Qualitative Theory in Structural Mechanics

Dajun Wang · Qishen Wang
Beichang (Bert) He

Qualitative Theory in Structural Mechanics

Qualitative Properties and Existence of Solutions

北京大学出版社
PEKING UNIVERSITY PRESS

Springer

Dajun Wang
Department of Mechanics
and Engineering Science
Peking University
Beijing, China

State Key Laboratory for Turbulence
and Complex Systems
Peking University
Beijing, China

Qishen Wang
School of Physics and Electrical Engineering
Anqing Normal University
Anqing, Anhui, China

Beichang (Bert) He
HBC Consulting
Seattle, WA, USA

ISBN 978-981-13-1375-2 ISBN 978-981-13-1376-9 (eBook)
https://doi.org/10.1007/978-981-13-1376-9

Jointly published with Peking University Press
The print edition is not for sale in China Mainland. Customers from China Mainland please order the print book from: Peking University Press

Library of Congress Control Number: 2018954600

This Springer imprint is published by the registered company Springer Nature Singapore Pte Ltd.
The registered company address is: 152 Beach Road, #21-01/04 Gateway East, Singapore 189721, Singapore

Preface

With enormous advances in structural engineering, the analytical, numerical, and experimental methods for solving static and dynamic problems in Structural Mechanics have reached a high level of maturity, and there is ongoing effort to enhance and perfect them. Strictly speaking, this remarkable progress is largely confined to the area of the quantitative theory in Structural Mechanics.

However, there is another important branch of Structural Mechanics, which we refer to as the qualitative theory in Structural Mechanics. Interestingly, this part of Structural Mechanics has been less understood by many of our colleagues working on research and application. After 50–60 years of quiet development since the middle of the twentieth century, a lot has been accomplished in the study of the qualitative theory. Consequently, it is now the right time for the qualitative theory of Structural Mechanics to be in the spotlight of engineering application, and become part of the students' classroom training.

The qualitative theory in Structural Mechanics primarily deals with static deformation and vibrational modes of linear elastic structures, and mainly cover subjects such as qualitative properties and the existence of solutions. Qualitative properties are shared by one type of structures, are at the system level and of clear regularity, and often result from analytical derivation and logical reasoning. A better understanding of qualitative properties can assist works in both numerical computation and experimental studies. It also promotes the development of better dynamic designs for structures. As for discussions on the existence of solutions, the validity of models in structural theories, etc., they help address fundamental issues in Structural Mechanics, and have far-reaching implication for engineering applications. A firm grasp of the theory of the existence of solutions and related subjects can aid in quantitative analysis, and help researchers build the theoretical background essential to their work.

Since the mid-1980s, the three authors of this book have conducted a systematic and extensive study on the qualitative theory in Structural Mechanics, and obtained some important results. We feel it will be beneficial for academic research, engineering application, and educational training if we summarize the existing work on the qualitative theory in Structural Mechanics, and publish a book dedicated to the subject. Our monograph in Chinese, *The Qualitative Theory in Structural*

Mechanics, was published in 2014. This book is the outcome of multiple years of additional revision and enhancement afterwards.

The qualitative theory covered in this book are highly structured and elegant. Furthermore, it provides some useful guidance to quantitative analysis and engineering application. We believe researchers, engineers, and students will become interested in reading the subject, and benefit from the study. However, derivation of certain qualitative properties involves relatively complex and difficult mathematics. Some readers may only want to learn the precise description of a few qualitative properties, and are less concerned about ways to obtain them. Therefore, Chap. 1 is tasked with the mission to introduce key conclusions of this book as well as present an overview of the main theories and methods used to establish them. Detailed explanation and derivation will be provided gradually in Chaps. 2–8. This approach should help avoid the pitfall of burying the practical and beautiful results of qualitative properties in highly involved mathematical proofs. Thus, the readers can easily pick portions of the book and spend more time on them, based on their own need and interest.

This book may serve as a reference book for researchers in the field of theoretical mechanics as well as structural and mechanical engineers working on strength and vibration testing, numerical analysis, and engineering design. It can also be used as the textbook for a graduate course in Solid Mechanics and Structural Engineering programs.

While preparing the manuscript for publication, the authors have benefited from the steadfast support and valuable feedback of the following academicians and professors: Zhemin Zheng, *Institute of Mechanics, Chinese Academy of Science*; Tongxi Yu, *the Hong Kong University of Science and Technology*; Zhongsheng (Simon) Liu, *National Research Council Canada*; Renhuai Liu, *Jinan University*; Bohua Sun, *Xian University of Architecture and Technology*; Haiyan Hu and Chunyan Zhou, *Beijing Institute of Technology*; Quan Wang, *Southern University of Science and Technology*; Huaiqiao Ying, *China Orient Institute of Noise and Vibration*; Guangji Qu, *China Academy of Space Technology*; Gongqing Zhang, Maozheng Guo, Jike Wu, Mingzhong Wang, Xianyue Su, Pu Chen, and Shaoqiang Tang, *Peking University*; and Zijun Zheng, *Chongqing University of Technology*. We would like to express our deep gratitude to all of them.

The authors feel fortunate to have worked with very experienced project teams at Springer Nature and Peking University Press. We wish to thank them for their professional guidance and timely assistance during the entire publication process of this book. We are also grateful to the reviewers, whose encouraging support means a lot to us.

Finally, we would appreciate notification of errors and omissions in our writing as well as comments on further improvement of this book from colleagues and readers.

Beijing, China — Dajun Wang
Anqing, China — Qishen Wang
Seattle, USA — Beichang (Bert) He
August 2019

Contents

Chapter 1
Overview

This chapter gives an introduction to the qualitative theory in Structural Mechanics, providing a brief history of its development, describing the content and methods of the study, and explaining the significance of the theory in research and application. In addition, this chapter presents main results of the qualitative theory in Structural Mechanics covered in the book, which makes it easier for readers to select portions of the book for further study based on their personal need and interest.

Structural Mechanics consists of multiple structural theories, dealing with Euler–Bernoulli beams, Timoshenko beams, thin plates, thick plates, shells, membrane shells, composite structures, assembly structures, etc. These theories can be applied to solve several types of problems in Solid Mechanics, such as static deformation, vibrational modes, and dynamic response. It falls within the scope of the quantitative theory to create analytical, numerical, and experimental methods for acquiring quantitative results of the aforementioned problems as well as to promote application of the quantitative results obtained; while it belongs to the realm of the qualitative theory to develop analytical and inferential methods for obtaining qualitative results as well as to study the significance of qualitative properties.

This book is dedicated to the qualitative theory. Our discussion consists of two parts. The first one is about qualitative properties of static deformation and vibration in linear elastic structures. Notice that our focus in this part will be qualitative properties of vibrational modes, including both natural frequencies and mode shapes (also known as normal modes), that are of regular pattern and at the system level. The second part of our discussion is on the existence of solutions of static deformation, vibrational modes, and dynamic response, within the context of linear elastic structures. This part also includes other subjects of fundamental theories such as validity of models in structural theories.

D. Wang et al., *Qualitative Theory in Structural Mechanics*,
https://doi.org/10.1007/978-981-13-1376-9_1

1.1 Brief History of the Development of the Qualitative Theory in Structural Mechanics

The history of studies on qualitative properties of structural vibration could be traced back to the 1830s. As pointed out by Rayleigh in his monograph *The Theory of Sound*, Sturm and Liouville did research at that time on nodal distribution of mode shapes of a bar using differential equations. In the 1950s, Courant and Hilbert, in their well-known book *Methods of Mathematical Physics*, established the qualitative relationship between natural frequencies and several parameters, including the mass, stiffness, constraints, size of the region, and boundary conditions, for a general system in vibration by applying the extremum principle in variational methods. They also provided some insightful and important discussion on nodal properties of mode shapes in the book.

Gantmacher and Krein's seminal work in the middle of the twentieth century started a new direction for the study of qualitative properties. They developed what they referred to as the theory of oscillatory matrices and kernels, and derived the general qualitative properties of static deformation and vibration for both discrete and continuous systems of one-dimensional structures, including strings, bars, and beams. In 1941, they published their monograph in Russian. In 1950, the second and expanded edition was in print. Later in 1961, the English version of this book, *Oscillation Matrices and Kernels and Small Vibrations of Mechanical Systems*, appeared. Gantmacher and Krein's book played a critical role in establishing the theoretical foundation of important qualitative properties, specifically, oscillatory properties in static deformation and vibration, of one-dimensional structures.

In the 1980s, Gladwell systematically studied inverse problems in vibration, where structural parameters are constructed from natural frequencies and mode shapes. The work led to the publication of his monograph *Inverse Problems in Vibration*. The second edition was released in 2004. Gladwell's book and many of his research papers helped advance the study of qualitative properties of vibration.

Since the 1980s, the three authors of this book as well as Zijun Zheng and Pu Chen have taken on qualitative properties of structural vibration as a separate research subject, and conducted a systematic and extensive study on modal qualitative properties. Additionally, the three authors published a monograph in Chinese, *Qualitative Theory in Structural Mechanics*, in 2014. These efforts have further increased and enhanced our understanding of this area.

The study on the existence of solutions in structural theories also involves a long development process. In the early days, some mathematicians employed the classical method of differential equations to analyze the existence of solutions of relatively simple mathematical equations such as the Sturm–Liouville equation. Advances in the theory of Hilbert functional spaces during the twentieth century enabled systematic study on the existence of solutions to equations in mathematical physics. Significant progress has been made. For instance, theories of Hilbert and Sobolev spaces were applied to establish existence conditions for solutions of static deformation, vibrational modes, and dynamics response in Elasticity. Details of the work can be found in several books, including *Some Applications of Functional Analysis in Mathematical Physics* by Sobolev in 1950, *The Problem of the Minimum of a Quadratic Functional* by Mikhlin in 1952, as well as *Existence Theorems in Elasticity* by Fichera in 1972. Based on multidimensional singular potentials and singular integral equations, Kupradze developed an alternative method to prove the existence of solutions in Elasticity. See his book *Potential Methods in the Theory of Elasticity*. Dajun Wang and Haichang Hu published a series of research papers during the period from 1982 to 1985, where they provided a unified proof for the existence of solutions to problems of static deformation and vibrational modes in Structural Mechanics by using a method that integrates the theories Mechanics and Functional Analysis. As an extension of this work, Dajun Wang later developed a theory on the validity of models in structural theories.

1.2 Areas Covered in the Study of the Qualitative Theory in Structural Mechanics

In this section, we will discuss areas covered by the qualitative theory in Structural Mechanics. Let us begin by introducing some conventions that are applicable throughout the book. A bar refers to a straight bar or rod. For the bar and its associated discrete systems, positive directions of their coordinate systems, forces, and displacements are set up to be identical and along the direction of the center line of the bar. A beam should always be interpreted as a straight Euler–Bernoulli beam. It deforms within the vertical symmetry plane of the beam. Moreover, forces and displacements of the beam and its discrete counterparts share the same positive direction that is perpendicular to the neutral axis of the beam. Finally, for membranes, plates, and shells, their forces and displacements also have the same positive direction.

Next, since some readers may not be familiar with the subject, we would like to present a few typical problems addressed in the qualitative theory. Hopefully, this will help stimulate the readers' interest and catch their attention.

Consider the static displacement of a beam, clamped at both ends and under the action of n (a positive integer) concentrated static forces. What rules are followed by the number of sign reversals of the displacement?

Think of a beam with a single span and subject to an arbitrary set of boundary constraints. Does it have a repeated natural frequency?

How many nodes does the i-th mode shape of the aforementioned beam have, where $i = 1, 2, \ldots$? Furthermore, how do nodes of this mode shape relate to those of a successive mode shape?

We know that the continuous system of a bar has an infinite set of modes, which include both natural frequencies and their corresponding mode shapes. But how many of them are "independent", i.e., how many modes are required to uniquely determine the system and derive all other modes?

As for the static deformation and forced vibration of a rotationally periodic structure, can they be decoupled into a series of smaller problems involving one substructure only?

Look at a circular plate clamped at the boundary and connected perpendicularly to a bar at the center. If the plate is modeled with the thin-plate theory, can we find solutions of static deformation and vibrational modes for the structure?

Also, for the aforementioned structure, if the plate is modeled with the Mindlin plate theory instead, do solutions to problems of static deformation and vibrational modes exist?

Attach a lumped mass, or add a concentrated constraint that is either elastic or rigid to a shell. Or, making it even simpler, apply one of these concentrated parameters to a membrane. Are these theoretical models valid? Can we find their vibrational modes?

All these questions can be answered with the qualitative theory in Structural Mechanics. The scope of the study of the qualitative theory is very broad. As of now, a lot of work has been done in the following areas:

(1) the existence of solutions of static deformation, vibrational modes, and dynamic response, as well as the validity of models in structural theories;
(2) geometric characteristics of static deformation of structures, such as oscillatory properties in static deformation of one-dimensional structures;
(3) effects on static deformation and vibrational modes of structures due to small changes in structural parameters;
(4) distribution of natural frequencies of a structure, such as existence of repeated frequencies and interlacement of frequencies associated with a structure subject to various boundary constraints;
(5) geometric characteristics of mode shapes of a structure, such as the distribution pattern of nodes, nodal lines, and nodal surfaces;
(6) determination of a structure using the specified modal data as well as the number of independent modes for a string, bar, or beam; and
(7) for repetitive structures, "repetitive" properties of static deformation, vibrational modes, and response to forced vibration, as well as their application in quantitative analysis.

We will introduce the fundamental theory of static deformation and vibrational modes of one dimensional structures, i.e., the theory of oscillatory matrices and kernels, in Chap. 2; discuss details of qualitative properties of static deformation and vibrational modes associated with strings, bars, beams, and multiple repetitive structures in Chaps. 3–7; and cover topics such as the existence of solutions in structural theories and the validity of structural models in Chap. 8.

Finally, we would like to conclude this section with two remarks. First, a lot of fruitful research has been conducted to determine the impact of small changes in structural parameters on vibrational modes. See Refs. [1, 5, 8, 9, 22]. Second, regarding the distribution of one-dimensional nodes, two-dimensional nodal lines, and three-dimensional nodal surfaces, Courant established an exquisite theorem in

1923 that has stimulated many follow-up studies. See Refs. [2, 6, 10, 11]. Due to space limitation, the content of these two subjects is not included in this book.

1.3 Key Theoretical Results and the Approach to Establish Them

To help the reader better grasp the essence of theories elaborated in this book, we are going to give a brief overview of key theoretical results as well as methods for establishing them in this section.

1.3.1 Oscillatory Properties in Static Deformation and Vibration of One-Dimensional Single-Span Structures

In their monograph [3], *namely, Oscillation Matrices and Kernels and Small Vibrations of Mechanical Systems*, Gantmacher and Krein expounded their theory of oscillatory properties in static deformation and vibration associated with discrete and continuous systems of one-dimensional single-span structures. We, the authors of this book, referred to their work as "the theory of oscillatory properties of one-dimensional single-span structures," and summarized its key components and the method for establishing the theory in a few words as "three properties and two links." The three properties are given in the next a few paragraphs.

Property 1 Oscillatory System *The flexibility matrix of the discrete system of a one-dimensional single-span structure is an oscillatory matrix; while the kernel, also known as the Green's function, of the related continuous system is an oscillatory kernel.*

Property 2 Oscillatory Properties in Static Deformation *When a concentrated static force is applied to the continuous or discrete system of a one-dimensional single-span structure, the displacement of the system is nonzero and in the direction of the force acting on the structure. Furthermore, when the system is under the action of an arbitrary set of* n *concentrated static forces, the number of sign reversals of the resulting static displacement or u-line does not exceed* $n-1$*, where* n *is a positive integer.*

Property 3 Oscillatory Properties in Vibration

(1) Natural frequencies of the system are distinct. For a continuous system, these frequencies have infinity as their limit point, and can be arranged in an increasing order as

$$0 < f_1 < f_2 < \cdots < f_n < \cdots;$$

while for a discrete system with n degrees of freedom, the n natural frequencies can also be sorted from small to large as

$$0 < f_1 < f_2 < \cdots < f_n.$$

(2) *The i-th mode shape $u_i(x)$ has precisely $i - 1$ nodes. Notice that the first mode shape does not have any node.*
(3) *Nodes of successive mode shapes interlace.*
(4) *If a vibration can be expressed as a linear combination of the p-th to the q-th mode shapes, i.e.,*

$$u(x, t) = \sum_{r=p}^{q} c_r(t)u_r(x),$$

where $1 \leq p \leq q$, then the node count of the vibration is no less than $p - 1$ and its zero count is no greater than $q - 1$ at any instant.

Notice that Property 3(4) is also applicable to the static deformation, in which case $u(x, t)$ and $\{c_r(t)\}$ in the above expression are all independent of the time variable t.

The authors of this book assign the following short description to the above four oscillatory properties in vibration: (1) distinctiveness of natural frequencies, (2) sequentiality of nodes of mode shapes, (3) interlacement of nodes between mode shapes, and (4) characteristics of nodes of a vibration.

Next, the two links in "three properties and two links" are as follows:

Link 1 *Property 1 is the necessary and sufficient condition of Property 2.*

Link 2 *Property 1 is a sufficient condition of Property 3.*

Gantmacher and Krein have shown that single-span bars and beams as well as some of their discrete systems are oscillatory systems, have oscillatory properties in static deformation, and possess oscillatory properties in vibration.

The theory of oscillatory properties of one-dimensional structures has amply demonstrated the depth and charm of integration between Mathematics and Mechanics. However, there were some shortcomings in the approach taken by Gantmacher and Krein, and thus some follow-up work is needed. These are highlighted in the two paragraphs below:

(1) Because the concepts of oscillatory kernels and matrices as well as oscillatory properties in static deformation are only applicable to systems without rigid-body motion, Gantmacher and Krein's theory does not cover systems with rigid-body movement that are commonplace in engineering applications. By the way, in this book, we refer to systems with and without rigid-body motion as under-constrained and well-constrained systems, respectively.
(2) Link 2 mentioned above shows the following: The sufficient condition for a system to hold oscillatory properties in vibration is for its flexibility matrix to

be oscillatory if the system is discrete, or its Green's function to be an oscillatory kernel if the system is continuous. But it was not clear whether this condition should also be necessary.

Since 1990s, there has been effort to address these issues. Wang, Wang, and He solved the first problem in Refs. [18, 19, 21] by employing the technique of conjugate systems. They demonstrated that the under-constrained systems also possess oscillatory properties in vibration. This work helped enhance Gantmacher and Krein's theory, and expand its application in engineering. As for the second problem mentioned above, Zheng, Chen, and Wang proved in Refs. [23, 24] that for a discrete system, its flexibility matrix being oscillatory is indeed the necessary condition for the system to possess oscillatory properties in vibration. Furthermore, they showed that for a continuous system, the Green's function being an oscillatory kernel is also the necessary condition for the system to hold oscillatory properties in vibration.

By summarizing the discussion above, we may reach a perfect conclusion. Specifically, for a one-dimensional single-span system, whether it is discrete or continuous, each of the following three properties is a necessary and sufficient condition of the other two: the system having oscillatory properties in static deformation, either its Green's function or flexibility matrix being oscillatory, and the system possessing oscillatory properties in vibration. This result has revealed qualitative properties of this type of structures at a much deeper level, and provided us with an alternative way to verify whether a structure possesses certain oscillatory properties.

We will prove in Chaps. 3–6 that the following one-dimensional systems have oscillatory properties in vibration: continuous and several discrete systems of both strings and bars with or without elastic foundations; continuous and finite difference systems of beams; two-node hybrid finite element systems of beams, with lumped mass and constructed by using the Hellinger–Reissner principle; and finite element systems of beams, derived from the principle of minimum potential energy as well as with lumped mass, the third-order Hermite shape function for the displacement, and some additional constraints on stiffness distribution within an element. See Refs. [23, 24]

Lastly, we should point out that not all one-dimensional structures have the oscillatory properties in vibration. Here are two examples: a beam resting on an elastic foundation and a Rayleigh beam with its kinetic energy of section rotation being taken into account. As another example, consider the finite element system of a beam built with the principle of minimum potential energy and third-order Hermite polynomials as displacement shape functions. Such a system may not have oscillatory properties in vibration when its bending stiffness and mass are distributed in certain way.

1.3.2 Other Important Qualitative Properties

Besides oscillatory properties presented above, this book also covers a few other important qualitative properties. These are summarized below.

(1) *Number of Independent Modes of a One-Dimensional Single-Span Structure* We know that the continuous system of a one-dimensional single-span structure has an infinite set of modes, including both natural frequency f_i and mode shape $u_i(x)(i = 1, 2, \ldots)$. When the system under consideration is discrete, counts of frequencies and mode shapes are equal to the number of degrees of freedom of the discrete system. Besides mode shapes being orthogonal to each other, what may other important relationship exist among them? Qishen Wang and Dajun Wang have proved in Ref. [19] that for the bar with a single-span, its tensile stiffness and mass density can be reconstructed from two modes while all other modes may be derived from them. This implies that there are only two independent vibrational modes for a bar, and they can be any two modes. Consequently, only two modes can be specified in the dynamic design of a bar. The related conclusion for a string with a single-span is that there is only one independent vibrational mode, which can be any one of the modes.
Zijun Zheng has proved in Ref. [23] that a single-span beam has two independent modes. In addition, he showed that, although untrue with some mode pairs, two modes are often sufficient to determine a beam.
For the spring–mass discrete system of a bar as well as the finite difference model of a beam, authors of this book have demonstrated in Refs. [16, 17] that these systems can be uniquely determined with two vibrational modes.

(2) *"Hearing" a Structure* Gladwell showed in his monograph [4] that, given three sets of interlacing spectra of a beam subject to different boundary constraints, such as clamped–free, clamped–pinned, and clamped–sliding, the bending stiffness and mass density of the beam can be determined. Since the spectra can be measured through acoustic means, the process to determining the beam is compared to "hearing" its physical parameters.

(3) *Qualitative Properties of Non-Displacement Mode Shapes of Bars and Beams* Authors of this book derived qualitative properties of strain mode shapes of a bar. They also uncovered qualitative properties of mode shapes associated with the slope, strain, and shear force of a beam. The work helps further reveal physical characteristics of structures, and is beneficial for engineering application.

(4) *Qualitative Properties of Continuous and Discrete Systems of Beams with Multiple Supports* A number of results on vibrational modes have been obtained through introduction of mathematical transformation systems.

(5) *Qualitative Properties of Continuous Systems of Repetitive Structures* Previous research on repetitive structures often targeted discrete models and was from the perspective of simplifying numerical solution of linear algebraic equations and matrix eigenvalue problems, while the emphasis of this book is to directly study eigenvalue and static deformation problems of differential equations of the continuous systems. The approach helps uncover special qualitative properties of vibration and static deformation associated with symmetric, rotationally periodic, linearly periodic, chain, and axisymmetric structures. In addition, this allows for more accurate revelation of the physical nature of these properties, and also makes it easier to apply knowledge of the qualitative theory to engineering practice.

1.3.3 Existence of Solutions in Structural Theories

There are two problems of hallmark significance in the study of fundamental theories of Solid Mechanics: One is the existence of solutions in Elasticity, including solutions of static deformation, vibrational modes, and dynamic response; while the other is the existence of solutions in structural theories. In the 1950s, Friedrichs, Mikhlin, et al. proved the existence of elasticity solutions by applying the theory of Hilbert function space.

As for the corresponding problem in structural theories, some mathematicians and researchers managed to prove, on a case-by-case basis, the existence of solutions in one- and two-dimensional structural theories. In the 1980s, Wang and Hu developed a method in [12–15] that extends the existence of solutions from Elasticity to structural theories. This was accomplished by deeply integrating Mechanics and Mathematics, taking advantage of connections in strain energies and kinetic energies between structural theories and Elasticity, applying the transferability of both boundedness and compactness of operators, and setting up a functional analysis framework. The theory is further enhanced in this book. In addition, it is applied to determining the validity of models of structures in structural theories, particularly those related to assembly structures.

1.4 Theoretical and Practical Significance of the Qualitative Theory in Structural Mechanics

1.4.1 Significance of Qualitative Properties of Structural Vibration and Static Deformation

Qualitative properties of structural vibration and static deformation not only form an important part of the theory of Structural Mechanics but may also play a guiding role in engineering applications. Below are some examples demonstrating their significance:

1. Fast Estimate Or Evaluation of Quantitative Analysis and Method

By taking advantage of their qualitative properties, we may produce quantitative estimates of static deformation and vibrational modes, prior to conducting detailed numerical analysis or lab testing. After the analysis or testing is completed, we can use qualitative properties to screen the quantitative results. More importantly, it helps us evaluate our analytical methods and experimental techniques. Specifically, if a quantitative result is inconsistent with the related qualitative properties, then either the quantitative method or quantitative result is wrong; while if the two are in agreement, then the result might be correct.

For instance, suppose the static deformation of a beam, clamped at both ends and under the action of 3 concentrated forces, has been numerically calculated, and is shown in Fig. 1.1a. Since the static deformation has 3 nodes, we know for sure that the result is wrong, and it is even reasonable to question the validity of the

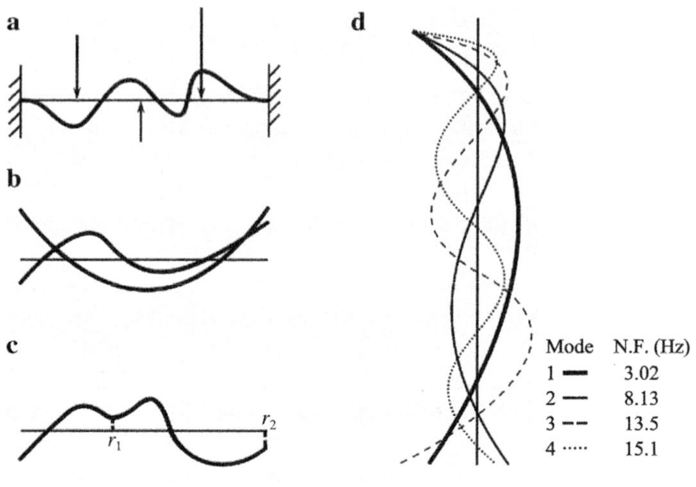

Fig. 1.1 Verification of static deformation and vibrational modes of beams

analysis method used. Our conclusion is based on the qualitative property that when 3 concentrated forces are applied to this beam, the number of sign reversals of its static deflection should be less than 3.

As another example, consider the vibration of an unconstrained beam. Suppose its 3rd and 4th mode shapes from numerical analysis are graphed in Fig. 1.1b. Although node counts for both mode shapes are correct, at least one of them has some excessive error that causes another qualitative property to be violated, which requires nodes of two successive mode shapes to interlace. Furthermore, the mode shape of this beam depicted in Fig. 1.1c is also wrong, because its shape does not satisfy qualitative properties at two locations, labeled r_1 and r_2. First, since the mode shape function reaches an extremum at r_1, the displacement u and the curvature u'' at this point ought to be of opposite signs, rather than share the same sign. Second, r_2 is located right at a free boundary, and thus the displacement u and the slope u' there need to be of identical sign, instead of having different signs.

As for Fig. 1.1d, it illustrates mode shapes and natural frequencies of a rocket from lab testing. Since the 1st and 2nd mode shapes represent rigid-body translation and rotation, respectively, the results shown should correspond to the 3rd, 4th, 5th, and 6th mode shapes of the rocket, which is modeled as a beam in the free–free configuration. Several observations can be made based on Fig. 1.1d: First, all non-zero natural frequencies (N.F.) are distinct; second, the i-th mode shape has $i - 1$ nodes, where $i = 3, 4, 5, 6$; third, nodes of successive mode shapes interlace; and fourth, the shape of each mode shape does not violate other main qualitative properties. Therefore, this set of test results has successfully passed our screening based on qualitative properties, and are likely to be correct.

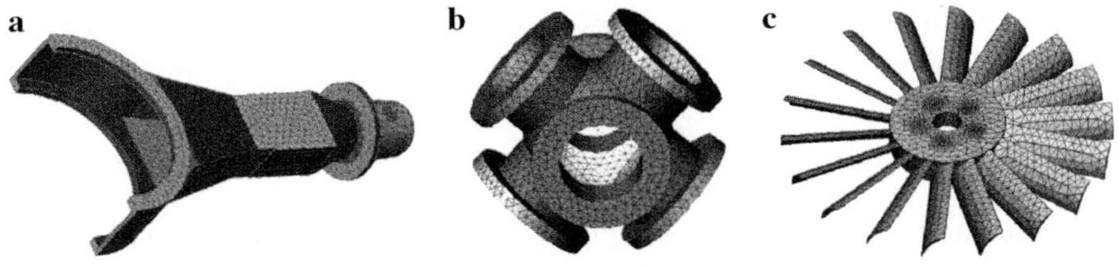

Fig. 1.2 Symmetric and rotationally periodic structures

As demonstrated in the above examples, qualitative properties may not only help screen quantitative results, but also serve as a basis for evaluating analytical, numerical, and experimental methods.

2. *Work Load Reduction in Quantitative Analyses of Repetitive Structures*

In regard to various repetitive structures, qualitative properties of static deformation and vibrational modes can be applied to simplify quantitative analyses in both numerical and experimental studies. This should lead to a dramatic reduction in the work load.

Let us look at a symmetric structure as an example. Since its mode shapes can be divided into two groups, one symmetric while the other anti-symmetric, we only need to perform numerical analysis on one half of the structure separated by the symmetry plane in order to compute natural frequencies and mode shapes of the entire structure. If the total number of degrees of freedom of the symmetric structure is N, then the amount of calculation required to solve its eigenvalue problem as a whole will be of the order N^3. However, by taking advantage of symmetry, we only need to solve two eigenvalue problems for a structure half of the original size. Thus, the amount of calculation is reduced to about $N^3/4$. In addition, if the structure has two symmetry planes, such as the one shown in Fig. 1.2a, we only need to solve 4 eigenvalue problems for 1/4 of the structure, and the amount of calculation is of the order $N^3/16$. Finally, if the structure has 3 symmetry planes, as in the case shown in Fig. 1.2b, the work load drops down further to only $N^3/64$.

For a rotationally periodic structure with n sectors, we may significantly reduce the computational effort by taking advantage of the periodicity. When n becomes large, the numerical computation involved is decreased to roughly $4/n^2$ of the work needed to analyze the structure as a whole. For the mechanical component depicted in Fig. 1.2c, $n = 18$ and the computational work required is only 1/81 of that needed to analyze the entire structure.

3. *Valid Specification of Modal Data*

Modal qualitative properties help ensure that a valid specification of natural frequencies and mode shapes is given in the inverse problem of vibration and dynamic design of structures. For instance, it is impossible to design a beam that has a nodal line in any mode shape. This is because, according to modal qualitative properties, the mode shape of a beam may have nodes only, and no nodal line should ever exist. As another example, at most two sets of vibrational modes can be specified in the design of a beam.

4. Validity of Discrete Model

Modal qualitative properties can be deployed to check the validity of discrete models of a structure, based on the fact that the main qualitative properties associated with vibrational modes of the simplified discrete system should be consistent with those of the original continuous system. For instance, the continuous system of a beam should have oscillatory properties in vibration pertaining to continuous systems, and we expect various discrete models of this beam to also have related oscillatory properties in vibration associated with discrete systems. If, on the contrary, a particular discrete model does not possess these oscillatory properties applicable to discrete systems, then it is not valid.

5. Easy Solution to Some Problems

Some problems can be solved quickly with the knowledge of qualitative properties, and thus complicated quantitative analysis may be avoided. For instance, sometimes we only need to know the general trend of the relationship between changes in structural parameters and the behavior of frequencies and mode shapes. In this case, qualitative properties from the qualitative theory should be sufficient to produce a satisfactory answer.

1.4.2 Significance of the Existence of Solutions in Structural Theories as Well as Other Subjects Within Fundamental Theories

Structural Mechanics deals with many types of problems: static deformation, vibrational modes, dynamic response, stability, and so on. Solution to all these problems involves building physics and mathematics models for the underling engineering objects and then solving the resulting equations of mathematical physics. Since we have to resort to approximation when working with most problems in Structural Mechanics, it is necessary for us to understand requirements on structural parameters, border shape, external forces, initial conditions, and boundary constraints that ensure existence of a solution. In addition, we need to know in what sense the solution exists. Clearly, knowledge of these should be a precondition of any problem-solving effort. Unfortunately, there were cases in the literature of structural vibration, where people blindly sought after solutions and completely ignored the fundamental question about their existence. For example, Leung et al. once pointed out errors in some research papers, where the authors had attempted to compute, using the method of series expansion, modes of either membranes or Mindlin plates with concentrated masses attached to them (see Ref. [7]).

Next, an important criterion for the validity of a model in the structural theory is the existence of a solution. If no solution exists, then the model of the structural theory is invalid. To find approximate solutions to structural problems, the Ritz method and other energy-based methods are highly effectual. But some careful studies are needed to evaluate the solutions obtained in this way. Specifically, we need to understand under what input conditions, including structural parameters, external loads, initial states, and boundary constraints, these methods may produce converged solutions; and in what sense the convergence can be achieved. The issues raised here not only reflect the requirement of mathematical integrity, but is of clear physical significance as well. They belong to the realm of fundamental theories in Structural Mechanics.

Before wrapping up this section, we would like to emphasize a few points and also make some additional comments:

Quantitative and qualitative theories are two aspects of the complete theory of Structural Mechanics. With the quantitative theory, we primarily go after the accuracy of analytical, numerical, and experimental methods as well as their coverage and efficiency in application; while with the qualitative theory, we primarily pursue the breadth and depth in our understanding of qualitative properties.

In engineering applications, we can get more accurate answers faster by utilizing problem-solving techniques in the quantitative theory together with qualitative properties in the qualitative theory. In the study of Structural Mechanics, we can foster further growth in this field by promoting interaction between quantitative and the qualitative theories.

Furthermore, it should benefit engineering students tremendously, if elements of the qualitative theory can be incorporated into the teaching of various structural theories, particularly, into the course of Structural Dynamics. It should increase the breadth of their knowledge and enhance their thinking capability. Authors of the book feel strongly that it is high time for us to take concrete steps in this direction.

The main qualitative properties of vibrational modes associated with bars, beams, and repetitive structures are presented in Sects. 1.5–1.7, respectively; while Sect. 1.8 is dedicated to some key results on the existence of solutions in structural theories. To get details beyond this brief introduction and to find proofs of various theorems, readers may decide, based on their needs and interests, to carefully study subsequent chapters of this book.

Regarding static deformation, the same qualitative property, i.e., Property 2 in Sect. 1.3.1, is shared by bars and beams. As for repetitive structures, results of their qualitative properties of static deformation are given in various subsections of Sect. 1.7.

1.5 Overview of Qualitative Properties of Vibration Associated with Bars

1.5.1 Qualitative Properties of Vibration Associated with the Continuous System of a Bar

Consider a bar with a single span, having variable cross sections, and subject to an arbitrary set of boundary conditions, as shown in Fig. 3.2a of Chap. 3. Depending on values of boundary parameters h and H, different types of end constraints may be imposed. For instance, $h = 0$ or $H = 0$ corresponds to a free end, while $h \to \infty$ or $H \to \infty$ leads to a fixed end. Suppose natural frequencies of the bar have been arranged in an ascending order. Let f_i denote the i-th natural frequency and $u_i(x)$ represent the corresponding mode shape, where $i = 1, 2, \ldots$. The main qualitative properties of these modes are summarized in the rest of this subsection.

(1) *Distinctiveness of Natural Frequencies* All natural frequencies are distinct, and the system has no repeated frequency. Thus, the frequencies can be sorted as:

$$0 \leq f_1 < f_2 < \cdots < f_n < \cdots.$$

(2) *Sequentiality of Nodes of Mode Shapes*

(a) The i-th displacement mode shape $u_i(x)$ has $i - 1$ nodes and no other zero. This may be expressed as

$$S_{u_i} = i - 1, \quad i = 1, 2, \ldots.$$

(b) The node count, i.e., the number of sign reversals, $S_{u'_i}$ of the i-th strain mode shape $u'_i(x)$ is fixed, which is listed in Table 5.1 in Chap. 5. Notice that the node count S_{u_i} of the i-th displacement mode shape $u_i(x)$ is also included in the table.

(3) *Interlacement of Nodes Between Mode Shapes*

(a) Nodes of successive displacement mode shapes $u_i(x)$ and $u_{i+1}(x)$ interlace.
(b) Nodes of successive mode shapes of the strain $u'_i(x)$ and $u'_{i+1}(x)$ interlace.
(c) Nodes of the strain mode shape $u'_i(x)$ interlace with those of the displacement mode shape $u_i(x)$, where both mode shapes are related to the same natural frequency f_i.

(4) *Interlacement of Natural Frequencies* Natural frequencies of a bar subject to various boundary constraints satisfy the following inequalities:

$$f_i^{\text{ff}} < f_i^{\text{cf}} < \left(f_i^{\text{cc}}, f_{i+1}^{\text{ff}}\right) < f_{i+1}^{\text{cf}}, \quad i = 1, 2, \ldots.$$

where superscripts ff, cf, and cc represent, respectively, free–free, fixed–free, and fixed–fixed boundary constraints at two ends of the bar.
It needs to be pointed out that all well-constrained Sturm–Liouville systems, including a bar resting on an elastic foundation, should have properties (1), (2)(a), (3)(a), and (4) given above.

(5) *Number of Independent Modes* There are only two independent modes. Given two positive numbers f_i and f_j as well as two functions $u_i(x)$ and $u_j(x)$ defined on the interval [0, l], if $u_i(x)$ and $u_j(x)$ individually satisfy the sign reversal conditions listed in Table 5.1 and moreover, $u_i(x)$, $u_j(x)$, $\lambda_i = (2\pi f_i)^2$, and $\lambda_j = (2\pi f_j)^2$ together meet the compatibility conditions specified in Sect. 5.4.2 of Chap. 5, then we can create the continuous system of a bar with $(f_i, u_i(x))$ and $(f_j, u_j(x))$ as its two modes.
The result shows that, among the infinite set of modes of the continuous system of a bar, only two are independent. This implies that the system is determined completely after the two modes are specified, and all other modes can be derived. Therefore, in dynamic design, at most two modes can be assigned to a bar.
In comparison, a string has only one independent mode.

(6) *Characteristics of a Mode Shape* These include:

  (a) If the intersection between a mode shape $u(x)$ and the x-axis, which is the center axis of the bar, is an internal zero, then it has to be a node, and cannot be other type of zeros or a zero line.
  (b) An internal zero of a strain mode shape must also be a node.
  (c) At the point x_r where an extremum is reached, we have

$$u_i'(x_r) = 0 \quad \text{and} \quad u_i(x_r)u_i''(x_r) < 0.$$

  (d) At the boundary, we get

$$u_i(0)u_i'(0) \geq 0 \quad \text{and} \quad u_i(l)u_i'(l) \leq 0,$$

  where the equality holds in the above inequalities when h or H either equals 0 or approaches ∞.

(7) *Characteristics of Nodes of a Vibration* If the displacement of a vibrating bar is produced by the superposition of the p-th to the q-th mode shapes, i.e.,

$$u(x,t) = \sum_{r=p}^{q} c_r(t)u_r(x),$$

where $1 \leq p \leq q$, then, at any moment, the node count of the vibration is no less than $p - 1$ while its number of zeroes is no greater than $q - 1$.

Notice that Property (7) is also applicable to the static deformation, in which case $u(x,t)$ and $\{c_r(t)\}$ in the above expression are all independent of the time variable t.

As indicated earlier, Properties (1), (2)(a), (3)(a), and (7) are collectively referred to as "oscillatory properties in vibration" of bars.

Details of this subsection can be found in Chap. 5.

1.5.2 Qualitative Properties of Vibration of a Spring–Mass System

The spring–mass system shown in Fig. 3.2b can serve as the common discrete system of strings, bars, and shafts. It contains $n+1$ masses or point particles, and has N degrees of freedom. Here, $N = n+1$ when both h and H are of finite value, $N = n$ when only one of h and H approaches to ∞, or $N = n-1$ when h and H approach to ∞ simultaneously. Sort natural frequencies of this system from small to big into $f_1, f_2, \ldots, f_N$. The corresponding mode shapes are represented by $\boldsymbol{u}^{(1)}, \boldsymbol{u}^{(2)}, \ldots, \boldsymbol{u}^{(N)}$, respectively. The qualitative properties of vibration associated with this discrete system are almost identical to those of the continuous system of a bar. The only differences between the two are as follows: The former has a finite number of modes while the latter has infinitely many, the strain mode shape of the continuous system should be replaced by the mode shape of the spring deformation or relative displacement of the discrete system, and there is a little difference in the description of Property (6).

1.5.3 Qualitative Properties of Vibration Associated with Other Discrete Systems of a Bar

The spring–mass–spring system depicted in Fig. 3.3 is one physical discrete system of a string or bar resting on an elastic foundation; while Fig. 3.5 shows another physical discrete system of a bar, consisting of massless elastic bars and masses. Mathematical discrete systems include those obtained using the finite difference or finite element method. Depending on how the distributed mass is discretized, systems established with the finite element method may be further categorized as systems with lumped or consistent mass matrices.

Stiffness matrices of all systems mentioned above are sign-oscillatory, or their flexibility matrices are oscillatory. Thus, both natural frequencies and mode shapes of these systems have qualitative properties very similar to their counterparts of the continuous system of a bar, which have been introduced in Sect. 1.5.1.

Details of this subsection and Sect. 1.5.2 can be found in Chap. 3.

1.6 Overview of Qualitative Properties of Vibration Associated with Beams

1.6.1 Qualitative Properties of Vibration Associated with the Continuous System of a Single-Span Beam

Natural frequencies and mode shapes of the continuous system of the beam shown in Fig. 6.1 of Chap. 6 have major qualitative properties listed in the rest of this subsection.

(1) *Distinctiveness of Natural Frequencies* For a well-constrained beam, its natural frequencies are not only positive but distinct as well; while for an under-constrained beam, it has one or two zero frequencies, but other nonzero natural frequencies are still distinct. Thus, we have

$$0 \le f_1 \le f_2 < f_3 < \cdots.$$

(2) *Sequentiality of Nodes of Mode Shapes*

(a) The i-th displacement mode shape $u_i(x)$ has $i-1$ nodes and no other zero. Its node count may be expressed as

$$S_{u_i} = i - 1, \quad i = 1, 2, \ldots.$$

(b) Table 6.1 of Chap. 6 tabulates node counts of mode shapes of the angle of rotation $u'(x)$, bending moment $\tau(x) = EJu''(x)$, and shear force $(EJu''(x))' = \phi(x)$. Notice that the node count S_{u_i} of the i-th displacement mode shape is also included in the table.

(3) *Interlacement of Nodes Between Mode Shapes* This property shows that nodes of each of the seven pairs of mode shapes given below interlace:

(a) successive mode shapes of the displacement,
(b) successive mode shapes of the angle of rotation,
(c) successive mode shapes of the bending moment,
(d) successive mode shapes of the shear force,
(e) the mode shape of the displacement and the mode shape of the angle of rotation corresponding to the same frequency,
(f) the mode shape of the angle of rotation and the mode shape of the bending moment corresponding to the same frequency, and
(g) the mode shape of the bending moment and the mode shape of the shear force corresponding to the same frequency.

(4) *Interlacement of Natural Frequencies* Natural frequencies of a beam subject to various boundary conditions have the following relationship:

$$f_i^{\mathrm{cf}} < f_i^{\mathrm{cs}} < f_i^{\mathrm{ca}} < f_i^{\mathrm{cp}} < \left(f_i^{\mathrm{cc}}, f_{i+1}^{\mathrm{cf}}\right) < f_{i+1}^{\mathrm{cs}}, \quad i = 1, 2, \ldots,$$

where superscripts cf, cs, ca, cp, and cc represent, respectively, boundary constraints on the beam at its left and right ends that are clamped–free, clamped–sliding, clamped–anti-resonant, clamped–pinned, and clamped–clamped.

(5) *Number of Independent Modes* A beam has two independent modes. Under most circumstances, the beam can be uniquely determined by two modes. However, there are cases where the beam cannot be uniquely defined by certain pair of modes.

(6) *Characteristics of a Mode Shape* These include:

(a) At the point x_r where an extremum is reached, i.e., $u'(x_r) = 0$, we have $u(x_r)u''(x_r) < 0$.

(b) When the left end of the beam is free,

$$u_i(0)u_i'(x) < 0, \quad 0 \le x \le \xi_1, \quad i = 1, 2, \ldots;$$

when the right end is free,

$$u_i(l)u_i'(x) > 0, \quad \xi_{i-1} \le x \le l, \quad i = 1, 2, \ldots;$$

when the left end is sliding,

$$u_i(0)u_i'(x) < 0, \quad 0 < x \le \xi_1, \quad i = 1, 2, \ldots;$$

and when the right end is sliding,

$$u_i(l)u_i'(x) > 0, \quad \xi_{i-1} \le x < l, \quad i = 1, 2, \ldots;$$

where ξ_1 and ξ_{i-1} represent, respectively, the first and last nodes of the i-th displacement mode shape $u_i(x)$.

(7) *Characteristics of Nodes of a Vibration* If the displacement of a vibrating beam can be expressed as a linear combination of the p-th to the q-th mode shapes, namely,

$$u(x, t) = \sum_{r=p}^{q} c_r(t)u_r(x),$$

where $1 \le p \le q$, then, at any moment, the node count of this vibration is no less than $p - 1$ while its zero count is no more than $q - 1$.

Notice that Property (7) is also applicable to the static deformation, in which case $u(x, t)$ and $\{c_r(t)\}$ in the above expression are all independent of the time variable t.

As indicated before, Properties (1), (2)(a), (3)(a), and (7) are collectively referred to as "oscillatory properties in vibration" of beams.

1.6.2 Qualitative Properties of Vibration Associated with Continuous Systems of Overhang Beams

Some main properties of natural frequencies and mode shapes are given below for two- and three-span overhang beams that are shown in Fig. 6.6.

(1) Natural frequencies are distinct. The systems do not have any repeated frequency.
(2) The first displacement mode shape $u_1(x)$ has p nodes, where $p = 1$ for a two-span beam while $p = 2$ for a three-span beam. Furthermore, the mode shape function is of the same sign within a span, but has opposite signs between adjacent spans.
(3) The i-th displacement mode shape $u_i(x)$ has $i + p - 1$ $(p = 1, 2)$ zeroes. An internal zero not overlapping with any internal support must be a node, while a zero located right at an internal support can be either a node or a null anti-node, where a null anti-node is counted as two distinct zeroes.
(4) Nodes of two successive mode shapes $u_i(x)$ and $u_{i+1}(x)$ may no longer interlace.
(5) Node counts of mode shapes of the angle of rotation, bending moment, and shear force are given by, respectively,

$$S_{u_i'} = i, \quad S_{\tau_i} = i - 1, \quad S_{\phi_i} = i, \quad i = 1, 2, \ldots.$$

Details of results listed in this and the previous subsection can be found in Chap. 6.

1.6.3 Qualitative Properties of Vibration Associated with the Finite Difference System of a Beam

Consider the system consisting of springs, masses, and rigid bars, as shown in Fig. 4.1b. The system has $n + 1$ masses, and its number of degrees of freedom is N. Here, $N = n + 1$ when the two ends of the system are neither clamped nor pinned, $N = n$ when only one end is either clamped or pinned, or $N = n - 1$ when both ends are either clamped or pinned. We obtain this system after discretizing the continuous system of a beam with the finite difference scheme. Sort its natural frequencies in an ascending order, and represent them by $f_1, f_2, \ldots, f_N$. The corresponding mode shapes are denoted by $\boldsymbol{u}^{(i)}$ $(i = 1, 2, \ldots, N)$. The qualitative properties of this system are almost identical to those of the continuous system of the beam, except that the discrete system only has a finite number of modes, as opposed to infinitely many modes in the case of the continuous system. In addition, the description of Property (6) is a little different.

1.6.4 Qualitative Properties of Vibration Associated with the Discrete System of a Multi-span Beam

Consider a single-span system consisting of springs, masses, and rigid bars. It can be converted to the discrete system of a multi-span beam by adding p pinned supports at the interior. Natural frequencies and mode shapes of this multi-span discrete system have major qualitative properties summarized in the subsequent paragraphs.

(1) Natural frequencies are all distinct, and the system has no repeated frequency.
(2) The first displacement mode shape $\boldsymbol{u}^{(1)}$ has p nodes. Moreover, components of the mode shape are of the same sign within a span, while they have opposite signs between adjacent spans.
(3) Node count of the i-th displacement mode shape $\boldsymbol{u}^{(i)}$ is

$$S_{\boldsymbol{u}^{(i)}} = i - 1 + p - 2s, \quad s \leq \min(i - 1, p),$$

where s is a parameter and its description can be found in Sect. 4.8.5.
(4) Nodes of successive mode shapes $\boldsymbol{u}^{(i)}$ and $\boldsymbol{u}^{(i+1)}$ may no longer interlace.
(5) For finite difference systems of two- and three-span beams with overhangs, their natural frequencies and mode shapes have additional properties listed below:

(a) Let $\bar{\boldsymbol{u}}^{(i)}$ be the i-th displacement mode shape that includes zero components at the internal supports. Its zero count is $i + p - 1$, where $p = 1, 2$. Among these zeroes, the one that is not located right at any internal support is a node; while the one overlapping with an internal support may be either a node or a null anti-node, where a null anti-node is counted as two distinct zeroes.
(b) Node counts of mode shapes associated with the angle of rotation, bending moment, and shear force are, respectively,

$$S_{\boldsymbol{\theta}^{(i)}} = i, \quad S_{\boldsymbol{\tau}^{(i)}} = i - 1, \quad \text{and} \quad S_{\boldsymbol{\phi}^{(i)}} = i,$$

where $i = 1, 2, \ldots, n - 1$.

1.6.5 Qualitative Properties of Vibration Associated with Other Discrete Systems of a Beam

Other discrete systems of a beam may include:

System 1 The system, depicted in Fig. 4.4, consists of masses and massless elastic beams.
System 2 This is the finite element system constructed by using the Hellinger–Reissner principle and a lumped mass matrix.

System 3 This is the finite element system consisting of elements built with third-order Hermite displacement shape functions and the principle of minimum potential energy. It also involves a lumped mass matrix. The bending stiffness within each element needs to satisfy the condition

$$\int_0^1 EJ(\xi)(9\xi^2 - 9\xi + 2)\mathrm{d}\xi > 0.$$

Modes of all these three systems have the oscillatory properties in vibration pertaining to discrete systems.

Details of Sects. 1.6.3–1.6.5 can be found in Chap. 4.

1.7 Overview of Qualitative Properties of Vibration and Static Deformation Associated with Repetitive Structures

Continuous systems of repetitive structures and their valid discrete counterparts have consistent qualitative properties, so far as vibrational modes and static deformation are concerned. As a result, we primarily discuss continuous systems here.

1.7.1 Structures with Mirror Symmetry

The shape, physical properties, and boundary conditions of a structure with mirror symmetry, also referred to as the symmetric structure in short, are all symmetric about one or multiple planes. The main property of their mode shapes is as follows: For each plane of symmetry, the mode shapes can be divided into two groups, one symmetric while the other antisymmetric with respect to the symmetry plane.

As for the static deformation, the most important qualitative property is that both the external forces and the static deformation can be decomposed into two groups with respect to each plane of symmetry, one symmetric while the other antisymmetric.

Consequently, to obtain vibrational modes and static deformation of a structure with one plane of symmetry, we only need to solve two smaller problems involving half of the original structure. Moreover, if the external loads are themselves symmetric or antisymmetric with respect to the same plane of symmetry, then we only need to solve the static deformation problem involving half of the structure.

If the simplified discrete system of a structure with mirror symmetry is valid, then its stiffness and mass matrices are both real symmetric matrices. In addition, the static deformation and mode shapes of this discrete system have the same qualitative properties as their counterparts associated with the respective continuous system.

1.7.2 Rotationally and Linearly Periodic Structures

The shape, physical properties, and boundary conditions of a rotationally periodic structure with n sectors are all periodic functions of the angle of rotation ψ about the center axis, and they have the same period $\psi = 2\pi/n$, where n is a positive integer. As the angle of rotation ψ assumes the values $2\pi k/n$ $(k = 1, 2, \ldots, n)$ one after the other, we obtain n sectors or substructures, and label them with numbers k $(k = 1, 2, \ldots, n)$ accordingly. These n substructures fill the entire circumferential space, and form the complete structure.

The displacement of the k-th substructure $\boldsymbol{u}_k = \boldsymbol{u}_k(\boldsymbol{x})$ may be a one-, two-, or three-dimensional function of $\boldsymbol{x}$, which represents the one-, two-, or three-dimensional spatial coordinate of the substructure, respectively. Thus, the displacement $\boldsymbol{u}$ of the entire structure can be expressed as a vector consisting of n components that denote displacements of the n substructures:

$$\boldsymbol{u} = (\boldsymbol{u}_1, \boldsymbol{u}_2, \ldots, \boldsymbol{u}_n)^{\mathrm{T}}.$$

The main qualitative properties of vibration associated with this type of structures are summarized in the next a few paragraphs.

Consider the r-th group of mode shapes

$$\boldsymbol{u}^{(r)} = (\boldsymbol{u}_1^{(r)}, \boldsymbol{u}_2^{(r)}, \ldots, \boldsymbol{u}_n^{(r)})^{\mathrm{T}}.$$

Mode shape components of adjacent substructures are related to each other by the following equation:

$$\boldsymbol{u}_{k+1}^{(r)} = \mathrm{e}^{\mathrm{i}r\psi}\boldsymbol{u}_k^{(r)}, \quad r = 1, 2, \ldots, n. \tag{1.7.1}$$

There are several scenarios based on Eq. (1.7.1):

(1) Mode shape components of adjacent substructures are identical, which implies mode shape components of all substructures to be the same. This corresponds to the case of $r = n$ in Eq. (1.7.1), i.e.,

$$\boldsymbol{u}^{(n)} = (\boldsymbol{q}_n, \boldsymbol{q}_n, \ldots, \boldsymbol{q}_n)^{\mathrm{T}},$$

where $\boldsymbol{q}_n$ is the eigenfunction obtained by solving the eigenvalue problem of a single substructure. Components of $\boldsymbol{u}^{(n)}$ from adjacent substructures should have the same value and sign, and so are other variables that might appear at the interface conditions.

(2) When n is an even number, mode shape components of adjacent substructures are of the same magnitude but opposite signs. This corresponds to the case of $r = n/2$ in Eq. (1.7.1), i.e.,

$$\boldsymbol{u}^{(n/2)} = (\boldsymbol{q}_{n/2}, -\boldsymbol{q}_{n/2}, \ldots, \boldsymbol{q}_{n/2}, -\boldsymbol{q}_{n/2})^{\mathrm{T}},$$

where $\boldsymbol{q}_{n/2}$ is also the eigenfunction obtained by solving the eigenvalue problem of one substructure. Components of $\boldsymbol{u}^{(n/2)}$ from adjacent substructures should have the same value but opposite signs at the two interfaces of the substructure, and so are other variables that might appear at the interface conditions.

(3) When $r \neq n$ and $r \neq n/2$ (n is an even number), the frequency is repeated, and there exist two sets of mode shapes for each distinct frequency:

$$\boldsymbol{v}_1^{(r)}, \boldsymbol{v}_2^{(r)}, \ldots \boldsymbol{v}_n^{(r)} \quad \text{and} \quad \boldsymbol{w}_1^{(r)}, \boldsymbol{w}_2^{(r)}, \ldots, \boldsymbol{w}_n^{(r)},$$

where $r = 1, 2, \ldots, (n-2)/2$ if n is an even number or $(n-1)/2$ if n is odd. Mode shapes of a substructure are related to those of the adjacent substructure through the following equations:

$$\begin{pmatrix} \boldsymbol{v}_{k+1}^{(r)} \\ \boldsymbol{w}_{k+1}^{(r)} \end{pmatrix} = \begin{pmatrix} \cos r\psi & -\sin r\psi \\ \sin r\psi & \cos r\psi \end{pmatrix} \begin{pmatrix} \boldsymbol{v}_k^{(r)} \\ \boldsymbol{w}_k^{(r)} \end{pmatrix}, \quad k = 1, 2, \ldots, n-1, \tag{1.7.2}$$

where $\boldsymbol{v}_1^{(r)}$ and $\boldsymbol{w}_1^{(r)}$ are eigenfunctions obtained by solving a coupled eigenvalue problem of just one substructure, which is twice the size of the decoupled eigenvalue problem we see earlier when $r = n$ or $r = n/2$ (n is even). The related boundary conditions at the two interfaces of the substructure are derived from Eq. (1.7.2). Details can be found in Sect. 7.2.

As for the static deformation of a rotationally periodic structure, we can discuss its qualitative properties by considering three scenarios for its external loads and static deformation: (1) Components of the external forces on one substructure are identical to those on any other substructure, and so are the components of the corresponding static deformation. There is only one group of such solution. (2) If n is an even number, then components of the external forces on a substructure are identical in magnitude but opposite in sign to those on any adjacent substructure, and so are the components of the respective static deformation. There is only one group of such solution, too. (3) One pair of external loads is applied to each substructure, and they are related to their counterparts at the adjacent substructure in a way similar to Eq. (1.7.2); while the corresponding pair of static deformation, say $\boldsymbol{v}^{(r)}$ and $\boldsymbol{w}^{(r)}$ ($r = 1, 2, \ldots, N$), satisfies Eq. (1.7.2), where $N = (n-2)/2$ if n is even and $N = (n-1)/2$ if n is odd. There are N groups of solutions of this type.

Following the process outlined above, we may obtain solutions of vibrational modes and static deformation at the substructure level. We still need to assemble them in order to get the corresponding solutions of the entire structure. The equation for assembling mode shapes of substructures is given in Sect. 7.2.2, while the static deformation of substructures can be assembled in a similar fashion.

Stiffness and mass matrices of the valid discrete system of a rotationally periodic structure are both circulant matrices. Qualitative properties of vibrational modes and static deformation associated with the discrete system parallel their counterparts of the respective continuous system.

Qualitative properties of vibrational modes and static deformation associated with some linearly periodic structures are similar to those of their corresponding rotationally periodic structures.

1.7.3 Chain Structures

A chain structure may be obtained by uniformly distributing a number of identical substructures, joining them with the same elastical and/or rigid connections, and then fixing the two external ends.

Consider a chain structure consisting of n substructures. Label these substructures sequentially as $1, 2, \ldots, k, \ldots, n$. Let $\boldsymbol{u}_k$ denote the displacement function of the k-th substructure. The displacement function of the entire structure may be represented by a vector $\boldsymbol{u}$ consisting of n components, each of which represents the displacement at one substructure, i.e.,

$$\boldsymbol{u} = (\boldsymbol{u}_1, \boldsymbol{u}_2, \ldots, \boldsymbol{u}_n)^{\mathrm{T}}.$$

Thus, the r-th group of mode shapes of the entire structure is of the form

$$\boldsymbol{u}^{(r)} = (\boldsymbol{u}_1^{(r)}, \boldsymbol{u}_2^{(r)}, \ldots, \boldsymbol{u}_n^{(r)})^{\mathrm{T}} = (\sin r\psi \boldsymbol{I}, \sin 2r\psi \boldsymbol{I}, \ldots, \sin nr\psi \boldsymbol{I})^{\mathrm{T}} \boldsymbol{q}_r, \quad (1.7.3)$$

where $r = 1, 2, \ldots, n$, $\boldsymbol{I}$ designates the identity matrix, $\psi = 2\pi/(n+1)$, and $\boldsymbol{q}_r$ is the eigenfunction of the decoupled eigenvalue problem involving one substructure only.

As for the static deformation, the external loads applied to a chain structure consisting of n substructures can be decomposed into n external forces, each of which acts on a single substructure, in a way similar to Eq. (1.7.3). Thus, we may solve for the static deformation of just one substructure, and then use Eq. (1.7.3) to get the static deformation of the entire structure.

Stiffness and mass matrices of the valid discrete system of a chain structure are both block tri-diagonal matrices. Furthermore, for each of these matrices, the submatrices along the diagonal are identical, and those along the subdiagonals are also the same. Qualitative properties of vibrational modes and static deformation associated with the discrete system parallel those of the corresponding continuous system.

1.7.4 Axisymmetric Structures

For an axisymmetric structure, its shape, physical properties, and boundary conditions are all independent of the angle of rotation about a straight line, which is referred to as the axis of rotation. Examples of three- and two-dimensional axisymmetric structures are, respectively, cylindrical solids and revolving shells, both of which are with uniform mass density and elastic parameters as well as subject to axisymmetric boundary conditions.

Mode shapes and static deformation of an axisymmetric structure can be expressed as a series of sine and cosine functions along the circumferential direction. This helps reduce the size of the problem of solving for mode shapes and static deformation by one dimension.

Before ending this section, we would like to mention that details of the content presented here can be found in Chap. 7. Furthermore, some other subjects related to repetitive structures are also discussed in Chap. 7, including methods for decoupling forced vibration and reducing the dimension in vibration control.

1.8 Overview of the Existence of Solutions in Elasticity and Structural Theories as Well as Other Subjects in Fundamental Theories

1.8.1 Types of Solutions in Elasticity and Structural Mechanics

There are three categories of problems in Elasticity and Structural Mechanics: Find static deformation caused by a given set of static external forces, solve for vibrational modes, and determine the dynamic response to the specified dynamic external loads as well as initial conditions of the displacement and velocity.

In addition, there are two types of solutions associated with each category of the problems outlined above:

(1) A classical solution is one that satisfies the governing differential equation as well as the specified boundary and, possibly, initial conditions.
(2) A generalized solution refers to the solution satisfying, in the context of generalized differentiation, the variational equation that is derived from the principle of minimum potential energy, minimization of the Rayleigh quotient, etc.

These solutions have dissimilar differentiability. Specifically, for a differential equation of the $2k$-th order, its classical solution should have, at least, $2k$-th classical derivatives; while the generalized solution only needs to have k-th generalized derivatives. Our discussion on the existence of solutions will be conducted within the context of generalized solutions. For a specific problem, we can determine the actual highest-order generalized derivatives of its generalized solution, based on the differentiability of coefficients of the governing equations, boundary shapes, external forces, as well as boundary conditions.

1.8.2 Existence Theorems for Solutions in Elasticity

The governing equations for static deformation and vibrational modes in Elasticity are, respectively,

$$\begin{cases} \boldsymbol{A}_{\mathrm{e}}\boldsymbol{u} = -\sum\limits_{i,k,l,m=1}^{3} \dfrac{\partial}{\partial x_i}(c_{iklm}\varepsilon_{lm}(\boldsymbol{u}))\boldsymbol{x}_k^{(0)} = \boldsymbol{f}(\boldsymbol{x}) & \text{in } \Omega, \\ \boldsymbol{B}_{\mathrm{e}}\boldsymbol{u} = \boldsymbol{0} & \text{on } \partial\Omega, \end{cases} \tag{1.8.1}$$

and

$$\begin{cases} \boldsymbol{A}_{\mathrm{e}}\boldsymbol{u} = \omega^2\rho\boldsymbol{u} & \text{in } \Omega, \\ \boldsymbol{B}_{\mathrm{e}}\boldsymbol{u} = \boldsymbol{0} & \text{on } \partial\Omega, \end{cases} \tag{1.8.2}$$

where $\boldsymbol{u}$ denotes the three-dimensional displacement vector, ε_{lm} is a strain component, $\boldsymbol{x}_k^{(0)}$ represents the unit vector along the coordinate axis x_k ($k = 1, 2, 3$), the elasticity operator $\boldsymbol{A}_{\mathrm{e}}$ is a second-order differential operator, and $\boldsymbol{B}_{\mathrm{e}}$ stands for the boundary differential operator.

Numbers of elastic coefficients c_{iklm} for anisotropic and isotropic objects are, respectively, 21 and 2. Additionally, for nonhomogeneous materials, their elastic coefficients are functions of the spatial coordinate $\boldsymbol{x}$; while, for homogeneous materials, they are all constants. The mass density ρ is either a positive number or a bounded positive function. Suppose an elastic object occupies a bounded domain Ω with the boundary $\partial\Omega$. There are six types of common boundary conditions: (1) Fixed boundary, i.e., $\boldsymbol{u}|_{\partial\Omega} = 0$. (2) Free boundary, i.e., $\boldsymbol{t}(\boldsymbol{u})|_{\partial\Omega} = 0$. (3) contact boundary, i.e., $\boldsymbol{u}_{(\nu)}|_{\partial\Omega} = 0$ and $\boldsymbol{t}(\boldsymbol{u})_{(\mathrm{s})}|_{\partial\Omega} = 0$, where ν and s denote normal and tangential directions at the boundary, respectively. (4) The displacement is unconstrained in the normal direction but fixed along the tangential direction, i.e., $\boldsymbol{u}_{(\mathrm{s})}|_{\partial\Omega} = 0$ and $\boldsymbol{t}(\boldsymbol{u})_{(\nu)}|_{\partial\Omega} = 0$. (5) Hybrid boundary, formed by combining the previous four types of conditions. (6) Elastic boundary, i.e., $\boldsymbol{t}(\boldsymbol{u})|_{\partial\Omega} + \boldsymbol{K}\boldsymbol{u}|_{\partial\Omega} = 0$, where $\boldsymbol{K}$ is a 3×3 diagonal matrix, with positive diagonal elements.

Define

$$(\boldsymbol{u},\, \boldsymbol{v}) = \int \boldsymbol{u} \cdot \boldsymbol{v}\, \mathrm{d}\Omega$$

to be the inner product of the square-integrable space L_2. Thus, $(\boldsymbol{A}_\mathrm{e}\boldsymbol{u}, \boldsymbol{u})$ denotes twice of the magnitude of the strain energy associated with the displacement $\boldsymbol{u}$, while $(\rho\boldsymbol{u}, \boldsymbol{u})$ represents two times the coefficient of the magnitude of the kinetic energy. Integrate $(\boldsymbol{A}_\mathrm{e}\boldsymbol{u}, \boldsymbol{u})$ by parts and apply the boundary conditions. The resulting Hilbert space is called the space with the strain energy norm and represented by H_{A_e}. Next, the Hilbert space with the norm $(\rho\boldsymbol{u}, \boldsymbol{u})$ is referred to as the space with the kinetic energy norm and denoted by H_ρ. Finally, the mapping from the space with the strain energy norm to the space with the kinetic energy norm is known as the energy embedding operator.

Theorem on the Positive Definiteness of the Elasticity Operator *If elastic coefficients* $\{c_{iklm}\}$ *of an elastic solid have piecewise continuous first derivatives in a domain* Ω *that is bounded and with piecewise smooth border, and any one set of boundary conditions (1)–(5) given above is satisfied; then the elasticity operator* $\mathbf{A}_\mathrm{e}$ *is positive definite, i.e.,*

$$(\boldsymbol{A}_\mathrm{e}\boldsymbol{u}, \boldsymbol{u}) \geq \gamma^2 \|\boldsymbol{u}\|^2,$$

where γ *is a positive number.*

In Physics, the positive definiteness of an elasticity operator indicates the ratio of the strain energy norm to the kinectic energy norm to be a positive number; while, in Mathematics, it implies that the embedding operator from the space with the strain energy norm to the space with the kinetic energy norm is bounded.

Theorem on the Compactness of the Energy Embedding Operator in Elasticity *If elastic coefficients* $\{c_{iklm}\}$ *of an elastic solid have piecewise continuous first derivatives in a domain* Ω *that is bounded and with piecewise smooth border, the mass density* ρ *is a bounded and positive function and* $\rho \in L_2$*, and any one set of the aforementioned boundary conditions (1)–(5) is satisfied; then the energy embedding operator in Elasticity is compact.*

It needs to be pointed out that there is an important comment in Sect. 8.4.2 with regard to certain boundary conditions allowing rigid-body motion of the elastic solid. In this situation, additional requirements must be imposed to ensure applicability of the above two theorems.

Based on the theory of Hilbert Space, we may obtain the following existence theorems for solutions in Elasticity:

Existence Theorem for the Solution of Static Deformation in Elasticity *If elastic coefficients* $\{c_{iklm}\}$ *of an elastic solid have piecewise continuous first derivatives in a domain* Ω *that is bounded and with piecewise smooth border, the external force function belongs to* $L_2(\Omega)$*, and any one set of the aforementioned boundary conditions (1)–(5) is satisfied; then there exists a unique generalized solution to the static deformation problem in Elasticity.*

Existence Theorem for Modal Solutions in Elasticity *If elastic coefficients* $\{c_{iklm}\}$ *of an elastic solid have piecewise continuous first derivatives in a domain* Ω *that is bounded and with piecewise smooth border, the mass density* ρ *is a bounded and positive function and* $\rho \in L_2$, *and any one set of the aforementioned boundary conditions (1)–(5) is satisfied; then there exist generalized solutions to the modal problem in Elasticity. Specifically, there are countable but infinitely many natural frequencies, and the infinity is their only limit point; while the respective mode shapes form a complete and orthogonal set in both the space with the kinetic energy norm and the space with the strain energy norm.*

1.8.3 Existence Theorems for Solutions in Structural Theories

There are a variety of structural theories in Structural Mechanics, and they originated from different structural models. These theoretical models cover not only components such as bars, beams, plates, and shells, but also the more sophisticated assembly structures that have complex shapes and are made of advanced materials. Although they involve diverse physical characteristics and mathematical representations, all these models may be collectively viewed as derivatives of the elasticity model of a three-dimensional elastic solid by going through three types of simplifications: (1) simplification of the deformation, realized primarily by incorporating displacement constraints; (2) simplification of the stress, generally accomplished through relaxation of certain stress components; and (3) simplification of the mass distribution, normally achieved via mass lumping.

The governing equation and boundary condition for the static deformation in a structural theory are of the form:

$$\begin{cases} \boldsymbol{A}_s \boldsymbol{w}(\boldsymbol{x}) = \boldsymbol{f}(\boldsymbol{x}) & \text{in } \Omega, \\ \boldsymbol{B}_s \boldsymbol{w}(\boldsymbol{x}) = \boldsymbol{0} & \text{on } \partial\Omega, \end{cases} \tag{1.8.3}$$

while the frequency and mode shape in the structural theory satisfy the following equation and boundary condition:

$$\begin{cases} \boldsymbol{A}_s \boldsymbol{w}(\boldsymbol{x}) - \omega^2 m \boldsymbol{w}(\boldsymbol{x}) = \boldsymbol{0} & \text{in } \Omega, \\ \boldsymbol{B}_s \boldsymbol{w}(\boldsymbol{x}) = \boldsymbol{0} & \text{on } \partial\Omega, \end{cases} \tag{1.8.4}$$

where $\boldsymbol{A}_s$ stands for the structural theory operator, $\boldsymbol{B}_s$ represents the boundary differential operator, $\boldsymbol{w}(\boldsymbol{x})$ denotes the generalized displacement in the structural theory, and m is the mass density of the structure. Similar to their counterparts in Elasticity, $(\boldsymbol{A}_s\boldsymbol{w}, \boldsymbol{w})$ and $(m\boldsymbol{w}, \boldsymbol{w})$ are, respectively, twice of the magnitude of the strain energy and two times the coefficient of the magnitude of the kinetic energy in the structural theory. Integrate $(\boldsymbol{A}_s\boldsymbol{w}, \boldsymbol{w})$ by parts and apply the boundary conditions. The resulting Hilbert space is called the space with the strain energy norm in the structural theory.

Furthermore, the Hilbert space with the norm $(m\boldsymbol{w}, \boldsymbol{w})$ is referred to as the space with the kinetic energy norm in the structural theory.

Given a structure and its boundary conditions, we may identify the equivalent three-dimensional elastic solid and related boundary conditions. The latter is referred to as the elastic solid corresponding to the given structure, or simply, the corresponding elastic solid. Furthermore, if simplification of the deformation is applied to this elastic solid, then the resulting object is called the constrained elastic solid corresponding to the original structure, or the corresponding constrained elastic solid in short.

Theorem on the Positive Definiteness of the Structural Theory Operator *For an elastic structure with given elastic parameters, geometric shape, and boundary conditions, if*

(1) *the corresponding elastic solid of the structure and its boundary conditions ensure the positive definiteness of the elasticity operator* $\mathbf{A}_e$, *and*
(2) *there is an equivalence of norms between the two H spaces with the strain energy norm, one associated with the original structure while the other related to the corresponding constrained elastic solid,*

then the structural theory operator $\mathbf{A}_e$ *of this structure is positive definite.*

Theorem on the Compactness of the Energy Embedding Operator in the Structural Theory *For an elastic structure with specified elastic parameters, mass density, geometric shape, and boundary conditions. If*

(1) *the corresponding elastic solid of the structure and its boundary conditions guarantee the energy embedding operator in Elasticity to be compact,*
(2) *there is an equivalence of norms between the two H spaces with the strain energy norm, one associated with the original structure while the other related to the corresponding constrained elastic solid, and*
(3) *there also exists an equivalence of norms between the two H spaces with the kinetic energy norm, one associated with the original structure while the other related to the corresponding constrained elastic solid,*

then the energy embedding operator in the structural theory, which is associated with the original structure, is compact.

The above two theorems are graphically illustrated in Fig. 8.2a, b in Chap. 8.

Existence Theorem for the Solution of Static Deformation in Structural Theory *For an elastic structure with given elastic parameters, geometric shape, and boundary conditions, if*

(1) *the corresponding elastic solid of the structure and its boundary conditions ensure the positive definiteness of the elasticity operator* $\mathbf{A}_e$,
(2) *there is an equivalence of norms between the two H spaces with the strain energy norm, one associated with the original structure while the other related to the corresponding constrained elastic solid, and*
(3) *the external force applied to the structure belongs to the* L_2 *space,*

then there exists a unique generalized solution to the static deformation problem of the structure.

On the one hand, this result may be referred to as the preservation theorem that preserves the existence of the solution of static deformation from Elasticity to structural theories, if the structural theories are regarded as some simplification of the more sophisticated three-dimensional elasticity theory. On the other hand, it may be called an extension theorem that extends the existence of the solution of static deformation from Elasticity to the structural theories, if the more multifarious and practical structural theories are viewed as certain extension of Elasticity.

Existence Theorem for Modal Solutions in Structural Theory *For an elastic structure with specified elastic parameters, mass density, geometric shape, and boundary conditions, if*

(1) *the corresponding elastic solid of the structure and its boundary conditions guarantee the positive definiteness of the elasticity operator* $\mathbf{A}_e$ *as well as the compactness of the energy embedding operator in Elasticity,*
(2) *there is an equivalence of norms between the two H spaces with the strain energy norm, one associated with the original structure while the other related to the corresponding constrained elastic solid, and*
(3) *there is also an equivalence of norms between the two H spaces with the kinetic energy norm, one associated with the original structure while the other related to the corresponding constrained elastic solid,*

then there exist generalized solutions to the modal problem of the structure. Specifically, there is a countable but infinite set of natural frequencies, with the infinity as the sole limit point; while the corresponding mode shapes formed a complete and orthogonal set in both the space with the kinetic energy norm and the space with the strain energy norm.

As with the previous theorem, this result is also known as, either the preservation theorem that preserves or the extension theorem that extends, modal solutions from Elasticity to Structural Theories, depending on a person's perspective.

1.8.4 Validity of Models of Structures in Structural Theories

If the generalized solution of a structure exists, then the structure is considered to be valid. Otherwise, it is not valid.

For an individual structure, existence of a solution to its structural problem involves two aspects: (1) validity of the particular theoretical model used, such as the Euler–Bernoulli beam, Timoshenko beam, thin plate, Mindlin plate, etc.; (2) validity of specific structural parameters, such as, in the case of a plate, the shape of the boundary, constraints at the border, distribution of the bending stiffness, etc.

From theorems in Sect. 1.8.3 on the existence of solutions of static deformation and vibrational modes in structural theories, the theoretical model of a structure is valid if it has the following properties: (1) there is an equivalence of norms between the two spaces with the strain energy norm, one associated with the theoretical model of the structure while the other related to the respective constrained elastic solid; and (2) there is also an equivalence of norms between the spaces with the kinetic energy norm, one associated with the theoretical model of the structure while the other related to the corresponding constrained elastic solid. For a specific structure with some valid theoretical model, if the corresponding elastic solid has a positive definite elasticity operator while its energy embedding operator is compact, then the structure itself is valid.

Existing models in structural theories can be classified into three types: (1) Many effective and widely used structural models are valid, including the beam, thin plate, thin shell, etc. The materials involved can be either conventional or composite. (2) If a concentrated parameter is applied to a structure whose Green's function has a singularity, then the structure could be invalid. This may occur when a concentrated support or mass is attached to the singular point on a Mindlin plate, shell, or elastic solid. (3) Assembly structures deserve some special attention. Two things need to be checked on: First, is each component a valid structural model? Second, is there any compatibility issue at the joint of components? Only when answers to both questions are positive, can we conclude that the assembly structure itself is valid. Incompatibility at the joint occurs, when the joint is reduced to a point or line and the Green's function of one component happens to be singular at this location. Here are some examples: a beam with one end connected to a three-dimensional solid, a plate with one side attached to a three-dimensional solid, a circular beam wrapped around a cylindrical solid of the same radius, a bar with one end linked to a Mindlin plate, and a beam with one end affixed to the surface of a shell.

Details of materials presented in this section can be found in Chap. 8.

References

1. Courant R, Hilbert D (1962) Methods of mathematical physics, vol I, 1953, vol II. Interscience Publishers, New York
2. Faber G (1923) Beweis, dass unter allen homogenen membranen van gleicher fläche und gleicher spannung die kreisfömige den tiefsten grundton gibt. S B Math-Nat KI Bayer Akad Wiss, 169–172
3. Gantmacher FP, Krein MG (1961) Oscillation matrices and kernels and small vibrations of mechanical systems. U. S. Atomic Energy Commission, Washington
4. Gladwell GML (2004) Inverse Problems in Vibration. 2nd edn. Springer, Dordrecht (1986, 1st edn, Martinus Nijhoff Publishers, Dordrecht)
5. Hu HC, Liu ZS, Wang DJ (1996) Influence of small changes of support location on vibration mode shape. Acta Mech Sin 28(1):23–32 (in Chinese)
6. Krahn E (1924) Uber eine von Raylegh formulierte minimaleigenschaft des kreises. Math Ann 1924(94):97–100
7. Leung AYT, Wang DJ, Wang Q (2004) On concentrated mass and stiffness in structural theories. Int J Struct Stab Dyn 4:171–179

8. Liu ZS, Hu HC, Wang DJ (1994) Effect of small variation of support location on natural frequencies. In: Proceedings of international conference on vibration engineering, ICVE'94, Beijing. International Academic Publishers, Singapore, pp 9–12
9. Liu ZS, Hu HC, Wang DJ (1996) New method for deriving eigenvalue rate with respect to support location. AIAA J 34(4):864–865
10. Pleijel A (1956) Remarks on Courant's nodal line theorem. Comm Pure Appl Math 1956(4):543–550
11. Polterovich I (2009) Pleijel's nodal domain theorem for free membranes. Proc Am Math Soc 137(3):1021–1024
12. Wang DJ, Hu HC (1982) A unified proof for the positive definiteness and compactness of two kinds of operators in the theory of elastic structure. Acta Mech Sin 14(2):111–121 (in Chinese)
13. Wang DJ, Hu HC (1982) A unified proof of the general properties of the linear vibrations in the theory of elastic structures. J Vibr Shock, 1(1):6–16 (in Chinese)
14. Wang DJ, Hu HC (1983) A unified proof for the positive-definiteness and compactness of two kinds of operators in the theories of elastic structures. In: Proceedings of the China-France symposium on finite element methods. Science Press China, New York, pp 6–16
15. Wang DJ, Hu HC (1985) Positive definiteness and compactness of two kinds of operators in theory of elastic structures. Sci Sin Ser A 28(7):727–739
16. Wang DJ, He BC, Wang QS (1990) On the construction of the Euler-Bernoulli beam via two sets of modes and the corresponding frequencies. Acta Mech Sin 22(4):479–483 (in Chinese)
17. Wang QS, Wang DJ (1987) Construction of the discrete system for the rod by partial natural modes and frequencies data. J Vibr Eng 1987(1):83–87 (in Chinese)
18. Wang QS, He BC, Wang DJ (1990) Some qualitative properties of frequencies and modes of Euler beams. J Vibr Eng 3(4):58–66 (in Chinese)
19. Wang QS, Wang DJ, He BC (1992) Qualitative properties of frequencies and modes of discrete system of continuous second-order systems. J Vibration and Shock, 11(3):7–12 (in Chinese)
20. Wang QS, Wang DJ (1994) An inverse mode problem for continuous second-order systems. In: Proceedings of international conference on vibration engineering, ICVE'94, Beijing. International Academic Publishers, Singapore, pp 167–170
21. Wang QS, Wang DJ (1997) Qualitative properties of frequency spectrum and modes of arbitrary supported beams in vibration. Acta Mech Sin 29(5):540–547 (in Chinese)
22. Wang QS, Huang PC (2008) The influence for the frequency of beams by adding internal support. Mod Vibr Noise Technol 6:104–107
23. Zheng ZJ, Chen P, Wang DJ (2013) Oscillation property of the vibrations for finite element models of Euler beam. Qtr J Mech & Appl Math, 66(4):587–608
24. Zheng ZJ (2014) The qualitative vibrational property and modal inverse problems of rods and Euler beams [D]. Department of Mechanics and Engineering Science, College of Engineering, Peking University (in Chinese)

Chapter 2
Oscillatory Matrices and Kernels as Well as Properties of Eigenpairs

The theory of oscillatory matrices and kernels forms the mathematical foundation for the study of qualitative properties of natural frequencies and mode shapes of bars and beams. This chapter provides an introduction to the theory. The content is drawn largely from the monograph, *Oscillation Matrices and Kernels and Small Vibrations of Mechanical Systems*, written by creators of the theory, Gantmacher and Krein; but Sect. 2.11 and most of Sect. 2.10 are the original work by authors of this book as well as their collaborators Zijun Zheng and Pu Chen.

2.1 A Few Notations and Definitions

Denote a matrix consisting of $m \times n$ elements in a rectangular table of m rows and n columns as

$$\boldsymbol{A} = \begin{bmatrix} a_{11} & a_{12} & \cdots & a_{1n} \\ a_{21} & a_{22} & \cdots & a_{2n} \\ \vdots & \vdots & & \vdots \\ a_{m1} & a_{m2} & \cdots & a_{mn} \end{bmatrix}.$$

It is called an $m \times n$ matrix and denoted by $(a_{ij})_{m\times n}$ in short. The number a_{ij} is called the element at the i-th row and j-th column. An $n \times n$ matrix is referred to as a square matrix of order n, and may be written in an abbreviated form as (a_{ij}).

A column vector is designated by $\boldsymbol{a} = (a_1, a_2, \ldots, a_n)^{\mathrm{T}}$.

The determinant of a matrix $\boldsymbol{A}$ is represented by either $|\boldsymbol{A}|$ or $\det \boldsymbol{A}$. The p-th-order minor of $\boldsymbol{A}$ refers to the determinant formed by elements at the intersection of p rows and p columns of the matrix. It is written as

D. Wang et al., *Qualitative Theory in Structural Mechanics*,
https://doi.org/10.1007/978-981-13-1376-9_2

$$A\begin{pmatrix} i_1\, i_2 \cdots i_p \\ j_1\, j_2 \cdots j_p \end{pmatrix} = \begin{vmatrix} a_{i_1j_1} & a_{i_1j_2} & \cdots & a_{i_1j_p} \\ a_{i_2j_1} & a_{i_2j_2} & \cdots & a_{i_2j_p} \\ \vdots & \vdots & & \vdots \\ a_{i_pj_1} & a_{i_pj_2} & \cdots & a_{i_pj_p} \end{vmatrix}, \quad 1 \le \begin{matrix} i_1 < i_2 < \cdots < i_p \\ j_1 < j_2 < \cdots < j_p \end{matrix} \le n. \quad (2.1.1)$$

The minor is called the principal minor of the p-th-order, if $i_r = j_r$ $(r = 1, 2, \ldots, p)$. When $i_r = j_r = r$ $(r = 1, 2, \ldots, p)$, the principal minor is further classified as the leading principal minor of the p-th order, and is denoted by D_p. A quasi-principal minor of $\boldsymbol{A}$ refers to one with $|i_r - j_r| \le 1$ and $j_r \le i_{r+1}$ $(r = 1, 2, \ldots, p-1)$.

The first minor M_{ij} corresponding to the element a_{ij} in the matrix $\boldsymbol{A}$, also known as the (i, j) minor, is a minor formed by remaining elements of $\boldsymbol{A}$ after its i-th row and j-th column are removed; while $A_{ij} = (-1)^{i+j}M_{ij}$ is called the cofactor of the element a_{ij}, or the (i, j) cofactor.

Listed in the next a few paragraphs are several types of matrices commonly used in this book.

The truncated submatrix of $(a_{ij})_{n\times n}$ is defined by

$$(a_{ij})_q^p = \begin{bmatrix} a_{qq} & a_{q,q+1} & \cdots & a_{qp} \\ a_{q+1,q} & a_{q+1,q+1} & \cdots & a_{q+1,p} \\ \vdots & \vdots & & \vdots \\ a_{pq} & a_{p,q+1} & \cdots & a_{pp} \end{bmatrix}, \quad 1 \le q \le p \le n.$$

The matrix $\boldsymbol{A} = (a_{ij})$ is symmetric if $a_{ij} = a_{ji}$, where i and $j = 1, 2, \ldots, n$. The majority of matrices in engineering applications are symmetric. Obviously, a symmetric matrix has to be a square matrix.

The $n \times m$ matrix obtained by exchanging rows and columns of an $m \times n$ matrix $\boldsymbol{A}$ is called the transpose of $\boldsymbol{A}$, and is written as $\boldsymbol{A}^{\mathrm{T}}$.

The diagonal matrix, denoted by $\mathrm{diag}(a_{11}, a_{22}, \ldots, a_{nn})$, refers to a special type of square matrices whose elements a_{ij} vanish whenever $i \ne j$. The matrix $\mathrm{diag}(1, 1, \ldots, 1)$ is called an identity matrix and designated by $\boldsymbol{I}$.

If $\boldsymbol{AB} = \boldsymbol{I}$, then $\boldsymbol{B}$ is the right inverse matrix of $\boldsymbol{A}$; while if $\boldsymbol{BA} = \boldsymbol{I}$, then $\boldsymbol{B}$ is the left inverse of $\boldsymbol{A}$. When $\boldsymbol{A}$ is a symmetric matrix, its left and right inverse matrices are, apparently, the same and may be denoted by $\boldsymbol{A}^{-1}$. It is easy to prove: If the determinant of a matrix $\boldsymbol{A}$ satisfies $|\boldsymbol{A}| \ne 0$, i.e., $\boldsymbol{A}$ is nonsingular, then its inverse matrix is given by

$$\boldsymbol{A}^{-1} = (A_{ji})/|\boldsymbol{A}|.$$

A nonnegative matrix refers to one whose elements are all nonnegative, while a positive matrix is one whose elements are all positive. Furthermore, a totally nonnegative matrix refers to a square matrix whose minors are all nonnegative, while a totally positive matrix is a square matrix whose minors are always positive. Obviously, a totally positive matrix must be nonsingular, i.e., its determinant cannot be zero. Also, notice that a totally positive matrix differs from a positive definite matrix. The necessary and sufficient conditions for the latter are as follows: Its leading principal minors are all positive and the matrix must be symmetric.

2.2 Some Relationship Among Minors

To support subsequent discussions, a few important relationships among minors are introduced in this section. Due to space limitation, the results are presented without proof.

If $\boldsymbol{A}$ and $\boldsymbol{B}$ are $n \times m$ and $m \times n$ matrices, respectively, then $\boldsymbol{C} = \boldsymbol{AB}$ is a square matrix of order n, and its element c_{ij} can be calculated by

$$c_{ij} = \sum_{k=1}^{n} a_{ik} b_{kj}, \quad i \text{ and } j = 1, 2, \ldots, n.$$

Generalizing this result, we get the determinant of the matrix $\boldsymbol{C}$ when $n < m$,

$$C\begin{pmatrix} 1\ 2 \cdots n \\ 1\ 2 \cdots n \end{pmatrix} = \sum A\begin{pmatrix} 1\ 2 \cdots n \\ j_1\ j_2 \cdots j_n \end{pmatrix} B\begin{pmatrix} j_1\ j_2 \cdots j_n \\ 1\ 2 \cdots n \end{pmatrix}, \tag{2.2.1}$$

where the summation runs through all possible index combinations satisfying $1 \le j_1 < j_2 < \cdots < j_n < m$. In addition, when both $\boldsymbol{A}$ and $\boldsymbol{B}$ are square matrices of order n, we obtain, for any p where $1 \le p \le n$, the p-th-order minor of the matrix $\boldsymbol{C}$ is

$$C\begin{pmatrix} i_1\ i_2 \cdots i_p \\ k_1\ k_2 \cdots k_p \end{pmatrix} = \sum A\begin{pmatrix} i_1\ i_2 \cdots i_p \\ j_1\ j_2\ \vdots\ j_p \end{pmatrix} B\begin{pmatrix} j_1\ j_2 \cdots j_p \\ k_1\ k_2 \cdots k_p \end{pmatrix}, \tag{2.2.2}$$

where $1 \le i_1 < i_2 < \cdots < i_p \le n$, $1 \le k_1 < k_2 < \cdots < k_p \le n$, and the summation also exhausts all possible index combinations satisfying $1 \le j_1 < j_2 < \cdots < j_p \le n$. This is the famous Binet–Cauchy identity. Starting from this result, we may get the theorem below:

Theorem 2.1 *The product of totally nonnegative matrices remains totally nonnegative, while the product of a totally positive matrix and a nonsingular totally nonnegative matrix is a totally positive matrix.*

Theorem 2.2 *Consider a square matrix* $\boldsymbol{A} = (a_{ij})_{n \times n}$ *and set* $\widetilde{\boldsymbol{A}} = (M_{ij})$. *Then,*

$$\widetilde{A}\begin{pmatrix} i_1\ i_2 \cdots i_p \\ j_1\ j_2 \cdots j_p \end{pmatrix} = |\boldsymbol{A}|^{p-1} A\begin{pmatrix} k_1\ k_2 \cdots k_{n-p} \\ s_1\ s_2 \cdots s_{n-p} \end{pmatrix}, \tag{2.2.3}$$

where $1, 2, \ldots, n$ *is the full set of indices,* $k_1 < k_2 < \cdots < k_{n-p}$ *constitutes the complement of* $i_1 < i_2 < \cdots < i_p\,(1 \le i_1; i_p \le n)$*, and* $s_1 < s_2 < \cdots < s_{n-p}$ *represents the complement of* $j_1 < j_2 < \cdots < j_p\,(1 \le j_1; j_p \le n)$.

Corollary 1 *When* $p = n$, $\widetilde{A}\begin{pmatrix} 1\ 2 \cdots n \\ 1\ 2 \cdots n \end{pmatrix} = |\boldsymbol{A}|^{n-1}$.

Corollary 2 *If the square matrix* $\boldsymbol{A} = (a_{ij})$ *is a totally nonnegative (positive) matrix, then* $\widetilde{\boldsymbol{A}} = (M_{ij})$ *is also a totally nonnegative (positive) matrix.*

Corollary 3 *If* $\boldsymbol{A} = (a_{ij})_{n\times n}$ *is a nonsingular totally nonnegative (positive) matrix, then* $(\boldsymbol{A}^*)^{-1}$ *is also a totally nonnegative (positive) matrix, where* $\boldsymbol{A}^* = ((-1)^{i+j} a_{ij})_{n\times n}$ *and is known as the sign-reverse matrix of* $\boldsymbol{A}$.

Theorem 2.3 *(Sylvester's Identity) Let* $\boldsymbol{A} = (a_{ij})_{n\times n}$ *be an arbitrary square matrix. For any* p *satisfying* $1 \le p < n$*, set*

$$b_{rs} = A\begin{pmatrix} 1\ 2 \cdots p\ r \\ 1\ 2 \cdots p\ s \end{pmatrix} \quad \text{and } \boldsymbol{B} = (b_{rs})_{p+1}^{n}.$$

Then, we have

$$|\boldsymbol{B}| = |\boldsymbol{A}|\left\{A\begin{pmatrix} 1\ 2 \cdots p \\ 1\ 2 \cdots p \end{pmatrix}\right\}^{n-p-1}. \tag{2.2.4}$$

The previous two theorems are applicable to all square matrices, while the next three theorems target totally nonnegative matrices only.

Theorem 2.4 *Let* $\boldsymbol{A}$ *be a totally nonnegative matrix of order* n*. If one of its leading principal minors* $D_q\,(1 \le q < n)$ *satisfies* $D_q = 0$*, then all leading principal minors* $D_r\,(q \le r \le n)$ *containing elements of* D_q *must also vanish.*

Theorem 2.5 *Suppose the square matrix* $\boldsymbol{A} = (a_{ij})_{n\times n}$ *is nonsingular and totally nonnegative. For any positive integer* p *satisfying* $1 \le p < n$*, we will have*

$$A\begin{pmatrix} 1\ 2 \cdots n \\ 1\ 2 \cdots n \end{pmatrix} \le A\begin{pmatrix} 1\ 2 \cdots p \\ 1\ 2 \cdots p \end{pmatrix} A\begin{pmatrix} p+1\ p+2 \cdots n \\ p+1\ p+2 \cdots n \end{pmatrix}.$$

It should be pointed out that the condition for $\boldsymbol{A}$ to be nonsingular in the above theorem is not necessary. In fact, the requirement may be relaxed to cover a wider range of situations. Nevertheless, the theorem, in its current form, is sufficient to support our need.

Theorem 2.6 *Let* $\boldsymbol{A} = (a_{ij})_{m\times n}$ *be a totally nonnegative matrix.*[1] *If a set of its rows with row numbers* $i_1, i_2, \ldots, i_p\,(1 \le i_1 < i_2 < \cdots < i_p \le m)$ *is linearly dependent*

[1]For a non-square matrix $(a_{ij})_{m\times n}$, it is totally nonnegative if all of its p-th-order minors are non-negative where p is any positive integer satisfying $p \le \min(m, n)$.

but the first and last $p-1$ rows in the set are both linearly independent, then the rank of $\boldsymbol{A}$ *is* $p-1$.

Corollary *Let* $\boldsymbol{A}=(a_{ij})_{m\times n}$ *be a totally nonnegative matrix. If one of its minors satisfies*

$$A\begin{pmatrix} i_1 & i_2 & \cdots & i_{p-1} & i_p \\ k_1 & k_2 & \cdots & k_{p-1} & k_p \end{pmatrix}=0,$$

while

$$A\begin{pmatrix} i_1 & i_2 & \cdots & i_{p-1} \\ k_1 & k_2 & \cdots & k_{p-1} \end{pmatrix} A\begin{pmatrix} i_2 & \cdots & i_{p-1} & i_p \\ k_2 & \cdots & k_{p-1} & k_p \end{pmatrix} \neq 0,$$

then the rank of $\boldsymbol{A}$ *is* $p-1$.

2.3 Jacobian Matrix

If a square matrix (a_{ij}) satisfies the condition $a_{ij}=0$ whenever $|i-j|>1$ (i and $j=1,2,\ldots,n$), then we get a tri-diagonal matrix, which is referred to as the Jacobian matrix and commonly denoted by $\boldsymbol{J}$. Set $a_i=a_{ii}$ ($i=1,2,\ldots,n$), $b_i=-a_{i,i+1}$, $c_i=-a_{i+1,i}$ ($i=1,2,\ldots,n-1$), and then $\boldsymbol{J}$ may be expressed as

$$\boldsymbol{J}=\begin{bmatrix} a_1 & -b_1 & 0 & \cdots & 0 & 0 & 0 \\ -c_1 & a_2 & -b_2 & \cdots & 0 & 0 & 0 \\ 0 & -c_2 & a_3 & \cdots & 0 & 0 & 0 \\ \vdots & \vdots & \vdots & & \vdots & \vdots & \vdots \\ 0 & 0 & 0 & \cdots & -c_{n-2} & a_{n-1} & -b_{n-1} \\ 0 & 0 & 0 & \cdots & 0 & -c_{n-1} & a_n \end{bmatrix}. \tag{2.3.1}$$

For any real number λ, introduce the following polynomial sequence:

$$D_0(\lambda)\equiv 1, \quad D_k(\lambda)=\det(a_{ij}-\lambda\delta_{ij})_{k\times k}, \quad k=1,2,\ldots,n.$$

It is not difficult to verify that this set of polynomials satisfy the recurrence relation below:

$$D_k(\lambda)=(a_k-\lambda)D_{k-1}(\lambda)-b_{k-1}c_{k-1}D_{k-2}(\lambda), \quad k=2,3,\ldots,n. \tag{2.3.2}$$

It is obvious from the above equation that b_k and c_k must appear as a product $b_k c_k$ in the expanded expression of $D_k(\lambda)$. Thus, from this point on, we only consider Jacobian matrices satisfying $b_k c_k > 0\,(k = 1, 2, \ldots, n-1)$.

Definition 2.1 If b_k and c_k $(k = 1, 2, \ldots, n-1)$ are all positive, the corresponding Jacobian matrix is referred to as the standard Jacobian matrix.

2.3.1 *Jacobian Matrix and Sturm Sequence*

Let Sequence A be the following polynomial sequence:

$$D_m(\lambda), D_{m-1}(\lambda), \ldots, D_1(\lambda), D_0(\lambda) \quad (m \leq n).$$

Starting from the recurrence relationship in Eq. (2.3.2), we may show that Sequence A has the following two properties associated with the so-called Sturm sequence:

Property 1 *$D_0(\lambda)$ never changes sign. In fact, we always have $D_0(\lambda) \equiv 1$.*

Property 2 *At a zero of $D_k(\lambda)$, $D_{k+1}(\lambda)$ and $D_{k-1}(\lambda)$ cannot vanish and must be of opposite sign, where $1 \leq k < m$.*

Before proceeding with discussion on properties of Sequence A, let us introduce the concept of the number of sign reversals or interchanges. Consider a sequence of real numbers $a_1, a_2, \ldots, a_n$. If $a_k a_{k+1} < 0$, we say that the sequence has a sign reversal in between two adjacent members a_k and a_{k+1}. If there exists an a_k such that $a_k = 0$, we may assign either a positive or negative sign to a_k. In this case, the sequence has only one sign reversal among a_{k-1}, a_k, and a_{k+1} if $a_{k-1}a_{k+1} < 0$, while it may have none or two sign reversals if $a_{k-1}a_{k+1} > 0$. The number or count of sign reversals of the entire sequence is defined as the sum of all sign reversals with its members. According to the above rules on sign reversals, when there is a zero member in the sequence, it has different minimum and maximum numbers of sign reversals, which may be denoted by S^- and S^+, respectively. Only when $S^- = S^+$, can a sequence have a fixed number of sign reversals, which is represented by S. If a sequence has a fixed number of sign reversals, then its first and last members must be nonzero numbers and, moreover, it cannot happen that two or more adjacent members in the sequence vanish simultaneously. Here are some examples of the number of sign reversals:

$$S(1, -1, 1, -1) = 3; \quad S(1, -1, 0, 1) = 2;$$
$$S^-(1, -1, 0, -1) = 1, \ S^+(1, -1, 0, -1) = 3;$$
$$S^-(0, 1, -1, -1) = 1, \ S^+(0, 1, -1, -1) = 2;$$
$$S^-(1, 0, 0, -1) = 1, \quad S^+(1, 0, 0, -1) = 3.$$

None of the last three sequences given above has a fixed number of sign reversals. Additionally, it should be pointed out that in subsequent discussion we will also talk about the number of sign reversals of a vector, which refers to the number of sign reversals of the sequence comprised of components of the vector.

Now let us examine the influence of a continuously increasing parameter λ on the sign reversal of Sequence A. Due to continuity, there could be a change in the number of sign reversals only if λ, as it sweeps through in the interval $[\alpha, \beta]$, passes a zero of $D_k(\lambda)$ $(1 \leq k \leq m)$. Furthermore, based on the continuity of polynomials and Property 2, we know that the number of sign reversals of Sequence A will not be affected when λ moves past a zero of $D_k(\lambda)$ in the middle of the sequence, i.e., $1 \leq k < m$. In fact, suppose λ_0 is a zero of $D_k(\lambda)$. Property 2 indicates that the subsequence consisting of D_{k+1}, D_k, and D_{k-1} always generates exactly one sign reversal for Sequence A, regardless of $D_k(\lambda)$ changing from positive to negative or negative to positive as λ passes through λ_0. Therefore, we may conclude that as λ continuously increases from α to β, the number of sign reversals of Sequence A may change only when λ goes past a zero of $D_m(\lambda)$. This observation leads to the theorem below.

Theorem 2.7 *Suppose $D_m(\alpha) \neq 0$ and $D_m(\beta) \neq 0$. As λ sweeps through in interval (α, β) and passes the zeroes of $D_m(\lambda)$ from left to right, the change in the number of sign reversals of Sequence A equals the count of roots of $D_m(\lambda)$ that make the product $D_m D_{m-1}$ switch from positive to negative minus the count of roots of $D_m(\lambda)$ that causes the same product to vary from negative to positive.*

Notice that $D_k(-\infty) > 0$ while $(-1)^k D_k(\infty) > 0$, where $k = 0, 1, \ldots, n$. Thus, Sequence A does not change sign as $\lambda \to -\infty$, while it has m sign reversals as $\lambda \to +\infty$. By applying Theorem 2.7 in the interval $(-\infty, +\infty)$, we may derive other properties of a Sturm sequence.

Property 3 *Polynomial $D_m(\lambda)$ has m distinct real roots, where $m = 1, 2, \ldots, n$.*

Property 4 *As λ continuously increases and passes through a zero of $D_m(\lambda)$, the sign of the product $D_m D_{m-1}$ changes from positive to negative.*

As a direct corollary of Theorem 2.7 and Property 4, we have the following property:

Property 5 *The root count of $D_m(\lambda)$ in the interval (α, β) equals the difference between numbers of sign reversals of Sequence A at $\lambda = \beta$ and $\lambda = \alpha$, where $m = 1, 2, \ldots, n$.*

Properties 3 and 4 as well as the continuity of the polynomial $D_m(\lambda)$ further lead to the following property:

Property 6 *There is exactly one real root of $D_{m-1}(\lambda)$ between each pair of adjacent real roots of $D_m(\lambda)$, where $m = 2, 3, \ldots, n$.*

Property 7 *Suppose $\lambda_1, \lambda_2, \ldots, \lambda_n$ are eigenvalues of the matrix $\boldsymbol{J}$ and $\lambda_1 < \lambda_2 < \ldots < \lambda_n$. In addition, let Sequence B represent the following polynomial sequence:*

$$D_{n-1}(\lambda), D_{n-2}(\lambda), \ldots, D_0(\lambda).$$

Then, Sequence B has precisely $i - 1$ sign reversals when $\lambda = \lambda_i$.

In fact, since the roots of $D_m(\lambda)$ and $D_{m-1}(\lambda)$ ($m = 2, 3, \ldots, n$) interlace, $D_{n-1}(\lambda)$ has $i - 1$ roots in the interval $(-\infty, \lambda_i]$, and there is exactly one root in each of the subintervals $(\lambda_1, \lambda_2), (\lambda_2, \lambda_3), \ldots, (\lambda_{i-1}, \lambda_i)$. Thus, as λ sweeps through from $-\infty$ to λ_i, the increase in the number of sign reversals of Sequence B should be $i - 1$. But because Sequence B does not have any sign reversal as $\lambda \to -\infty$, all of these $i - 1$ sign reversals must occur at $\lambda = \lambda_i$.

2.3.2 Eigenvalues and Eigenvectors of a Standard Jacobian Matrix

Suppose λ and $\boldsymbol{u} = (u_1, u_2, \ldots, u_n)^{\mathrm{T}}$ to be an eigenvalue of the matrix $\boldsymbol{J}$ and the corresponding eigenvector, respectively. They satisfy the following vector form of equation:

$$\boldsymbol{Ju} - \lambda \boldsymbol{u} = 0,$$

which may be rewritten in component form as below:

$$-c_{k-1}u_{k-1} + (a_k - \lambda)u_k - b_k u_{k+1} = 0, \quad k = 1, 2, \ldots, n-1; \ c_0 = 0, \tag{2.3.3}$$

$$-c_{n-1}u_{n-1} + (a_n - \lambda)u_n = 0. \tag{2.3.4}$$

When λ is an eigenvalue of the matrix $\boldsymbol{J}$, $D_{n-1}(\lambda) \neq 0$. Thus, the $n - 1$ equations in (2.3.3) are linearly independent, and Eq. (2.3.4) is a linear combination of equations in (2.3.3). Introduce the following transformation:

$$v_1 = u_1 \quad \text{and} \quad v_k = b_1 b_2 \cdots b_{k-1} u_k, \quad k = 2, 3, \ldots, n.$$

Then, Eq. (2.3.3) may be rewritten as

$$v_{k+1} = (a_k - \lambda)v_k - b_{k-1}c_{k-1}v_{k-1}, \quad k = 1, 2, \ldots, n-1.$$

Notice that this recurrence relation is consistent with the one shown in Eq. (2.3.2). Thus, we conclude,

$$v_k = CD_{k-1}(\lambda), \quad k = 1, 2, \ldots, n,$$

where C is an arbitrary nonzero constant. This leads to

$$u_k = CD_{k-1}(\lambda)/(b_1 b_2 \cdots b_{k-1}), \quad k = 1, 2, \ldots, n. \tag{2.3.5}$$

In particular, components of the eigenvector $\boldsymbol{u}^{(i)}$ corresponding to λ_i can be expressed as

$$u_{ki} = CD_{k-1}(\lambda_i)/(b_1 b_2 \cdots b_{k-1}), \quad k \text{ and } i = 1, 2, \ldots, n. \tag{2.3.6}$$

Based on previous discussion, it is not difficult to obtain Theorem 2.8.

Theorem 2.8 *Eigenpairs of a standard Jacobian matrix have the following properties:*

(1) All eigenvalues are real and distinct, i.e., $\lambda_1 < \lambda_2 < \cdots < \lambda_n$.
(2) Eigenvalues of successive leading principal submatrices interlace.
(3) The sequence consisting of components of Eigenvector $\boldsymbol{u}^{(i)}$*, which corresponds to the eigenvalue* λ_i*, has precisely* $i - 1$ *sign reversals. This result is represented by* $S_{\boldsymbol{u}^{(i)}} = i - 1$.

Corollary *For a positive definite standard Jacobian matrix, not only does it have Properties (1)–(3) given in Theorem 2.8, but its eigenvalues are all positive as well, i.e.,*

$$0 < \lambda_1 < \lambda_2 < \cdots < \lambda_n.$$

2.3.3 The Concept of *u*-Lines and Their Nodes

To support future needs, we introduce the concept of u-lines in this subsection.

Consider a vector $\boldsymbol{u} = (u_1, u_2, \ldots, u_n)^{\mathrm{T}}$. Its u-line refers to the curve $P_1 P_2 \cdots P_n$ comprised of broken lines with vertices $P_k = (k, u_k)\,(k = 1, 2, \ldots, n)$ in a planar Cartesian coordinate system xoy. When the u-line crosses the x-axis, the intersection point is called a node. In this book, we may also refer to this point directly as a node of the original vector $\boldsymbol{u}$. Apparently, in order for any point where the u-line touches the x-axis to be a node, the necessary and sufficient conditions are as follows: If it happens that $u_k = 0$, then there must be $u_{k-1}u_{k+1} < 0\,(1 < k < n)$; while if the intersection lies within the segment between u_{k-1} and u_k, then $u_{k-1}u_k < 0\,(1 < k < n)$.

Without loss of generality, let us now set $C = 1$ in Eq. (2.3.5). Then, components of the vector

$$\boldsymbol{u}(\lambda) = (u_1(\lambda), u_2(\lambda), \ldots, u_n(\lambda))^{\mathrm{T}},$$

as functions of λ, may be expressed as

$$u_1(\lambda) \equiv 1, \quad u_k(\lambda) = D_{k-1}(\lambda)/(b_1 b_2 \cdots b_{k-1}), \quad k = 2, 3, \ldots, n.$$

The corresponding u-line of the vector assumes the following form, when $k - 1 < x < k$:

$$y(x, \lambda) = (k - \lambda)u_{k-1}(\lambda) + (x - k + 1)u_k(\lambda), \quad k = 2, 3, \ldots, n. \tag{2.3.7}$$

Next, we will study the shape of the $u(\lambda)$-line as well as movement of its nodes in response to a change in λ.

First, from Property 2 of a Sturm sequence, every intersection point between the $u(\lambda)$-line and x-axis in the interval $(1, n)$ is a node. Second, according to Theorem 2.7, when $\lambda = \lambda_i$, the $u(\lambda)$-line has exactly $i - 1$ nodes. Under the general condition that λ assumes an arbitrary value, the nodal count of the $u(\lambda)$-line is equal to the number of roots of $D_{n-1}(\lambda)$ in $(-\infty, \lambda)$, per Property 5 of a Sturm sequence.

Let us further examine the effect of symmetrification of $\boldsymbol{J}$ by substituting b_k and c_k with $\sqrt{b_k c_k}$. In this case, $u_k(\lambda)$ is replaced by

$$\begin{aligned}\tilde{u}_k(\lambda) &= D_{k-1}(\lambda)/\sqrt{b_1 b_2 \cdots b_{k-1} c_1 c_2 \cdots c_{k-1}} \\ &= \sqrt{b_1 b_2 \cdots b_{k-1} c_1^{-1} c_2^{-1} \cdots c_{k-1}^{-1}}\, u_k(\lambda), \quad k = 1, 2, \ldots, n.\end{aligned}$$

It is easy to see that the $\tilde{u}(\lambda)$-line, which results from symmetrification of $\boldsymbol{J}$, has a shape different from the original $u(\lambda)$-line. However, there are still a few important commonalities between the two:

(1) Their numbers of nodes are always identical.
(2) If there is a node in the interval $(k - 1, k)$ for one of them, then the other must also have a node in the same interval.
(3) Their corresponding nodes move to the left or right in sync, as the value of λ varies.

Because of these properties, we need only to deal with the case of $\boldsymbol{J}$ being symmetric, i.e., $b_k = c_k$ $(k = 1, 2, \ldots, n - 1)$, in the discussion below. With this restriction, the equation satisfied by $u_k(\lambda)$ becomes

$$-b_{k-1}u_{k-1}(\lambda) + a_k u_k(\lambda) - b_k u_{k+1}(\lambda) = \lambda u_k(\lambda).$$

With the above preparation, we are ready to introduce the following lemma and theorem.

Lemma 2.1 *If $\lambda < \mu$, then there is at least one node of the $u(\mu)$-line between every two adjacent nodes of the $u(\lambda)$-line.*

Theorem 2.9 *Let $\boldsymbol{u}^{(i)}$ and $\boldsymbol{u}^{(i+1)}$ denote two successive eigenvectors of a Jacobian matrix, corresponding to eigenvalues λ_i and λ_{i+1}, respectively. Their nodes, i.e., nodes of $u(\lambda_i)$- and $u(\lambda_{i+1})$-lines, must interlace.*

Detailed proofs for Lemma 2.1 and Theorem 2.9 can be found in Ref. [3] or [4].

2.4 Oscillatory Matrices

2.4.1 *Definition of an Oscillatory Matrix and Its Identification Criteria*

Let us begin with the definition of an oscillatory matrix.

Definition 2.2 If the square matrix $A = (a_{ij})\,(i, j = 1, 2 \cdots, n)$ is a totally nonnegative matrix and there exists a positive integer s such that A^s becomes a totally positive matrix, then A is called an oscillatory matrix while the minimal value of s is referred to as the oscillation index of the oscillatory matrix.

If $A = (a_{ij})$ is an oscillatory matrix with the oscillation index s, then it has the following properties:

Property 1 *Since* $|A^s| = |A|^s > 0$, *an oscillatory matrix must be nonsingular, i.e.,* $|A| > 0$.

Property 2 *Since* $(A^p)^s = (A^s)^p$, *an oscillatory matrix to the n-th power remains to be an oscillatory matrix, where n is any positive integer.*

Property 3 *Any truncated submatrix* $B = (a_{ij})_q^r\,(1 \le q \le r \le n)$ *of an oscillatory matrix* A *remains to be an oscillatory matrix.*

Proof of Property 3 can be found in Ref. [3].

As a direct corollary of Property 3, we get the following property:

Property 4 *Diagonal and subdiagonal elements of an oscillatory matrix are all positive.*

In fact, since the truncated matrix of order 1 of an oscillatory matrix is also oscillatory, we may immediately conclude that the main diagonal elements satisfy $a_{ii} > 0\,(i = 1, 2, \ldots, n)$. As for the subdiagonal elements $a_{i,i+1}$ and $a_{i+1,i}$, if there exists a zero element $a_{i+1,i} = 0$, then the corresponding truncated matrix of order 2 ought to be of the form

$$\begin{bmatrix} a_i & a_{i,i+1} \\ 0 & a_{i+1,i+1} \end{bmatrix}.$$

It is supposed to be an oscillatory matrix, but the bottom-left element of this matrix to any power will remain to be zero. This contradiction shows that we must have $a_{i+1,i} > 0\,(i = 1, 2, \ldots, n-1)$. By the same token, $a_{i,i+1} > 0\,(i = 1, 2, \ldots, n-1)$ must also be true.

Because of the special role that oscillatory matrices play in this book, it is necessary to study the criteria for an oscillatory matrix before we discuss in depth properties of its eigenvectors and eigenvalues.

Theorem 2.10 *(Criteria for an Oscillatory Matrix) The necessary and sufficient conditions for a square matrix* $\boldsymbol{A} = (a_{ij})$ *to be oscillatory are as follows:* $\boldsymbol{A}$ *is nonsingular and totally nonnegative, while its subdiagonal elements are all positive, i.e.,* $a_{i+1,i} > 0$ *and* $a_{i,i+1} > 0$ *where* $i = 1, 2, \ldots, n-1$.

Proof The necessary part of this theorem comes straight out of Properties 1 and 4 introduced earlier. We only need to prove the sufficient part and will proceed in two steps:

First, let us prove that under the conditions given, all quasi-principal minors of $\boldsymbol{A}$ are positive.

Per conditions of the theorem, $a_{i+1,i} > 0$ and $a_{i,i+1} > 0$. In addition, according to Theorem 2.5 in Sect. 2.2, we have

$$0 < \det \boldsymbol{A} \leq A\begin{pmatrix} 1\ 2 \cdots n-1 \\ 1\ 2 \cdots n-1 \end{pmatrix} \cdot a_{nn}.$$

Since the minor of the totally nonnegative matrix $\boldsymbol{A}$,

$$A\begin{pmatrix} 1\ 2 \cdots n-1 \\ 1\ 2 \cdots n-1 \end{pmatrix},$$

is nonnegative, there must be $a_{nn} > 0$. Because the submatrices of $\boldsymbol{A} = (a_{kj})\,(k, j = 1, 2, \ldots, i)$ remain to be totally nonnegative, by applying mathematical induction, we may reach the conclusion that all diagonal elements satisfy $a_{ii} > 0$. As a result, the first-order quasi-principal minors are all positive.

Now let us assume quasi-principal minors of orders less than p to be all positive, and we will show that the p-th-order quasi-principal minors are also greater than zero. This is accomplished by using the method of proof by contradiction. Suppose there exists a p-th-order quasi-principal minor satisfying $A\begin{pmatrix} i_1\ i_2 \cdots i_p \\ k_1\ k_2 \cdots k_p \end{pmatrix} = 0$. The induction hypothesis yields

$$A\begin{pmatrix} i_1\ i_2 \cdots i_{p-1} \\ k_1\ k_2 \cdots k_{p-1} \end{pmatrix} \cdot A\begin{pmatrix} i_2\ i_3 \cdots i_p \\ k_2\ k_3 \cdots k_p \end{pmatrix} > 0.$$

Thus, based on the corollary of Theorem 2.6, the rank of Matrix $\boldsymbol{A}$ should be $p-1$, which contradicts the condition that $\boldsymbol{A}$ is nonsingular.

Again, from mathematical induction, we know that all quasi-principal minors of Matrix $\boldsymbol{A}$ to be positive.

Second, let us prove the square matrix A^{n-1} to be totally positive. Starting from the Binet–Cauchy identity given in Eq. (2.2.2), we get the following result:

$$A^{n-1}\begin{pmatrix} i_1 & i_2 & \cdots & i_p \\ k_1 & k_2 & \cdots & k_p \end{pmatrix} = \sum A\begin{pmatrix} i_1 & i_2 & \cdots & i_p \\ \alpha'_1 & \alpha'_2 & \cdots & \alpha'_p \end{pmatrix} \cdot A\begin{pmatrix} \alpha'_1 & \alpha'_2 & \cdots & \alpha'_p \\ \alpha''_1 & \alpha''_2 & \cdots & \alpha''_p \end{pmatrix}$$
$$\cdots A\begin{pmatrix} \alpha_1^{(n-2)} & \alpha_2^{(n-2)} & \cdots & \alpha_p^{(n-2)} \\ k_1 & k_2 & \cdots & k_p \end{pmatrix}. \tag{2.4.1}$$

Notice that all terms in the summation on the right-hand side of Eq. (2.4.1) are nonnegative. Thus, in order to show that the left-hand side is positive, we only need to demonstrate that there is at least one term in the summation on the right-hand side to be greater than zero. In fact, it will be shown in the next paragraph that there always exists one family of dummy index sets in the summation, specifically,

$$\alpha'_1, \alpha'_2, \ldots, \alpha'_p; \quad \alpha''_1, \alpha''_2, \ldots \alpha''_p; \quad \ldots; \quad \alpha_1^{(n-2)}, \alpha_2^{(n-2)}, \ldots, \alpha_p^{(n-2)}$$

such that all factors of the respective term in the summation of Eq. (2.4.1) are quasi-principal minors, and thus this particular term is of value greater than zero. Steps to obtain this family of index sets are explained in the paragraph below.

Compare the indices $i_1, i_2, \ldots, i_p$ with an index k_r, where $r = 1, 2, \ldots, p$, and then assign elements in the index set $i_1, i_2, \ldots, i_p$ to one of the three groups: (1) positive group if $i_r < k_r$, (2) negative group if $i_r > k_r$, and (3) zero group if $i_r = k_r$. Add 1 to the last index in the positive group, subtract 1 from the first index in the negative group, and keep all other indices unchanged. Denote the resulting indices as $\alpha'_1, \alpha'_2, \ldots, \alpha'_p$. Next, divide $\alpha'_1, \alpha'_2, \ldots, \alpha'_p$ into positive, negative, and zero groups based on conditions $\alpha'_r < k_r$, $\alpha'_r > k_r$, and $\alpha'_r = k_r$, respectively. Increase each of the last two indices in the positive group by 1, decrease each of the first two indices in the negative group by 1, and retain values for other indices. This yields the indices $\alpha''_1, \alpha''_2, \ldots, \alpha''_p$. Let this process continue. In general, to derive the index set $\alpha_1^{(t+1)}, \alpha_2^{(t+1)}, \ldots, \alpha_p^{(t+1)}$ from $\alpha_1^{(t)}, \alpha_2^{(t)}, \ldots, \alpha_p^{(t)}$, we always begin by splitting the index set $\alpha_1^{(t)}, \alpha_2^{(t)}, \ldots, \alpha_p^{(t)}$ into positive, negative, and zero groups based on conditions $\alpha_r^{(t)} < k_r$, $\alpha_r^{(t)} > k_r$, and $\alpha_r^{(t)} = k_r$, respectively. Then, we increase each of the last t indices in the positive group by 1, reduce each of the first t indices in the negative group by 1, and keep other indices unchanged. Here, if t is greater than the number of indices in a group, just modify all indices in the group. These steps will produce the desired new indices $\alpha_1^{(t+1)}, \alpha_2^{(t+1)}, \ldots, \alpha_p^{(t+1)}$. Apparently, the family of index sets obtained in this process should correspond to one of the summation terms on the right-hand side of Eq. (2.4.1). Moreover, all of the factors of this term, except the last one, are guaranteed to be quasi-principal minors. As for the last factor, notice that in the expression on the left-hand side of Eq. (2.4.1), $r \le i_r, k_r \le n-p+r$ $(r = 1, 2, \ldots, p)$. This leads to $|i_r - k_r| \le n - p$, where $r = 1, 2, \ldots, p$. Furthermore, if there is a pair of indices such that $i_r \ne k_r$, then the procedure outlined above would allow the index i_r to remain unchanged for at most $p-1$ times before being moved toward k_r by 1 with each index modification. Continuing modification on the indices for $n-2$ times, we will get $\left|\alpha_r^{(n-2)} - k_r\right| \leq 1$, which implies that the last factor must also be a quasi-principal minor. ■

It is clear from the above argument that the oscillation index of an $n \times n$ oscillatory matrix is no greater than $n-1$. Furthermore, base on Theorem 2.10, an oscillatory matrix may be defined alternatively as a nonsingular and totally nonnegative matrix with positive subdiagonal elements. Theorem 2.10 also leads to the property below:

Property 5 *The product of a nonsingular totally nonnegative matrix and an oscillatory matrix is an oscillatory matrix.*

In fact, the product of a nonsingular totally nonnegative matrix and an oscillatory matrix is obviously a nonsingular totally nonnegative matrix. Moreover, if $\boldsymbol{A}$ is a nonsingular and totally nonnegative matrix, then $a_{ii} > 0\,(i = 1, 2, \ldots, n)$; while if $\boldsymbol{B}$ is an oscillatory matrix, then $b_{i,i+1} > 0$ and $b_{i+1,i} > 0\,(i = 1, 2, \ldots, n)$. When $\boldsymbol{C} = \boldsymbol{AB}$,

$$c_{i,i+1} = \sum_{k=1}^{n} a_{ik} b_{k,i+1} \geq a_{ii} b_{i,i+1} > 0 \quad \text{and} \quad c_{i+1,i} = \sum_{k=1}^{n} a_{i+1,k} b_{ki} \geq a_{i+1,i+1} b_{i+1,i} > 0.$$

Thus, $\boldsymbol{C}$ is an oscillatory matrix. The conclusion remains valid when $\boldsymbol{A}$ is an oscillatory matrix while $\boldsymbol{B}$ is a nonsingular and totally nonnegative matrix.

2.4.2 Sign-Oscillatory Matrix

Sign-Oscillatory matrices are closely related to oscillatory matrices. They are defined as follows:

Definition 2.3 Consider a square matrix $\boldsymbol{A} = (a_{ij})_{n\times n}$. If the sign-reverse matrix of $\boldsymbol{A}$, i.e., $\boldsymbol{A}^* = ((-1)^{i+j} a_{ij})_{n\times n}$, is an oscillatory matrix, then $\boldsymbol{A}$ is referred to as a sign-oscillatory matrix.

Based on this definition as well as the criteria for an oscillatory matrix, the necessary and sufficient conditions for $\boldsymbol{A}$ to be a sign-oscillatory matrix are as follows: $\boldsymbol{A}$ is nonsingular; its subdiagonal elements are all negative numbers, i.e., $a_{i+1,i} < 0$ and $a_{i,i+1} < 0\,(i = 1, 2, \ldots, n-1)$; and A^* is totally nonnegative.

We may also get the following corollary, based on the above necessary and sufficient conditions as well as Properties 3 and 5 of an oscillatory matrix.

Corollary *Truncated submatrices of a sign-oscillatory matrix remain to be sign-oscillatory. The product of a positive definite diagonal matrix, such as the inverse of a diagonal mass matrix, and a sign-oscillatory matrix is also a sign-oscillatory matrix.*

In addition, the following theorem can be established.

Theorem 2.11 *The inverse of an oscillatory matrix is a sign-oscillatory matrix, while the inverse of a sign-oscillatory matrix is an oscillatory matrix.*

Proof Suppose the square matrix $\boldsymbol{A} = (a_{ij})_{n\times n}$ to be oscillatory. We get from Theorem 2.10,

$$\det \boldsymbol{A} > 0.$$

Thus, the inverse of the matrix $\boldsymbol{A}$ exists, and $\boldsymbol{A}^{-1} = (\widetilde{A}_{ji})_{n\times n}/|\boldsymbol{A}|$ while $\det \boldsymbol{A}^{-1} > 0$. Moreover, it is known from Corollary 3 of Theorem 2.2 that $(\boldsymbol{A}^*)^{-1}$ is a totally nonnegative matrix. Notice that the diagonal and subdiagonal elements of the matrix $(\boldsymbol{A}^{-1})^* = (\boldsymbol{A}^*)^{-1}$ happen to be the quasi-principal minors of $\boldsymbol{A}$, and are thus all positive. As a result, $\boldsymbol{A}^{-1}$ is a sign-oscillatory matrix.

By the same token, it can be shown that if $\boldsymbol{A}$ is sign-oscillatory then $\boldsymbol{A}^{-1}$ must be oscillatory. ■

2.4.3 Examples of Oscillatory and Sign-Oscillatory Matrices

Example 1 Consider a matrix $\boldsymbol{L} = (l_{ij})_{n\times n}$ with elements

$$l_{ij} = \begin{cases} \varphi_i\psi_j, & i \le j, \\ \varphi_j\psi_i, & i > j, \end{cases} \quad 1 \le i \text{ and } j \le n,$$

where $\varphi_1, \varphi_2, \ldots, \varphi_n$ and $\psi_1, \psi_2, \ldots, \psi_n$ are all real constants. Matrix $\boldsymbol{L}$ has three properties, which will be introduced in the rest of this example.

First, for any p-th–order minor ($p \ge 2$) of the matrix $\boldsymbol{L}$, if

$$1 \le i_1, j_1 < i_2, j_2 < \cdots < i_p, j_p \le n, \tag{2.4.2}$$

then

$$L\begin{pmatrix} i_1 & i_2 & \cdots & i_p \\ j_1 & j_2 & \cdots & j_p \end{pmatrix} = \varphi_{\alpha_1}\begin{vmatrix} \psi_{\beta_1} & \varphi_{\beta_1} \\ \psi_{\alpha_2} & \varphi_{\alpha_2} \end{vmatrix}\begin{vmatrix} \psi_{\beta_2} & \varphi_{\beta_2} \\ \psi_{\alpha_3} & \varphi_{\alpha_3} \end{vmatrix}\cdots\begin{vmatrix} \psi_{\beta_{p-1}} & \varphi_{\beta_{p-1}} \\ \psi_{\alpha_p} & \varphi_{\alpha_p} \end{vmatrix}\psi_{\beta_p}, \tag{2.4.3}$$

where

$$\alpha_r = \min(i_r, j_r) \quad \text{and} \quad \beta_r = \max(i_r, j_r).$$

Second, for any p-th–order minor ($p \ge 2$) of the matrix $\boldsymbol{L}$, if Eq. (2.4.2) is not satisfied, then the minor is zero, i.e.,

$$L\begin{pmatrix} i_1 & i_2 & \cdots & i_p \\ j_1 & j_2 & \cdots & j_p \end{pmatrix} = 0.$$

In fact, since Matrix $\boldsymbol{L}$ is symmetric, we may assume, without loss of generality, that $i_2 \le j_2$. Thus,

$$L\begin{pmatrix} i_1 & i_2 & \cdots & i_p \\ j_1 & j_2 & \cdots & j_p \end{pmatrix} = \begin{vmatrix} \varphi_{\alpha_1}\psi_{\beta_1} & \varphi_{i_1}\psi_{j_2} & \varphi_{i_1}\psi_{j_3} & \cdots & \varphi_{i_1}\psi_{j_p} \\ \varphi_{j_1}\psi_{i_2} & \varphi_{i_2}\psi_{j_2} & \varphi_{i_2}\psi_{j_3} & \cdots & \varphi_{i_2}\psi_{j_p} \\ \vdots & \vdots & \vdots & & \vdots \end{vmatrix}.$$

Subtracting the product of the second row and $\varphi_{i_1}/\varphi_{i_2}$ from the first row in the determinant on the right-hand side of the above equation, we obtain

$$\begin{aligned} L\begin{pmatrix} i_1 & i_2 & \cdots & i_p \\ j_1 & j_2 & \cdots & i_p \end{pmatrix} &= (\varphi_{\alpha_1}\psi_{\beta_1} - \varphi_{j_1}\psi_{i_2}(\varphi_{i_1}/\varphi_{i_2}))L\begin{pmatrix} i_2 & i_3 & \cdots & i_p \\ j_2 & j_3 & \cdots & j_p \end{pmatrix} \\ &= \frac{\varphi_{\alpha_1}}{\varphi_{\alpha_2}}\begin{vmatrix} \psi_{\beta_1} & \varphi_{\beta_1} \\ \psi_{\alpha_2} & \varphi_{\alpha_2} \end{vmatrix} L\begin{pmatrix} i_2 & i_3 & \cdots & i_p \\ j_2 & j_3 & \cdots & j_p \end{pmatrix}, \end{aligned}$$

where the obvious relation $\varphi_{i_1}\varphi_{j_1} = \varphi_{\alpha_1}\varphi_{\beta_1}$ has been used. Continuing this process and also taking into account $L\begin{pmatrix} i_p \\ j_p \end{pmatrix} = \varphi_{\alpha_p}\psi_{\beta_p}$, we derive the first property.

As for the second property, let us assume $i_{r+1} \le j_r$, as an example. Then, the first two rows of the minor $L\begin{pmatrix} i_r & i_{r+1} & \cdots & i_p \\ j_r & j_{r+1} & \cdots & j_p \end{pmatrix}$ are proportional. Thus, the minor must be zero.

Starting from the above properties, we may get another property.

Third, the necessary and sufficient conditions for $\boldsymbol{L}$ to be a totally nonnegative matrix are: $\varphi_1, \varphi_2, \ldots, \varphi_n$ and $\psi_1, \psi_2, \ldots, \psi_n$ are all nonzero; they have the same positive or negative sign; and the following inequalities are satisfied

$$\frac{\varphi_1}{\psi_1} \le \frac{\varphi_2}{\psi_2} \le \cdots \le \frac{\varphi_n}{\psi_n}. \tag{2.4.4}$$

Additionally, the rank of Matrix $\boldsymbol{L}$ is equal to one plus the number of times that an equality does not hold in (2.4.4). If no equality is satisfied in Eq. (2.4.4), then $\boldsymbol{L}$ is a totally positive matrix. In other words, it is an oscillatory matrix with an oscillation index of 1.

Example 2 A positive definite standard Jacobian matrix is sign-oscillatory.

In fact, a positive definite standard Jacobian matrix must be non-singular and also have negative subdiagonal elements. Thus, it needs only to prove that J^* is a totally nonnegative matrix. Consider the following minor:

$$J^*\begin{pmatrix} i_1 & i_2 & \cdots & i_p \\ j_1 & j_2 & \cdots & j_p \end{pmatrix}, \quad 1 \le p < n. \tag{2.4.5}$$

There are three possible scenarios related to this minor. These will be discussed in the remaining part of this example.

First, when $i_r = j_r\,(r = 1, 2, \ldots, p)$ for all indices, the expression (2.4.5) defines a principal minor that is, of course, greater than zero.

Now consider the case where a row index differs from its corresponding column index, specifically, $i_r \neq j_r$. If $i_r < j_r$, then

$$a_{i_\alpha j_\beta} = 0, \quad \alpha = 1, 2, \ldots, r; \quad \beta = r + 1, r + 2, \ldots, p.$$

If $i_r > j_r$, then

$$a_{i_\alpha j_\beta} = 0, \quad \alpha = r + 1, r + 2, \ldots, p; \quad \beta = 1, 2, \ldots, r.$$

Thus, the minor in the expression (2.4.5) can be factored as

$$J\begin{pmatrix} i_1 & i_2 & \cdots & i_p \\ j_1 & j_2 & \cdots & j_p \end{pmatrix} = J\begin{pmatrix} i_1 & i_2 & \cdots & i_r \\ j_1 & j_2 & \cdots & j_r \end{pmatrix} \cdot J\begin{pmatrix} i_{r+1} & i_{r+2} & \cdots & i_p \\ j_{r+1} & j_{r+2} & \cdots & j_p \end{pmatrix}. \tag{2.4.6}$$

Based on Eq. (2.4.6), we can easily deal with the other two scenarios.

Second, suppose there is at least one row index to be different from its corresponding column index. Further assume, for this group of index pairs, the difference between the row and respective column index satisfies $|i_r - j_r| = 1$. Then, the minor in the expression (2.4.5) equals the product of a principal minor and b_i, and is thus positive.

Third, suppose we can find index pairs such that $i_r \neq j_r$ and among them there exists at least one pair satisfying $|i_r - j_r| > 1$. Because $a_{i_\alpha j_\beta} = 0\,(\alpha = 1, 2, \ldots, r)$ or $a_{i_\alpha j_\beta} = 0\,(\beta = 1, 2, \ldots, r)$, the minor (2.4.5) must be zero.

2.5 Perron's Theorem and Compound Matrices

2.5.1 *Perron's Theorem*

In order to derive oscillatory properties of an oscillatory matrix, we introduce a theorem on eigenvalues and eigenvectors of positive matrices of the general form.

Notice that the matrices under consideration may not be symmetric. The theorem is known as Perron's Theorem. It plays a critical role in the subsequent discussion.

Definition 2.4 If all components of a vector $\boldsymbol{x} = (x_1, x_2, \dots, x_n)^{\mathrm{T}}$ satisfy $x_i \geq 0 (> 0)$, where $i = 1, 2, \dots, n$, then $\boldsymbol{x}$ is called a nonnegative (positive) vector and denoted by $\boldsymbol{x} \geq 0 (> 0)$. If two vectors $\boldsymbol{x}$ and $\boldsymbol{y}$ satisfy $\boldsymbol{y} - \boldsymbol{x} \geq 0$, then the relationship is represented by $\boldsymbol{y} \geq \boldsymbol{x}$.

Theorem 2.12 *(Perron's Theorem) A positive square matrix has a unique eigenvalue whose absolute value is the greatest among all its eigenvalues. Furthermore, this eigenvalue is both positive and distinct, while the corresponding eigenvector can be taken as a positive vector, whose components are all greater than zero.*

Proof of this theorem can be found in Ref. [3] or [4].

2.5.2 *Compound Matrices*

To apply Perron's theorem to oscillatory matrices, we still need to introduce a new concept, the compound matrix. Here is its definition: The p-th compound matrix of $\boldsymbol{A}$ refers to a matrix $\underline{\boldsymbol{A}}_p$ composed of all p-th-order minors of $\boldsymbol{A}$ arranged in a lexicographic order. Specifically, if $\boldsymbol{A} = (a_{ij})_{n \times n}$, then its compound matrix $\underline{\boldsymbol{A}}_p = (a_{st}^{(p)})_{N \times N}$ is of the order $N = C_n^p$ and

$$a_{st}^{(p)} = A\begin{pmatrix} i_1\, i_2 \cdots i_p \\ j_1\, j_2 \cdots j_p \end{pmatrix}.$$

Here, the index numbers $s = s(i_1, i_2, \dots, i_p)$ and $t = s(j_1, j_2, \dots, j_p)$, as functions of row and column indices, respectively, are defined through the following process: Arbitrarily select p different indices $i_1, i_2, \dots, i_p$ from the set of positive integers $1, 2, \dots, n$, and make sure they satisfy $i_1 < i_2 < \cdots < i_p$. There are a total of C_n^p different ways to construct the index set $i_1, i_2, \dots, i_p$. Now arrange these index sets into a sequence and the relative position of any two index sets $i_1', i_2', \dots, i_p'$ and $i_1'', i_2'', \dots, i_p''$ in the sequence may be determined by checking differences between their corresponding indices

$$i_1' - i_1'', \quad i_2' - i_2'', \dots, \quad i_p' - i_p''.$$

If the first a few of these differences are all zeros and the one immediately after them is negative, then place the index set $i_1', i_2', \dots, i_p'$ ahead of $i_1'', i_2'', \dots, i_p''$. This process ensures that each index set $i_1, i_2, \dots, i_p$ has a fixed position in the sequence. Thus, we may assign a unique sequence number $s(i_1, i_2, \dots, i_p)$ to the index set $i_1, i_2, \dots, i_p$. For example, when $n = 5$ and $p = 3$, the C_5^3 index combinations are

$$123\ 124\ 125\ 134\ 135\ 145\ 234\ 235\ 245\ 345.$$

Thus, $s(1, 2, 4) = 2$, $s(2, 3, 4) = 7$, and so on.

The following result can be easily derived from the definition of compound matrices as well as the Binet–Cauchy identity in Eq. (2.2.2):

Theorem 2.13 *If* $\boldsymbol{C} = \boldsymbol{AB}$, *then the respective compound matrices also have a similar relationship* $\underline{\boldsymbol{C}}_p = \underline{\boldsymbol{A}}_p\underline{\boldsymbol{B}}_p$.

Proof Since

$$C\begin{pmatrix} i_1\ i_2 \cdots i_p \\ j_1\ j_2 \cdots j_p \end{pmatrix} = \sum A\begin{pmatrix} i_1\ i_2 \cdots i_p \\ k_1\ k_2 \cdots k_p \end{pmatrix} B\begin{pmatrix} k_1\ k_2 \cdots k_p \\ j_1\ j_2 \cdots j_p \end{pmatrix},$$

we have $c_{st}^{(p)} = \sum_{k=1}^{N} a_{sk}^{(p)} b_{kt}^{(p)}$, which is just the result we are looking for. ■

The following corollaries can be obtained from the theorem above.

Corollary 1 *The p-th compound matrix of the inverse of a nonsingular matrix* $\boldsymbol{A}$ *is the same as the inverse of the p-th compound matrix of* $\boldsymbol{A}$.

Proof If $\boldsymbol{B} = \boldsymbol{A}^{-1}$, then $\boldsymbol{AB} = \boldsymbol{I}$. Thus, $\underline{\boldsymbol{A}}_p\underline{\boldsymbol{B}}_p = \underline{\boldsymbol{I}}_p$. However, $\underline{\boldsymbol{I}}_p$ remains to be an identity matrix. This concludes the proof. ■

Corollary 2 *Suppose* $\lambda_1, \lambda_2, \ldots, \lambda_n$ *are eigenvalues of an* $n \times n$ *matrix* $\boldsymbol{A}$. *Then, eigenvalues of the p-th compound matrix of* $\boldsymbol{A}$ *are of the form* $\lambda_{i_1}\lambda_{i_2}\cdots\lambda_{i_p}$, *where the index set* $i_1, i_2, \ldots, i_p$ *represents any possible subset of* $1, 2, \ldots, n$ *and satisfies the condition* $1 \le i_1 < i_2 < \cdots < i_p \le n$.

Proof Under the condition given in this corollary, there exists an invertible matrix $\boldsymbol{T}$ such that

$$\boldsymbol{A} = \boldsymbol{T\Lambda T}^{\mathrm{T}}, \quad \boldsymbol{TT}^{\mathrm{T}} = \boldsymbol{I},$$

where $\boldsymbol{\Lambda}$ is an upper triangular matrix and its main diagonal elements happen to be $\lambda_1, \lambda_2, \ldots, \lambda_n$. Thus, we get

$$\underline{\boldsymbol{A}}_p = \underline{\boldsymbol{T}}_p\underline{\boldsymbol{\Lambda}}_p\underline{\boldsymbol{T}}_p^{\mathrm{T}}.$$

It is not difficult to verify that $\underline{\boldsymbol{\Lambda}}_p = (b_{st})_{N\times N}$ is also an upper triangular matrix. In fact, when $s > t$ there must exists an index q such that $i_r = k_r\ (r = 1, 2, \ldots, q-1)$ and $i_q > k_q$. Thus,

$$b_{st} = A\begin{pmatrix} i_1 & i_2 & \cdots & i_p \\ k_1 & k_2 & \cdots & k_p \end{pmatrix} = \begin{vmatrix} \lambda_1 & * & \cdots & * & * & \cdots & * \\ & \lambda_2 & \cdots & * & * & \cdots & * \\ & & \ddots & \vdots & \vdots & & \vdots \\ & & & \lambda_{q-1} & * & \cdots & * \\ & 0 & & & 0 & \cdots & * \\ & & & & & \ddots & * \\ & & & & & & 0 \end{vmatrix} = 0,$$

$$b_{ss} = A\begin{pmatrix} i_1 & i_2 & \cdots & i_p \\ i_1 & i_2 & \cdots & i_p \end{pmatrix} = \lambda_{i_1}\lambda_{i_2}\cdots\lambda_{i_p},$$

where $*$ indicates an element not affecting the value of the determinant while $i_1, i_2, \ldots, i_p$ represents any combination taken from $1, 2, \ldots, n$ and satisfies $1 \le i_1 < i_2 < \cdots < i_p \le n$. Apparently, b_{ss} is an eigenvalue of $\underline{\boldsymbol{A}}_p$. ■

For a matrix $\boldsymbol{A}$ with n distinct eigenvalues, the previous result can be further extended. In fact, let $\boldsymbol{U}$ denote the eigenvector matrix of $\boldsymbol{A}$ and, in this case, $\boldsymbol{\Lambda} = \mathrm{diag}(\lambda_1, \lambda_2, \ldots, \lambda_n)$. Then, we have

$$\boldsymbol{A} = \boldsymbol{U}\boldsymbol{\Lambda}\boldsymbol{U}^{\mathrm{T}}.$$

The corresponding compound matrices satisfy

$$\underline{\boldsymbol{A}}_p = \underline{\boldsymbol{U}}_p\underline{\boldsymbol{\Lambda}}_p\underline{\boldsymbol{U}}_p^{\mathrm{T}}.$$

It is easy to see that $\underline{\boldsymbol{\Lambda}}_p$ is still a diagonal matrix with the product $\lambda_{i_1}\lambda_{i_2}\cdots\lambda_{i_p}$ as the element along the main diagonal. This leads to the corollary below.

Corollary 3 *For a fixed index set $k_1, k_2, \ldots, k_p$ and a variable index set $i_1, i_2, \ldots, i_p$, the minor*

$$U\begin{pmatrix} i_1 & i_2 & \cdots & i_p \\ k_1 & k_2 & \cdots & k_p \end{pmatrix}, \quad 1 \le p \le n$$

is the s-th component of the eigenvector of $\underline{\boldsymbol{A}}_p$ corresponding to the eigenvalue $\lambda_{i_1}\lambda_{i_2}\cdots\lambda_{i_p}$, where $s = s(i_1, i_2, \ldots, i_p)$. In particular, $U\begin{pmatrix} i_1 & i_2 & \cdots & i_p \\ 1 & 2 & \cdots & p \end{pmatrix}$ is the s-th component of the eigenvector of $\underline{\boldsymbol{A}}_p$ associated with the greatest eigenvalue.

2.6 Eigenpairs of an Oscillatory Matrix

Based on Perron's Theorem and results of compound matrices, we will derive a number of theorems related to eigenvalues and eigenvectors of an oscillatory matrix in this section. Notice that alternative proofs of some of the theorems can be found in Refs. [4].

Theorem 2.14 *All eigenvalues of an oscillatory matrix are positive and distinct. They may be sorted from large to small as follows:*

$$\lambda_1 > \lambda_2 > \cdots > \lambda_n > 0.$$

Proof Suppose the square matrix $\boldsymbol{A} = (a_{ij})_{n\times n}$ is oscillatory and its eigenvalues are $\lambda_1, \lambda_2, \ldots, \lambda_n$. Per the definition of an oscillatory matrix, there exists a positive integer s, which is the oscillation index of $\boldsymbol{A}$, such that $\boldsymbol{B} \equiv \boldsymbol{A}^m$ is a totally positive matrix when $m \geq s$ and m is a positive integer also. Apparently, $\mu_i = \lambda_i^m$ is the eigenvalue of $\boldsymbol{B}$. Let $\underline{\boldsymbol{B}}_q$ be the q-th compound matrix of $\boldsymbol{B}$, where $1 \leq q \leq n$. Obviously, it is a positive matrix, and thus Perron's theorem is applicable to $\underline{\boldsymbol{B}}_q$. If eigenvalues of $\boldsymbol{B}$ are sorted by their absolute values from big to small, i.e., $|\mu_1| \geq |\mu_2| \geq \cdots \geq |\mu_n|$, then the eigenvalue of $\underline{\boldsymbol{B}}_q$ having the greatest absolute value should be $\mu_1\mu_2\cdots\mu_q$ and we get from Perron's theorem

$$\mu_1\mu_2\cdots\mu_q > 0 \quad \text{and} \quad \mu_1\mu_2\cdots\mu_q > \left|\mu_1\mu_2\cdots\mu_{q-1}\mu_{q+1}\right|.$$

The first inequality above leads to $\mu_q > 0$, i.e., $\lambda_q^m > 0$ or $\lambda_q > 0\,(q = 1, 2, \ldots, n)$; while the second inequality yields $\mu_q > \mu_{q+1}$, i.e., $\lambda_q^m > \lambda_{q+1}^m$ or $\lambda_q > \lambda_{q+1}$ $(q = 1, 2, \ldots, n)$. ■

Theorem 2.15 *Suppose that eigenvalue of an oscillatory matrix* $\boldsymbol{A}$ *have been sorted from large to small as* $\lambda_1 > \lambda_2 > \cdots > \lambda_n$ *and the eigenvector corresponding to* λ_i *is denoted by* $\boldsymbol{u}^{(i)} = (u_{1i}, u_{2i}, \ldots, u_{ni})^{\mathrm{T}}$. *Then, for an arbitrary set of real numbers* $c_i\,(i = p, p+1, \ldots, q)$ *that are not zero simultaneously, the number of sign reversals of the following vector*

$$\boldsymbol{u} = \sum_{i=p}^{q} c_i \boldsymbol{u}^{(i)}, \quad 1 \leq p \leq q \leq n$$

is between $p-1$ *and* $q-1$. *Namely,*

$$p - 1 \leq S_{\boldsymbol{u}}^- \leq S_{\boldsymbol{u}}^+ \leq q - 1. \tag{2.6.1}$$

In particular, the number of sign reversals of $\boldsymbol{u}^{(i)}$ *is exactly* $i-1$, *i.e.,*

$$S_{\boldsymbol{u}^{(i)}}^- = S_{\boldsymbol{u}^{(i)}}^+ = S_{\boldsymbol{u}^{(i)}} = i - 1.$$

Proof For an arbitrary integer $m \geq 1, \boldsymbol{u}^{(i)}$ is an eigenvector of both $\boldsymbol{A}$ and $\boldsymbol{A}^m$ corresponding to λ_i and λ_i^m, respectively. In addition, when $m \geq s_0$, where s_0 is the oscillation index of Matrix $\boldsymbol{A}$, $\boldsymbol{A}^m$ becomes a totally positive matrix. Therefore, we may assume, without loss of generality, $\boldsymbol{A}$ itself to be a totally positive matrix. Because eigenvalues of $\boldsymbol{A}$ are distinct, each of its eigenvector $\boldsymbol{u}^{(i)}$ is uniquely determined within a constant factor.

Let us first prove the last part of Eq. (2.6.1). Without loss of generality, we may set $p = 1$. Per Corollary 3 of Theorem 2.13, the minor

$$U\begin{pmatrix} i_1\ i_2 \cdots i_q \\ 1\ 2 \cdots q \end{pmatrix}$$

is a component of the eigenvector of $\underline{\boldsymbol{A}}_q$, which is associated with its greatest eigenvalue. Thus, regardless of the composition of the index set $i_1, i_2, \ldots, i_q$, the above minor is always of same positive or negative sign, which may be denoted by ε_q. Then, by multiplying eigenvectors $\boldsymbol{u}^{(1)}, \boldsymbol{u}^{(2)}, \ldots, \boldsymbol{u}^{(n)}$ of $\boldsymbol{A}$ and factors $\varepsilon_1, \varepsilon_2/\varepsilon_1, \ldots, \varepsilon_n/\varepsilon_{n-1}$, respectively, we can make

$$U\begin{pmatrix} i_1\ i_2 \cdots i_q \\ 1\ 2 \cdots q \end{pmatrix} > 0, \quad q = 1, 2, \ldots, n.$$

We are now ready to prove the last inequality in (2.6.1) using the method of proof by contradiction. Let us assume $S_{\boldsymbol{u}}^+ > q - 1$. Then, there exist some components of $\boldsymbol{u}$ such that

$$u_{i_r} u_{i_r+1} \leq 0, \quad r = 1, 2, \ldots, q.$$

Here, $u_{i_1}, u_{i_2}, \ldots, u_{i_q}$ cannot vanish simultaneously. Otherwise, the following system of homogenous equations with c_k as unknowns

$$\sum_{k=1}^{q} c_k u_{i_r k} = u_{i_r} = 0, \quad r = 1, 2, \ldots, q,$$

would only have zero solution, because the determinant of its coefficients

$$U\begin{pmatrix} i_1\ i_2 \cdots i_q \\ 1\ 2 \cdots q \end{pmatrix}$$

is nonzero, But this contradicts the condition that c_i $(i = 1, 2, \ldots, q)$ are not all zero. Expanding the determinant

$$\begin{vmatrix} u_{i_1 1} & u_{i_1 2} & \cdots & u_{i_1 q} & u_{i_1} \\ u_{i_2 1} & u_{i_2 2} & \cdots & u_{i_2 q} & u_{i_2} \\ \vdots & \vdots & & \vdots & \vdots \\ u_{i_{q+1} 1} & u_{i_{q+1} 2} & \cdots & u_{i_{q+1} q} & u_{i_{q+1}} \end{vmatrix},$$

whose value is guaranteed to be zero, by the last column, we get

$$\sum_{s=1}^{q+1} (-1)^{s+q+1} u_{i_s} U \begin{pmatrix} i_1 \; i_2 \cdots i_{s-1} & i_{s+1} & i_{s+2} & \cdots & i_{q+1} \\ 1 \;\; 2 \cdots s-1 & s & s+1 & \cdots & q \end{pmatrix} = 0.$$

But this is impossible, because at least one term in the summation on the left-hand side of the equation is nonzero, while other terms either share the same sign with the nonzero term or vanish. As a result, the above equation is invalid. This contradiction leads to the conclusion that the last part of Eq. (2.6.1) must hold true.

Next, let us prove the first part of Eq. (2.6.1), i.e., $S_{\boldsymbol{u}}^- \geq p-1$. To this end, set $B = (A^*)^{-1}$, where A^* is the sign-reverse matrix of $\boldsymbol{A}$. Also set $\boldsymbol{\Lambda} = \mathrm{diag}(\lambda_1, \lambda_2, \ldots, \lambda_n)$ and let $\boldsymbol{U}$ denote the matrix formed by eigenvectors of $\boldsymbol{A}$. Then, $\boldsymbol{AU} = \boldsymbol{U\Lambda}$ and accordingly,

$$\boldsymbol{B U}^* = \boldsymbol{U}^* \boldsymbol{\Lambda}^{-1}. \tag{2.6.2}$$

Thus, $(\boldsymbol{u}^{(k)})^* = (u_{1k}, -u_{2k}, \ldots, (-1)^{n-1} u_{nk})^{\mathrm{T}}$ is the eigenvector of $\boldsymbol{B}$ corresponding to the eigenvalue λ_k^{-1}. Now apply the same procedure used to prove the last part of Eq. (2.6.1) to the vector below

$$\boldsymbol{u}^* = c_p(\boldsymbol{u}^{(p)})^* + c_{p+1}(\boldsymbol{u}^{(p+1)})^* + \cdots + c_q(\boldsymbol{u}^{(q)})^*.$$

Since $\boldsymbol{B}$ is also a totally positive matrix (see Corollary 3 of Theorem 2.2 in Sect. 2.2), $S_{\boldsymbol{u}^*}^+ \leq n-p$. But because $S_{\boldsymbol{u}^*}^+ + S_{\boldsymbol{u}}^- = n-1$, we must have

$$S_{\boldsymbol{u}}^- \geq p-1.$$

Finally, it is obviously true that the vector $\boldsymbol{u}^{(i)}$ has $i-1$ sign reversals. ■

Starting from this theorem, we may get the following theorem.

Theorem 2.16 *Eigenvalues of a sign-oscillatory matrix* $\boldsymbol{A}$ *are both positive and distinct. Arrange these eigenvalues from small to large as* $0 < \lambda_1 < \lambda_2 < \cdots < \lambda_n$, *and denote the eigenvector corresponding to the eigenvalue* λ_i *as* $\boldsymbol{u}^{(i)}$. *Then, for any set of real numbers* c_i $(i = p, p+1, \ldots, q$ *and* $1 \leq p \leq q \leq n)$ *that do not vanish simultaneously, the count of sign reversals of the vector*

$$\boldsymbol{u} = c_p \boldsymbol{u}^{(p)} + c_{p+1} \boldsymbol{u}^{(p+1)} + \cdots + c_q \boldsymbol{u}^{(q)}$$

satisfies $p - 1 \le S_{\boldsymbol{u}}^- \le S_{\boldsymbol{u}}^+ \le q - 1$. *In particular,* $\boldsymbol{u}^{(i)}$ *has precisely* $i - 1$ *sign reversals, where* $i = 1, 2, \ldots, n$.

Proof Let λ_i and $\boldsymbol{u}^{(i)}$ $(i = 1, 2, \ldots, n)$ be eigenpairs of a sign-oscillatory matrix $\boldsymbol{A}$, i.e.,

$$\boldsymbol{A}\boldsymbol{u}^{(i)} = \lambda_i \boldsymbol{u}^{(i)}.$$

Rewrite the above equation into the following form:

$$\boldsymbol{B}\boldsymbol{u}^{(i)} = \mu_i \boldsymbol{u}^{(i)},$$

where $\boldsymbol{B} = \boldsymbol{A}^{-1}$ and $\mu_i = \lambda_i^{-1}$. Because $\boldsymbol{A}$ is sign-oscillatory, $\boldsymbol{B}$ is oscillatory. Thus, $\boldsymbol{u}^{(i)}$ $(i = 1, 2, \ldots, n)$ should also be eigenvectors of the oscillatory matrix $\boldsymbol{B}$. Notice that λ_i and μ_i have the same sequence number. When $\{\mu_i\}$ are arranged from large to small, $\{\lambda_i\}$ are listed from small to large. As a result, $\boldsymbol{u}^{(i)}$ $(i = 1, 2, \ldots, n)$ is the i-th eigenvector of both the sign-oscillatory matrix $\boldsymbol{A}$ and the oscillatory matrix $\boldsymbol{B}$. Thus, Theorem 2.16 can be easily established, based on Theorem 2.15. ■

Theorem 2.17 *When eigenvalues of an oscillatory matrix are arranged in a descending order (in the case of a sign-oscillatory matrix, the eigenvalues are organized in an ascending order), nodes of two successive eigenvectors* $\boldsymbol{u}^{(i)}$ *and* $\boldsymbol{u}^{(i+1)}$ *associated with* λ_i *and* λ_{i+1}, *respectively, interlace, where* $i = 2, 3, \ldots, n - 1$.

Proof Based on definitions of the u-line and node given in Sect. 2.3 as well as the last result of Theorem 2.15, we may immediately conclude that the u-line of the eigenvector $\boldsymbol{u}^{(i)}$ associated with the i-th eigenvalue has exactly $i - 1$ nodes and no other zero.

In order to prove the theorem, we look at the vector

$$\boldsymbol{u} = c\boldsymbol{u}^{(i)} + d\boldsymbol{u}^{(i+1)}. \tag{2.6.3}$$

Let $u(x)$, $u^{(i)}(x)$, and $u^{(i+1)}(x)$ denote u-line functions corresponding to vectors $\boldsymbol{u}$, $\boldsymbol{u}^{(i)}$, and $\boldsymbol{u}^{(i+1)}$, respectively. For any set of real numbers c and d that are not simultaneously zero, we get by applying Theorem 2.15 to $\boldsymbol{u}$:

$$i - 1 \le S_{\boldsymbol{u}}^- \le S_{\boldsymbol{u}}^+ \le i. \tag{2.6.4}$$

Set $d = -1$ and assume α and β to be two adjacent nodes of $\boldsymbol{u}^{(i)}$, i.e.,

$$\boldsymbol{u} = c\boldsymbol{u}^{(i)} - \boldsymbol{u}^{(i+1)}, \quad u^{(i)}(\alpha) = 0, \quad u^{(i)}(\beta) = 0. \tag{2.6.5}$$

First, we use the method of proof by contradiction to show $u^{(i+1)}(\alpha) \neq 0$. If this is not true, we must have $u(\alpha) = 0$. Pick a point γ such that $\alpha < \gamma \le [\alpha + 1]$, where the notation $[t]$ represents the integer portion of the number t. Thus, by setting

$$c = u^{(i+1)}(\gamma)/u^{(i)}(\gamma),$$

we get $u(\gamma) = 0$. Due to the linearity of the u-line on the interval $([\alpha], [\alpha + 1])$, this part of u-line must lie flat on the x-axis. In other words, two successive components of the vector $\boldsymbol{u}$ must be zero. Consequently, we should have

$$S_{\boldsymbol{u}}^{+} - S_{\boldsymbol{u}}^{-} \geq 2,$$

which is contradictory to Eq. (2.6.4). By the same token, we can prove $u^{(i+1)}(\beta) \neq 0$.

Next, suppose that the $u^{(i+1)}$-line has no node in the interval (α, β) and, consequently, does not change sign in this interval. Without loss of generality, we may assume

$$u^{(i)}(x) > 0 \quad \text{and} \quad u^{(i+1)}(x) > 0, x \in (\alpha, \beta). \tag{2.6.6}$$

Let us again look at the vector $\boldsymbol{u}$ defined in Eq. (2.6.5). For a sufficiently large constant c, we should have $u(x) > 0\,(\alpha < x < \beta)$ based on the assumptions in Eq. (2.6.6). By gradually reducing c to a smaller number c_0, this part of the u-line must, at least, touch the x-axis at one point while its two ends stay on the same side of the x-axis. But this implies that the u-line has a zero, while the two components of $\boldsymbol{u}$ adjacent this zero must be of the same positive sign. But this also leads to $S_u^{+} - S_u^{-} \geq 2$, and is contradictory to Eq. (2.6.4). Therefore, there has to be at least one node of the $u^{(i+1)}$-line in the interval (α, β). Furthermore, as explained earlier, the $u^{(i)}$-line has and only has $i - 1$ nodes. Consequently, in each of the small interval (α_k, α_{k+1}) formed by nodes of the $u^{(i)}$-line, where $k - 0, 1, \ldots, i - 1$ and $\alpha_0 = 0 = \alpha_i$, there can be one and only one node of the $u^{(i+1)}$-line. In other words, nodes of $u^{(i)}$- and $u^{(i+1)}$-lines must interlace.

The previous argument is, apparently, applicable to the sign-oscillatory matrix, too. ■

The discussion above shows that qualitative properties of eigenpairs associated with both oscillatory and sign-oscillatory matrices parallel the oscillatory properties in vibration presented in Sect. 1.3. Because of this, the theory of oscillatory matrices forms the mathematical basis for the study of qualitative properties of the vibration of discrete systems in this book.

2.7 Integral Equations with Symmetric Kernels and the Concept of Oscillatory Kernels

With the help of the Green's function, we can express the natural vibration problem of any one-dimensional structure as the following eigenvalue problem of the integral equation:

$$u(x) = \lambda \int_a^b G(x, s)u(s)\rho(s)ds. \quad (2.7.1)$$

By setting

$$\varphi(x) = u(x)\sqrt{\rho(x)} \ \text{ and } \ K(x, s) = G(x, s)\sqrt{\rho(x)\rho(s)}, \quad (2.7.2)$$

we may transform Eq. (2.7.1) into a symmetrical form

$$\varphi(x) = \lambda \int_a^b K(x, s)\varphi(s)ds. \quad (2.7.3)$$

In some classical books, such as Ref. [1], it has been proved that both eigenvalues and eigenfunctions exist when the kernel in the integral equation is real, symmetric, continuous, and positive definite. Additionally, they have been shown to possess the following properties:

(1) Eigenvalues are all positive.
(2) When $K(x, s)$ is a degenerated kernel, i.e.,

$$K(x, s) = \alpha_0(x)\beta_0(s) + \alpha_1(x)\beta_1(s) + \cdots + \alpha_p(x)\beta_p(s),$$

there is a finite number of eigenvalues. For all other cases when $K(x, s)$ is not degenerated, there is countable but infinitely many eigenvalues, whose only limit point is the infinity. Thus, if there are repeated eigenvalues, their multiplicity has to be finite.
(3) The following integral transform of any piecewise continuous function $h(s)$

$$g(x) = \int_a^b K(x, s)h(s)ds$$

can be expanded into a generalized Fourier series of the eigenfunctions of $K(x, s)$. Furthermore, the series converges uniformly and absolutely. In this sense, the family of eigenfunctions of $K(x, s)$ constitutes a complete orthogonal system on the interval $[a, b]$.

As we all know, one-dimensional structures commonly seen in engineering applications, such as bars and beams, have eigenvalues that are not only real and positive but also isolated and distinct. This suggests that their Green's functions must have some special properties above and beyond those listed above. As a result, we are going to introduce one type of eigenvalue problems of integral equations that involves oscillatory kernels and is of practical significance. As with the oscillatory matrix, which may not be symmetric, an oscillatory kernel needs not to be symmetric.

Now let us define a point set I, the notation of which will be used not only in this section but also in subsequent discussion as well. The point set I includes (1) an open interval (a, b), (2) the end point a if $K(a, a) \neq 0$, and (3) the end point b if $K(b, b) \neq 0$. In Structural Dynamics, there is a clear physical interpretation for I: It represents the set consisting of all movable points in the closed interval $[a, b]$.

We are now ready to introduce the concept of an oscillatory kernel.

Definition 2.5 A continuous bivariate function $K(x, s)\,(a \leq x, s \leq b)$ is called an oscillatory kernel if it satisfies the following three conditions:

$$(1) \quad K(x, s) > 0, \quad x, s \in I \quad \text{and} \quad (x, s) \neq (a, b)^2; \tag{2.7.4}$$

$$(2) \quad K\begin{pmatrix} x_1\ x_2 \cdots x_n \\ s_1\ s_2 \cdots s_n \end{pmatrix} \geq 0, \quad a \leq \begin{matrix} x_1 < x_2 < \cdots < x_n \\ s_1 < s_2 < \cdots < s_n \end{matrix} \leq b, \ \text{and}\ n = 1, 2, \ldots; \tag{2.7.5}$$

$$(3) \quad K\begin{pmatrix} x_1\ x_2 \cdots x_n \\ x_1\ x_2 \cdots x_n \end{pmatrix} > 0, \quad x_1 < x_2 < \cdots < x_n \in I, \ \text{and}\ n = 1, 2, \ldots; \tag{2.7.6}$$

where

$$K\begin{pmatrix} x_1\ x_2 \cdots x_n \\ s_1\ s_2 \cdots s_n \end{pmatrix} = \begin{vmatrix} K(x_1, s_1) & K(x_1, s_2) & \cdots & K(x_1, s_n) \\ K(x_2, s_1) & K(x_2, s_2) & \cdots & K(x_2, s_n) \\ \vdots & \vdots & & \vdots \\ K(x_n, s_1) & K(x_n, s_2) & \cdots & K(x_n, s_n) \end{vmatrix}.$$

Starting from inequalities in Definition 2.5 and also using the criteria for an oscillatory matrix (Theorem 2.10), we may derive the theorem below.

Theorem 2.18 *A continuous function $K(x, s)$ defined on the rectangular region $a \leq x, s \leq b$ is an oscillatory kernel if and only if the matrix $(K(x_i, x_j))$ is oscillatory for any n and one-dimensional point set $\{x_i\}_1^n$, where $x_1 < x_2 < \cdots < x_n$, $x_i \in I$ $(i = 1, 2, \ldots, n)$, and at least one of the points is an internal point.*

Proof Equation (2.7.4) implies that both $K(x_i, x_{i+1})$ and $K(x_{i+1}, x_i)$ are positive, Eq. (2.7.6) indicates that the matrix $\boldsymbol{K} = (K(x_i, x_j))$ is nonsingular, and Eq. (2.7.5) shows that $\boldsymbol{K}$ is a totally nonnegative matrix. Thus, $\boldsymbol{K}$ must be an oscillatory matrix.

Conversely, when $\boldsymbol{K} = (K(x_i, x_j))$ is an oscillatory matrix, we may arrive at the following conclusions by applying Theorem 2.10 (the criteria for an oscillatory matrix): (1) $(K(x_i, x_j))$ is a nonsingular and totally nonnegative matrix, which leads to Eq. (2.7.6). (2) Eq. (2.7.4) results from $K(x_i, x_{i+1}) > 0$ as well as the arbitrariness

[2]Here, $(x, s) \neq (a, b)$ means that $x = a$ and $s = b$ cannot occur simultaneously, and neither can $x = b$ and $s = a$. Thus, as an example, when $x = a$, we should have $s \neq b$.

of x_i and x_{i+1}. (3) For any two point sets $x_1 < x_2 < \cdots < x_n$ and $s_1 < s_2 < \cdots < s_n$, it is always possible to combine them into one set and sort it from small to large as

$$x'_1 < x'_2 < \cdots < x'_m, \quad n \leq m \leq 2n.$$

Since $(K(x'_i, x'_j))_{m\times m}$ is still an oscillatory matrix, any of its minors is nonnegative. This leads to Eq. (2.7.5). ■

Starting from this theorem, we may derive a corollary that is very useful in engineering applications.

Corollary *Suppose $G(x, s)$ is an oscillatory kernel while $f(x)$ is a positive function. Then*

$$K(x, s) = G(x, s)f(x)f(s)$$

is also an oscillatory kernel.

In fact, given an oscillatory kernel $G(x, s)$, any integer n, and an arbitrary set of real numbers $x_1 < x_2 < \cdots < x_n$, if $x_i \in I\,(i = 1, 2, \ldots, n)$ and at least one of the points is an internal point, then, based on Theorem 2.18, Matrix $(G(x_i, x_j))$ is oscillatory. As a special case of Property 5 of oscillatory matrices, the corresponding matrix

$$(K(x_i, x_j)) = \begin{bmatrix} G(x_1, x_1)f(x_1)f(x_1) & G(x_1, x_2)f(x_1)f(x_2) & \cdots & G(x_1, x_n)f(x_1)f(x_n) \\ G(x_2, x_1)f(x_2)f(x_1) & G(x_2, x_2)f(x_2)f(x_2) & \cdots & G(x_2, x_n)f(x_2)f(x_n) \\ \vdots & \vdots & & \vdots \\ G(x_n, x_1)f(x_n)f(x_1) & G(x_n, x_2)f(x_n)f(x_2) & \cdots & G(x_n, x_n)f(x_n)f(x_n) \end{bmatrix}$$
$$= \mathrm{diag}(f(x_1), f(x_2), \ldots, f(x_n)) \cdot (G(x_i, x_j)) \cdot \mathrm{diag}(f(x_1), f(x_2), \ldots, f(x_n))$$

obviously remains to be an oscillatory matrix. Consequently,

$$K(x, s) = G(x, s)f(x)f(s)$$

is also an oscillatory kernel.

2.8 Perron's Theorem for Integral Equations and the Concept of Compound Kernels

In order to derive oscillatory properties of the eigenvalue problem associated with an integral equation having an oscillatory kernel, we first present Perron's theorem for an integral equation having a continuous kernel, and then introduce the concept of compound kernels as well as some of their simple properties. The approach parallels that of Sect. 2.5.

Theorem 2.19 *(Perron's Theorem) If a kernel $K(x, s)$ is continuous on the rectangular region $a \le x, s \le b$ and also satisfies conditions below:*

$$K(x, s) \ge 0 \quad and \quad K(x, x) > 0, \quad for\ x\ and\ s \in I, \tag{2.8.1}$$

then there exists a unique eigenvalue λ_1 for the integral equation (2.7.3), which has the smallest absolute value among all eigenvalues and is both positive and distinct. In addition, the eigenfunction associated with λ_1 does not change sign in the interval I.

The proof of this theorem can be found in Ref. [3] or [4].
Next, let us introduce the concept of compound kernels.

Definition 2.6 The p-th compound kernel $\underline{K}_p(X, S)$ of $K(x, s)$ is defined as

$$\underline{K}_p(X, S) = K\begin{pmatrix} x_1 & x_2 & \cdots & x_p \\ s_1 & s_2 & \cdots & s_p \end{pmatrix}, \tag{2.8.2}$$

where $X = (x_1, x_2, \ldots, x_p)$ and $S = (s_1, s_2, \ldots, s_p)$ exhaust all points of the p-dimensional simplex M^p defined by the inequality

$$a \le x_1 \le x_2 \le \cdots \le x_p \le b.$$

When none of the equalities holds in the inequality above, the corresponding point X is called an internal point of M^p.

In the subsequent discussion, we will also deal with iterated kernels. Here is their definition: The q-th iterated kernel of $K(x, s)$ refers to

$$K^{(q)}(x, s) = \int_a^b K^{(q-1)}(x, t)K(t, s)\mathrm{d}t.$$

Specifically, the second iterated kernel is

$$K^{(2)}(x, s) = \int_a^b K(x, t)K(t, s)\mathrm{d}t.$$

Now we would like to present two important properties of a compound kernel.

Property 1 *If three kernels $K(x, s)$, $L(x, s)$, and $M(x, s)$ are related by*

$$M(x, s) = \int_a^b K(x, t)L(t, s)dt, \tag{2.8.3}$$

then

$$M\begin{pmatrix} x_1 \ x_2 \cdots x_p \\ s_1 \ s_2 \cdots s_p \end{pmatrix}$$
$$= \int_a^b \int_a^{t_p} \cdots \int_a^{t_2} K\begin{pmatrix} x_1 \ x_2 \cdots x_p \\ t_1 \ t_2 \cdots t_p \end{pmatrix} L\begin{pmatrix} t_1 \ t_2 \cdots t_p \\ s_1 \ s_2 \cdots s_p \end{pmatrix} \cdot dt_1 dt_2 \cdots dt_p,$$

i.e.,

$$\underline{M}_p(X, S) = \int_{M^p} \underline{K}_p(X, T)\underline{L}_p(T, S)\mathrm{d}T. \tag{2.8.4}$$

In fact, Eq. (2.8.4) may be derived from the following integral identity:

$$\int_{M^p} \Delta\begin{pmatrix} \psi_1 \ \psi_2 \cdots \psi_p \\ t_1 \ t_2 \cdots t_p \end{pmatrix} \Delta\begin{pmatrix} \chi_1 \ \chi_2 \cdots \chi_p \\ t_1 \ t_2 \cdots t_p \end{pmatrix} \cdot \mathrm{d}t_1 \mathrm{d}t_2 \cdots \mathrm{d}t_p$$
$$= \frac{1}{p!} \int_a^b \int_a^b \cdots \int_a^b \Delta\begin{pmatrix} \psi_1 \ \psi_2 \cdots \psi_p \\ t_1 \ t_2 \cdots t_p \end{pmatrix} \Delta\begin{pmatrix} \chi_1 \ \chi_2 \cdots \chi_p \\ t_1 \ t_2 \cdots t_p \end{pmatrix} \cdot \mathrm{d}t_1 \mathrm{d}t_2 \cdots \mathrm{d}t_p$$
$$= \det\left(\int_a^b \psi_i(t)\chi_j(t)\mathrm{d}t \right)_{p\times p} \tag{2.8.5}$$

by setting $\psi_i = K(x_i, t)$ and $\chi_i = L(t, s_i)$, where

$$\Delta\begin{pmatrix} \psi_1 \ \psi_2 \cdots \psi_p \\ t_1 \ t_2 \cdots t_p \end{pmatrix} = \begin{vmatrix} \psi_1(t_1) & \psi_1(t_2) & \cdots & \psi_1(t_p) \\ \psi_2(t_1) & \psi_2(t_2) & \cdots & \psi_2(t_p) \\ \vdots & \vdots & & \vdots \\ \psi_p(t_1) & \psi_p(t_2) & \cdots & \psi_p(t_p) \end{vmatrix}. \tag{2.8.6}$$

Property 2 *The p-th compound kernel of the q-th iterated kernel of $K(x, s)$ equals the q-th iterated kernel of the p-th compound kernel of $K(x, s)$, namely,*

$$[\underline{K}_p(X, S)]^{(q)} = \underline{K}_p^{(q)}(X, S). \tag{2.8.7}$$

To prove Property 2, set $L(x, s) = K(x, s)$ in Eq. (2.8.3) . Then, Eq. (2.8.4) yields when $q = 2$,

$$[\underline{K}_p(X, S)]^{(2)} = \underline{K}_p^{(2)}(X, S).$$

We can complete the proof by applying the method of mathematical induction to cases of $q > 2$.

Starting from these two properties, we may determine how eigenvalues and eigenfunctions of a continuous kernel $K(x, s)$ relate to their counterparts of the corresponding p-th compound kernel $\underline{K}_p(X, S)$.

Theorem 2.20 *Suppose the integral equation (2.7.3) has a continuous kernel and the sequence of its eigenpairs is represented by* $\{\lambda_i, \varphi_i(x)\}_1^\infty$. *Then, the eigenvalue and the associated eigenfunction of the integral equation*

$$\Phi(X) = \Lambda \int\limits_{M^p} \underline{K}_p(X, S)\Phi(S)dS$$

are, respectively, $\lambda_{i_1}\lambda_{i_2}\cdots\lambda_{i_p}$ *and*

$$\Delta\begin{pmatrix} \varphi_{i_1} & \varphi_{i_2} & \cdots & \varphi_{i_p} \\ x_1 & x_2 & \cdots & x_p \end{pmatrix},$$

where $i_1, i_2, \ldots, i_p$ *are any p different indices taken from* $1, 2, \ldots, n, \ldots$ *and satisfy the condition* $i_1 < i_2 < \cdots < i_p$.

Proof Set $\psi_k(t) = K(x_k, t)$ and $\chi_k(t) = \varphi_{i_k}(t)$ in Eq. (2.8.5). Also notice that

$$\varphi_{i_k}(x) = \lambda_i \int\limits_a^b K(x, t)\varphi_{i_k}(t)\mathrm{d}t.$$

We get

$$\Delta\begin{pmatrix} \varphi_{i_1} & \varphi_{i_2} & \cdots & \varphi_{i_p} \\ x_1 & x_2 & \cdots & x_p \end{pmatrix} = \lambda_{i_1}\lambda_{i_2}\cdots\lambda_{i_p}\left|\int\limits_a^b K(x_j, t)\varphi_{i_j}(t)\mathrm{d}t\right|_{p\times p}$$
$$= \lambda_{i_1}\lambda_{i_2}\cdots\lambda_{i_p}\int\limits_{M^p} K\begin{pmatrix} x_1 & x_2 & \cdots & x_p \\ t_1 & t_2 & \cdots & t_p \end{pmatrix}$$
$$\cdot\,\Delta\begin{pmatrix} \varphi_{i_1} & \varphi_{i_2} & \cdots & \varphi_{i_p} \\ t_1 & t_2 & \cdots & t_p \end{pmatrix} \cdot \mathrm{d}t_1\mathrm{d}t_2\cdots\mathrm{d}t_p.$$

■

2.9 Eigenpairs of the Integral Equation with an Oscillatory Kernel

In this section, we first introduce the concept, criteria, and main properties of a family of oscillatory functions known as the sequence of Markov functions, and then discuss oscillatory properties of the integral equation with an oscillatory kernel.

2.9.1 Family of Oscillatory Functions

Let us begin with a few definitions related to the family of oscillatory functions.

Definition 2.7 Consider a function $f(x)$ defined on the interval $[a, b]$. If $f(c) = 0\,(a \leq c \leq b)$, then the point c is called a zero of $f(x)$. If there exists a subinterval $J \subset [a, b]$ such that $f(c) = 0$ for any point c in J, then J is referred as a zero subinterval of $f(x)$. Furthermore, if c is a zero of $f(x)$ and

$$f(c+\varepsilon)f(c-\varepsilon) < 0$$

for an arbitrarily small positive number ε, then c is referred to as a node of $f(x)$; while if c is a zero of $f(x)$ and

$$f(c+\varepsilon)f(c-\varepsilon) > 0$$

for an arbitrarily small positive number ε, then c is called a null anti-node of $f(x)$.

Definition 2.8 Suppose the function $u(x)$ is defined on the interval $[a, b]$. If there exists a point set $\{x_r\}_0^k$ in $[a, b]$ such that

$$u(x_r)u(x_{r+1}) < 0, \quad r = 0, 1, \ldots, k-1,$$

and we cannot find $k + 2$ points that have the aforementioned property, then the function $u(x)$ has k sign reversals in the interval $[a, b]$, which is denoted by $S_u = k$. Apparently, if $u(x)$ is continuous on $[a, b]$, then the function undergoing k sign reversals in $[a, b]$ is equivalent to the function having and only having k nodes within the same interval.

Definition 2.9 Consider a family of functions $\varphi_i(x)\,(i = 1, 2, \ldots, n)$ that are continuous on the interval $[a, b]$. Given an arbitrary set of real numbers $c_i\,(i = 1, 2, \ldots, n)$ that do not vanish simultaneously, if the number of zeroes of the function

$$\varphi(x) = c_1\varphi_1(x) + c_2\varphi_2(x) + \cdots + c_n\varphi_n(x)$$

does not exceed $n - 1$ in $I \subset [a, b]$, then $\{\varphi_i(x)\}_1^n$ $(i = 1, 2, \ldots, n)$ form a family of Chebyshev functions on the point set I.

Definition 2.10 Consider an infinite sequence of continuous functions $\varphi_i(x)\,(i = 1, 2, \ldots, n, \ldots)$ defined on the interval $[a, b]$. If $\varphi_i(x)\,(i = 1, 2, \ldots, n)$ form a family of Chebyshev functions on $I \subset [a, b]$ for any given $n\,(n = 1, 2, \ldots)$, then we refer to $\{\varphi_i(x)\}_1^\infty$ as a sequence of Markov functions.

The following criteria may be established for a sequence of Markov functions:

Theorem 2.21 *Let $\varphi_i(x)\,(i = 1, 2, \ldots, n, \ldots)$ represent a sequence of continuous functions on the interval $[a, b]$. They constitute a sequence of Markov functions if and only if, for any $n = 1, 2, \ldots$, the determinant*

$$\Phi(X) \equiv \Delta\begin{pmatrix} \varphi_1 & \varphi_2 & \cdots & \varphi_n \\ x_1 & x_2 & \cdots & x_n \end{pmatrix} \tag{2.9.1}$$

always has strictly fixed sign ε_n, where $x_i\,(i = 1, 2, \ldots, n)$ are an arbitrary set of points in I satisfying $a \le x_1 < x_2 < \cdots < x_n \le b$.

Proof Let us first prove the necessary part of the theorem by employing the method of proof by contradiction. For any given n, suppose there exists a set of points $x_1 < x_2 < \cdots < x_n \in I$ such that

$$\Delta\begin{pmatrix} \varphi_1 & \varphi_2 & \cdots & \varphi_n \\ x_1 & x_2 & \cdots & x_n \end{pmatrix} = 0.$$

Then the following system of equations

$$c_1\varphi_1(x_r) + c_2\varphi_2(x_r) + \cdots + c_n\varphi_n(x_r) = 0, \quad r = 1, 2, \ldots, n \tag{2.9.2}$$

has a nontrivial solution. Thus, the function $\varphi(x) = c_1\varphi_1(x) + c_2\varphi_2(x) + \cdots + c_n\varphi_n(x)$ has n zeroes, which contradicts the definition for the sequence of Markov functions. Furthermore, because Φ is a continuous function on the n-dimensional simplex M^n, the fact that the value Φ is not zero at any point $X = (x_1, x_2, \ldots, x_n)$ in M^n implies it has strictly fixed sign in the region. This completes the proof of the necessary part of the theorem.

It needs to be pointed out: We have only proved in the previous paragraph that, for a given n, the determinant (2.9.1) is of fixed sign at any point $X = (x_1, x_2, \ldots, x_n)$ in the simplex M^n and the sign may be denoted by ε_n. However, when n varies, the determinant in Eq. (2.9.1) may not preserve the same positive or negative sign.

Now let us prove the sufficient part of the theorem. In fact, the condition that the determinant (2.9.1) has strictly fixed sign implies $\Phi(X) \neq 0$, where n is any positive integer and $x_i\,(i = 1, 2, \ldots, n)$ are any point set in I satisfying $a \le x_1 < x_2 < \cdots < x_n \le b$. This indicates that the system of equations (2.9.2) should have a trivial solution only when the number of equations exceeds $n - 1$. In other words, the condition for the system of equations (2.9.2) to have a nontrivial solution is as follows: Its number of equations must be no more than $n-1$, i.e., the number of zeroes of $\varphi(x)$ is no greater than $n - 1$. Thus, the family of functions $\varphi_i(x)\,(i = 1, 2, \ldots)$ constitutes a sequence of Markov functions. ■

In the definition of a sequence of Markov functions, it involves some accounting work on the number of zeroes of a function. Thus, we introduce the corollary below.

Corollary *Suppose $\varphi_i(x)\,(i = 1, 2, \ldots)$ comprise a sequence of Markov functions on the interval $[a, b]$. If the function*

$$\varphi(x) = \sum_{i=1}^{n} c_i\varphi_i(x), \quad \sum_{i=1}^{n} c_i^2 > 0 \quad \text{and} \quad n = 1, 2, \ldots$$

has r distinct zeroes in $[a, b]$, which include p nodes and q null anti-nodes, then $r + q \leq n - 1$. This indicates that a null anti-node should be treated like two single zeroes when the total number of zeroes is counted.

Proof To facilitate the discussion, let us first introduce Property Z of a point set: Given a function $\varphi(x)$ on $[a, b]$, the s points $x_k \in I\,(k = 1, 2, \ldots, s)$, where $x_1 < x_2 < \cdots < x_s$, have Property Z if there exists an integer h such that

$$(-1)^{k+h}\varphi(x_k) \geq 0, \quad k = 1, 2, \ldots, s.$$

Suppose nodes of $\varphi(x)$ to be $\{\alpha_i\}_1^p$. Per the definition of a node, it is not difficult to see that there exist points $x_k \in (\alpha_k, \alpha_{k+1})\,(k = 0, 1, \ldots, p; \alpha_0 = a, \alpha_{p+1} = b)$ such that

$$(-1)^{k+h}\varphi(x_k) > 0,$$

i.e., $\{x_k\}_0^p$ have Property Z. Moreover, for each of the null anti-nodes $\beta_m\,(m = 1, 2, \ldots, q)$ of $\varphi(x)$, it is always possible to find two points in the neighborhood of β_m such that the new point set formed by adding these new points to the original set $\{x_k\}_0^p$ continues to have Property Z. In fact, let ε denote a positive number that is sufficiently small. If $x_k < \beta_m < \alpha_{k+1}$, we may set $x_m^- = \beta_m$ and $x_m^+ = \beta_m + \varepsilon$; while if $\alpha_k < \beta_m < x_k$, we can choose $x_m^- = \beta_m - \varepsilon$ and $x_m^+ = \beta_m$. Thus, we get a set consisting of $p + 2q + 1$ points:

$$\{x_k\}_0^p \cup \{x_m^-, x_m^+\}_1^q.$$

The new set $\{s_k\}_1^{r+q+1}$ obtained by rearranging the point set given above in an increasing order, where $p + q = r$, must possess Property Z.

We are now ready to prove the theorem through proof by contradiction. Assume that $r + q > n - 1$. Expanding the determinant

$$\Delta\begin{pmatrix} \varphi_1 & \varphi_2 & \cdots & \varphi_n & \varphi \\ s_1 & s_2 & \cdots & s_n & s_{n+1} \end{pmatrix},$$

which is identically zero, by the last row, we get

$$\sum_{k=1}^{n}(-1)^{n+k+1}\varphi(s_k)\Delta\begin{pmatrix}\varphi_1 & \varphi_2 & \cdots & \cdots & \cdots & \cdots & \varphi_n\\ s_1 & s_2 & \cdots & s_{k-1} & s_{k+1} & \cdots & s_{n+1}\end{pmatrix}=0.$$

Since all terms in the summation on the left-hand side of the equation are of the same positive or negative sign, all of $\varphi(s_k)\,(k=1,2,\ldots,n)$ have to vanish for the equation to be valid. But this is obviously impossible. ■

The sequence of Markov functions has a series of important properties, which are summarized in the next theorem.

Theorem 2.22 *Suppose $\varphi_i(x)\,(i=1,2,\ldots)$ are a sequence of Markov functions on the interval $[a,b]$. It is further assumed that they are orthogonal with respect to the weight function $\rho(x)$ on the same interval. Then, the following properties exist:*

(1) $\varphi_1(x)$ has no zero in the point set $I\subset(a,b)$.
(2) $\varphi_i(x)$ has $i-1$ nodes and no other zeroes in $I\subset(a,b)$, where $i=2,3,\ldots$.
(3) In the point set $I\subset(a,b)$, the function

$$\varphi(x)=\sum_{k=p}^{q}c_k\varphi_k(x),\quad 0\le p\le q\ \ and\ \ \sum_{i=q}^{p}c_i^2>0$$

has at least $p-1$ nodes and no more than $q-1$ zeroes. In particular, if $\varphi(x)$ has $q-1$ different zeroes, then all of them are nodes.
(4) Nodes of two successive functions $\varphi_i(x)$ and $\varphi_{i+1}(x)$ interlace.

Proof Notice that Properties (1) and (2) are both special cases of Property (3). In addition, it is obviously true that $\varphi(x)$ has no more than $q-1$ zeroes, as shown in Property (3), based on the corollary of Theorem 2.21. Thus, we only need to demonstrate that the node count of $\varphi(x)$ is no less than $p-1$ here. To this end, let $\xi_1<\xi_2<\cdots<\xi_{r-1}$ denote the nodes of $\varphi(x)$ and introduce the following function:

$$\psi(x)=\Delta\begin{pmatrix}\varphi_1 & \varphi_2 & \cdots & \varphi_{r-1} & \varphi_r\\ \xi_1 & \xi_2 & \cdots & \xi_{r-1} & x\end{pmatrix}.$$

We know from the definition of a sequence of Markov functions as well as Theorem 2.21: $\psi(x)\neq 0$ when $x\neq\xi_i\,(i=1,2,\ldots,r-1)$; $\psi(x)$ retains its positive or negative sign when x varies within the interval $(\xi_i,\xi_{i+1})\,(i=0,1,\ldots,r-1;\xi_0=a,\xi_r=b)$; and $\psi(x)$ undergoes sign reversal only when x passes through ξ_i. These behaviors indicate that $\xi_i\,(i=0,1,\ldots,r-1)$ are also nodes of $\psi(x)$. Therefore,

$$(\rho\varphi,\psi)\neq 0.$$

Notice that φ and ψ are, respectively, linear combinations of $\varphi_i(x)\,(i=p,p+1,\ldots,q)$ and $\varphi_i(x)\,(i=1,2,\ldots,r)$. Thus, the previous inequality implies that there has to be some overlapping between the two sets of basis functions, $\{\varphi_i(x)\}_p^q$ and $\{\varphi_i(x)\}_1^r$, which implies $r \geq p$. This completes the proof of Property (3).

In order to prove Property (4), let us define the following function:

$$\psi(x) = \varphi_i(x)/\varphi_{i+1}(x).$$

Suppose nodes of $\varphi_{i+1}(x)$ are $\{\alpha_k\}_1^i$. They divide the interval $[a, b]$ into $i + 1$ subintervals (α_k, α_{k+1}) $(k = 0, 1, \ldots, i; \alpha_0 = a, \alpha_{i+1} = b)$. We are now ready to prove the result in two steps.

First, let us show that $\psi(x)$ is monotonic within each of the subintervals given above. If this is not true and $\psi(x)$ is not a monotonic function in a subinterval (α_k, α_{k+1}), then there exist three points $x_1 < x_2 < x_3$ in this subinterval such that

$$(\psi(x_2) - \psi(x_1))(\psi(x_2) - \psi(x_3)) > 0.$$

Without loss of generality, we may assume $\psi(x_2) > \psi(x_1)$, which leads to $\psi(x_2) > \psi(x_3)$. Hence, $\psi(x)$ must reach the maximum $\psi(x_0)$ at an internal point x_0 within the closed interval $[x_1, x_3]$. Consider the function

$$\varphi(x) = \varphi_{i+1}(x)(\psi(x) - \psi(x_0)) = \varphi_i(x) - \psi(x_0)\varphi_{i+1}(x).$$

On the one hand, it is known from Property (3) that the node count of $\varphi(x)$ is no less than $i - 1$, while its zero count is no more than i on $[a, b]$; and consequently, all zeroes of $\varphi(x)$ have to be its nodes in this interval. On the other hand, since

$$\psi(x) - \psi(x_0) \leq 0, \quad x \in [x_1, x_3],$$

x_0 must be a null anti-node of $\varphi(x)$. These results produce a contradiction, which shows that $\psi(x)$ has to be a monotonic function.

Next, we will prove that $\psi(x)$ varies monotonically either from $-\infty$ to $+\infty$ or from $+\infty$ to $-\infty$ within each of the subintervals (α_k, α_{k+1}) $(k = 1, 2, \ldots, i-1)$ given above. To this end, we only need to exclude possibilities that any of the following limits can be of finite value:

$$\lim_{x \to \alpha_k - 0} \psi(x) = c_k', \quad \lim_{x \to \alpha_k + 0} \psi(x) = c_k''.$$

We will again proceed with the method of proof by contradiction. Suppose at least one of the limits, for instance, c_k' to be of finite value, which is possible only when α_k is also a node of $\varphi_i(x)$. There are two possibilities in this situation: (1) $c_k' \neq c_k''$, but they share a common sign and c_k'' can be either of finite value or infinity; (2) $c_k' = c_k''$. The four possible shapes of $\psi(x)$ in the neighborhood of α_k are depicted in Fig. 2.1,

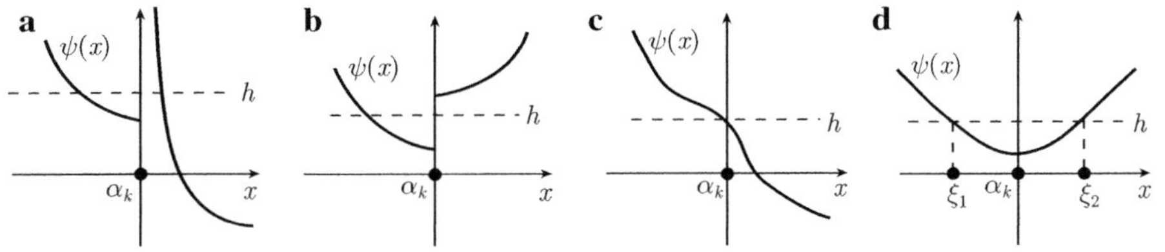

Fig. 2.1 Four possible shapes of $\psi(x)$ in the neighborhood of α_k

where it has been assumed that, without loss of generality, $\psi(x)$ is monotonically decreasing in the subinterval (α_{k-1}, α_k). For the first three cases a, b, and c in Fig. 2.1, there exists a number h such that $\psi(x) - h$ reverses sign as x passes through α_k from left to right. Thus, α_k should be a null anti-node of the function

$$\varphi(x) = \varphi_{i+1}(x)(\psi(x) - h) = \varphi_i(x) - h\varphi_{i+1}(x), \tag{2.9.3}$$

which is impossible according to Property (3). As for the case d in Fig. 2.1, choose an h such that $h > c'_k$. The equation $\psi(x) = h$ has two roots denoted by ξ_1 and ξ_2 that ought to be nodes of the function $\varphi(x)$ defined in Eq. (2.9.3). As a result, $\varphi(x)$ would have two more nodes than $\varphi_{i+1}(x)$ does. But this is also impossible. These contradictions indicate that c'_k cannot be of finite value. By the same token, c''_k cannot be a finite number, too.

By summarizing the above discussion, we may conclude: $\psi(x)$ either monotonically increases from $-\infty$ to $+\infty$ or monotonically decreases from $+\infty$ to $-\infty$ in (α_k, α_{k+1}) $(k = 1, 2, \ldots, i-1)$. Thus, $\varphi_i(x)$ has one and only one node in the subinterval (α_k, α_{k+1}). Since $\varphi_i(x)$ has a total of $i - 1$ nodes only, there should be no node of the function in subintervals (a, α_1) and (α_i, b). This completes proof of interlacing properties of nodes of $\varphi_i(x)$ and $\varphi_{i+1}(x)$. ■

2.9.2 Eigenpairs of Integral Equations with Oscillatory Kernels

With preparation in the previous section, we are now ready to prove the following theorem on properties of an oscillatory kernel.

Theorem 2.23 *Consider the following integral equation with a continuous oscillatory kernel:*

$$\varphi(x) = \lambda \int_a^b K(x, s)\varphi(s)ds. \tag{2.9.4}$$

Its eigenvalues are both positive and distinct. In addition, if these eigenvalues are sorted in an ascending order $0 < \lambda_1 < \lambda_2 < \cdots$, *then the corresponding eigenfunctions form a sequence of Markov functions on the interval* $[a, b]$, *and thus have the four properties of the sequence of Markov functions described in Theorem 2.22.*

Proof Arrange eigenvalues of Eq. (2.9.4) in the order such that their absolute values are increasing:

$$|\lambda_1| \leq |\lambda_2| \leq |\lambda_3| \leq \cdots.$$

Denote the p-th compound kernel of $K(x, s)$ as $\underline{K}_p(X, S)$. When $K(x, s)$ is a continuous oscillatory kernel, $\underline{K}_p(X, S)$ satisfies conditions of Theorem 2.19 (Perron's Theorem). Thus, the integral equation

$$\Phi(X) = \Lambda \int\limits_{M^p} \underline{K}_p(X, S)\Phi(S)\mathrm{d}S$$

has an eigenvalue $\lambda_1\lambda_2\cdots\lambda_p$ whose absolute value is the smallest among all eigenvalues, and

$$\lambda_1\lambda_2\cdots\lambda_p > 0, \quad \lambda_1\lambda_2\cdots\lambda_p < \left|\lambda_1\lambda_2\cdots\lambda_{p-1}\lambda_{p+1}\right|.$$

The first inequality above shows that all eigenvalues are positive, while the second one indicates $\lambda_p < \lambda_{p+1} (p = 1, 2, \ldots)$. Furthermore, the eigenfunction

$$\Delta\begin{pmatrix} \varphi_1 & \varphi_2 & \cdots & \varphi_p \\ x_1 & x_2 & \cdots & x_p \end{pmatrix}$$

of $\underline{K}_p(X, S)$ corresponding to the eigenvalue $\lambda_1\lambda_2\cdots\lambda_p$ has strictly fixed sign. From Theorem 2.21, $\varphi_i(x)\,(i = 1, 2, \ldots)$ form a sequence of Markov functions on the interval $[a, b]$, and thus possess all properties described in Theorem 2.22. ■

Finally, we would like to mention that Eq. (2.7.4) in the definition of an oscillatory kernel does not appear as a condition in Theorem 2.19 (Perron's Theorem) and consequently, is not used in the proof of Theorem 2.23. Kernels satisfying only Eqs. (2.7.5) and (2.7.6) are referred to as Kellogg kernels. Gantmacher and Krein pointed in Ref. [3] that all known Kellogg kernels as of now are oscillatory.

2.10 Relationship Among Oscillatory Properties in Static Deformation, Flexibility Functions (Matrices) Being Oscillatory Kernels (Matrices), and Oscillatory Properties in Vibration

Definition 2.11 Consider the discrete system modeling a one-dimensional continuous system. The discrete system has oscillatory properties in vibration, if following requirements are met: (1) All of its natural angular frequencies are distinct, i.e., $0 < \omega_1 < \omega_2 < \cdots < \omega_n$. (2) The i-th mode shape $\boldsymbol{u}^{(i)}$ has $i-1$ sign reversals. (3) For an arbitrary set of real numbers c_i $(i = p, p+1, \ldots, q$ and $1 \le p \le q \le n)$ that do not vanish simultaneously, counts of sign reversals of the vector

$$\boldsymbol{u} = c_p\boldsymbol{u}^{(p)} + c_{p+1}\boldsymbol{u}^{(p+1)} + \cdots + c_q\boldsymbol{u}^{(q)}$$

satisfy the condition $p-1 \le S_{\boldsymbol{u}}^- \le S_{\boldsymbol{u}}^+ \le q-1$. (4) Nodes of two successive mode shapes interlace.

Definition 2.12 A one-dimensional continuous system has oscillatory properties in vibration, if the following conditions are met: (1) All of its natural angular frequencies are distinct, i.e., $0 < \omega_1 < \omega_2 < \cdots < \omega_n < \cdots$. (2) The i-th mode shape $u_i(x)$ has $i-1$ sign reversals in the point set $I \subset [0, l]$. (3) For an arbitrary set of real numbers c_i $(i = p, p+1, \ldots, q$ and $1 \le p \le q)$ that do not vanish simultaneously, the number of nodes of the function

$$u(x) = \sum_{i=p}^{q} c_i u_i(x)$$

is no less than $p-1$, while its zero count is no more than $q-1$. (4) Nodes of two successive mode shapes interlace.

Although the theory of oscillatory matrices and kernels, developed by Gantmacher and Krein and briefly introduced in Sects. 2.3–2.9, are very elegant and fairly comprehensive, there are two remaining questions that deserve further investigation and contemplation:

First, it is very difficult to determine, directly from the definition, whether the Green's function of a structure to be oscillatory, or if the flexibility (or stiffness) matrix of the discrete system of the structure to be oscillatory (or sign-oscillatory). Second, it has been proved in Sect. 2.9 that its Green's function being an oscillatory kernel is a sufficient condition for a well-constrained one-dimensional continuous system to have oscillatory properties in vibration. Similarly, we have concluded that its flexibility matrix being oscillatory is a sufficient condition for the discrete system modeling the well-constrained one-dimensional continuous system to have oscillatory properties in vibration. However, it is not clear whether the respective conditions should also be necessary.

The good news is that we now have fairly complete answers to both of these questions. They will be discussed in greater details in this section.

2.10.1 Relationship Between Oscillatory Properties in Static Deformation and Flexibility Functions (Matrices) Being Oscillatory Kernels (Matrices)

For a one-dimensional continuous system, Gantmacher and Krein pointed out in Ref. [3] that Conditions (2.7.4) and (2.7.5) of the oscillatory kernel are equivalent, respectively, to the following two properties:

Property A *If a concentrated force is applied to a movable point on a one-dimensional continuous system, the deflection at every movable point of the system is nonzero and in the direction of the applied force.*

Property B *Under the action of n concentrated forces, the deflection $u(x)$ of the one-dimensional continuous system has at most $n-1$ sign reversals. Namely, $S_u \le n-1$.*

We refer to Properties A and B collectively as oscillatory properties in static deformation of a one-dimensional continuous system. In addition, we will follow the convention introduced at the beginning of Sect. 1.2, when dealing with the direction of the displacement or deflection induced by concentrated forces applied to the one-dimensional object.

Based on the definition and symmetry of the Green's function, the mathematical representation of Property A is Eq. (2.7.4). In order to establish the relationship between Property B and Condition (2.7.5) of an oscillatory kernel, we need to use Lemmas 2.2 and 2.3 to be introduced later.

But before getting to these lemmas, we present a formula without proof: Suppose $\{\varphi_i(x)\}_1^n$ and $m(x,s)$ are, respectively, continuous functions in the interval $[0,l]$ and the square region $0 \le x, s \le l$. In addition, set

$$\psi_i(x) = \int_0^l m(x,s)\varphi_i(s)\mathrm{d}s, \quad i = 1,2,\ldots,n.$$

Then, for any $i, j = 1,2,\ldots,n$, we have

$$\begin{aligned} \Psi\begin{pmatrix} x_1 & x_2 & \cdots & x_n \\ 1 & 2 & \cdots & n \end{pmatrix} &= \det\big(\psi_i(x_j)\big) \\ &= \int\int_V \cdots \int M\begin{pmatrix} x_1 & x_2 & \cdots & x_n \\ s_1 & s_2 & \cdots & s_n \end{pmatrix} \\ &\quad \cdot \Phi\begin{pmatrix} s_1 & s_2 & \cdots & s_n \\ 1 & 2 & \cdots & n \end{pmatrix} \cdot \mathrm{d}s_1\mathrm{d}s_2\cdots\mathrm{d}s_n. \end{aligned} \tag{2.10.1}$$

where V is a simplex and its interior point satisfies the inequality $0 \le s_1 < s_2 < \cdots < s_n \le l$ while

$$M\begin{pmatrix} x_1 & x_2 & \cdots & x_n \\ s_1 & s_2 & \cdots & s_n \end{pmatrix} = \det(m(x_i, s_j)), \quad 0 \le \begin{matrix} x_1 < x_2 < \cdots < x_n \\ s_1 < s_2 < \cdots < s_n \end{matrix} \le l,$$

and

$$\Phi\begin{pmatrix} s_1 & s_2 & \cdots & s_n \\ 1 & 2 & \cdots & n \end{pmatrix} = \det(\varphi_i(s_j)), \quad i, j = 1, 2, \ldots, n.$$

Lemma 2.2 *Suppose $\varphi(x)$ is a function that is continuous, not identically zero, and with no more than $n-1$ sign reversals in the interval $[0, l]$. Set*

$$\psi(x) = \int_0^l m(x, s)\varphi(s)ds, \quad 0 \le x \le l,$$

where $m(x, s)$ is continuous on $0 \le x, s \le l$ and satisfies the following condition:

$$M\begin{pmatrix} x_1 & x_2 & \cdots & x_n \\ s_1 & s_2 & \cdots & s_n \end{pmatrix} > 0, \quad 0 \le \begin{matrix} x_1 < x_2 < \cdots < x_n \\ s_1 < s_2 < \cdots < s_n \end{matrix} \le l.$$

Then, $\psi(x)$ vanishes no more than $n-1$ times in $I \subset [0, l]$. Here, I is defined in a way similar to that in Sect. 2.7, where we only need to replace values of the kernel $K(x, s)$ at end points, specifically, $K(a, a)$ and $K(b, b)$, with the corresponding values of $\psi(x)$ at end points, specifically, $\psi(0)$ and $\psi(l)$.

Proof According to the conditions given, there exists a set of points $\{\xi_r\}_0^n$, where $0 = \xi_0 < \xi_1 < \cdots < \xi_n = l$, such that $\varphi(x)$ retains its sign and is not identically zero in each of the subintervals $[\xi_{r-1}, \xi_r]$ $(r = 1, 2, \ldots, n)$. Set

$$\psi_i(x) = \int_{\xi_{i-1}}^{\xi_i} m(x, s)\varphi(s)ds, \quad i = 1, 2, \ldots, n, \tag{2.10.2}$$

Apparently,

$$\psi(x) = \psi_1(x) + \psi_2(x) + \cdots + \psi_n(x).$$

Thus, for any point set $\{x_i\}_1^n$ satisfying $0 \le x_1 < x_2 < \cdots < x_n \le l$, we have

$$\Psi\begin{pmatrix} x_1 & x_2 & \cdots & x_n \\ 1 & 2 & \cdots & n \end{pmatrix} = \int_{\xi_{n-1}}^{\xi_n} \cdots \int_{\xi_0}^{\xi_1} M\begin{pmatrix} x_1 & x_2 & \cdots & x_n \\ s_1 & s_2 & \cdots & s_n \end{pmatrix}$$

$$\cdot\, \varphi(s_1)\varphi(s_2)\cdots\varphi(s_n)\mathrm{d}s_1\mathrm{d}s_2\cdots\mathrm{d}s_n.$$

Since each factor in the integrand has a fixed sign and is not identically zero, the determinant on the left-hand side of the above equation must be of strictly fixed sign for the point set $\{x_i\}_1^n$. Thus, per Theorem 2.21, $\psi_i(x)\,(i = 1, 2, \ldots, n)$ form a family of Chebyshev functions on the interval $I \subset [0, l]$ and consequently, the number of zeroes of $\psi(x)$ in $I \subset [0, l]$ does not exceed $n - 1$. ■

The most important kernel satisfying conditions of Lemma 2.2 is the one in Thermodynamics:

$$m_t(x, s) = \frac{2}{\sqrt{\pi t}}\exp\left\{\frac{-(x-s)^2}{t^2}\right\}.$$

This kernel is well behaved. It has been proved in Refs. [1, 4] that $\delta(x - s)$ is the weak limit of the kernel as $t \to 0$. Thus, if

$$\psi(x, t) = \int_0^l m_t(x, s)\varphi(s)\mathrm{d}s, \quad 0 \le x \le l,$$

where $\varphi(x)$ is a continuous function on interval $[0, l]$, then

$$\begin{aligned}\lim_{t\to 0}\psi(x, t) &= \lim_{t\to 0}\int_0^l m_t(x, s)\varphi(s)\mathrm{d}s \\ &= \int_0^l \delta(x - s)\varphi(s)\mathrm{d}s = \varphi(x), \quad 0 \le x \le l.\end{aligned} \tag{2.10.3}$$

By taking advantage of this kernel $m_t(x, s)$, we may prove the lemma below.

Lemma 2.3 *Suppose $\varphi_i(x)\,(i = 1, 2, \ldots, n)$ are a family of linearly independent continuous functions on $[0, l]$. Let $c_i\,(i = 1, 2, \ldots, n)$ be a set of real numbers that are not all zero. The necessary and sufficient condition for the number of sign reversals of the function*

$$\varphi(x) = \sum_{i=1}^{n} c_i\varphi_i(x)$$

to satisfy the inequality $S_\varphi \le n - 1$ on $[0, l]$ is as follows: For any set of $x_i \in I\,(i = 1, 2, \ldots, n)$ such that $0 \le x_1 < x_2 < \cdots < x_n \le l$, the determinant

$$\Phi\begin{pmatrix} x_1 & x_2 & \cdots & x_n \\ 1 & 2 & \cdots & n \end{pmatrix} = \det(\varphi_i(x_j))_{n\times n} \tag{2.10.4}$$

has a fixed sign, i.e., it retains the same positive or negative sign whenever the value of the expression (2.10.4) differs from zero.

Proof Since $\varphi_i(x)\,(i = 1, 2, \ldots, n)$ are linearly independent on $[0, l]$, the determinant (2.10.4) is not identically zero. Set

$$\psi(x, t) = \int_0^l m_t(x, s)\varphi(s)\mathrm{d}s, \quad 0 \le x \le l, \tag{2.10.5}$$

and

$$\psi_i(x, t) = \int_0^l m_t(x, s)\varphi_i(s)\mathrm{d}s, \quad i = 1, 2, \ldots, n; 0 \le x \le l. \tag{2.10.6}$$

Thus,

$$\psi(x, t) = \int_0^l m_t(x, s)\sum_{i=1}^n c_i\varphi_i(s)\mathrm{d}s = \sum_{i=1}^n c_i\psi_i(x, t).$$

According to Lemma 2.2, $S_\varphi \le n - 1$ is equivalent to the requirement that the function $\psi(x, s)$ assumes zero value for no more than $n - 1$ times on $I \subset [0, l]$; while, according to Theorem 2.21, the necessary and sufficient condition for the latter is that the determinant

$$\Psi\begin{pmatrix} x_1 & x_2 & \cdots & x_n \\ 1 & 2 & \cdots & n \end{pmatrix} \tag{2.10.7}$$

has strictly fixed sign for any set of $x_i \in I(i = 1, 2, \ldots, n)$ satisfying $x_1 < x_2 < \cdots < x_n$. It is known from Eq. (2.10.3) that Determinant (2.10.4) has a fixed sign for any set of $x_i \in I\,(i = 1, 2, \ldots, n)$ satisfying $x_1 < x_2 < \cdots < x_n$. Conversely, when we choose $m(x, s) = m_t(x, s)$, Eq. (2.10.1) indicates that as long as Determinant (2.10.4) has fixed sign for any set of $x_i \in I\,(i = 1, 2, \ldots, n)$ satisfying $x_1 < x_2 < \cdots < x_n$, Determinant (2.10.7) should have strictly fixed sign for the same set of $x_i \in I\,(i = 1, 2, \ldots, n)$. ■

Theorem 2.24 *For a one-dimensional continuous system, the oscillatory property in static deformation B is the necessary and sufficient condition of Eq. (2.7.5) associated with an oscillatory kernel.*

Proof In fact, under the action of n concentrated forces $F_1, F_2, \ldots$, and F_n at points $(0 \le)s_1 < s_2 < \cdots < s_n(\le l)$, the deflection of the one-dimensional continuous system can be expressed as

$$y(x) = \sum_{i=1}^{n} F_i \varphi_i(x),$$

where $\varphi_i(x) = K(x, s_i)\,(i = 1, 2, \ldots, n)$ and $K(x, s)$ represents the Green's function of the continuous system. In this case,

$$\Phi\begin{pmatrix} x_1 & x_2 & \cdots & x_n \\ 1 & 2 & \cdots & n \end{pmatrix} = K\begin{pmatrix} x_1 & x_2 & \cdots & x_n \\ s_1 & s_2 & \cdots & s_n \end{pmatrix}. \tag{2.10.8}$$

On the one hand, this implies, based on Lemma 2.3, that Property B is equivalent to the condition: Determinant (2.10.8), wherever its value is nonzero, has a fixed sign for all $x_i \in I\,(i = 1, 2, \ldots, n)$ such that $x_1 < x_2 < \cdots < x_n$. On the other hand, if the system has Property B, the positive definiteness of the following bilinear form

$$V = \frac{1}{2}\sum K(s_i, s_j) F_i F_j \tag{2.10.9}$$

leads to the conclusion that Determinant (2.10.8) is positive at $x_i = s_i\,(i = 1, 2, \ldots, n)$. This indicates that Determinant (2.10.8) is always nonnegative, namely,

$$K\begin{pmatrix} x_1 & x_2 & \cdots & x_n \\ s_1 & s_2 & \cdots & s_n \end{pmatrix} \geq 0, \quad 0 \leq \begin{matrix} x_1 < x_2 < \cdots < x_n \\ s_1 < s_2 < \cdots < s_n \end{matrix} \leq l.$$

The reverse is also true. ■

When proving the Green's function of a beam to be an oscillatory kernel in Chap. 6, we will see that it is a practical approach to first check whether the system has the oscillatory properties in static deformation and then, based on Theorem 2.24, decide if its Green's function is oscillatory. When dealing with discrete systems, we are also in need of this approach. For instance, in the discussion of oscillatory properties associated with the finite element system of a beam, we are faced with the problem of proving the flexibility matrix of the discrete system to be oscillatory. In the rest of this subsection, we will focus on establishing the results applicable to discrete systems. Let us begin by introducing the corresponding oscillatory properties in static deformation as well as the discrete form of Lemma 2.3.

Property C *If a concentrated force is applied to a movable discretization node of the discrete system modeling a one-dimensional continuous system, then the displacement at any movable discretization node in the system will be nonzero and in the direction of the force applied.*

Property D *If the discrete system of a one-dimensional continuous system is subject to n concentrated forces at some movable discretization nodes where $n \leq N$ and N is the number of degrees of freedom of the discrete system, then the u-line consists of displacements $\{u_i\}_1^N$ at the movable discretization nodes has no more than $n - 1$ sign reversals, i.e., $S_u \leq n - 1$.*

Properties C and D are collectively referred to as oscillatory properties in static deformation of the discrete system modeling a one-dimensional continuous system.

Lemma 2.4 *Consider a family of linearly independent m-dimensional vectors* $\{\boldsymbol{u}^{(i)}\}_1^n$, *where* $\boldsymbol{u}^{(i)} = (u_{1i}, u_{2i}, \ldots, u_{mi})^{\mathrm{T}}$. *Any of their nontrivial combination*

$$\boldsymbol{u} = \sum_{i=1}^{n} c_i \boldsymbol{u}^{(i)}, \quad \sum_{i=1}^{n} c_i^2 > 0 \tag{2.10.10}$$

has no more than $n - 1$ *sign reversals, if and only if, for any set of integers* $\{j_i\}_1^n$ *satisfying* $1 \le j_1 < j_2 < \ldots < j_n \le m$, *the following determinant*

$$\det \begin{bmatrix} u_{1j_1} & u_{1j_2} & \cdots & u_{1j_n} \\ u_{2j_1} & u_{2j_2} & \cdots & u_{2j_n} \\ \vdots & \vdots & & \vdots \\ u_{nj_1} & u_{nj_2} & \cdots & u_{nj_n} \end{bmatrix} \tag{2.10.11}$$

is either always nonpositive or always nonnegative.

Proof Represent the $u^{(i)}$-line as $\varphi_i(x)$. Then, $\{\varphi_i(x)\}_1^n$ forms a family of linearly independent continuous functions defined on the interval $[1, m]$.

Let us begin by proving the necessary part of the theorem. In this case, the number of sign reversals of any linear combination of $\{\varphi_i(x)\}_1^n$, i.e.,

$$\varphi(x) = \sum_{i=1}^{n} c_i \varphi_i(x)$$

does not exceed $n - 1$. Additionally, set $x_i = j_i$. Then, according to Lemma 2.3, as the integer set $\{j_i\}$ changes, the determinant (2.10.11) is either always nonpositive or always nonnegative. In other words, the determinant (2.10.11) does not change its positive or negative sign.

Next, we prove the sufficient part of the theorem. According to Lemma 2.3, we only need to show: If the determinant (2.10.11) does not change sign for any integer set of $\{j_i\}$, then the determinant (2.10.4) also does not change sign for any $x_i \in I\,(i = 1, 2, \ldots, n)$ such that $x_1 < x_2 < \cdots < x_n$. When $\{x_i\}$ are all integers, Determinant (2.10.4) may be represented by Determinant (2.10.11). Thus, the two should share the same behavior to remain either positive or negative. Now look at the case when there is only one non-integer x_i among all points $x_1 < x_2 < \cdots < x_n$. Suppose $x_i \in (k_i, k_i + 1)$, where k_i is an integer no greater than m. Since $\varphi_i(x)$ consists of line segments, we have

$$\Phi\begin{pmatrix} x_1 & x_2 & \cdots & x_i & \cdots & x_n \\ 1 & 2 & \cdots & i & \cdots & n \end{pmatrix}$$

$$= (k_i + 1 - x_i)\Phi\begin{pmatrix} x_1 & x_2 & \cdots & k_i & \cdots & x_n \\ 1 & 2 & \cdots & i & \cdots & n \end{pmatrix}$$

$$+ (x_i - k_i)\Phi\begin{pmatrix} x_1 & x_2 & \cdots & k_i + 1 & \cdots & x_n \\ 1 & 2 & \cdots & i & \cdots & n \end{pmatrix}. \qquad (2.10.12)$$

Based on the assumption, the two terms on the right-hand side of the above equation have the same sign. In addition, their signs would not change as $x_1, x_2, \ldots, x_n$, excluding x_i, assume any different integer values. Therefore, when there is only one x_i not to be an integer among all points $x_1 < x_2 < \cdots < x_n$, the Determinant (2.10.4) always retains its positive or negative sign. Using the method of mathematical induction and repeating decomposition similar to Eq. (2.10.12), we may reach the conclusion that for any set of $x_i \in [1, m]$ satisfying $x_1 < x_2 < \cdots < x_n$, the determinant (2.10.4) stays either positive or negative all the time. This completes the proof on the sufficient part. ■

Definition 2.13 Suppose there is a set of linearly independent m-dimensional vectors $\{\boldsymbol{u}^{(i)}\}_1^n$, where $\boldsymbol{u}^{(i)} = (u_{1i}, u_{2i}, \ldots, u_{mi})^{\mathrm{T}}$. If, for any set of real numbers $c_i\,(i = 1, 2, \ldots, n)$ that are not all zero, the number of sign reversals of the vector

$$\boldsymbol{u} = \sum_{i=1}^{n} c_i \boldsymbol{u}^{(i)}, \quad \sum_{i=1}^{n} c_i^2 > 0$$

does not exceed $n - 1$, then these vectors are referred to as the family of Chebyshev vectors.

Theorem 2.25 *The positive definite $N \times N$ matrix $\boldsymbol{A} = (a_{ij})_{N\times N}$ is oscillatory, if and only if*

(1) all of its subdiagnal elements are positive, i.e., $a_{i,i+1} > 0$ and $a_{i+1,i} > 0$ for every possible i; and
(2) for any positive integer $n<N$, $\{\boldsymbol{a}^{(j_i)} | 1 \le j_1 < j_2 < \cdots < j_n \le N\}$ form a family of Chebyshev vectors, where $\boldsymbol{a}^{(j_i)}$ is the j_i-th column vector of the matrix $\boldsymbol{A}$.

Proof The positive definite matrix $\boldsymbol{A}$ must be nonsingular. From Lemma 2.4 and Definition 2.13, The two things, Condition (2) and the positive definiteness of $\boldsymbol{A}$, are equivalent to the statement of Matrix $\boldsymbol{A}$ being totally nonnegative. By further taking into account Condition (1), we know from Theorem 2.10 that Matrix $\boldsymbol{A}$ is an oscillatory matrix. ■

Thus, for the discrete system of a one-dimensional continuous object, the oscillatory property C in static deformation implies that subdiagonal elements of its flexibility matrix are all positive, while the oscillatory property D in static deformation indicates that column vectors of the flexibility matrix constitute a family of Chebyshev vectors. From Theorem 2.25, we conclude that, for the discrete system modeling a one-dimensional continuous object, having oscillatory properties in static deformation is the necessary and sufficient condition for the flexibility matrix of the system to be oscillatory.

2.10.2 Relationship Between Flexibility Functions (Matrices) Being Oscillatory Kernels (Matrices) and Oscillatory Properties in Vibration

We now turn to the second problem mentioned at the beginning of this section, i.e., the necessary condition for a system to possess the oscillatory properties in vibration. See Refs. [7, 8].

Theorem 2.26 *For the discrete system modeling a well-constrained one-dimensional continuous object and having a lumped mass matrix of any form, if the system has oscillatory properties in vibration, then its flexibility matrix is oscillatory.*

Proof For the discrete system with N degrees of freedom, obtained by discretizing the displacement field of a one-dimensional continuous object, the governing equation for its free vibration can be written as

$$\boldsymbol{K}\boldsymbol{u} = \lambda \boldsymbol{M}\boldsymbol{u}, \tag{2.10.13}$$

where $\boldsymbol{K}$ denotes the stiffness matrix, $\boldsymbol{M}$ represents the mass matrix, $\boldsymbol{u}$ is the nodal displacement vector, $\lambda = \omega^2$ stands for the eigenvalue, and ω designates the natural angular frequency of the system. In terms of the flexibility matrix, the above equation may be expressed as

$$\boldsymbol{u} = \lambda \boldsymbol{R}\boldsymbol{M}\boldsymbol{u}, \tag{2.10.14}$$

where the flexibility matrix $\boldsymbol{R} = (r_{ij})$ is the inverse of the stiffness matrix $\boldsymbol{K}$. Since it has been assumed that the system has oscillatory properties in vibration, it is known from the property (1) in Definition 2.11 that the flexibility matrix $\boldsymbol{R}$ is nonsingular. Now let us prove $\boldsymbol{R}$ to be a totally nonnegative matrix. This can be accomplished through proof by contradiction. Suppose $\boldsymbol{R}$ is not totally nonnegative. Then, according to Lemma 2.4, there must be k $(1 < k < N)$ column vectors in the flexibility matrix $\boldsymbol{R}$, say columns $j_1 < j_2 < \cdots < j_k$, such that within the space spanned by them there exists a vector whose number of sign reversals is no less than k.

Now let us construct a sequence of diagonal mass matrices $\{\boldsymbol{M}^{(i)}\}$, whose diagonal elements satisfy

$$M_{tt}^{(i)} = \begin{cases} \left(\frac{1}{2}\right)^i, & t \notin \{j_p\}, \\ 1, & t \in \{j_p\}. \end{cases}$$

Denote the limit of this sequence as $\boldsymbol{M}^{(\infty)}$. Obviously, $\boldsymbol{M}^{(\infty)}$ is a diagonal matrix with the rank k and diagonal elements

$$M_{tt}^{(\infty)} = \begin{cases} 0, & t \notin \{j_p\}, \\ 1, & t \in \{j_p\}. \end{cases}$$

Next, we will prove the following result: Consider the two systems

$$\mu^{(i)}\boldsymbol{u}_i = \boldsymbol{R}\boldsymbol{M}^{(i)}\boldsymbol{u}_i \tag{2.10.15}$$

and

$$\mu^{(\infty)}\boldsymbol{u}_\infty = \boldsymbol{R}\boldsymbol{M}^{(\infty)}\boldsymbol{u}_\infty. \tag{2.10.16}$$

Suppose eigenvalues of each system have been sorted from large to small. Then the subspace spanned by the first k eigenvectors $\boldsymbol{u}_i^{(1)}, \boldsymbol{u}_i^{(2)}, \ldots, \boldsymbol{u}_i^{(k)}$ of the system (2.10.15) converges to the subspace spanned by the first k eigenvectors $\boldsymbol{u}_\infty^{(1)}, \boldsymbol{u}_\infty^{(2)}, \ldots, \boldsymbol{u}_\infty^{(k)}$ of the system (2.10.16).

When i is sufficiently large, System (2.10.15) may be reagarded as a perturbed system infinitely close to System (2.10.16). From the $\sin\theta$ theorem on matrix perturbation (see Ref. [5]), we get

$$\frac{\left|\Delta\mu^{(\infty)}\right| \cdot \left|\sin\theta^{(i)}\right|}{2} \leq \left\|\boldsymbol{Y}^{(i)}\right\|. \tag{2.10.17}$$

Here, $\left\|\boldsymbol{Y}^{(i)}\right\|$ is the residual of Eq. (2.10.16) as the System (2.10.16) is perturbed to the System (2.10.15), and it obviously converges to zero as i increases; $\theta^{(i)}$ stands for the angle between subspaces spanned by vectors $\boldsymbol{u}_\infty^{(1)}, \boldsymbol{u}_\infty^{(2)}, \ldots, \boldsymbol{u}_\infty^{(k)}$ and vectors $\boldsymbol{u}_i^{(1)}, \boldsymbol{u}_i^{(2)}, \ldots, \boldsymbol{u}_i^{(k)}$; and $\left|\Delta\mu^{(i)}\right|$ represents the minimal absolute value of the differences between the first k and last $N-k$ eigenvalues of Eq. (2.10.16). Apparently, System (2.10.16) happens to have k nonzero eigenvalues, and these are also eigenvalues of the matrix

$$\begin{bmatrix} r_{j_1j_1} & r_{j_1j_2} & \cdots & r_{j_1j_k} \\ r_{j_2j_1} & r_{j_2j_2} & \cdots & r_{j_2j_k} \\ \vdots & \vdots & & \vdots \\ r_{j_kj_1} & r_{j_kj_2} & \cdots & r_{j_kj_k} \end{bmatrix}. \tag{2.10.18}$$

However, Matrix (2.10.18) is positive definite, and all of its eigenvalues need to be positive. Consequently, the first k eigenvalues of System (2.10.16) are all greater than zero and thus, $\left|\Delta\mu^{(\infty)}\right| = \Delta\mu^{(\infty)} > 0$. From Eq. (2.10.17), the angle between two subspaces converges to zero as i increases. Hence, the subspace spanned by $\boldsymbol{u}_i^{(1)}, \boldsymbol{u}_i^{(2)}, \ldots, \boldsymbol{u}_i^{(k)}$ converges to the subspace spanned by $\boldsymbol{u}_\infty^{(1)}, \boldsymbol{u}_\infty^{(2)}, \ldots, \boldsymbol{u}_\infty^{(k)}$, namely,

$$\lim_{i\to\infty} \operatorname{span}(\boldsymbol{u}_i^{(1)}, \boldsymbol{u}_i^{(2)}, \ldots, \boldsymbol{u}_i^{(k)}) = \operatorname{span}(\boldsymbol{u}_\infty^{(1)}, \boldsymbol{u}_\infty^{(2)}, \ldots, \boldsymbol{u}_\infty^{(k)}). \tag{2.10.19}$$

Based on the assumption, System (2.10.15) has the oscillatory property (2), i.e., there is no vector in the subspace spanned by $\boldsymbol{u}_i^{(1)}, \boldsymbol{u}_i^{(2)}, \ldots, \boldsymbol{u}_i^{(k)}$ whose number of sign reversals exceeds $k-1$. As a result, in the subspace that is the limit of the sequence of subspaces, a vector with the count of sign reversals exceeding $k-1$ cannot exist, too.

Furthermore, it is easy to verify

$$\begin{aligned}
&[\boldsymbol{R}^{(j_1)}, \boldsymbol{R}^{(j_2)}, \ldots, \boldsymbol{R}^{(j_k)}] = [\boldsymbol{u}_\infty^{(1)}, \boldsymbol{u}_\infty^{(2)}, \ldots, \boldsymbol{u}_\infty^{(k)}] \cdot \operatorname{diag}(\lambda_1, \lambda_2, \ldots, \lambda_k) \cdot \bar{\boldsymbol{u}}^{-1},\\
&\bar{\boldsymbol{u}} = [\boldsymbol{e}^{(j_1)}, \boldsymbol{e}^{(j_2)}, \ldots, \boldsymbol{e}^{(j_k)}]^{\mathrm{T}} \cdot [\boldsymbol{u}_\infty^{(1)}, \boldsymbol{u}_\infty^{(2)}, \ldots, \boldsymbol{u}_\infty^{(k)}],\\
&\boldsymbol{e}^{(j_n)} = \begin{pmatrix} 0 \cdots 0 & \underset{j_n-th}{1} & 0 \cdots 0 \end{pmatrix}^{\mathrm{T}}.
\end{aligned} \tag{2.10.20}$$

This indicates that the subspace spanned by $\boldsymbol{u}_\infty^{(1)}, \boldsymbol{u}_\infty^{(2)}, \ldots, \boldsymbol{u}_\infty^{(k)}$ happens to be a subspace spanned by column vectors of the flexibility matrix $\boldsymbol{R}^{(j_1)}, \boldsymbol{R}^{(j_2)}, \ldots, \boldsymbol{R}^{(j_k)}$, i.e.,

$$\operatorname{span}(\boldsymbol{R}^{(j_1)}, \boldsymbol{R}^{(j_2)}, \ldots, \boldsymbol{R}^{(j_k)}) = \operatorname{span}(\boldsymbol{u}_\infty^{(1)}, \boldsymbol{u}_\infty^{(2)}, \ldots, \boldsymbol{u}_\infty^{(k)}). \tag{2.10.21}$$

In fact, Eq. (2.10.21) is one of the theoretical bases for the method of modal superposition widely used in engineering (see Ref. [2]). There is no vector on the right-hand side of Eq. (2.10.21) whose number of sign reversals exceeds $k-1$, while there exists a vector on the left-hand side of this equation whose count of sign reversals is at least k. This creates a contradiction that showes $\boldsymbol{R}$ has to be a totally nonnegative matrix.

Now let us check on subdiagonal elements of the matrix $\boldsymbol{R}$. Assume that there is a subdiagonal element to be nonpositive. According to the definition of a totally nonnegative matrix, this element has to be zero. Without loss of generality, suppose there exists an index i such that $r_{i+1,i} = r_{i,i+1} = 0$. Then, for any $j \le i$, we must have $r_{i+1,j} = r_{j,i+1} = 0$. Otherwise, the minor

$$R\begin{pmatrix} i & i+1 \\ j & i \end{pmatrix} = 0 - r_{ii}r_{i+1,j} < 0,$$

which would contradict the condition of Matrix $\boldsymbol{R}$ being totally nonnegative. By the same token, for any $i < p$, we get $r_{ip} = r_{pi} = 0$. Furthermore, for any $j \le i < p$, we must have $r_{jp} = r_{pj} = 0$. Otherwise, the minor

$$R\begin{pmatrix} i+1 & p \\ j & i+1 \end{pmatrix} < 0.$$

This implies that the matrix $\boldsymbol{R}$ may be decoupled into two parts as below:

$$\begin{bmatrix} r_{11} & r_{12} & \cdots & r_{1i} & 0 & 0 & \cdots & 0 \\ r_{21} & r_{22} & \cdots & r_{2i} & 0 & 0 & \cdots & 0 \\ \vdots & \vdots & & \vdots & \vdots & \vdots & & \vdots \\ r_{i1} & r_{i2} & \cdots & r_{ii} & 0 & 0 & \vdots & 0 \\ 0 & 0 & \cdots & 0 & r_{i+1,i+1} & r_{i+1,i+2} & \cdots & r_{i+1,N} \\ 0 & 0 & \cdots & 0 & r_{i+2,i+1} & r_{i+2,i+2} & \cdots & r_{i+2,N} \\ \vdots & \vdots & & \vdots & \vdots & \vdots & & \vdots \\ 0 & 0 & \cdots & 0 & r_{N,i+1} & r_{N,i+2} & \cdots & r_{NN} \end{bmatrix} = \begin{bmatrix} \boldsymbol{R}_1 & \boldsymbol{0} \\ \boldsymbol{0} & \boldsymbol{R}_2 \end{bmatrix}.$$

Modes associated with these two parts are independent. By adjusting components of the mass matrix $\boldsymbol{M}$ corresponding to the two decoupled submatrices $\boldsymbol{R}_1$ and $\boldsymbol{R}_2$, we will certainly be able to make two local modes share the same eigenvalue. Thus, this system must have a repeat eigenvalue, but this contradicts the first oscillatory property in vibration, given in Definition 2.11. Based on the criteria for oscillatory matrices (Theorem 2.10), we know $\boldsymbol{R}$ is an oscillatory matrix. This completes the proof of the theorem. ■

Starting from Theorem 2.26, we may derive, with relative ease, the next theorem applicable to continuous systems.

Theorem 2.27 *Consider a well-constrained, one-dimensional, and continuous system. Its mass distribution can be of any form, including the lumped mass. If the system has oscillatory properties in vibration, then its Green's function is an oscillatory kernel.*

Proof Let l be the length of the one-dimensional continuous system and $G(x, s)$ denote its Green's function, where x and $s \in \boldsymbol{I} \subset [0, l]$. For any integer n, arbitrarily select points $\{x_i\}_1^n$ such that

$$0 < x_1 < x_2 < \cdots < x_n < l.$$

The flexibility matrix of the corresponding discrete system is

$$\boldsymbol{R} = (r_{ij})_{n\times n} = (G(x_i, x_j))_{n\times n}.$$

Suppose an arbitrary set of concentrated masses $\{m_i\}_1^n$ has been attached to the points $\{x_i\}_1^n$ of the discrete system. Since the original continuous system has oscillatory properties in vibration associated with continuous systems, the discrete system should also have oscillatory properties in vibration associated with discrete systems. Based on Theorem 2.26 given above, the flexibility matrix $\boldsymbol{R}$ of this discrete system has to be an oscillatory matrix. Furthermore, since both n and the point set $\{x_i\}_1^n$ are arbitrarily chosen, it is known from Theorem 2.18 that the Green's function of the original continuous system must be an oscillatory kernel. ■

Thus, we may reach the following conclusions: For the well-constrained system of a one-dimensional continuous object, its Green's function being an oscillatory kernel is the necessary and sufficient condition for the system to have oscillatory properties in vibration. In addition, for the discrete system modeling the well-constrained system of a one-dimensional continuous object, if the lumped mass matrix is used, its flexibility matrix being an oscillatory matrix (or its stiffness matrix being a sign-oscillatory matrix) is the necessary and sufficient condition for the discrete system to have oscillatory properties in vibration.

Combining discussion in the previous two subsections, we obtain a perfect result, which is captured in the theorem below.

Theorem 2.28 *For a well-constrained one-dimensional structure, regardless of it being modeled as a continuous system or its discrete counterpart with a lumped mass matrix, each of the following three things is the necessary and sufficient condition of the other two: the system has oscillatory properties in static deformation, the Green's function/flexibility matrix is an oscillatory kernel/matrix, and the system possesses oscillatory properties in vibration.*

Thus, in order to determine whether a well-constrained system has oscillatory properties in vibration, it only needs to verify if the system has oscillatory properties in static deformation or its Green's function/flexibility matrix is an oscillatory kernel/matrix.

2.11 From Oscillatory Matrices to Oscillatory Kernels

In previous sections, we separately introduced the definition, identification criteria, and properties of related eigenvalue problems for oscillatory matrices and kernels. However, because of the significant similarity between the two, it is natural to ask what kind of connection exists between oscillatory matrices and oscillatory kernels, and whether it is possible to determine an oscillatory kernel directly from the theory of oscillatory matrices. We have introduced one type of matrices derived from a kernel in Theorem 2.18. If the kernel is oscillatory, then the derived matrix will be oscillatory. The converse is also true. Next, we look at another situation, where the kernel is obtained by letting elements of the matrix approach their respective limits. As an example, the flexibility coefficients of the finite difference model of a bar or beam are transformed into the corresponding Green's function in the limit. From here, we can prove the following: If the flexibility matrix of the finite difference model is oscillatory, its limit is a Kellogg kernel. See Ref. [6].

Theorem 2.29 *Consider the following algebraic eigenvalue problem:*

$$\boldsymbol{u} = \lambda \boldsymbol{R}\boldsymbol{M}\boldsymbol{u}. \tag{2.11.1}$$

Here, $\boldsymbol{u}$ is a column vector obtained from its continuous counterpart at the discretization nodes $0 = x_0 < x_1 < \cdots < x_N = l$. If the component corresponding to an end point is zero, then it is removed from the vector $\boldsymbol{u}$. Moreover, $\boldsymbol{M} = diag(m_0, m_1, \ldots, m_N)$. Suppose the continuous function $G(x, s)$ becomes the limit of the element r_{ij} of $\boldsymbol{R}$ as $N \to \infty$ and $\max\limits_{1 \le r \le N} \Delta x_r \to 0$. Then, $G(x, s)$ is a Kellogg kernel when $\boldsymbol{R}$ is an oscillatory matrix.

Proof Based on the definition of a Kellogg kernel given in Sect. 2.9, we only need to show that the following inequalities hold true for any given point sets $\{\xi_r\}_1^n \in I \subset [0, l]$ and $\{s_r\}_1^n \in I \subset [0, l]$:

$$G\begin{pmatrix} \xi_1 \ \xi_2 \ \cdots \ \xi_n \\ s_1 \ s_2 \ \cdots \ s_n \end{pmatrix} = \det(G(\xi_i, s_j)) \ge 0, \quad 0 \le \begin{matrix} \xi_1 < \xi_2 < \cdots < \xi_n \\ s_1 < s_2 < \cdots < s_n \end{matrix} \le l, \tag{2.11.2}$$

$$G\begin{pmatrix} \xi_1 \ \xi_2 \ \cdots \ \xi_n \\ \xi_1 \ \xi_2 \ \cdots \ \xi_n \end{pmatrix} > 0, \quad 0 \le \xi_1 < \xi_2 < \cdots < \xi_n \le l. \tag{2.11.3}$$

To prove the above inequalities, consider a new set $\{\eta_r\}_1^m$ obtained by rearranging the point set $\{\xi_r\}_1^n \cup \{s_r\}_1^n$ from small to large. Notice that if it occurs that points from the two set overlap, for instance, $\xi_r = s_r$, then we need to keep just one point. Consequently, regarding the size of the new point set, we should have $m \le 2n$. Starting from $\{\eta_r\}_1^m$, we may insert new points and build yet another set $\{x_r\}_0^N$ $(N > m)$ that satisfies conditions of the theorem. The related Eq. (2.11.1) may be viewed as the system of equations governing the vibration of a discrete system. Because $\boldsymbol{R}$ is an oscillatory matrix, its minor satisfies

$$R\begin{pmatrix} i_1 \ i_2 \ \cdots \ i_n \\ j_1 \ j_2 \ \cdots \ j_n \end{pmatrix} \ge 0, \tag{2.11.4}$$

where i_r and j_r are, respectively, indices of ξ_r and s_r in the point set $\{x_r\}_0^N$. As $N \to \infty$ and $\max\limits_{1 \le r \le N} \Delta x_r \to 0$, $r_{ij} \to G(x, s)$. Thus, after taking the limit of Eq. (2.11.4), we get

$$\lim_{N \to \infty, \delta \to 0} R\begin{pmatrix} i_1 \ i_2 \ \cdots \ i_n \\ j_1 \ j_2 \ \cdots \ j_n \end{pmatrix} = G\begin{pmatrix} \xi_1 \ \xi_2 \ \cdots \ \xi_n \\ s_1 \ s_2 \ \cdots \ s_n \end{pmatrix} \ge 0.$$

In order to prove Eq. (2.11.3), notice again that Eq. (2.11.1) may be viewed as the equation of vibration of a one-dimensional discrete system, and thus $\boldsymbol{R}$ can be viewed as the flexibility matrix of the system. Then, based on the concept of flexibility coefficients, the physical interpretation of its limit $G(x, s)$ shows that this should be the Green's function of the corresponding continuous system.

Now let us look at the strain energy of the system. Imagine there is a concentrated force F_r acting on each interior point ξ_r, where $r = 1, 2, \ldots, N$. Then, from the definition of the Green's function, the displacement at ξ_i and the strain energy of the system are, respectively,

$$u_i = \sum_{i=1}^{N} G(\xi_i, \xi_j)F_j, \quad V = \sum_{i=1}^{N} G(\xi_i, \xi_j)F_iF_j/2. \tag{2.11.5}$$

Thus, as long as the system is well-constrained, we will have $V > 0$, which indicates that the right-hand side of the second equation in (2.11.5) is a positive definite bilinear form. Therefore, Eq. (2.11.3) is valid. ■

It needs to be pointed out that the theorem above only shows that $G(x, s)$ is a Kellogg kernel. But this is sufficient to support our application, because, as mentioned at the end of Sect. 2.9, only Eqs. (2.7.5) and (2.7.6) are needed to derive oscillatory properties of eigenvalues and eigenfunctions of an oscillatory kernel.

References

1. Courant R, Hilbert D (1962) Methods of mathematical physics, vol I, 1953, vol II. InterScience Publishers, New York
2. Davis C, Kahan WM (1970) The rotation of eigenvectors by a perturbation. III. SIAM J Num Anal 7(1):1–46
3. Gantmacher FP, Krein MG (1961) Oscillation matrices and kernels and small vibrations of mechanical systems. U. S. Atomic Energy Commission, Washington
4. Gladwell GML (2004) Inverse Problems in Vibration, 2nd edn. Springer, Dordrecht (1986, 1st edn, Martinus Nijhoff Publishers, Dordrecht)
5. Shigley JE, Mischke CR, Budynas RG (2004) Mechanical engineering design. McGraw-Hill Professional
6. Wang QS, Wang DJ (1997) United proof for qualitative properties of discrete and continuous systems of vibrating rod and beam. Acta Mech Sin 29(1):99–102 (in Chinese)
7. Zheng ZJ, Chen P, Wang DJ (2013) Oscillation property of the vibrations for finite element models of Euler beam. Q J Mech Appl Mech 66(4):587–608
8. Zheng ZJ (2014) The qualitative vibrational property and modal inverse problems of rods and Euler beams [D]. Department of Mechanics and Engineering Science, College of Engineering, Peking University (in Chinese)

Chapter 3
Qualitative Properties of Vibration and Static Deformation Associated with Discrete Systems of Strings and Bars

In this chapter, we will study qualitative properties of natural frequencies and mode shapes of discrete models of second-order continuous systems, such as strings in lateral vibration, bars in longitudinal vibration, and shafts in torsional vibration. We will also discuss qualitative properties of static deformation associated with some of these discrete systems.

3.1 Discrete Systems of Strings and Bars

3.1.1 Physical Discrete Systems of Strings and Bars

The continuous system of a string is shown in Fig. 3.1a. The string is subject to a tension force T and elastically supported in the lateral direction by springs with spring constants h and H at the two ends. Here, $h = 0$ and $H = 0$ indicate that the string is unconstrained or free at boundaries $x = 0$ and $x = l$, respectively; while $h \to \infty$ and $H \to \infty$ represent, respectively, fixed ends at $x = 0$ and $x = l$, i.e., $u(0) = 0$ and $u(l) = 0$.

Now let us convert distributed parameters to concentrated or lumped ones. Select a set of discrete points $\{x_i\}_0^n$ on the string. See Fig. 3.1b. Suppose that the distance between two adjacent points, i.e., the length of a string segment, is $l_r = x_r - x_{r-1}$, where $r = 1, 2, \ldots, n$. Next, set the linear density of the string x_r to be $\rho(x_r) = \rho_r$. Then, lump the distributed mass of the string around the point x_r together in a concentrated mass m_r using formulae below:

$$\begin{cases} m_0 = \rho_0 l_1/2, \\ m_r = \rho_r(l_r + l_{r+1})/2, \quad r = 1, 2, \ldots, n-1, \\ m_n = \rho_n l_n/2. \end{cases} \tag{3.1.1}$$

D. Wang et al., *Qualitative Theory in Structural Mechanics*,
https://doi.org/10.1007/978-981-13-1376-9_3

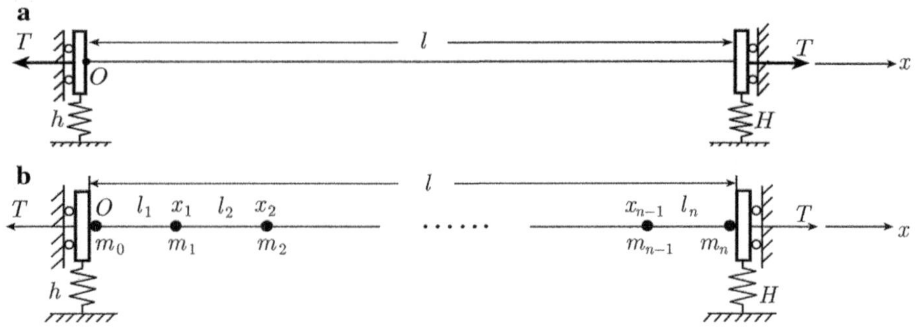

Fig. 3.1 **a** The continuous system of a string; **b** the corresponding discrete system of the string, consisting of massless string segments and concentrated masses

After these steps, the continuous system in Fig. 3.1a is transformed into a discrete system depicted in Fig. 3.1b, which may be referred to as a massless string–concentrated mass system.

Set $k_r = T/l_r$ $(r = 1,\ 2,\ \ldots,\ n)$ and let the amplitude of the displacement of the r-th mass be u_r. Then, the modal equations of the system in a natural vibration can be written as:

$$\begin{cases} (k_1 + h)u_0 - k_1 u_1 = \omega^2 m_0 u_0, \\ -k_r u_{r-1} + (k_r + k_{r+1})u_r - k_{r+1} u_{r+1} = \omega^2 m_r u_r, \quad r = 1, 2, \ldots, n-1, \\ -k_n u_{n-1} + (k_n + H)u_n = \omega^2 m_n u_n. \end{cases} \tag{3.1.2}$$

Equation (3.1.2) may also be recast in the matrix form:

$$\boldsymbol{K u} = \lambda \boldsymbol{M u}. \tag{3.1.3}$$

Here, the mode shape $\boldsymbol{u} = (u_0, u_1, \ldots, u_n)^{\mathrm{T}}$; $\lambda = \omega^2$, where ω is the natural angular frequency and is related to the natural frequency through $f = \omega/2\pi$; and the mass and stiffness matrices $\boldsymbol{M}$ and $\boldsymbol{K}$ are, respectively, diagonal and symmetric triangular matrices

$$\boldsymbol{M} = \mathrm{diag}\,(m_0, m_1, \ldots, m_n) \tag{3.1.4}$$

and

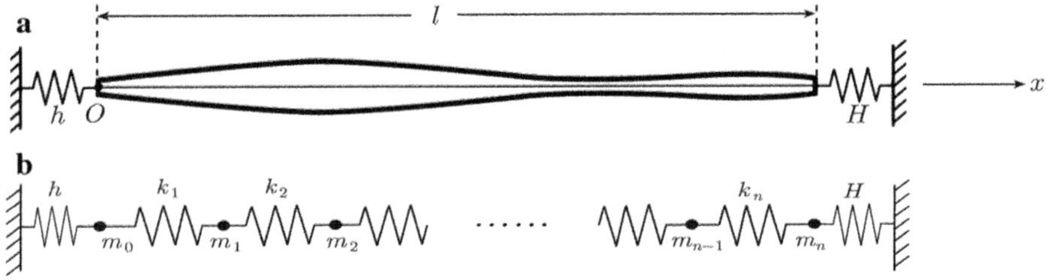

Fig. 3.2 **a** The continuous system of a bar; **b** the corresponding physical discrete system of the bar, consisting of springs and masses

$$\boldsymbol{K} = \begin{bmatrix} k_1 + h & -k_1 & 0 & \cdots & 0 & 0 & 0 \\ -k_1 & k_1 + k_2 & -k_2 & \cdots & 0 & 0 & 0 \\ \vdots & \vdots & \vdots & & \vdots & \vdots & \vdots \\ 0 & 0 & 0 & \cdots & -k_{n-1} & k_{n-1} + k_n & -k_n \\ 0 & 0 & 0 & \cdots & 0 & -k_n & k_n + H \end{bmatrix}. \tag{3.1.5}$$

The continuous system of a bar in longitudinal vibration is shown in Fig. 3.2a. It can be discretized in a way similar to the string. Lump the distributed mass along the bar near the point x_r together in a concentrated mass m_r by using Eq. (3.1.1). Then, convert the distributed stiffness along the bar segment l_r to a linear spring, with the spring constant

$$k_r = \frac{EA(x_r) + EA(x_{r-1})}{2l_r}, \quad r = 1, 2, \ldots, n. \tag{3.1.6}$$

Thus, the continuous system illustrated in Fig. 3.2a is transformed into the discrete system depicted in Fig. 3.2b, where h and H are elastic constants of the springs at left and right ends of the bar, respectively, while k_r ($r = 1, 2, \ldots, n$) represent elastic constants of internal springs. The mass and stiffness matrices of this system are also of the forms shown in Eqs. (3.1.4) and (3.1.5), respectively.

Thus, the continuous systems of strings in lateral vibration, bars in longitudinal vibration, shafts in torsional vibration, and even short bars in shear vibration can all be transformed to the same discrete system, which is the spring–mass system of finite degrees of freedom shown in Fig. 3.2b. Natural frequencies and mode shapes of the discrete system satisfy Eq. (3.1.2) or Eqs. (3.1.3)–(3.1.5), which define an eigenvalue problem.

When a string, bar, or shaft vibrates on an elastic foundation, its behavior can be determined mathematically by solving the vibration problem of a Sturm–Liouville system. The distributed elastic coefficient $q(x)$ (≥ 0) of the foundation can be converted to elastic constants of a set of springs in the discrete system as follows:

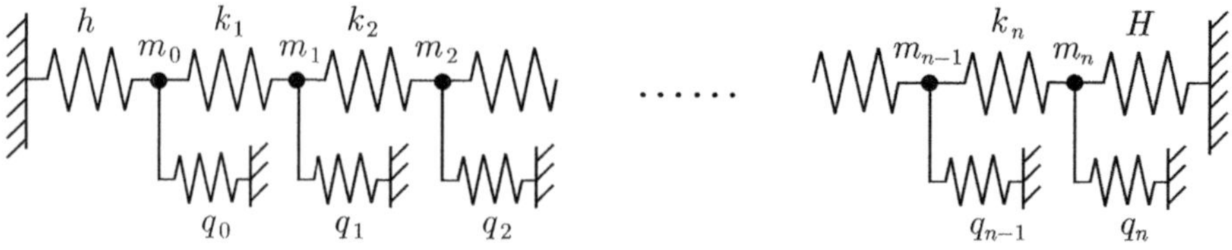

Fig. 3.3 The physical discrete system of a string or bar on an elastic foundation

$$\begin{cases} q_0 = q(0)l_1/2, \\ q_r = q(x_r)(l_r + l_{r+1})/2, \; r = 1, 2, \ldots, n-1, \\ q_n = q(l)l_n/2. \end{cases}$$

The corresponding discrete system consisting of twin springs and concentrated masses is sketched in Fig. 3.3. Frequencies and mode shapes of the system satisfy Eq. (3.1.3), the mass matrix is of the form shown in Eq. (3.1.4), and the stiffness matrix now assumes the following form:

$$\boldsymbol{K} = \begin{bmatrix} k_1 + q_0 + h & -k_1 & 0 & \cdots & 0 & 0 \\ -k_1 & k_1 + k_2 + q_1 & -k_2 & \cdots & 0 & 0 \\ \vdots & \vdots & \vdots & & \vdots & \vdots \\ 0 & 0 & 0 & \cdots & -k_n & k_n + q_n + H \end{bmatrix}. \tag{3.1.7}$$

Although the physical discrete system derived above for strings, bars, and shafts, i.e., the spring–mass system, appears to be rather simple, it is in fact highly representative. We will see in subsequent subsections that the physical discrete system also covers the mathematical discrete models of strings, bars, and shafts.

Finally, we would like to mention boundary conditions applied to the spring–mass system. There are three types of boundary constraints. Taking the left end as an example, we can have:

(1) Fixed end: $h \to \infty$, which yields $u_0 = 0$. Thus, the number of degrees of freedom of the system drops by one.
(2) Free or unconstrained end: $h = 0$.
(3) Elastic support: $0 < h < \infty$, in which case $u_0 \neq 0$.

3.1.2 The Finite Difference System of a String in Lateral Vibration

The natural angular frequency ω and mode shape $u(x)$ of a vibrating string satisfy the differential equation

$$-\rho(x)\omega^2 u(x) = Tu''(x), \quad 0 < x < l \tag{3.1.8}$$

as well as boundary conditions

$$Tu'(0) - hu(0) = 0 = Tu'(l) + Hu(l), \tag{3.1.9}$$

where T represents the tension, h and H are spring constants, ρ denotes the linear density, and l designates the length of the string.

By approximating derivatives with finite differences, we may discretize the continuous system. Introduce a partition $0 = x_0 < x_1 < \cdots < x_n = l$ and set $l_r = x_r - x_{r-1}$ $(r = 1, 2, \ldots, n)$. Also let u_r, u_r', and u_r'' represent values of the displacement, slope, and approximate curvature functions at x_r, respectively, where $r = 0, 1, \ldots, n$. Then, from the Taylor series expansion with the second-order accuracy, i.e.,

$$\begin{cases} u_{r+1} = u_r + u_r' l_{r+1} + u_r'' l_{r+1}^2/2 + \mathrm{O}(l_{r+1}^3), & r = 0, 1, \ldots, n-1, \\ u_{r-1} = u_r - u_r' l_r + u_r'' l_r^2/2 + \mathrm{O}(l_r^3), & r = 1, 2, \ldots, n, \end{cases} \tag{3.1.10}$$

we get the second-order central finite difference formulae as follows:

$$\begin{cases} u_r' = -u_{r-1}\dfrac{l_{r+1}}{l_r(l_r + l_{r+1})} + u_r\dfrac{l_{r+1} - l_r}{l_r l_{r+1}} \\ \qquad + u_{r+1}\dfrac{l_r}{l_{r+1}(l_r + l_{r+1})}, & r = 1, 2, \ldots, n-1, \\ u_r'' = \dfrac{2}{l_r + l_{r+1}}\left(\dfrac{u_{r-1} - u_r}{l_r} - \dfrac{u_r - u_{r+1}}{l_{r+1}}\right). \end{cases} \tag{3.1.11}$$

Next, we may discretize Eq. (3.1.8) and obtain the system of equations below, by applying the first equation in (3.1.10) and the boundary condition (3.1.9) to the left end, the second equation in (3.1.11) to internal discretization points, as well as the second equation in (3.1.10) and the boundary condition (3.1.9) to the right end:

$$\begin{cases} \rho_0\omega^2 u_0 = -\dfrac{2T}{l_1}\dfrac{u_1 - u_0}{l_1} + \dfrac{2hu_0}{l_1}, \\ \rho_r\omega^2 u_r = \dfrac{2T}{l_r + l_{r+1}}\left(\dfrac{u_r - u_{r-1}}{l_r} - \dfrac{u_{r+1} - u_r}{l_{r+1}}\right), & r = 1, 2, \ldots, n-1, \\ \rho_n\omega^2 u_n = \dfrac{2T}{l_n}\dfrac{u_n - u_{n-1}}{l_n} + \dfrac{2Hu_n}{l_n}. \end{cases} \tag{3.1.12}$$

After setting

$$\begin{cases} m_0 = \rho_0 l_1/2, \ m_n = \rho_n l_n/2, \\ m_r = \rho_r(l_r + l_{r+1})/2, & r = 1, 2, \ldots, n-1, \\ k_r = T/l_r, & r = 1, 2, \ldots, n, \end{cases} \tag{3.1.13}$$

we can convert the modal equation (3.1.12) into the form of either Eq. (3.1.2) or (3.1.3)–(3.1.5). This shows that the finite difference model of a string is equivalent to the spring–mass system introduced earlier.

3.1.3 *The Finite Difference System of a Bar in Longitudinal Vibration*

The natural angular frequency ω and mode shape $u(x)$ of a bar in longitudinal vibration satisfy the differential equation

$$(p(x)u'(x))' + \lambda\rho(x)u(x) = 0, \quad 0 < x < l \tag{3.1.14}$$

and boundary conditions

$$p(0)u'(0) - hu(0) = 0, \;\; p(l)u'(l) + Hu(l) = 0. \tag{3.1.15}$$

Here, $p(x) = EA(x)$ is the tension stiffness of the bar, where E represents the elastic modulus and A stands for the area of the cross section; ρ denotes the linear density; l designates the length of the bar; and h and H are, respectively, spring constants at the left and right ends.

Similar to the string problem, we may discretize the continuous system by approximating derivatives with finite differences. To this purpose, rewrite Eq. (3.1.14) as

$$p(x)u''(x) + p'(x)u'(x) = -\lambda\rho(x)u(x), \quad 0 < x < l. \tag{3.1.16}$$

Introduce the partition $0 = x_0 < x_1 < \cdots < x_n = l$; set $l_r = x_r - x_{r-1}$ ($r = 1, 2, \ldots, n$); and let p_r, p'_r, u_r, u'_r, and u''_r denote values of the respective functions at points x_r ($r = 0, 1, \ldots, n$).

At the interior points x_r ($r = 1, 2, \ldots, n-1$), we get

$$\begin{aligned} -\lambda\rho_r u_r = {} & \frac{2p_r}{l_r + l_{r+1}}\left(\frac{u_{r-1} - u_r}{l_r} - \frac{u_r - u_{r+1}}{l_{r+1}}\right) \\ & + p'_r\left[-u_{r-1}\frac{l_{r+1}}{l_r(l_r + l_{r+1})} + u_r\frac{l_{r+1} - l_r}{l_r l_{r+1}} + u_{r+1}\frac{l_r}{l_{r+1}(l_r + l_{r+1})}\right], \end{aligned}$$

after substituting Eq. (3.1.11) into Eq. (3.1.16). At the boundary points $r = 0$ and n, we obtain, respectively,

$$-\lambda\rho_0 u_0 = \frac{2p_0}{l_1^2}(u_1 - u_0) - \left(\frac{2}{l_1} - \frac{p'_0}{p_0}\right)hu_0$$

and

$$-\lambda\rho_n u_n = \frac{2p_n}{l_n^2}(u_{n-1} - u_n) - \left(\frac{2}{l_n} + \frac{p_n'}{p_n}\right)Hu_n$$

from the Taylor series expansion (3.1.10) as well as the boundary condition (3.1.15).

Set

$$\begin{cases} a_r = b_{r+1} + c_r, & r = 0, 1, \ldots, n, \\ b_1 = \dfrac{p_0}{l_1},\ b_{n+1} = 0,\ b_{r+1} = \dfrac{2p_r + p_r' l_r}{2l_{r+1}}, & r = 1, 2, \ldots, n-1, \\ c_0 = 0,\ c_n = \dfrac{p_n}{l_n},\ c_r = \dfrac{2p_r - p_r' l_{r+1}}{2l_r}, & r = 1, 2, \ldots, n-1, \\ m_0 = \dfrac{\rho_0 l_1}{2},\ m_r = \rho_r \dfrac{l_r + l_{r+1}}{2}, & r = 1, 2, \ldots, n-1, \\ m_n = \rho_n l_n / 2. \end{cases} \tag{3.1.17}$$

Then, the above-discretized form of the differential equation, Eq. (3.1.14), is reformatted to

$$\begin{cases} a_0 u_0 - b_1 u_1 + Q_0 = \lambda m_0 u_0, \\ -c_r u_{r-1} + a_r u_r - b_{r+1} u_{r+1} = \lambda m_r u_r,\ r = 1, 2, \ldots, n-1, \\ -c_n u_{n-1} + a_n u_n + Q_n = \lambda m_n u_n, \end{cases} \tag{3.1.18}$$

where parameters

$$Q_0 = \left(1 - \frac{l_1}{2}\frac{p_0'}{p_0}\right)hu_0 = q_{00}u_0 \text{ and } Q_n = \left(1 + \frac{l_n}{2}\frac{p_n'}{p_n}\right)Hu_n = q_{n0}u_n \tag{3.1.19}$$

represent boundary forces at the two ends of the bar. It is clear that, under general circumstances, this system of equations differs from that of a spring–mass system. However, Eq. (3.1.18) is covered under the standard Jacobian system introduced in Sect. 2.3. We will see that the spring–mass system and the Jacobian system share identical qualitative properties.

3.2 Basic Qualitative Properties of Vibration and Static Deformation of Spring–Mass Systems

3.2.1 Oscillatory Properties in Vibration of a Spring–Mass System

Suppose the spring–mass system shown in Fig. 3.2b is in a natural vibration. We will study its characteristics directly from Eqs. (3.1.3)–(3.1.5). If the spring–mass system is elastically supported on two ends, then the system of modal equations (3.1.3) can be rewritten in the matrix form as

$$\boldsymbol{A}\boldsymbol{u} = \lambda \boldsymbol{u}, \tag{3.2.1}$$

where $\lambda = \omega^2$ is the eigenvalue, $\boldsymbol{u} = (u_0, u_1, \dots, u_n)^{\mathrm{T}}$ stands for the displacement mode shape, and the matrix

$$\boldsymbol{A} = \boldsymbol{M}^{-1}\boldsymbol{K} = \begin{bmatrix} \frac{k_1+h}{m_0} & -\frac{k_1}{m_0} & 0 & \cdots & 0 & 0 \\ -\frac{k_1}{m_1} & \frac{k_1+k_2}{m_1} & -\frac{k_2}{m_1} & \cdots & 0 & 0 \\ \vdots & \vdots & \vdots & & \vdots & \vdots \\ 0 & 0 & 0 & \cdots & -\frac{k_n}{m_n} & \frac{k_n+H}{m_n} \end{bmatrix}$$

happens to be a standard Jacobian matrix.

Let $\boldsymbol{A}_1$ denote the truncated submatrix formed by removing the first row and first column of the tridiagonal matrix $\boldsymbol{A}$, and $\boldsymbol{A}_{1n}$ represent the truncated submatrix created by deleting the first and last rows as well as the first and last columns of $\boldsymbol{A}$. Then, the modal equation of the spring–mass system with a fixed left end and a free right end becomes

$$\lambda \boldsymbol{u}^{\mathrm{cf}} = \boldsymbol{A}_1 \boldsymbol{u}^{\mathrm{cf}}, \tag{3.2.2}$$

while the equation of the system fixed at both ends assumes the form

$$\lambda \boldsymbol{u}^{\mathrm{cc}} = \boldsymbol{A}_{1n} \boldsymbol{u}^{\mathrm{cc}}. \tag{3.2.3}$$

Here, superscripts cf and cc represent fixed–free and fixed–fixed boundary conditions, respectively, and

$$\begin{aligned} &\boldsymbol{u}^{\mathrm{cf}} = (u_1, u_2, \dots, u_n)^{\mathrm{T}}, \quad \boldsymbol{u}^{\mathrm{cc}} = (u_1, u_2, \dots, u_{n-1})^{\mathrm{T}}, \\ &\boldsymbol{A}_1 = (\boldsymbol{M}^{\mathrm{cf}})^{-1}\boldsymbol{K}^{\mathrm{cf}}, \quad \boldsymbol{A}_{1n} = (\boldsymbol{M}^{\mathrm{cc}})^{-1}\boldsymbol{K}^{\mathrm{cc}}, \end{aligned}$$

where

$$\boldsymbol{M}^{\mathrm{cf}} = \mathrm{diag}(m_1, m_2, \dots, m_n), \quad \boldsymbol{M}^{\mathrm{cc}} = \mathrm{diag}(m_1, m_2, \dots, m_{n-1}),$$

$$\boldsymbol{K}^{\mathrm{cf}} = \begin{bmatrix} k_1+k_2 & -k_2 & 0 & \cdots & 0 & 0 \\ -k_2 & k_2+k_3 & -k_3 & \cdots & 0 & 0 \\ \vdots & \vdots & \vdots & & \vdots & \vdots \\ 0 & 0 & 0 & \cdots & -k_n & k_n \end{bmatrix},$$

and $\boldsymbol{K}^{cc}$ is a square matrix of order $n-1$ obtained by removing the last row and column of Matrix $\boldsymbol{K}^{cf}$.

It is easy to verify that both $\boldsymbol{A}_1$ and $\boldsymbol{A}_{1n}$ are positive definite standard Jacobian matrices. In addition, $\boldsymbol{A}$ is a positive definite standard Jacobian matrix when $h+H>0$, while it becomes a positive semidefinite standard Jacobian matrix when $h=0=H$. Thus, we may get the following theorem based on oscillatory properties of eigenpairs associated with the standard Jacobin matrix, which have been introduced in Sect. 2.3.

Theorem 3.1 *Natural frequencies f_i ($i=1,2,\ldots,N$) of a spring–mass system are distinct and can be arranged in an ascending order as*

$$0 \le f_1 < f_2 < \cdots < f_N,$$

where the equality holds only when the system is unconstrained, $N=n+1$ when the system is either unconstrained or elastically supported at both ends, $N=n$ when the system is subject to fixed–free or fixed–elastic constraints, and $N=n-1$ when the system is fixed at both ends.

Notice that this interpretation of N is applicable in the rest of this chapter.

Theorem 3.2 *When natural frequencies of a spring–mass system are sorted in an ascending order, the following properties may be established:*

(1) *The displacement mode shape $\boldsymbol{u}^{(i)}$ corresponding to f_i has exactly $i-1$ sign reversals, where $i=1,2,\ldots,N$.*
(2) *Nodes of two successive mode shapes of the displacement $\boldsymbol{u}^{(i)}$ and $\boldsymbol{u}^{(i+1)}$ ($i=2,3,\ldots,N-1$) interlace.*
(3) *Consider the vibration expressed as a linear combination of the p-th to q-th mode shapes, where $p<q$. The node count of this vibration is in between $p-1$ and $q-1$ at any moment. In other words, if*

$$\boldsymbol{u} = \sum_{i=p}^{q} c_i \boldsymbol{u}^{(i)},$$

then the minimum and maximum numbers of sign reversals $S_{\boldsymbol{u}}^-$ and $S_{\boldsymbol{u}}^+$ of $\boldsymbol{u}$ satisfy the condition

$$p-1 \le S_{\boldsymbol{u}}^- \le S_{\boldsymbol{u}}^+ \le q-1.$$

It needs to be pointed out that the basis for Theorem 3.2 is oscillatory properties associated with the positive definite standard Jacobian matrix. However, for a spring–mass system unconstrained at both ends, its stiffness matrix $\boldsymbol{A}$ becomes a positive semidefinite standard Jacobian matrix, and is no longer a sign-oscillatory matrix. Consequently, Property (3) in Theorem 3.2 has to be separately proved for a spring–mass system unconstrained at both ends. This will be done at the end of Sect. 3.3.1.

3.2.2 Oscillatory Properties in Static Deformation of a Spring–Mass System

We have pointed out in Example 2 of Sect. 2.4 within Chap. 2 that a positive definite standard Jacobian matrix is sign-oscillatory. Thus, the stiffness matrix of a well-constrained spring–mass system is sign-oscillatory. Furthermore, it is known from Theorem 2.28 of Chap. 2 that the static deformation of a well-constrained spring—mass system has the following properties:

(1) When a concentrated force is applied to a movable mass in a well-constrained spring–mass system, the displacement of any mass in the system must be nonzero and should be along the direction of the force applied.
(2) When n concentrated forces are applied to some movable masses in the system, where $n \leq N$ and N is the number of degrees of freedom of the well-constrained spring–mass system, the u-line determined by displacement components $\{u_i\}$ of the system should have no more than $n - 1$ sign reversals, i.e., $S_{\boldsymbol{u}} \leq n - 1$.

3.2.3 Interlacement of Natural Frequencies of Spring–Mass Systems

Theorem 3.3 *Consider three spring–mass systems: The first is unconstrained at two ends, the second is subject to fixed–free constraints, and the third is fixed at both ends. Suppose they share identical internal springs and masses, with parameters* $\{k_r\}$ *and* $\{m_r\}$*, respectively. Natural frequencies of the three systems have the following relationship:*

$$f_i^{\mathrm{ff}} < f_i^{\mathrm{cf}} < (f_i^{\mathrm{cc}}, f_{i+1}^{\mathrm{ff}}) < f_{i+1}^{\mathrm{cf}}, \quad i = 1, 2, \ldots, n-1, \tag{3.2.4}$$

where the superscripts cf *and* cc *have been defined before, while* ff *indicates the free –free or unconstrained boundary conditions.*

Proof We know, from Property 6 of the sequence A

$$D_m(\lambda), D_{m-1}(\lambda), \ldots D_1(\lambda), D_0(\lambda)$$

given in Sect. 2.3, that there is exactly one real root of $D_{m-1}(\lambda)$ in between each pair of successive real roots of $D_m(\lambda)$, where $m = 2, 3, \ldots, n$. From this, we conclude there is one and only one natural frequency of the spring–mass system in the fixed–free configuration between two successive natural frequencies of the spring–mass system unconstrained at both ends, namely,

$$f_i^{\mathrm{ff}} < f_i^{\mathrm{cf}} < f_{i+1}^{\mathrm{ff}}, \quad i = 1, 2, \ldots, n. \tag{3.2.5}$$

By the same token, there is one and only one natural frequency of the spring–mass system fixed at both ends between two successive natural frequencies of the spring–mass system subject to fixed–free constraints, i.e.,

$$f_i^{\mathrm{cf}} < f_i^{\mathrm{cc}} < f_{i+1}^{\mathrm{cf}}, \quad i = 1, 2, \ldots, n-1. \tag{3.2.6}$$

After combining the above two sets of inequalities, we get Eq. (3.2.4). ■

Based on this theorem, it was pointed out in Ref. [1] that we can determine all physical parameters $\{m_r\}$ and $\{k_r\}$ of a spring–mass system in the fixed–free configuration after knowing its n natural frequencies f_i^{cf} $(i = 1, 2, \ldots, n)$ and also the $n-1$ natural frequencies f_i^{cc} $(i = 1, 2, \ldots, n-1)$ of the same system in the fixed–fixed configuration. Obviously, the latter system may be obtained from the first one by replacing its free end with a fixed end. The algorithm developed by Golub and Boley in 1977 was incorporated in the numerical implementation (see Ref. [3]). Interested readers are referred to Ref. [1].

In summary, we have obtained, with the help of the theory of Jacobian matrices, oscillatory properties of spring–mass systems as well as the interlacement relationship among their natural frequencies. However, we need to develop further understanding on other qualitative properties of such systems to support theoretical research and engineering application, such as solution of inverse problems. Thus, we are going to discuss additional properties of frequencies and mode shapes associated with spring–mass systems in the next section.

3.3 Necessary and Sufficient Conditions for Mode Shapes of a Spring–Mass System

3.3.1 Further Properties of Mode Shapes of a Spring–Mass System

The primary focus of this section is to discuss further properties of the mode shapes of a spring–mass system. To this end, let us introduce the following $n \times (n+1)$ matrix:

$$\boldsymbol{E} = \begin{bmatrix} 1 & -1 & & & 0 \\ & 1 & -1 & & \\ & & \ddots & \ddots & \\ & 0 & & 1 & -1 \end{bmatrix}$$

and $\underline{\boldsymbol{K}} = \mathrm{diag}(k_1, k_2, \cdots, k_n)$. See Ref. [6]. When the system is unconstrained at both ends, i.e., $h = 0 = H$, Eq. (3.1.3) can be rewritten as

$$\lambda \boldsymbol{M} \boldsymbol{u} = \boldsymbol{E}^{\mathrm{T}} \underline{\boldsymbol{K}} \boldsymbol{E} \boldsymbol{u}. \tag{3.3.1}$$

Let $w_r = u_r - u_{r-1}$ denote the elongation of the r-th spring and $\sigma_r = k_r w_r$ represent the corresponding spring force, where $r = 1, 2, \ldots, n$. Then,

$$\boldsymbol{w} = (w_1, w_2, \ldots, w_n)^{\mathrm{T}} = -\boldsymbol{E}\boldsymbol{u},\ \boldsymbol{\sigma} = (\sigma_1, \sigma_2, \ldots, \sigma_n)^{\mathrm{T}} = \underline{\boldsymbol{K}} \boldsymbol{w}. \tag{3.3.2}$$

In the subsequent discussion, we refer to $\boldsymbol{w}$ as the mode shape of the spring deformation or relative displacement, and call $\boldsymbol{\sigma}$ the mode shape of the spring force. For clarity, $\boldsymbol{u}$ is now referred to as the displacement mode shape or the mode shape of the displacement. With these notations, Eq. (3.3.1) can be further transformed into the following format:

$$\lambda \underline{\boldsymbol{K}}^{-1} \boldsymbol{\sigma} = \boldsymbol{E} \boldsymbol{M}^{-1} \boldsymbol{E}^{\mathrm{T}} \boldsymbol{\sigma}. \tag{3.3.3}$$

By comparing Eqs. (3.3.1) and (3.3.3), we can easily find two differences: First, M and $\underline{\boldsymbol{K}}$ in Eq. (3.3.1) are replaced by $\underline{\boldsymbol{K}}^{-1}$ and $\boldsymbol{M}^{-1}$ in Eq. (3.3.3), respectively. Second, there is one fewer equation in (3.3.3) than in (3.3.1), but the coefficient matrix $\boldsymbol{E}\boldsymbol{M}^{-1}\boldsymbol{E}^{\mathrm{T}}$ on the right-hand side of Eq. (3.3.3) remains to be a standard Jacobian matrix. Thus, we may interpret Eq. (3.3.3) as the governing equation of another spring–mass system with parameters $\{\bar{m}_r = k_r^{-1}\}_1^n$ and $\{\bar{k}_r = m_r^{-1}\}_0^n$. Such a system is referred to as the conjugate of the original spring–mass system.

Additionally, it is not difficult to verify several observations: First, when the original system is unconstrained at both ends, the respective conjugate system is fixed at both ends, and the number of degree of freedom of the latter is reduced by one from that of the former. Second, if the original system is fixed at both ends, then the corresponding conjugate system will be free at both ends, and the latter has one more degree of freedom than the former. Finally, if the original system is in a fixed–free configuration, then the conjugate system will be subject to free–fixed constraints, and the numbers of degrees of freedom of the two systems are identical. Thus, we may extend the discussion in the previous section and obtain a theorem.

Theorem 3.4 *Suppose natural frequencies of a spring–mass system have been sorted in an ascending order. Let S_w and S_σ denote, respectively, the numbers of sign reversals associated with its mode shapes of the spring deformation $\boldsymbol{w}^{(i)}$ and spring force $\boldsymbol{\sigma}^{(i)}$, both of which correspond to the natural frequency f_i. Then, S_w and S_σ satisfy the following conditions:*

$$S_{w^{\mathrm{ff}}} = S_{\sigma^{\mathrm{ff}}} = i - 2,\ i = 2, 3, \ldots, n + 1,$$

$$S_{w^{\mathrm{cc}}} = S_{\sigma^{\mathrm{cc}}} = i \ \textit{and}\ S_{w^{\mathrm{cf}}} = S_{\sigma^{\mathrm{cf}}} = i - 1, i = 1, 2, \ldots, N.$$

Proof For a spring-mass system in the free-free configuration, its conjugate is the spring-mass system in the fixed-fixed configuration. Let f_i^* and $\boldsymbol{\sigma}_*^{(i)}$ represent the natural frequency and the corresponding "displacement" mode shape of the conjugate system, respectively. Obviously, we have $f_i^* = f_{i+1}$ $(i = 1, 2, \ldots, n)$ while $f_1 = 0$. Thus, $\boldsymbol{\sigma}_*^{(i)}$ is, in fact, $\boldsymbol{\sigma}^{(i+1)}$ of the original system. The result $S_{\sigma_*} = i - 1$ leads to $S_{\sigma^{\mathrm{ff}}} = i - 2$, where $i = 2, 3, ..., n + 1$.

Next, the conjugate of the a spring-mass system fixed at both ends is one unconstrained at both ends. Here is the relationship between natural frequencies f_i^* and f_i of the conjugate system and the original spring-mass system fixed at both ends, respectively: $f_1^* = 0$ and $f_i^* = f_{i-1}$ $(i = 2, 3, \ldots, n)$. Since the number of sign reversals of the "displacement" mode shape $\boldsymbol{\sigma}_*^{(i)}$ associated with the conjugate system satisfies the equation $S_{\sigma_*} = i - 1$, we can derive $S_{\sigma^{\mathrm{cc}}} = i$, where $i = 1, 2, ..., n - 1$.

Finally, the conjugate of a spring-mass system in the fixed-free configuration is a system in the free-fixed configuration. The two systems should have identical natural frequencies. As a result, the "displacement" mode shape $\boldsymbol{\sigma}_*^{(i)}$ of the conjugate system and the spring force mode shape $\boldsymbol{\sigma}^{(i)}$ of the original system should share the same sequence number. Based on this and the result $S_{\sigma_*} = i - 1$, we may obtain $S_{\sigma^{\mathrm{cf}}} = i - 1$, where $i = 1, 2, \ldots, n$. ∎

A number of important corollaries can be derived from this theorem and Theorem 3.2 regarding displacement mode shapes of a spring–mass system.

Corollary 1 *Consider any displacement mode shape of a spring–mass system. None of its two adjacent components can be zero simultaneously. In addition, it is impossible for any of its three consecutive components to be identical.*

The validity of this corollary can be easily established through proof by contradiction. In fact, if there exist two adjacent components of a displacement mode shape that are both zero, then the number of sign reversals of the displacement mode shape is not determinate, which contradicts Theorem 3.2. Furthermore, if there are three consecutive components of a displacement mode shape to be identical, then the number of sign reversals associated with the mode shape of the spring deformation is undetermined, which contradicts Theorem 3.4.

Corollary 2 *Except for the first mode shape involving the free–free configuration, the following conclusions are valid for all other mode shapes of a spring–mass system:*

(1) *If the left end is free, then* $u_0 w_1 < 0$*; while if the right end is free, then* $u_n w_n > 0$.
(2) *If the left end is fixed, then* $u_1 w_1 > 0$*; while if the right end is fixed, then* $u_{n-1} w_n < 0$.

Here, the first two inequalities are obtained from the first and last equations in (3.1.2), i.e.,

$$-k_1 w_1 = \omega^2 m_0 u_0, \; k_n w_n = \omega^2 m_n u_n;$$

while the latter two inequalities can be established by using the definition of $\boldsymbol{w}$, specifically,

$$w_1 = u_1 - u_0,\ w_n = u_n - u_{n-1}.$$

Corollary 3 *If u_r is a local maximum of the displacement mode shape of a spring–mass system, then there must be $u_r > 0$.*

In fact, the modal equation of the mass corresponding to u_r is

$$k_r(u_r - u_{r-1}) - k_{r+1}(u_{r+1} - u_r) = \lambda m_r u_r, \quad 0 < r < n. \tag{3.3.4}$$

According to the definition of a local maximum,

$$u_r - u_{r-1} \geq 0 \text{ and } u_r - u_{r+1} \geq 0.$$

But only one of the two inequalities above can be equality, per Corollary 1. Since λ, k_r, and m_r are all positive numbers, we must have $u_r > 0$. It is easy to see that this result remains valid in the case of a free end.

Similarly, if u_r is a local minimum of the spring–mass system, then $u_r < 0$.

Corollary 4 *If u_r is an extremum, then it is possible to have either $u_r = u_{r+1}$ or $u_{r-1} = u_r$. Conversely, if either $u_r = u_{r+1}$ or $u_{r-1} = u_r$, then u_r must be an extremum.*

In fact, if u_r is a local maximum, then $u_r \geq u_{r+1}$ and $u_r \geq u_{r-1}$. According to Corollary 1, it is impossible for the inequalities to become equalities simultaneously. However, per Eq. (3.3.4), it is permissible for one of these inequalities to be an equality. The same argument can be applied to the situation where u_r is a local minimum. We can establish the rest of this corollary through proof by contradiction. If $u_r = u_{r+1}$ and u_r is not an extremum, then we must have

$$(u_r - u_{r-1})(u_{r+2} - u_r) > 0.$$

But this contradicts the fact that $\boldsymbol{w}$ has a fixed number of sign reversals.

By combining the previous results, we conclude that components of the displacement mode shape of a spring–mass system can be divided into i segments of the form

$$u_{\alpha_r}, u_{\alpha_r+1}, \cdots, u_{\beta_r-1}, u_{\beta_r}, \quad r = 1, 2, \ldots, i,$$

such that components within each segment share the same sign while components of adjacent segments are of opposite signs. In other words, the following conditions are satisfied:

$$u_{\alpha_r} u_{\alpha_{r+1}} < 0,\ r = 1, 2, \ldots, i-1;$$
$$u_{\alpha_r} u_j > 0,\ j = \alpha_r + 1,\ \alpha_r + 2, \ldots, \beta_r.$$

It needs to be pointed out that starting and ending indices α_1 and β_i respectively in the partition of $\boldsymbol{u}^{(i)}$ depend on boundary conditions. Specifically, $\alpha_1 = 0$ if the left end is either free or elastically supported, while $\alpha_1 = 1$ if the left end is fixed. Also, $\beta_i = n - 1$ if the right end is fixed, while $\beta_i = n$ if the right end is either free or elastically supported. Furthermore, depending on the existence of a zero component of $\boldsymbol{u}^{(i)}$ between two adjacent segments, we have the following relationship between related indices: If $u_{\beta_r+1} = 0$, then $\alpha_{r+1} - \beta_r = 2$; while if $u_{\beta_r+1} \neq 0$, then $\alpha_{r+1} - \beta_r = 1$.

In each segment that shares a common sign, the displacement mode shape has one and only one extremum that is either a local maximum in a positive segment or a local minimum in a negative segment. Finally, within two adjacent extrema, the associated *u-line* is monotonically increasing or decreasing and consequently, there is one and only one node. These observations lead to the following corollary:

Corollary 5 *If mode shapes of the displacement* $\boldsymbol{u}^{(i)}$ *and the spring deformation* $\boldsymbol{w}^{(i)}$ *correspond to the same natural frequency, then nodes of the respective u-line and w-line interlace.*

In addition, we have:

Corollary 6 *Consider a spring–mass system fixed at the left end but free at the right end. Suppose the system has n degrees of freedom. Set the n-th component of* $\boldsymbol{u}^{(i)}$ *to be positive, i.e.,* $u_{ni} > 0$*, where* $i = 1, 2, \ldots, n$*. Then, components of the first mode shape are all positive and monotonically increasing. In addition, the last segment of any u-line of the system must be monotonically increasing and concave upward.*

Theorem 2.17 is concerned with the interlacement of nodes of two successive eigenvectors $\boldsymbol{u}^{(k)}$ and $\boldsymbol{u}^{(k+1)}$, associated with either an oscillatory or sign-oscillatory matrix. In order to prove this theorem, it only needs to use Theorem 2.15, namely, Properties (1) and (3) of Theorem 3.2 in this chapter. Thus, we can further draw the following conclusion:

Corollary 7 *Let* $\boldsymbol{w}^{(i)}$ *and* $\boldsymbol{w}^{(i+1)}$ *be two successive spring deformation mode shapes of a spring–mass system subject to an arbitrary set of boundary constraints. Then, nodes of* $\boldsymbol{w}^{(i)}$ *and* $\boldsymbol{w}^{(i+1)}$ *interlace.*

In the remainder of this subsection, we will finish the leftover work of proving Property (3) of Theorem 3.2, i.e., to show that the property remains valid for a spring–mass system unconstrained at both ends. To this end, we first introduce a proposition. Also see Ref. [8].

Proposition 3.1 *Consider two vectors* $\boldsymbol{x} = (x_0, x_1, \ldots, x_n)^{\mathrm{T}}$ *and* $\boldsymbol{y} = (y_1, y_2, \ldots, y_n)^{\mathrm{T}}$ *that are related by the equation* $\boldsymbol{y} = -\boldsymbol{E}\boldsymbol{x}$*. If* $S_x^- \geq j$*, then*

$$S_y^- \geq j + 1 - H(x_0) - H(x_n);$$

while if $S_y^+ \leq j$*, then*

$$S_x^+ \leq j + 1.$$

Here,

$$H(t) = \begin{cases} 0, & t = 0, \\ 1, & t \neq 0. \end{cases}$$

Proof Let us begin by proving the first half of Proposition 3.1. From the condition $S_x^- \geq j$ as well as the comment after Corollary 4, components of the vector $\boldsymbol{x}$ can be grouped into at least $j+1$ segments, within each of which components share the same positive or negative sign. When $x_0 = 0$ and $x_n = 0$, each of these component segments of $\boldsymbol{x}$ should have at least one extremum, and on the two sides of each extremum the vector $\boldsymbol{y}$ has one sign reversal. Thus, we get $S_y^- \geq j+1$. When $x_0 \neq 0$, the first component segment of $\boldsymbol{x}$ may not cause $\boldsymbol{y}$ to have a sign reversal. By the same token, when $x_n \neq 0$, the last component segment of $\boldsymbol{x}$ may not trigger a sign reversal in $\boldsymbol{y}$. Consequently,

$$S_{\boldsymbol{y}}^- \geq j + 1 - H(x_0) - H(x_n).$$

Next, let us prove the second half of Proposition 3.1 using the method of proof by contradiction. Suppose $S_x^+ \geq j+2$ when $S_y^+ \leq j$. Based on the concept of the number of sign reversals given in Sect. 2.3, there exist some components

$$x_{i_r},\ r = 1, 2, \ldots, j+2 \quad \text{and} \quad 0 \leq i_1 < i_2 < \cdots < i_{j+2} \leq n$$

such that

$$x_{i_r} \cdot x_{i_r+1} \leq 0, \quad r = 1, 2, \ldots, j+2,$$

and

$$x_{i_r} \cdot x_{i_{r+1}} \leq 0, \quad r = 1, 2, \ldots, j+1.$$

Without loss of generality, let us assume $x_{i_1} > 0$. Then,

$$(-1)^r y_{i_r} = (-1)^r (x_{i_r+1} - x_{i_r}) \geq 0, \ r = 1, 2, \ldots, j+2.$$

But this would lead to $S_{\boldsymbol{y}}^+ \geq j+1$, which contradicts the condition given. Thus, the second half of Proposition 3.1 must also hold true. ■

Now let us return to the subject of proving Property (3) of Theorem 3.2, for the case of an unconstrained spring–mass system. From the discussion at the beginning of this section, the conjugate of a spring–mass system unconstrained at two ends is the system fixed at both ends. The "displacement" mode shape of the conjugate system happens to be the mode shape of the spring force $\boldsymbol{\sigma}_*^{(i)}$ associated with the original system. Thus, Property (3) of Theorem 3.2 is applicable to $\boldsymbol{\sigma}_*^{(i)}$. Since $\boldsymbol{\sigma}_*^{(i)}$ is related to the spring deformation $\boldsymbol{w}_*^{(i)}$ by the second equation in (3.3.2), we know

$w_*^{(i)}$ should have Property (3) of Theorem 3.2, too. Specifically, for any $1 \le p \le q \le n$ and a set of real constants $c_p, c_{p+1}, \ldots, c_q$ not vanishing simultaneously, the minimum and maximum counts of sign reversals S_w^- and S_w^+, associated with the vector $w = c_p w_*^{(p)} + c_{p+1} w_*^{(p+1)} + \ldots + c_q w_*^{(q)}$, satisfy the condition

$$p - 1 \le S_w^- \le S_w^+ \le q - 1; \tag{3.3.5}$$

while the minimum and maximum numbers of sign reversals S_σ^- and S_σ^+, related to the vector $\sigma = c_p \sigma_*^{(p)} + c_{p+1} \sigma_*^{(p+1)} + \cdots + c_q \sigma_*^{(q)}$, satisfy the inequality

$$p - 1 \le S_\sigma^- \le S_\sigma^+ \le q - 1. \tag{3.3.6}$$

Thus, for vectors u and w related by $w = -Eu$, we get the relationship below from the second half of Proposition 3.1 and Eq. (3.3.5):

$$S_u^+ \le q.$$

Furthermore, the modal equation of a spring–mass system unconstrained at both ends may be rewritten as

$$\lambda m_r u_r = \sigma_{r+1} - \sigma_r, \sigma_0 = 0, \sigma_{n+1} = 0, \quad \text{and} \quad r = 0, 1, \ldots, n.$$

It is clear from the above equation that, so far as the numbers of sign reversals are concerned, the relationship between vectors $\tilde{\sigma} = (\sigma_0, \sigma_1, \ldots, \sigma_{n+1})^{\mathrm{T}}$ and u is similar to that between x and y in Proposition 3.1. Consequently, we obtain the following result from the first half of Proposition 3.1 as well as Eq. (3.3.6):

$$S_u^- \ge p.$$

By combining the previous two results, we conclude

$$p \le S_u^- \le S_u^+ \le q. \tag{3.3.7}$$

Notice that while the "displacement" mode shape of the conjugate system is the mode shape of the spring force $\sigma_*^{(i)}$ associated with the original system, the corresponding natural frequency f_i^* of the conjugate system equals f_{i+1} of the original system, due to existence of a zero frequency in the original system. In other words,

$$w_*^{(i)} = -Eu^{(i+1)}, \ i = 1, 2, \ldots, n.$$

Thus, the vector u, corresponding to the vector $w = c_p w_*^{(p)} + c_{p+1} w_*^{(p+1)} + \cdots + c_q w_*^{(q)}$, is

$$u = c_p u^{(p+1)} + c_{p+1} u^{(p+2)} + \cdots + c_q w^{(q+1)}. \tag{3.3.8}$$

By comparing Eqs. (3.3.7) and (3.3.8), we finally obtain Property (3) of Theorem 3.2.

3.3.2 Necessary and Sufficient Conditions for One Mode Shape of a Spring–Mass System

By further generalizing the previous discussion, we may derive the next theorem.

Theorem 3.5 *Depending on the type of boundary constraints, the necessary and sufficient conditions for a vector* $\boldsymbol{u} = (u_0, u_1, \ldots, u_n)^{\mathrm{T}}$ *to be the displacement mode shape of a spring–mass system are:*

$$\begin{aligned} &S_{\boldsymbol{u}^{\mathrm{ff}}} = i-1,\ S_{\boldsymbol{w}^{\mathrm{ff}}} = i-2,\ i = 2, 3, \ldots, n+1; \\ &S_{\boldsymbol{u}^{\mathrm{cf}}} = i-1,\ S_{\boldsymbol{w}^{\mathrm{cf}}} = i-1,\ i = 1, 2, \ldots, n; \qquad \text{or} \\ &S_{\boldsymbol{u}^{\mathrm{cc}}} = i-1,\ S_{\boldsymbol{w}^{\mathrm{cc}}} = i, \qquad i = 1, 2, \ldots, n-1. \end{aligned} \tag{3.3.9}$$

Here, it is assumed that the vector $\boldsymbol{u}$ *satisfies the respective boundary conditions, which may include free–free, fixed–free, and fixed–fixed constraints;* $\boldsymbol{w} = -\boldsymbol{E}\boldsymbol{u}$ *denotes the mode shape of the spring deformation corresponding to* $\boldsymbol{u}$; $S_{\boldsymbol{u}^{\mathrm{ff}}}$ *and* $S_{\boldsymbol{w}^{\mathrm{ff}}}$ *designate numbers of sign reversals of mode shapes associated with the displacement and spring deformation, respectively, for the spring–mass system unconstrained at both ends; and other symbols representing numbers of sign reversals can be interpreted in a similar way.*

Proof Since the necessary part of the conditions is guaranteed by Theorems 3.2 and 3.4, we only need to prove the sufficient part.

Let us arbitrarily choose a positive number λ. We will then proceed to construct a spring–mass system such that $\boldsymbol{u}$ is one of its displacement mode shapes while $\sqrt{\lambda}$ is the respective angular frequency. Here, $\boldsymbol{u}$ and $\boldsymbol{w} = -\boldsymbol{E}\boldsymbol{u}$ satisfy one set of equations in (3.3.9) as well as the corresponding boundary conditions.

We begin by considering the case where the right end of the system is unconstrained, i.e., $u_n \neq 0$. The inequality $u_n w_n > 0$ can be derived directly from the given conditions on sign reversals, i.e., Equation (3.3.9). We may follow the approach introduced in Sect. 3.3.1 to divide components of $\boldsymbol{u}$ into several contiguous segments, within each of which the components share a common sign. Set

$$U_r = \sum_{j=r}^{n} \lambda m_j u_j.$$

Then, the modal equations in (3.1.2) can be rewritten as

$$k_r w_r = U_r, \quad r = n, n-1, \ldots, 1 \tag{3.3.10a}$$

when the left end is fixed; or they assume the form

$$k_r w_r = U_r, \quad r = n, n-1, \ldots, 1, \quad U_0 = 0 \tag{3.3.10b}$$

when the left end is unconstrained. Thus, by properly choosing positive values for m_r $(r = n, n-1, \ldots, 1)$ when $u_r \neq 0$ or by setting $m_r = 1$ when $u_r = 0$ $(0 < r < n)$, we can always make w_r and U_r $(r = 1, 2, \ldots, n)$ either be of the same sign or simultaneously vanish. Thus, we may calculate positive numbers k_r $(r = n, n-1, \ldots, 1)$ from Eq. (3.3.10a) or (3.3.10b) if $w_r \neq 0$, or assign any positive value to k_r if $w_r = 0$. When the system is also unconstrained at the left side, we need to select the positive number m_0 properly to ensure $U_0 = 0$.

Next, let us look at the case where the spring–mass system is fixed at both ends. Without loss of generality, we assume that $u_1 > 0$. Then, there should exist among all components of $\boldsymbol{u}$ at least one segment, in which all members are positive and we can find one positive local maximum u_s $(0 < s < n)$. There are two scenarios:

(1) If $u_s > u_{s-1}$ and $u_s > u_{s+1}$, set

$$k_s w_s = \lambda m_{s1} u_s, \quad -k_{s+1} w_{s+1} = \lambda m_{s2} u_s,$$

$$U_r = \lambda(m_{s1} u_s + m_{s-1} u_{s-1} + \cdots + m_r u_r), \quad r = s, s-1, \ldots, 1,$$

$$V_r = \lambda(m_{s2} u_s + m_{s+1} u_{s+1} + \cdots + m_r u_r), \quad r = s, s+1, \ldots, n-1,$$

where $m_s = m_{s1} + m_{s2}$. Similar to the case with an unconstrained end, we may rewrite the modal equations as

$$k_r w_r = U_r, r = s, \ s-1, \ldots, 1 \text{ and } -k_{r+1} w_{r+1} = V_r, r = s, s+1, \ldots, n-1.$$

By carefully selecting positive numbers m_{s1}, m_{s-1}, ..., m_1 and m_{s2}, m_{s+1}, ..., m_{n-1}, we can simultaneously achieve the following two objectives: First, make U_r and w_r $(r = s, s-1, \ldots, 1)$ either share the same sign or simultaneously be zero; and have V_r and w_{r+1} $(r = s, s+1, \ldots, n-1)$ either be of opposite signs or vanish together. Thus, k_r $(r = 1, 2, \ldots, n)$ calculated from the modal equations are all positive numbers.

(2) If $u_s = u_{s+1}$, $u_s > u_{s-1}$, and $u_s > u_{s+2}$, the proof follows the same process outlined above. But, since $V_s = 0$ in this case, we can set $m_{s2} = 0$ and allow k_{s+1} to be any positive number.

Thus, as long as the given vector $\boldsymbol{u}$ satisfies both boundary conditions and requirements on sign reversals, it will be one of the displacement mode shapes of a real spring–mass system. This concludes the proof of the sufficiency of the condition (3.3.9). ■

3.3.3 Compatibility of Two Mode Shapes and the Number of Independent Mode Shapes

The previous discussion primarily focused on qualitative properties of a single displacement mode shape associated with a spring–mass system. Before wrapping up this section, let us look at the relationship between two displacement mode shapes. See Refs. [1, 5]. Besides the well-known orthogonality requirement as well as the interlacing behavior among nodes mentioned in Sect. 3.2.1, the two displacement mode shapes of a spring–mass system should also satisfy some other important conditions.

Suppose ω_i represents the i-th natural angular frequency, while let $\boldsymbol{u}^{(i)} = (u_{\alpha_1 i}, u_{\alpha_1+1,i}, \ldots, u_{\beta_i i})^{\mathrm{T}}$ and $\boldsymbol{w}^{(i)} = (w_{1i}, w_{2i}, \ldots, w_{ni})^{\mathrm{T}}$ denote the corresponding mode shapes of the displacement and spring deformation, respectively, where $i = 1, 2, \ldots, N$ and the values of α_1 and β_i depend upon specific boundary conditions. For instance, if the spring–mass system is free at the left end, $\alpha_1 = 0$; while if it is fixed at the left end, $\alpha_1 = 1$. Similarly, if the right end of the system is free, $\beta_i = n$; while if its right end is fixed, $\beta_i = n - 1$. With these notations, we are ready to introduce some properties and theorems.

Property 1 *For an arbitrary integer* r $(1 \le r \le n-1)$ *as well as any fixed pair of integers* i *and* j $(1 \le i < j \le N)$, *we may calculate a triplet* $\{p_r, q_r, s_r\}$ *from the two given sets of modes as below:*

$$\begin{cases} p_r = \lambda_j u_{rj} w_{ri} - \lambda_i u_{ri} w_{rj}, \\ q_r = \lambda_j u_{rj} w_{r+1,i} - \lambda_i u_{ri} w_{r+1,j}, \\ s_r = w_{rj} w_{r+1,i} - w_{ri} w_{r+1,j}, \end{cases} \tag{3.3.11}$$

where $\lambda_i = \omega_i^2$ *and* $\lambda_j = \omega_j^2$ *are the respective eigenvalues. The calculated numbers* p_r, q_r, *and* s_r *are either of identical sign or equal to zero simultaneously, where* $r = 1, 2 \ldots, n-1$.

In fact, we have from modal equations of a spring–mass system (3.3.4):

$$k_r w_{ri} - k_{r+1} w_{r+1,i} = \lambda_i m_r u_{ri} \quad \text{and} \quad k_r w_{rj} - k_{r+1} w_{r+1,j} = \lambda_j m_r u_{rj}.$$

Solving the equations, we get

$$\frac{k_r}{m_r} = \frac{\lambda_j u_{rj} w_{r+1,i} - \lambda_i u_{ri} w_{r+1,j}}{w_{rj} w_{r+1,i} - w_{ri} w_{r+1,j}} = \frac{q_r}{s_r}, \tag{3.3.12}$$

$$\frac{k_{r+1}}{m_r} = \frac{\lambda_j u_{rj} w_{ri} - \lambda_i u_{ri} w_{rj}}{w_{rj} w_{r+1,i} - w_{ri} w_{r+1,j}} = \frac{p_r}{s_r}. \tag{3.3.13}$$

Since k_r, k_{r+1}, and m_r are all positive numbers, Property 1 is established. See Ref. [5].

Property 2 *If the spring–mass system is free at the left and right ends, then there are following results, respectively,*

$$-\frac{k_1}{m_0} = \lambda_i \frac{u_{0i}}{w_{1i}} = \lambda_j \frac{u_{0j}}{w_{1j}} < 0 \quad and \quad \frac{k_n}{m_n} = \lambda_i \frac{u_{ni}}{w_{ni}} = \lambda_j \frac{u_{nj}}{w_{nj}} > 0, \tag{3.3.14}$$

where $i \neq j$.

In fact, modal equations of a spring–mass system at the free ends are as follows:

$$-k_1 w_{1i} = \lambda_i m_0 u_{0i}, -k_1 w_{1j} = \lambda_j m_0 u_{0j}, k_n w_{ni} = \lambda_i m_n u_{ni}, k_n w_{nj} = \lambda_j m_n u_{nj}.$$

Thus, Eq. (3.3.14) obviously holds true.

Properties 1 and 2 introduced above are referred to as compatibility conditions of two different modes. Based on this discussion, we may obtain the following important theorem:

Theorem 3.6 *Consider two modes* $(f_i, \boldsymbol{u}^{(i)})$ *and* $(f_j, \boldsymbol{u}^{(j)})$ *of a spring–mass system with n degrees of freedom, where i and j are two different positive integers no greater than n. If they satisfy requirements on numbers of sign reversals given in Eq.* (3.3.9) *as well as the above compatibility conditions, then all parameters of the system, i.e.* $\{m_r\}$ *and* $\{k_r\}$ *may be determined. Specifically, if the spring–mass system does not have an unconstrained end, its parameters are computed using Eqs.* (3.3.12) *and* (3.3.13); *while if it has a free end, the parameters are calculated with Eqs.* (3.3.12)–(3.3.14). *Furthermore, all other modes of the system can be derived thereafter.*

Since the above result are of great significance in both Mechanics and Mathematics, we would like to rephrase the conclusion from a different perspective:

Theorem 3.7 *Among the n sets of displacement modes* $(f_i, \boldsymbol{u}^{(i)})$ $(i = 1, 2, \cdots, n)$ *of a spring–mass system with n degrees of freedom, there are only two sets that are independent. Furthermore, this pair of independent modes* $(f_{i_1}, \boldsymbol{u}^{(i_1)})$ *and* $(f_{i_2}, \boldsymbol{u}^{(i_2)})$ $(i_1 \neq i_2)$ *can be any two out of the n modes.*

The theorems indicate that if two modes (including two natural frequencies and the two corresponding mode shapes) of two *spring*–mass systems are identical, then the two systems must match precisely within a constant factor. The result also implies: We can design a spring–mass system to have one or two sets of given modes, where, of course, the modes need to satisfy the respective necessary and sufficient conditions for mode shapes as well as compatibility requirements. However, we may not be able to demand that the system possess more than two sets of specified modes in general situations. It needs to be pointed out that when dealing with a system with two free ends, the rigid-body modes should be excluded for discussion of this subsection to remain valid.

3.4 Modal Qualitative Properties of the Finite Difference System of a Bar

It has been shown in Sect. 3.1.3 that the finite difference model of a bar in longitudinal, shear, or torsional vibration is equivalent to a spring–mass or standard Jacobian system. Thus, modal qualitative properties of the spring–mass system derived in the previous sections are also applicable to the finite difference model of a bar in longitudinal, shear, or torsional vibration. These properties may be summarized as below:

(1) Natural frequencies and mode shapes of the finite difference model of a bar have the four oscillatory properties in vibration described in Theorems 3.1 and 3.2.
(2) The necessary and sufficient conditions for the vector $\boldsymbol{u} = (u_\alpha, u_{\alpha+1}, \ldots, u_\beta)^{\mathrm{T}}$ to be a mode shape of the finite difference system of a bar are: $\boldsymbol{u}$ and the derived vector $\boldsymbol{w} = -\boldsymbol{E}\boldsymbol{u}$ both have fixed numbers of sign reversals, and they satisfy

$$S_u = S_w - \gamma.$$

Here, α is 1 if the left end is fixed or 0 if it is free, while β assumes the value of $n-1$ if the right end is fixed or n if it is free; $\boldsymbol{w} = (w_1, w_2, \ldots, w_n)^{\mathrm{T}}$, where $w_r = u_r - u_{r-1}$ and $r = 1, 2, \ldots, n$; and the value of γ is set to 1 for a bar fixed at both ends, 0 for a bar in the fixed–free configuration, and -1 for an unconstrained bar. If it is further required that $\boldsymbol{u}$ be a mode shape associated with the i-th natural angular frequency ω_i, then numbers of sign reversals should satisfy

$$S_u = i - 1,\ S_w = i + \gamma - 1,\ i = 1, 2, \ldots, N.$$

(3) Nodes of two successive mode shapes of the relative displacement, $\boldsymbol{w}^{(i)}$ and $\boldsymbol{w}^{(i+1)}$, interlace, where $i = 2, 3, \ldots, n-1$; while nodes of the displacement mode shape $\boldsymbol{u}^{(i)}$ and the corresponding mode shape of the relative displacement $\boldsymbol{w}^{(i)}$, interlace, where $i = 2, 3, \ldots, N$.
(4) Components of two arbitrary mode shapes must satisfy the following compatibility conditions: For any integer r $(r = 1, 2, \ldots, n-1)$ and a pair of fixed integers i and j $(1 \le i < j \le N)$, the expressions

$$\begin{cases} p_r = \lambda_j u_{rj} w_{ri} - \lambda_i u_{ri} w_{rj}, \\ q_r = \lambda_j u_{rj} w_{r+1,i} - \lambda_i u_{ri} w_{r+1,j}, \\ s_r = w_{rj} w_{r+1,i} - w_{ri} w_{r+1,j} \end{cases} \tag{3.4.1}$$

must either be of the same sign or vanish simultaneously, where $\lambda_i = \omega_i^2$ and $\lambda_j = \omega_j^2$ are the corresponding eigenvalues. If the bar is free at the right end, then there is an additional result,

$$\lambda_i u_{ni}/w_{ni} = \lambda_i u_{nj}/w_{nj} > 0,\ i \neq j. \tag{3.4.2}$$

(5) Consider the finite difference system of a bar in the fixed-free configuration. Suppose it undergoes a forced vibration due to an external harmonic excitation at the right end with the frequency f. The displacement at this end is found to be

$$u_n = q_n \sum_{i=1}^{n} \frac{u_{ni}^2}{\lambda_i - \lambda},$$

where $\lambda = (2\pi f)^2$, $\lambda_i = \omega_i^2$, and u_{ni} is the last component of the mode shape $\boldsymbol{u}^{(i)}$ corresponding to the natural angular frequency ω_i. From this result, we may further derive the frequency equation of the bar fixed at both ends:

$$\sum_{i=1}^{n} \frac{u_{ni}^2}{\lambda_i - \lambda} = 0. \tag{3.4.3}$$

Based on the above equation, we conclude that natural frequencies of finite difference systems of the same bar in the fixed–fixed and fixed–free configurations should interlace.

In addition, we would like to point out that other results in Sects. 3.2 and 3.3 should also be applicable to the finite difference model of a bar. Due to space limitation, we have to skip the details here.

3.5 Modal Qualitative Properties of the Finite Element System of a Bar

Consider a bar of length l. Choose its left end as the origin O of the coordinate system and its centerline as the x-axis. Divide the bar into n elements with points $0 = x_0 < x_1 < \cdots < x_{n-1} < x_n = l$, as shown in Fig. 3.4. Let the x-displacements at two end points of the r-th element to be y_{r-1} and y_r respectively, where $r = 1, 2, \ldots, n$. The displacement within the element can be represented as a linear interpolation of those at the two end points as below:

$$y(x,t) = y_{r-1}(t)(1-\xi) + y_r(t)\xi, \; x_{r-1} < x < x_r, \tag{3.5.1}$$

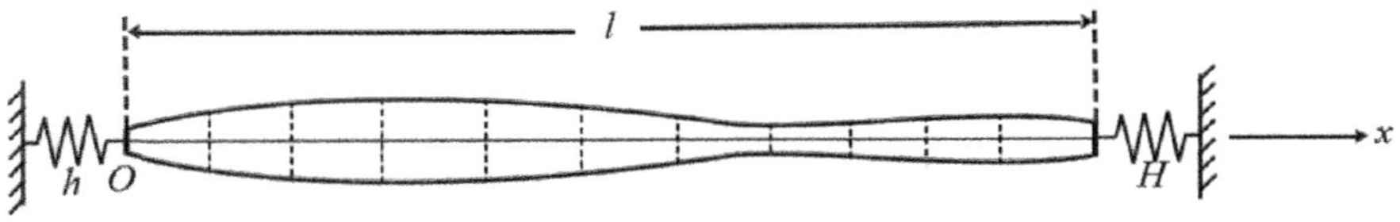

Fig. 3.4 The finite element model of a bar

where

$$\xi = \frac{x - x_{r-1}}{l_r}. \tag{3.5.2}$$

Apparently, the value of ξ ranges from 0 to 1. As r is set to $1, 2, \ldots, n$, Eq. (3.5.1) expresses the displacement at every point of the bar in terms of the $n + 1$ nodal displacements $\{y_r\}_0^n$.

Next, let us establish the finite element model of the bar using the well-known Lagrange's equation, namely,

$$\frac{\mathrm{d}}{\mathrm{d}t}\left(\frac{\partial L}{\partial \dot{q}_r}\right) - \frac{\partial L}{\partial q_r} = 0. \tag{3.5.3}$$

Here, the generalized displacement $q_r = y_r$, and the term related to the generalized force have been dropped, indicating that only free vibration of the bar is considered. In the rest of this section, we are going to discuss finite element models with lumped and distributed masses separately.

3.5.1 Finite Element Model with Lumped Masses

In this subsection, we assume that the mass of the bar has been physically lumped in the finite element nodes. In other words, there exist concentrated masses m_r $(r = 0, 1, \ldots, n)$ at the nodes, and they replace the original distributed mass of the bar. Thus, the kinetic and strain energies of the system become, respectively,

$$T = \frac{1}{2}\sum_{r=0}^{n} m_r \dot{y}_r^2, \tag{3.5.4}$$

$$V = \frac{1}{2}\int_0^l EA\left(\frac{\partial y}{\partial x}\right)^2 \mathrm{d}x, \tag{3.5.5}$$

where *EA* represents the tensile stiffness of the bar.

Substituting Eq. (3.5.1), which expresses the displacement in terms of nodal displacements, into Eq. (3.5.5), we get

$$V = \frac{1}{2}\sum_{r=1}^{n} k_r (y_r - y_{r-1})^2, \tag{3.5.6}$$

where

$$k_r = \frac{1}{l_r}\int_0^1 [EA(x_{r-1} + l_r\xi)]\mathrm{d}\xi. \tag{3.5.7}$$

Since $L = T - V$, we obtain the following equations of motion for the system, after substituting the above expressions of T and V into Lagrange's equation:

$$\begin{cases} m_0\ddot{y}_0 = k_1(y_1 - y_0), \\ m_r\ddot{y}_r = -k_r(y_r - y_{r-1}) + k_{r+1}(y_{r+1} - y_r), \quad r = 1, 2, \ldots, n-1, \\ m_n\ddot{y}_n = -k_n(y_n - y_{n-1}). \end{cases} \tag{3.5.8}$$

After inserting the following harmonic motion

$$y_r(t) = u_r \sin\omega t, r = 0, 1, \ldots, n$$

into Eq. (3.5.8) and eliminating the factor $\sin\omega t$, we get exactly the system of equations shown in (3.1.2).

As for boundary conditions, let us look at the left end of the bar only, because the right end can be dealt with in a similar way. If the bar is fixed at the left end, we obviously should set $y_0 = 0$ or make the corresponding amplitude vanish, i.e., $u_0 = 0$. If the bar is unconstrained at the left end, Eq. (3.5.1) yields $y'(0, t) \neq 0$, which is incorrect. But, this is all right, because invalid function values at a few isolated locations should not have a real impact on the overall results when the energy method is employed. In fact, the system of equations in (3.5.8) happens to be applicable, as is, to the unconstraint boundary. Finally, if the bar is elastically constrained at the left end, it is equivalent to a fixed end with an extra degree of freedom, just like how its finite difference counterpart has been treated.

Thus, we see that the finite element model with lumped masses of a bar is fully covered by our earlier discussion on the spring–mass system. Consequently, it has all the properties presented in Sect. 3.4.

3.5.2 Finite Element Model with Distributed Mass

In this subsection, the strain energy V is still given by Eq. (3.5.6) but the kinetic energy of the system takes a different form as follows:

$$
\begin{aligned}
T &= \frac{1}{2}\int_0^1 \rho A \dot{y}^2 \mathrm{d}x \\
&= \frac{1}{2}\sum_{r=1}^{n}\int_0^1 \rho A(x_{r-1}+l_r\xi)[(1-\xi)\dot{y}_{r-1}+\dot{y}_r\xi]^2 l_r \mathrm{d}\xi \\
&= \frac{1}{2}\sum_{r=1}^{n}(a_r\dot{y}_{r-1}^2 + 2b_r\dot{y}_r\dot{y}_{r-1} + c_r\dot{y}_r^2),
\end{aligned}
\tag{3.5.9}
$$

where

$$
\begin{aligned}
a_r &= \int_0^1 \rho A(x_{r-1}+l_r\xi)(1-\xi)^2 l_r \mathrm{d}\xi, \\
b_r &= \int_0^1 \rho A(x_{r-1}+l_r\xi)(1-\xi)\xi l_r \mathrm{d}\xi, \quad r = 1, 2, \ldots, n. \\
c_r &= \int_0^1 \rho A(x_{r-1}+l_r\xi)\xi^2 l_r \mathrm{d}\xi,
\end{aligned}
\tag{3.5.10}
$$

After substituting Eqs. (3.5.6) and (3.5.9) into Lagrange's equation, we get the discretized equations of motion below:

$$
\begin{cases}
a_1\ddot{y}_0 + b_1\ddot{y}_1 = -k_1(y_0 - y_1), \\
b_r\ddot{y}_{r-1} + (c_r + a_{r+1})\ddot{y}_r + b_{r+1}\ddot{y}_{r+1} \\
\qquad = k_r(y_{r-1} - y_r) - k_{r+1}(y_r - y_{r+1}), \quad r = 1, 2, \ldots, n-1, \\
b_n\ddot{y}_{n-1} + c_n\ddot{y}_n = k_n(y_{n-1} - y_n).
\end{cases}
\tag{3.5.11}
$$

Set $y_r(t) = u_r \sin\omega t$, insert it into the above system of equations, and eliminate the factor $\sin\omega t$. We finally get

$$
\begin{cases}
\lambda(a_1u_0 + b_1u_1) = k_1(u_0 - u_1), \\
\lambda[b_ru_{r-1} + (c_r + a_{r+1})u_r + b_{r+1}u_{r+1}] \\
\qquad = -k_r(u_{r-1} - u_r) + k_{r+1}(u_r - u_{r+1}), \quad r = 1, 2, \ldots, n-1, \\
\lambda(b_nu_{n-1} + c_nu_n) = -k_n(u_{n-1} - u_n).
\end{cases}
\tag{3.5.12}
$$

Rewrite the above system of equations in the matrix form as

$$
\lambda \boldsymbol{A}\boldsymbol{u} = \boldsymbol{C}\boldsymbol{u}, \tag{3.5.13}
$$

where $\lambda = \omega^2$ and $\boldsymbol{u} = (u_0, u_1, \ldots, u_n)^{\mathrm{T}}$ are the eigenvalue and amplitude vector of the displacement, respectively, while mass and stiffness matrices are both tridiagonal and assume the following forms, respectively,

$$\boldsymbol{A} = \begin{bmatrix} a_1 & b_1 & 0 & \cdots & 0 & 0 & 0 \\ b_1 & c_1 + a_2 & b_2 & \cdots & 0 & 0 & 0 \\ \vdots & \vdots & \vdots & & \vdots & \vdots & \vdots \\ 0 & 0 & 0 & \cdots & b_{n-1} & c_{n-1} + a_n & b_n \\ 0 & 0 & 0 & \cdots & 0 & b_n & c_n \end{bmatrix}$$

and

$$\boldsymbol{C} = \begin{bmatrix} k_1 & -k_1 & 0 & \cdots & 0 & 0 & 0 \\ -k_1 & k_1 + k_2 & -k_2 & \cdots & 0 & 0 & 0 \\ \vdots & \vdots & \vdots & & \vdots & \vdots & \vdots \\ 0 & 0 & 0 & \cdots & -k_{n-1} & k_{n-1} + k_n & -k_n \\ 0 & 0 & 0 & \cdots & 0 & -k_n & k_n \end{bmatrix}.$$

It is not difficult to verify that $\boldsymbol{A}$ is nonsingular while $\boldsymbol{C}$ is singular, i.e., $\det \boldsymbol{C} = 0$. However, as long as at least one end of the bar is fixed, the corresponding stiffness and mass matrices will be positive definite Jacobian matrices. Specifically, the stiffness and mass matrices are: $\boldsymbol{A}_1$ and $\boldsymbol{C}_1$, if the left end is fixed; $\boldsymbol{A}_n$ and $\boldsymbol{C}_n$, if the right end is fixed; and $\boldsymbol{A}_{1n}$ and $\boldsymbol{C}_{1n}$, if two ends of the bar are fixed. All of these submatrices of $\boldsymbol{A}$ and $\boldsymbol{C}$ are positive definite Jacobian matrices. Thus, based on discussion in Sect. 2.4, they are oscillatory and sign-oscillatory matrices respectively. Furthermore, from the properties of the oscillatory matrix (Theorem 2.11), $\boldsymbol{C}_1^{-1}$, $\boldsymbol{C}_n^{-1}$, and $\boldsymbol{C}_{1n}^{-1}$ are oscillatory matrices, and consequently, $\boldsymbol{C}_1^{-1}\boldsymbol{A}_1$, $\boldsymbol{C}_n^{-1}\boldsymbol{A}_n$, and $\boldsymbol{C}_{1n}^{-1}\boldsymbol{A}_{1n}$ should also be oscillatory (see Property 5 in Sect. 2.4.1). This indicates that so far as the finite element model of a bar in a fixed–free, free–fixed, or fixed–fixed configuration is concerned, its natural vibration can be covered by the eigenvalue problem involving a single oscillatory coefficient matrix $\boldsymbol{B}$, i.e.,

$$(\lambda^{-1}\boldsymbol{I} - \boldsymbol{B})\boldsymbol{u} = 0. \tag{3.5.14}$$

Here,

$$\boldsymbol{B} = \boldsymbol{C}_1^{-1}\boldsymbol{A}_1,\ \boldsymbol{u} = (u_1, u_2, \ldots, u_n)^{\mathrm{T}}, \tag{3.5.15}$$

for a fixed–free bar; while

$$\boldsymbol{B} = \boldsymbol{C}_{1n}^{-1}\boldsymbol{A}_{1n},\ \boldsymbol{u} = (u_1, u_2, \ldots, u_{n-1})^{\mathrm{T}}, \tag{3.5.16}$$

for a bar fixed at both ends.

It is immediately clear from the theory of oscillatory matrices that when the finite element models of a bar are subject to any of the aforementioned two types of boundary constraints, its natural frequencies and mode shapes must have the four oscillatory properties in vibration, which have been described in Theorems 3.1 and 3.2. Furthermore, for a bar in the fixed–free configuration, we may derive some additional properties.

As in Sect. 3.3, introduce the matrix $\widetilde{\boldsymbol{E}}$

$$\widetilde{\boldsymbol{E}} = \begin{pmatrix} 1 & -1 & 0 & & \\ 0 & 1 & -1 & & 0 \\ & & \ddots & \ddots & \\ & 0 & & 1 & -1 \\ & & & & 1 \end{pmatrix},$$

Then, the modal equation of a fixed–free bar can be reformatted into

$$\lambda \boldsymbol{A}_1 \boldsymbol{u} = \widetilde{\boldsymbol{E}} \underline{\boldsymbol{K}} \widetilde{\boldsymbol{E}}^{\mathrm{T}} \boldsymbol{u}, \tag{3.5.17}$$

where $\underline{\boldsymbol{K}} = \mathrm{diag}(k_1, k_2, \ldots, k_n)$ and $\boldsymbol{u} = (u_1, u_2, \ldots, u_n)^{\mathrm{T}}$. Set

$$\boldsymbol{w} = (w_1, w_2, \ldots w_n)^{\mathrm{T}} = -\widetilde{\boldsymbol{E}}^{\mathrm{T}} \boldsymbol{u}. \tag{3.5.18}$$

Thus, Eq. (3.5.17) is transformed to

$$\underline{\boldsymbol{K}}^{-1} \widetilde{\boldsymbol{E}}^{-1} \boldsymbol{A}_1 (\widetilde{\boldsymbol{E}}^{-1})^{\mathrm{T}} \boldsymbol{w} = \lambda^{-1} \boldsymbol{w}, \tag{3.5.19}$$

where

$$\widetilde{\boldsymbol{E}}^{-1} = \begin{bmatrix} 1 & \cdots & 1 \\ & \ddots & \vdots \\ 0 & & 1 \end{bmatrix}.$$

Obviously, $\widetilde{E}$ is a totally nonnegative matrix and $\det \widetilde{\boldsymbol{E}}^{-1} = 1$. Hence, $\underline{\boldsymbol{K}}^{-1} \widetilde{\boldsymbol{E}}^{-1} \boldsymbol{A}_1 (\widetilde{\boldsymbol{E}}^{-1})^{\mathrm{T}}$ is an oscillatory matrix, and we may draw the conclusions below:

(1) The mode shape of the relative displacement $\boldsymbol{w}^{(i)}$ corresponding to the i-th natural frequency f_i has exactly $i - 1$ sign reversals, where $i = 1, 2, \ldots, n$.

Starting from results related to numbers of sign reversals, we may conclude that qualitative properties of displacement mode shapes, described in Corollaries 1 to 7 of Sect. 3.3.1, remain valid for the finite element model of a bar in the fixed–free configuration. Specifically, the finite element model of a bar in the fixed-free configuration has the following properties:

(2) $u_{1i}w_{1i} > 0$ and $u_{ni}w_{ni} > 0$, where $i = 1, 2, \ldots, n$.
(3) Components of the displacement mode shape $\boldsymbol{u}^{(i)}$ can be divided into i contiguous segments as

$$u_{\alpha_r i}, u_{\alpha_r+1, i}, \ldots, u_{\beta_r-1, i}, u_{\beta_r i}, r = 1, 2, \ldots, i,$$

and within each of these segments the members share a common sign, i.e.,

$$u_{\alpha_r i}u_{\alpha_{r+1} i} < 0, \quad u_{\alpha_r i}u_{ji} > 0, \quad j = \alpha_r + 1, \alpha_r + 2, \ldots, \beta_r,$$

where $r = 1, 2, \ldots, i-1$, $\alpha_1 = 1$, and $\beta_i = n$. Furthermore, in each segment, the displacement mode shape has one and only one extremum, which is a local maximum when all members are positive and a local minimum when all of them are negative. In between two adjacent extrema, the u-line is either monotonically increasing or decreasing, and thus has one and only one node.
(4) If we set $u_{ni} > 0\,(i = 1, 2, \ldots, n)$, then components of the first displacement mode shape are all positive and monotonically increasing. In addition, the last segment of any of its u-line must be monotonically increasing and concave upward.
(5) Nodes of two successive mode shapes of the relative displacements $\boldsymbol{w}^{(i)}$ and $\boldsymbol{w}^{(i+1)}$ $(i = 2, 3, \ldots, n-1)$ interlace.

The content of this section is primarily taken from Refs. [1, 2].

3.6 Modal Qualitative Properties of the System Consisting of a Massless Elastic Bar and a Number of Concentrated Masses

There is another type of physical discrete models for a vibrating bar of length l. Select a series of points $\{x_r\}_1^n$ on the bar that satisfy $0 = x_0 < x_1 < \ldots < x_{n-1} < x_n = l$, and let the length of the r-th segment be $l_r = x_r - x_{r-1}\,(r = 1, 2, \ldots, n)$. Lump the distributed mass of the bar in a few concentrated masses at the nodes using Eq. (3.1.1), but still keep the stiffness of the bar distributed. Then, we get a system consisting of a massless elastic bar and a number of concentrated masses, as shown in Fig. 3.5.

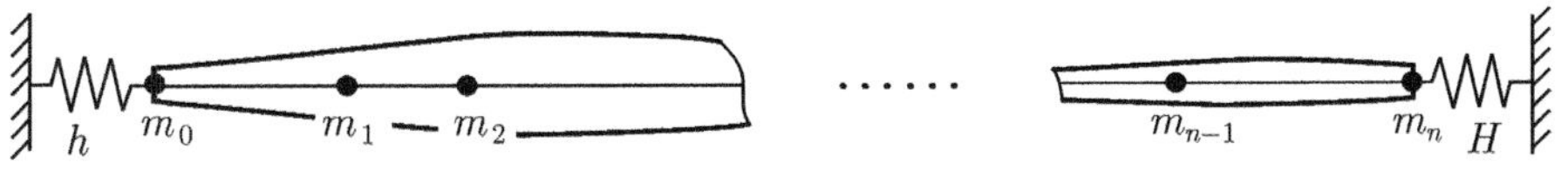

Fig. 3.5 A system consisting of a massless elastic bar and several concentrated masses

Suppose the flexibility coefficients of such a system are r_{ij} $(i, j = 0, 1, \ldots, n)$. The physical interpretation of r_{ij} is that it represents the displacement at x_i due to a unit force applied to the bar at x_j. Denote $\boldsymbol{R} = (r_{ij})$ $(i, j = 0, 1, \ldots, n)$ as the flexibility matrix. The governing equation for the free vibration of this system is

$$\boldsymbol{y}(t) = \boldsymbol{R}(-\boldsymbol{M}\ddot{\boldsymbol{y}}(t)),$$

where $\boldsymbol{y}\,(t) = (y_0(t), y_1(t), \ldots, y_n(t))^{\mathrm{T}}$ is the displacement vector while $\boldsymbol{M} = \operatorname{diag}(m_0, m_1, \ldots, m_n)$ represents the mass matrix. To obtain the natural angular frequency $\omega = \sqrt{\lambda}$ and the corresponding mode shape $\boldsymbol{u} = (u_0, u_1, \ldots, u_n)^{\mathrm{T}}$, set $y_r(t) = u_r \sin \omega t$ and substitute it into the governing equation of free vibration. This yields

$$\boldsymbol{u} = \lambda \boldsymbol{R}\boldsymbol{M}\,\boldsymbol{u}. \tag{3.6.1}$$

We need to derive the flexibility matrix $\boldsymbol{R}$ in order to study the qualitative properties of the system consisting of a massless elastic bar and some concentrated masses. For the continuous system of a bar, we will derive its Green's function in Chap. 5, which is of the form

$$G(x, s) = \begin{cases} \varphi(x)\psi(s), & x \le s, \\ \varphi(s)\psi(x), & x > s. \end{cases}$$

We are also going to prove there that functions $\varphi(x)$ and $\psi(x)$ have the following properties (refer to Theorem 5.3 in Chap. 5): (1) $\varphi(x)$ and $\psi(x)$ are of strictly fixed sign on the interval I, (2) $\varphi(x)/\psi(x)$ is strictly increasing on the interval I, and (3) $\varphi(x)\psi(x) > 0$ for any $x \in I$. Thus, without loss of generality, we may assume $\varphi(x) > 0$ and $\psi(x) > 0$ $(x \in I)$. Here, I refers to the set consisting of all movable points inside the closed interval $[0, l]$, i.e.,

$$I = \begin{cases} [0, l] \text{ if both } h \text{ and } H \text{ are of finite value}, \\ [0, l) \text{ if } h \text{ is of finite value while } H \to \infty, \\ (0, l] \text{ if } h \to \infty \text{ while } H \text{ is of finite value}, \\ (0, l) \text{ if both h and H approach to } \infty. \end{cases}$$

For the sake of convenience, let us use abbreviations $\varphi(x_i) = \varphi_i$ and $\psi(x_k) = \psi_k$. Then, the flexibility coefficients of the current discrete model can be expressed as

$$r_{ik} = G(x_i, x_k) = \begin{cases} \varphi(x_i)\psi(x_k), \ x_i \le x_k \\ \varphi(x_k)\psi(x_i), \ x_i > x_k \end{cases} = \begin{cases} \varphi_i \psi_k, \ i \le k, \\ \varphi_k \psi_i, \ i > k, \end{cases} \tag{3.6.2}$$

where

$$\varphi_0 \ge 0, \ \psi_n \ge 0, \ \varphi_i \ \text{ and } \ \psi_i > 0, \quad i = 1, 2, \cdots, n-1. \tag{3.6.3}$$

Notice that the first inequality in (3.6.3) becomes an equality only when $h \to \infty$, while the second one changes to an equality only if $H \to \infty$.

Thus, for the system consisting of a massless elastic bar and several concentrated masses as well as subject to fixed–fixed, fixed–free, and elastic–elastic ($0 < h,\ H < \infty$) constraints, we may establish, respectively, the following inequalities:

$$\frac{\varphi_1}{\psi_1} < \frac{\varphi_2}{\psi_2} < \cdots < \frac{\varphi_{n-1}}{\psi_{n-1}}, \frac{\varphi_1}{\psi_1} < \frac{\varphi_2}{\psi_2} < \cdots < \frac{\varphi_n}{\psi_n}, \text{ and } \frac{\varphi_0}{\psi_0} < \frac{\varphi_1}{\psi_1} < \cdots < \frac{\varphi_n}{\psi_n}.$$

According to Example 1 in Sect. 2.4, their respective flexibility matrices, specifically,

$$\begin{aligned} \boldsymbol{R}^{\mathrm{cc}} &= (r_{ij})_{(n-1)\times(n-1)}, & i, j &= 1, 2, \ldots, n-1, \\ \boldsymbol{R}^{\mathrm{cf}} &= (r_{ij})_{n\times n}, & i, j &= 1, 2, \ldots, n, \\ \boldsymbol{R}^{\mathrm{tt}} &= (r_{ij})_{(n+1)\times(n+1)}, & i, j &= 0, 1, \ldots, n, \end{aligned}$$

are all oscillatory matrices, where the suffix tt indicates the system is elastically constrained at both ends. Furthermore, it is known that the product of any of these matrices and a positive definite diagonal matrix remains to be oscillatory. As a result, we can establish the next theorem based on properties of eigenpairs of an oscillatory matrix.

Theorem 3.8 *Consider a system comprised of a massless elastic bar and several concentrated masses as shown in Fig.* 3.5. *When it is subject to fixed–fixed, fixed–free, or elastic–elastic* ($0 < h,\ H < \infty$) *constraints, natural frequencies and mode shapes of the system have the four oscillatory properties* in *vibration described in Theorems* 3.1 *and* 3.2.

3.7 Modal Qualitative Properties of Discrete Systems of Strings and Bars on Elastic Foundations

Systems given in Sect. 3.1 of strings, bars, and shafts vibrating on elastic foundations are, in fact, the most general form of the Sturm–Liouville system. Their physical discrete system is depicted in Fig. 3.3. Natural frequencies and mode shapes of this discrete system satisfy Eqs. (3.1.3), (3.1.4), and (3.1.7).

In comparison, natural frequencies and mode shapes of finite difference model of the aforementioned structural systems satisfies either Eqs. (3.1.3), (3.1.4), and (3.1.7) or the following system of equations:

$$\begin{cases} (a_0 + q_0)u_0 - b_1 u_1 = \lambda m_0 u_0, \\ -c_r u_{r-1} + (a_r + q_r)u_r - b_{r+1}u_{r+1} = \lambda m_r u_r, \quad r = 1, 2, \ldots, n-1, \\ -c_n u_{n-1} + (a_n + q_n)u_n = \lambda m_n u_n. \end{cases} \tag{3.7.1}$$

Thus, the physical discrete system of strings, bars, and shafts vibrating on an elastic foundation remains to be a positive definite standard Jacobian system, while the corresponding finite difference system can be a positive definite and either symmetric or nonsymmetric standard Jacobian system. Therefore, the introduction of an elastic foundation should not alter the oscillatory properties of the system. In other words, the qualitative properties established in previous sections for the spring–mass system and finite difference model of bars are readily applicable to both the physical discrete system and finite difference model of strings, bars, and shafts on elastic foundations.

The content of this section comes primarily from Refs. [4, 7].

References

1. Gladwell GML (2004) Inverse Problems in Vibration. 2nd edn. Springer, Dordrecht (1986, 1st edn, Martinus Nijhoff Publishers, Dordrecht)
2. Gladwell GML (1991) Qualitative properties of finite element models I: Sturm-Liouville systems. Q J Mech Appl Math 44(2):249–265
3. Golub GH, Boley D (1977) Inverse Eigenvalue Problems for Band Matrices. In: Watson GA (ed) Numerical analysis. Springer, Heidelberg, New York, pp 23–31
4. Tian X, Dai H (2007) Inverse vibration problem for the discrete system of a rod. J Shandong Inst Light Ind (Nat Sci Ed) 21(1):4–7 (in Chinese)
5. Wang QS, Wang DJ (1987) Construction of the discrete system for the rod by partial natural modes and frequencies data. J Vib Eng 1:83–87 (in Chinese)
6. Wang QS, Wang DJ, He BC (1992) Qualitative properties of frequencies and modes of discrete system of continuous second-order systems. J Vib Shock 11(3):7–12 (in Chinese)
7. Wang QS, Wang DJ (1997) Qualitative properties of frequencies and modes and inverse modes problems of discrete systems for a rod on the elastic bases. J. AQTC (Nat Sci Ed) 3(2):19–25 (in Chinese)
8. Wang QS, Wang DJ (2014) Supplementary proof of some oscillation property for discrete systems of rod and beam having rigid modes. Chin Q Mech 35(2):262–269 (in Chinese)

Chapter 4
Qualitative Properties of Vibration and Static Deformation Associated with Discrete Systems of Beams

The focus of the first six sections of this chapter is on qualitative properties of the finite difference model or the related physical model of a beam. We will set up the governing equations of motion and boundary conditions associated with the finite difference system; derive various modal qualitative properties of the discrete model under different boundary constraints, by applying the theory of oscillatory matrices and the concept of conjugate beams; and establish qualitative properties in static deformation of the finite difference system of a well-constrained beam. In the remaining sections of this chapter, we are going to introduce modal qualitative properties of two types of finite element systems of a beam, and discuss the modal qualitative properties of the discrete models of multi-span beams, in particular, beams with overhangs.

4.1 The Finite Difference Model of a Beam and the Corresponding Physical Model

Consider a beam with length l, linear density $\rho(x)$, and section bending stiffness $EJ(x)$. Its natural frequency $f = \omega/2\pi$ and mode shape $u(x)$ satisfy the modal equation

$$[EJu''(x)]'' = \omega^2\rho(x)u(x), \quad 0 < x < l, \tag{4.1.1}$$

as well as the following general form of boundary conditions

$$[EJu''(x)]'|_{x=0} + h_1u(0) = 0 = [EJu''(x)]'|_{x=l} - h_2u(l), \tag{4.1.2}$$

$$E(0)J(0)u''(0) - \beta_1u'(0) = 0 = E(l)J(l)u''(l) + \beta_2u'(l), \tag{4.1.3}$$

D. Wang et al., *Qualitative Theory in Structural Mechanics*,
https://doi.org/10.1007/978-981-13-1376-9_4

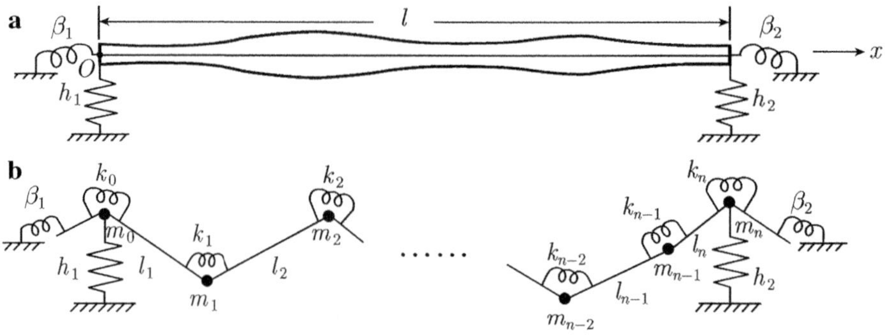

Fig. 4.1 **a** The continuous system of a beam; **b** Its physical discrete model consisting of springs, masses, and rigid bars

where h_1, h_2, β_1, and β_2 represent spring constants of the elastic supports and are thus all positive numbers. The beam is shown in Fig. 4.1a.

Gladwell introduced a physical discrete model for the beam that consists of springs, masses, and rigid bars (see Refs. [2, 3]). The model not only has a clear physical meaning but also is of practical significance. Its parameters can be determined from those of the corresponding continuous system with a procedure outlined in the next paragraph.

Take the centerline of the beam as the x-axis, and apply the partition $0 = x_0 < x_1 < \cdots < x_n = l$ to the interval $[0, l]$. Set $l_r = x_r - x_{r-1}$ $(r = 1, 2, \ldots, n)$, define $\rho_r = \rho(x_r)$ and $s_r = EJ(x_r)$ $(r = 0, 1, \ldots, n)$, and convert the distributed mass and stiffness into lumped parameters using the formulae below:

$$\begin{cases} m_0 = \dfrac{\rho_0 l_1}{2}, \ m_r = \rho_r \dfrac{l_r + l_{r+1}}{2} \quad (r = 1, 2, \ldots, n-1), \ m_n = \dfrac{\rho_n l_n}{2}; \\ k_0 = \dfrac{2s_0}{l_1}, \quad k_r = \dfrac{2s_r}{l_r + l_{r+1}} \quad (r = 1, 2, \ldots, n-1), \quad k_n = \dfrac{2s_n}{l_n}. \end{cases} \tag{4.1.4}$$

After these steps, the continuous system of the beam is discretized into the form shown in Fig. 4.1b. Here, the massless and rigid bars are hinged at the ends and constrained with rotational springs. There are also concentrated masses at the joint of two adjacent bars. In the figure, m_r $(r = 0, 1, \ldots, n)$ stand for the mass of concentrated masses, k_r $(r = 0, 1, \ldots, n)$ represent the stiffness of rotational springs that control relative rotation between two adjacent bars, and l_r $(r = 1, 2, \ldots, n)$ denote the length of massless rigid bars.

It is interesting to note that this physical discrete model is identical to the finite difference model of the beam. We will demonstrate this in greater detail in the next a few paragraphs. Also see Ref. [13].

Let us use the notation of Chap. 3. Consider the Taylor series expansion below for the function $u(x)$:

$$
\begin{aligned}
u_{r+1} &= u_r + u_r' l_{r+1} + u_r'' l_{r+1}^2/2 + O(l_{r+1}^3), \quad & r = 0, 1, \ldots, n-1, \\
u_{r-1} &= u_r - u_r' l_r + u_r'' l_r^2/2 + O(l_r^3), \quad & r = 1, 2, \ldots, n.
\end{aligned} \tag{4.1.5}
$$

The following second-order central finite difference formula for the second derivative was derived in Chap. 3:

$$
u_r'' = \frac{2}{l_r + l_{r+1}} \left(\frac{u_{r-1} - u_r}{l_r} - \frac{u_r - u_{r+1}}{l_{r+1}} \right). \tag{4.1.6}
$$

Substituting the above equation into Eq. (4.1.1), we obtain the finite difference equation of the beam at the r-th grid point in the interior domain:

$$
\begin{aligned}
\omega^2 \rho_r u_r = \frac{4}{l_r + l_{r+1}} \Bigg\{ & \frac{s_{r-1}}{l_{r-1} l_r (l_{r-1} + l_r)} u_{r-2} - \frac{1}{l_r^2} \left(\frac{s_{r-1}}{l_{r-1}} + \frac{s_r}{l_{r+1}} \right) u_{r-1} \\
& + \left[\frac{s_{r-1}}{l_r^2 (l_{r-1} + l_r)} + \frac{s_r}{l_r l_{r+1}} \left(\frac{1}{l_r} + \frac{1}{l_{r+1}} \right) + \frac{s_{r+1}}{l_{r+1}^2 (l_{r+1} + l_{r+2})} \right] u_r \\
& - \frac{1}{l_{r+1}^2} \left(\frac{s_r}{l_r} + \frac{s_{r+1}}{l_{r+2}} \right) u_{r+1} + \frac{s_{r+1}}{l_{r+1} l_{r+2} (l_{r+1} + l_{r+2})} u_{r+2} \Bigg\}, \quad r = 2, 3, \ldots, n-2.
\end{aligned} \tag{4.1.7}
$$

Next, we need to derive equations associated with finite difference grid points at the boundary. Here, the finite difference formula (4.1.6) is not always applicable. By combining the first equation in (4.1.5), Eq. (4.1.6), and boundary conditions at the left end of the beam given in Eqs. (4.1.2) and (4.1.3), we may obtain

$$
\omega^2 \rho_0 u_0 = \frac{4}{l_1^2} \left[\left(\frac{s_0}{l_1 + \bar{\beta}_1} + \frac{s_1}{l_1 + l_2} \right) \frac{u_0}{l_1} - \left(\frac{s_0}{l_1 + \bar{\beta}_1} + \frac{s_1}{l_2} \right) \frac{u_1}{l_1} + \frac{s_1}{l_1 + l_2} \frac{u_2}{l_2} \right] + \frac{2\phi_0}{l_1}, \tag{4.1.8}
$$

$$
\begin{aligned}
\omega^2 \rho_1 u_1 = \frac{4}{l_1 + l_2} \Bigg\{ & - \left(\frac{s_0}{l_1 + \bar{\beta}_1} + \frac{s_1}{l_2} \right) \frac{u_0}{l_1^2} + \left[\frac{s_0}{l_1^2 (l_1 + \bar{\beta}_1)} + \frac{s_1}{l_1 l_2} \left(\frac{1}{l_1} + \frac{1}{l_2} \right) \right. \\
& \left. + \frac{s_2}{l_2^2 (l_2 + l_3)} \right] u_1 - \left(\frac{s_1}{l_1} + \frac{s_2}{l_3} \right) \frac{u_2}{l_2^2} + \frac{s_2}{l_2 + l_3} \frac{u_3}{l_2 l_3} \Bigg\},
\end{aligned} \tag{4.1.9}
$$

where $\phi_0 = h_1 u_0$ and $\bar{\beta}_1 = 2s_0/\beta_1$. Likewise, by combining the second equation in (4.1.5), Eq. (4.1.6), as well as boundary conditions at the right end given in Eqs. (4.1.2) and (4.1.3), we get

$$
\begin{aligned}
\omega^2 \rho_{n-1} u_{n-1} = \frac{4}{l_{n-1} + l_n} \Bigg\{ & \frac{s_{n-2}}{l_{n-2} + l_{n-1}} \frac{u_{n-3}}{l_{n-2} l_{n-1}} - \left(\frac{s_{n-2}}{l_{n-2}} + \frac{s_{n-1}}{l_n} \right) \frac{u_{n-2}}{l_{n-1}^2} \\
& + \left[\frac{s_{n-2}}{l_{n-1}^2 (l_{n-2} + l_{n-1})} + \frac{s_{n-1}}{l_{n-1} l_n} \left(\frac{1}{l_{n-1}} + \frac{1}{l_n} \right) + \frac{s_n}{l_n^2 (l_n + \bar{\beta}_2)} \right] u_{n-1} \\
& - \left(\frac{s_{n-1}}{l_{n-1}} + \frac{s_n}{l_n + \bar{\beta}_2} \right) \frac{u_n}{l_n^2} \Bigg\},
\end{aligned} \tag{4.1.10}
$$

$$\omega^2 \rho_n u_n = -\frac{2\phi_{n+1}}{l_n} + \frac{4}{l_n^2}\left[\frac{s_{n-1}}{l_{n-1}+l_n}\frac{u_{n-2}}{l_{n-1}} - \left(\frac{s_{n-1}}{l_{n-1}} + \frac{s_n}{l_n+\bar{\beta}_2}\right)\frac{u_{n-1}}{l_n} + \left(\frac{s_{n-1}}{l_{n-1}+l_n} + \frac{s_n}{l_n+\bar{\beta}_2}\right)\frac{u_n}{l_n}\right], \tag{4.1.11}$$

where $\phi_{n+1} = -h_2 u_n$ and $\bar{\beta}_2 = 2s_n/\beta_2$. Thus, Eqs. (4.1.7)–(4.1.11) define a full set of modal equations of a beam in the finite difference form.

Set $\lambda = \omega^2$, define parameters m_r and k_r $(r = 0, 1, \ldots, n)$ by Eq. (4.1.4), and apply the transformation

$$k_0^* = \frac{k_0}{1 + 2s_0/\beta_1 l_1}, \quad k_n^* = \frac{k_n}{1 + 2s_n/\beta_2 l_n}. \tag{4.1.12}$$

We may then rewrite Eqs. (4.1.7)–(4.1.11) into the following form:

$$\lambda m_0 u_0 = \frac{k_0^* + k_1}{l_1^2}u_0 - \left[\frac{k_0^*}{l_1^2} + \frac{k_1}{l_1}\left(\frac{1}{l_1} + \frac{1}{l_2}\right)\right]u_1 + \frac{k_1}{l_1 l_2}u_2 + \phi_0, \tag{4.1.13}$$

$$\lambda m_1 u_1 = -\left[\frac{k_0^*}{l_1^2} + \frac{k_1}{l_1}\left(\frac{1}{l_1} + \frac{1}{l_2}\right)\right]u_0 + \left[\frac{k_0^*}{l_1^2} + \left(\frac{1}{l_1} + \frac{1}{l_2}\right)^2 k_1 + \frac{k_2}{l_2^2}\right]u_1 - \left[\frac{k_1}{l_2}\left(\frac{1}{l_1} + \frac{1}{l_2}\right) + \frac{k_2}{l_2}\left(\frac{1}{l_2} + \frac{1}{l_3}\right)\right]u_2 + \frac{k_2}{l_2 l_3}u_3, \tag{4.1.14}$$

$$\lambda m_r u_r = \frac{k_{r-1}}{l_{r-1}l_r}u_{r-2} - \left[\left(\frac{1}{l_{r-1}} + \frac{1}{l_r}\right)\frac{k_{r-1}}{l_r} + \left(\frac{1}{l_r} + \frac{1}{l_{r+1}}\right)\frac{k_r}{l_r}\right]u_{r-1} + \left[\frac{k_{r-1}}{l_r^2} + \left(\frac{1}{l_r} + \frac{1}{l_{r+1}}\right)^2 k_r + \frac{k_{r+1}}{l_{r+1}^2}\right]u_r - \left[\left(\frac{1}{l_r} + \frac{1}{l_{r+1}}\right)\frac{k_r}{l_{r+1}} + \left(\frac{1}{l_{r+1}} + \frac{1}{l_{r+2}}\right)\frac{k_{r+1}}{l_{r+1}}\right]u_{r+1} + \frac{k_{r+1}}{l_{r+1}l_{r+2}}u_{r+2}, \quad r = 2, 3, \ldots, n-2, \tag{4.1.15}$$

$$\lambda m_{n-1} u_{n-1} = \frac{k_{n-2}}{l_{n-2}l_{n-1}}u_{n-3} - \left[\left(\frac{1}{l_{n-2}} + \frac{1}{l_{n-1}}\right)\frac{k_{n-2}}{l_{n-1}} + \left(\frac{1}{l_{n-1}} + \frac{1}{l_n}\right)\frac{k_{n-1}}{l_{n-1}}\right]u_{n-2} + \left[\frac{k_{n-2}}{l_{n-1}^2} + \left(\frac{1}{l_{n-1}} + \frac{1}{l_n}\right)^2 k_{n-1} + \frac{k_n^*}{l_n^2}\right]u_{n-1} - \left[\left(\frac{1}{l_{n-1}} + \frac{1}{l_n}\right)\frac{k_{n-1}}{l_n} + \frac{k_n^*}{l_n^2}\right]u_n, \tag{4.1.16}$$

$$\lambda m_n u_n = \frac{k_{n-1}}{l_{n-1}l_n}u_{n-2} - \left[\left(\frac{1}{l_{n-1}} + \frac{1}{l_n}\right)\frac{k_{n-1}}{l_n} + \frac{k_n^*}{l_n^2}\right]u_{n-1} + \frac{k_{n-1} + k_n^*}{l_n^2}u_n - \phi_{n+1}. \tag{4.1.17}$$

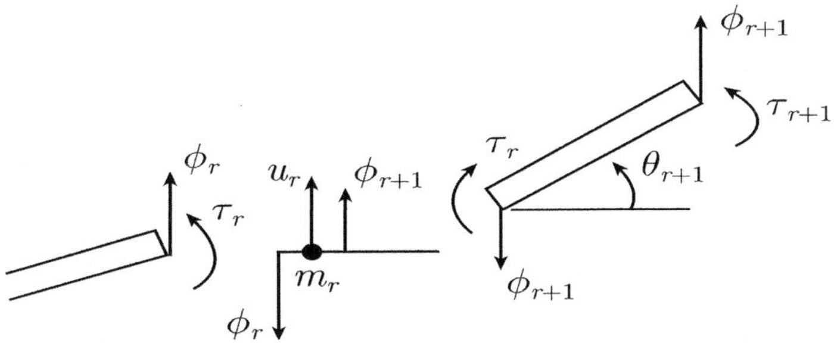

Fig. 4.2 Free-body diagram of the mass m_r

It is not difficult to verify that Eqs. (4.1.13)–(4.1.17) are none other than the equations of motion of the system depicted in Fig. 4.1b, which consists of springs, masses, and rigid bars.

In fact, let us look at the free body diagram in Fig. 4.2 of the r-th mass m_r of the system depicted in Fig. 4.1b. Below is the modal equation associated with the r-th mass after the time variable is separated:

$$\lambda m_r u_r = \phi_r - \phi_{r+1}, \quad r = 0, 1, \ldots, n, \tag{4.1.18}$$

where ϕ_r and ϕ_{r+1} represent shear forces applied to the mass m_r by its two adjacent rigid bars. Next, we get, from the equilibrium condition on the rigid bar connecting masses m_{r-1} and m_r,

$$\phi_r = (\tau_{r-1} - \tau_r)/l_r, \quad r = 1, 2, \ldots, n, \tag{4.1.19}$$

where τ_r denotes the moment generated by the rotational spring k_r and the relative rotation between two rigid bars connected to the mass m_r, i.e.,

$$\tau_r = k_r(\theta_{r+1} - \theta_r) = k_r w_r, \quad r = 0, 1, \ldots, n. \tag{4.1.20}$$

Furthermore, the angle of rotation of the rigid bar joining masses m_{r-1} and m_r satisfies

$$\theta_r = (u_r - u_{r-1})/l_r, \quad r = 1, 2, \ldots, n. \tag{4.1.21}$$

For all internal masses of the model shown in Fig. 4.1b, we can derive Eq. (4.1.15) by substituting Eqs. (4.1.19)–(4.1.21) into Eq. (4.1.18). As for the mass at the left end, we may obtain Eqs. (4.1.13) and (4.1.14) by introducing, on the left side of the system, the rotational angle θ_0 of an imaginary bar and the related reaction force ϕ_0 in the transverse direction. By the same token, we may establish Eqs. (4.1.16) and (4.1.17) by introducing the rotational angle θ_{n+1} and the reaction force ϕ_{n+1} on the right side of the system. This confirms that the system illustrated in Fig. 4.1b is indeed a physical model of the beam discretized with the finite difference method.

With the notation defined above, boundary conditions for various types of constraints may be expressed as:

(1) *Clamped* $u_0 = \theta_0 = 0$ or $u_n = \theta_{n+1} = 0$.
(2) *Pinned* $u_0 = \beta_1 = 0$ or $u_n = \beta_2 = 0$.
(3) *Sliding* $\theta_0 = \phi_0 = 0$ or $\theta_{n+1} = \phi_{n+1} = 0$.
(4) *Free* $\beta_1 = \phi_0 = 0$ or $\beta_2 = \phi_{n+1} = 0$.
(5) *Anti-resonant* (only anti-resonant conditions at the right end are given, as an example) $u_n = \phi_{n+1} = 0$ or $\theta_{n+1} = \beta_2 = 0$.

The beam model, which consists of springs, masses, and rigid bars, as well as its associated system of equations (4.1.13)–(4.1.17) is of simplistic form and has clear physical meaning. In addition, by starting from this model, we can take advantage of existing results easily. For these reasons, our discussion in the remaining part of this chapter will be based on the physical model.

Since Eqs. (4.1.13)–(4.1.17) occupy a large space, we will rewrite Eqs. (4.1.13)–(4.1.21) into a more compact vector form. This makes it easier to apply the theory of oscillatory matrices. Let us represent the displacement, angle of rotation, bending moment, and shear force of the spring–mass–rigid bar system as, respectively,

$$\begin{aligned} \boldsymbol{u} &= (u_0, u_1, \ldots, u_n)^{\mathrm{T}}, \quad \boldsymbol{\theta} = (\theta_1, \theta_2, \ldots, \theta_n)^{\mathrm{T}}, \\ \boldsymbol{\tau} &= (\tau_0, \tau_1, \ldots, \tau_n)^{\mathrm{T}}, \quad \boldsymbol{\phi} = (\phi_1, \phi_2, \ldots, \phi_n)^{\mathrm{T}}. \end{aligned} \tag{4.1.22}$$

Denote the mass matrix as $\boldsymbol{M} = \mathrm{diag}(m_0, m_1, \ldots, m_n)$ and the matrix of spring constants as $\boldsymbol{K} = \mathrm{diag}(k_0, k_1, \ldots, k_n)$, and set $\boldsymbol{L} = \mathrm{diag}(l_1, l_2, \ldots, l_n)$. Next, define the $n \times (n+1)$ matrix $\boldsymbol{E}$, the square matrix $\widetilde{\boldsymbol{E}}$ of order $n+1$, the $(n+1)$-dimensional vectors $\boldsymbol{e}^{(1)}$, and the $(n+1)$-dimensional vectors $\boldsymbol{e}^{(n+1)}$ as

$$\boldsymbol{E} = \begin{bmatrix} 1 & -1 & & & \\ & 1 & -1 & & 0 \\ & & \ddots & \ddots & \\ 0 & & & & \\ & & & 1 & -1 \end{bmatrix}, \quad \widetilde{\boldsymbol{E}} = \begin{bmatrix} 1 & -1 & & 0 \\ & \ddots & \ddots & \\ & & 1 & -1 \\ 0 & & & 1 \end{bmatrix}, \tag{4.1.23}$$

$$\boldsymbol{e}^{(1)} = (1, 0, \ldots, 0)^{\mathrm{T}}, \text{ and } \boldsymbol{e}^{(n+1)} = (0, \ldots, 0, 1)^{\mathrm{T}},$$

respectively. Then, Eqs. (4.1.19)–(4.1.21) become

$$\boldsymbol{\theta} = -\boldsymbol{L}^{-1}\boldsymbol{E}\boldsymbol{u}, \boldsymbol{w} = \boldsymbol{E}^{\mathrm{T}}\boldsymbol{\theta} - \theta_0\boldsymbol{e}^{(1)} + \theta_{n+1}\boldsymbol{e}^{(n+1)}, \boldsymbol{\tau} = \boldsymbol{K}\boldsymbol{w}, \boldsymbol{\phi} = \boldsymbol{L}^{-1}\boldsymbol{E}\boldsymbol{\tau}. \tag{4.1.24}$$

Furthermore, Eq. (4.1.18) may be written as

$$\begin{aligned}\lambda \boldsymbol{M}\boldsymbol{u} &= -\boldsymbol{E}^{\mathrm{T}}\boldsymbol{\phi} + \phi_0 \boldsymbol{e}^{(1)} - \phi_{n+1}\boldsymbol{e}^{(n+1)} \\ &= \boldsymbol{A}\boldsymbol{u} + \boldsymbol{E}^{\mathrm{T}}\boldsymbol{L}^{-1}\boldsymbol{E}(k_0\theta_0\boldsymbol{e}^{(1)} - k_n\theta_{n+1}\boldsymbol{e}^{(n+1)}) \\ &\quad + h_1 u_0 \boldsymbol{e}^{(1)} + h_2 u_n \boldsymbol{e}^{(n+1)},\end{aligned} \tag{4.1.25}$$

where

$$\boldsymbol{A} = \boldsymbol{E}^{\mathrm{T}}\boldsymbol{L}^{-1}\boldsymbol{E}\boldsymbol{K}\boldsymbol{E}^{\mathrm{T}}\boldsymbol{L}^{-1}\boldsymbol{E}. \tag{4.1.26}$$

From boundary conditions in (4.1.3), we get

$$\theta_0 = \frac{EJ(0)u''(0)}{\beta_1} = \frac{\bar{\beta}_1}{l_1}\frac{u_1 - u_0}{l_1 + \bar{\beta}_1} \text{ and } \theta_{n+1} = \frac{EJ(l)u''(l)}{\beta_2} = \frac{\bar{\beta}_2}{l_n}\frac{u_{n-1} - u_n}{l_n + \bar{\beta}_2}.$$

By substituting the above results into Eq. (4.1.25) and eliminating the second term at the right-hand side of the second equal sign, we may further simplify the modal equation of natural vibration into the following form:

$$\lambda \boldsymbol{M}\boldsymbol{u} = \underline{\boldsymbol{A}}\boldsymbol{u} + h_1 u_0 \boldsymbol{e}^{(1)} + h_2 u_n \boldsymbol{e}^{(n+1)}, \tag{4.1.27}$$

where the matrices $\boldsymbol{A}$ and $\underline{\boldsymbol{A}}$ differ only in that k_0 and k_n of the former have been replaced with k_0^* and k_n^* of the latter, responsively. Thus,

$$\underline{\boldsymbol{A}} = \boldsymbol{E}^{\mathrm{T}}\boldsymbol{L}^{-1}\boldsymbol{E}\underline{\boldsymbol{K}}\boldsymbol{E}^{\mathrm{T}}\boldsymbol{L}^{-1}\boldsymbol{E}, \tag{4.1.28}$$

where $\underline{\boldsymbol{K}} = \mathrm{diag}(k_0^*, k_1, \ldots, k_{n-1}, k_n^*)$. We will start from Eq. (4.1.27) to study the oscillatory properties of the discrete beam system in subsequent sections.

4.2 Qualitative Properties of Vibration and Static Deformation Associated with Finite Difference Systems of Well-Constrained Beams

4.2.1 Modal Qualitative Properties of Finite Difference Models of Well-Constrained Beams

First, we will demonstrate that stiffness matrices of well-constrained beams are sign-oscillatory. Let us begin by introducing a proposition.

Proposition 4.1 *If $\widetilde{\boldsymbol{E}}$ is defined in Eq. (4.1.23), then its sign-reverse matrix $\widetilde{\boldsymbol{E}}^*$ is totally nonnegative.*

Notice that any submatrix of $\widetilde{\boldsymbol{E}}^*$ must be one of the following three forms: a null matrix, an upper triangular matrix, or a lower triangular matrix. In addition, the main diagonal elements of this submatrix are either 0 or 1. Thus, Proposition 4.1 is obviously valid.

By the same token, $\boldsymbol{L}^{-1}$ should also be a totally nonnegative matrix.

Corollary *If* $\boldsymbol{B} = \boldsymbol{E}^{\mathrm{T}}\boldsymbol{L}^{-1}\boldsymbol{E}$, *then its sign-reverse matrix* $\boldsymbol{B}^*$ *is totally nonnegative.*

Proof On the one hand, $\boldsymbol{L}^{-1}$ is a totally nonnegative matrix. On the other hand, any minor of $\boldsymbol{E}^*$ and $(\boldsymbol{E}^{\mathrm{T}})^*$ must be a minor of the totally nonnegative matrix $\widetilde{\boldsymbol{E}}^*$ and $(\widetilde{\boldsymbol{E}}^*)^{\mathrm{T}}$, respectively. Thus, we may establish the corollary by applying the Binet—Cauchy formula given in Sect. 2.2. ■

Now, set $\underline{\boldsymbol{A}} = (a_{ij})_{(n+1)\times(n+1)}$, $\underline{\boldsymbol{A}}_1 = (a_{ij})_2^{n+1}$, $\underline{\boldsymbol{A}}_n = (a_{ij})_1^n$, and $\underline{\boldsymbol{A}}_{1n} = (a_{ij})_2^n$, where the last three matrices are all truncated submatrices of $\underline{\boldsymbol{A}}$. From the above corollary, $\underline{\boldsymbol{A}}^* = \boldsymbol{B}^*\underline{\boldsymbol{K}}\boldsymbol{B}^*$ is a totally nonnegative matrix. Since $\underline{\boldsymbol{A}}_1^*$, $\underline{\boldsymbol{A}}_n^*$, and $\underline{\boldsymbol{A}}_{1n}^*$ are all truncated submatrices of $\underline{\boldsymbol{A}}^*$, they are, apparently, totally nonnegative, too. It can be directly verified that

$$\det \underline{\boldsymbol{A}} = 0,\ \det \underline{\boldsymbol{A}}_{1n} > 0,\ \text{ and } a_{r,r+1} = a_{r+1,r} < 0.$$

Additionally, when one of the parameters k_0^* and k_n^* is greater than zero, we get $\det \underline{\boldsymbol{A}}_1 > 0$ and $\det \underline{\boldsymbol{A}}_n > 0$.

In order to further prove the stiffness matrix in Eq. (4.1.27) to be nonsingular, rewrite Eq. (4.1.27) as follows, by taking advantage of Eq. (4.1.28):

$$\lambda \boldsymbol{M}\boldsymbol{u} = \boldsymbol{C}\boldsymbol{u},$$

where elements of the stiffness matrix $\boldsymbol{C} = (c_{ij})_{(n+1)\times(n+1)}$ are related to those of $\underline{\boldsymbol{A}} = (a_{ij})_{(n+1)\times(n+1)}$ by the following equations:

$$c_{11} = a_{11} + h_1,\ c_{n+1,n+1} = a_{n+1,n+1} + h_2, \text{ and } c_{ij} = a_{ij} \text{ elsewhere.}$$

Thus, the sign-reverse matrix $\boldsymbol{C}^*$ of $\boldsymbol{C}$ is totally nonnegative. After expanding the determinant of $\boldsymbol{C}$ by the first and then last columns, we obtain

$$\det \boldsymbol{C} = \det \underline{\boldsymbol{A}} + h_1 \det \underline{\boldsymbol{A}}_1 + h_2 \det \underline{\boldsymbol{A}}_n + h_1 h_2 \det \underline{\boldsymbol{A}}_{1n}.$$

Based on this, we may consider the following scenarios:

(1) When $h_1 + h_2 > 0$ and $\beta_1 + \beta_2 > 0$, there must be $k_0^* + k_n^* > 0$. Hence, one of the following inequalities always holds:

$$h_1 \det \underline{\boldsymbol{A}}_1 > 0 \text{ or } h_2 \det \underline{\boldsymbol{A}}_n > 0.$$

This indicates $\det \boldsymbol{C} > 0$, i.e., $\boldsymbol{C}$ is a nonsingular matrix. Furthermore, Eqs. (4.1.13)–(4.1.17) show that subdiagonal elements of $\boldsymbol{C}$ are all of negative values. According to the criteria for a sign-oscillatory matrix (see Sect. 2.4.2), $\boldsymbol{C}$ is sign-oscillatory and the corresponding system is well-constrained.

(2) Now consider the case with $h_1 \cdot h_2 > 0$ and $\beta_1 = \beta_2 = 0$. Although $k_0^* = k_n^* = 0$ in this case, which leads to

$$\det \underline{A}_1 = 0 \quad \text{and} \quad \det \underline{A}_n = 0,$$

we still have

$$\det \boldsymbol{C} = h_1 h_2 \det \underline{A}_{1n} > 0,$$

i.e., $\boldsymbol{C}$ is a nonsingular matrix, Furthermore, Eqs. (4.1.13)–(4.1.17) show that subdiagonal elements of $\boldsymbol{C}$ are all negative numbers. Consequently, $\boldsymbol{C}$ continues to be a sign-oscillatory matrix and the corresponding system remains to be well-constrained.

(3) Finally, if one of h_1 and h_2 is zero while $\beta_1 = \beta_2 = 0$, or if both h_1 and h_2 are zero, then we must have $\det \boldsymbol{C} = 0$. This leads to the conclusion that the system has a zero eigenvalue.

In the discussion above, the property of sign oscillation has been established for the stiffness matrix of a beam subject to the general form of boundary constraints. We now turn our attention to a few special cases mentioned in Sect. 4.1, where the beams are under a combination of four types of boundary conditions: free, sliding, pinned, and clamped constraints. The stiffness and mass matrices for each of these cases are given below:

(1) *Pinned on both sides* $\boldsymbol{A}_{\mathrm{pp}} = \underline{A}_{1n}\big|_{k_0^*=k_n^*=0}$, $\boldsymbol{M}_{\mathrm{pp}} = \mathrm{diag}(m_1, m_2, \ldots, m_{n-1})$.

(2) *Clamped–free* $\boldsymbol{A}_{\mathrm{cf}} = \underline{A}_1\big|_{k_n^*=0}$, $\boldsymbol{M}_{\mathrm{cf}} = \mathrm{diag}(m_1, m_2, \ldots, m_n)$.

(3) *Pinned–sliding* $\boldsymbol{A}_{\mathrm{ps}} = \underline{A}_1\big|_{k_0^*=0}$, $\boldsymbol{M}_{\mathrm{ps}} = \boldsymbol{M}_{\mathrm{cf}}$.

(4) *Clamped at both ends* $\boldsymbol{A}_{\mathrm{cc}} = \underline{A}_{1n}$, $\boldsymbol{M}_{\mathrm{cc}} = \boldsymbol{M}_{\mathrm{pp}}$.

(5) *Clamped–pinned* $\boldsymbol{A}_{\mathrm{cp}} = \underline{A}_{1n}\big|_{k_n^*=0}$, $\boldsymbol{M}_{\mathrm{cp}} = \boldsymbol{M}_{\mathrm{pp}}$.

(6) *Clamped–sliding* $\boldsymbol{A}_{\mathrm{cs}} = \underline{A}_1$, $\boldsymbol{M}_{\mathrm{cs}} = \boldsymbol{M}_{\mathrm{cf}}$.

(7) *Free at both ends* $\boldsymbol{A}_{\mathrm{ff}} = \underline{A}\big|_{k_0^*=k_n^*=0}$, $\boldsymbol{M}_{\mathrm{ff}} = \boldsymbol{M} = \mathrm{diag}(m_0, m_1, \ldots, m_n)$.

(8) *Pinned–free* $\boldsymbol{A}_{\mathrm{pf}} = \underline{A}_1\big|_{k_0^*=k_n^*=0}$, $\boldsymbol{M}_{\mathrm{pf}} = \boldsymbol{M}_{\mathrm{cf}}$.

(9) *Sliding–free* $\boldsymbol{A}_{\mathrm{sf}} = \underline{A}\big|_{k_n^*=0}$, $\boldsymbol{M}_{\mathrm{sf}} = \boldsymbol{M}$.

(10) *Sliding on both sides* $\boldsymbol{A}_{\mathrm{ss}} = \underline{A}$, $\boldsymbol{M}_{\mathrm{ss}} = \boldsymbol{M}$.

Obviously, the beams are well-constrained with the first six types of boundary supports, while they allow one or two kinds of rigid-body motion under the latter four types of constraints listed above. Based on the definition of sign-oscillatory matrices as well as their identification criteria introduced in Sect. 2.4, we may derive the following lemma:

Lemma 4.1 *Stiffness matrices of finite difference models for beams in pinned–pinned, clamped–pinned, clamped–clamped, clamped–free, clamped–sliding, and pinned–sliding configurations are all sign-oscillatory.*

Notice that mass matrices associated with the six types of beams listed above are all nonsingular, diagonal, and totally nonnegative. Furthermore, the product of a sign-oscillatory matrix and a totally nonnegative matrix remains to be sign-oscillatory (see Property 5 in Sect. 2.4). By applying the theory of sign-oscillatory matrices, specifically, Theorems 2.16 and 2.17, to the aforementioned six types of beams, we can establish the theorem below:

Theorem 4.1 *Finite difference models of beams subject to pinned–pinned, pinned - clamped, clamped–clamped, clamped–free, clamped–sliding, or pinned–sliding constraints have the following oscillatory properties in vibration*:

(1) *Natural frequencies are all distinct. They can be arranged in an ascending order as*

$$0 < f_1 < f_2 < \cdots < f_N,$$

where $N = n - 1$ *for the first three types of constraints and* $N = n$ *for the latter three. Notice that this interpretation of* N *is also applicable to the next a few sections of this chapter.*

(2) *Let* $\boldsymbol{u}^{(i)} = (u_{1i}, u_{2i}, \ldots, u_{Ni})^{\mathrm{T}}$ *be the displacement mode shape corresponding to* f_i. *Then, the number of sign reversals in the sequence consisting of components of the mode shape* $\boldsymbol{u}^{(i)}$ *happens to be* $i - 1$, *i.e.,* $\boldsymbol{u}^{(i)}$ *has* $i - 1$ *nodes, which may be expressed as* $S_{\boldsymbol{u}^{(i)}} = i - 1$. *This further leads to the following results, based on the concept of the number of sign reversals in a sequence (see Sect. 2.3):*

    (a) *The first and last components of* $\boldsymbol{u}^{(i)}$ *must be nonzero, namely,* $u_{1i} \neq 0$ *and* $u_{Ni} \neq 0$, *where* $i = 1, 2, \ldots, N$.
    (b) *If the* r-th *component of the* i-th *mode shape satisfies* $u_{ri} = 0$, *then we must have* $u_{r-1,i}u_{r+1,i} < 0$, *where* $i = 1, 2, \ldots, N$ and $r = 2, 3, \ldots, N - 1$.

(3) *For any set of real numbers* c_i $(i = p, p + 1, \ldots, q)$ *that do not vanish simultaneously, the number of sign reversals of the displacement vector*

$$\boldsymbol{u} = c_p\boldsymbol{u}^{(p)} + c_{p+1}\boldsymbol{u}^{(p+1)} + \cdots + c_q\boldsymbol{u}^{(q)}, \quad 1 \leq p \leq q \leq n$$

is in between $p - 1$ *and* $q - 1$, *namely,*

$$p - 1 \leq S_{\boldsymbol{u}}^- \leq S_{\boldsymbol{u}}^+ \leq q - 1.$$

(4) *Nodes of two successive displacement mode shapes* $\boldsymbol{u}^{(i)}$ *and* $\boldsymbol{u}^{(i+1)}$ *interlace, where* $i = 2, 3, \ldots, N - 1$.

The content of this subsection is taken primarily from Ref. [13].

4.2.2 *Qualitative Properties of Static Deformation Associated with Finite Difference Models of Well-Constrained Beams*

We proved in the previous subsection that the stiffness matrices of well-constrained beams are sign-oscillatory. Thus, it is known from Theorem 2.28 in Chap. 2 that the static deformation of discrete systems of well-constrained beams has the following properties:

(1) When a concentrated force is applied to a discretization node of the discrete system of a well-constrained beam, the displacement at any discretization node of the discrete system is nonzero, and its direction is identical to that of the applied force.
(2) If n concentrated forces are applied to some discretization nodes of the discrete system of a well-constrained beam, where $n \le N$ and N is the number of degrees of freedom of the system, then the u-line formed by components u_i of the nodal displacement has no more than $n-1$ sign reversals, i.e., $S_{\boldsymbol{u}} \le n-1$.

4.3 Modal Qualitative Properties of the Finite Difference System of an Under-Constrained Beam

In order to study oscillation properties of the four types of beams with rigid-body motion, which were described in Sect. 4.2.1, we will borrow the concept of conjugate beams in Mechanics of Materials and introduce discretized "conjugate beams" for finite difference models of beams. To this end, let us take another look at Eq. (4.1.27). After limiting ourselves to only four types of end conditions, i.e., clamped, pinned, sliding, and free constraints, we have

$$k_0^*\theta_0 = 0, \quad k_n^*\theta_{n+1} = 0, \quad \phi_0 = h_1 u_0 = 0, \quad \phi_{n+1} = -h_2 u_{n+1} = 0.$$

Thus, Eq. (4.1.24) becomes

$$\boldsymbol{\theta} = -\boldsymbol{L}^{-1}\boldsymbol{E}\boldsymbol{u}, \quad \boldsymbol{w} = \boldsymbol{E}^{\mathrm{T}}\boldsymbol{\theta}, \quad \boldsymbol{\tau} = \boldsymbol{K}\boldsymbol{w}, \quad \boldsymbol{\phi} = \boldsymbol{L}^{-1}\boldsymbol{E}\boldsymbol{\tau}, \tag{4.3.1}$$

while the modal equation of natural vibration (4.1.27) is simplified to

$$\lambda \boldsymbol{M}\boldsymbol{u} = \boldsymbol{E}^{\mathrm{T}}\boldsymbol{L}^{-1}\boldsymbol{E}\boldsymbol{K}\boldsymbol{E}^{\mathrm{T}}\boldsymbol{L}^{-1}\boldsymbol{E}\boldsymbol{u} = -\boldsymbol{E}^{\mathrm{T}}\boldsymbol{L}^{-1}\boldsymbol{E}\boldsymbol{\tau}. \tag{4.3.2}$$

Multiplying the left and right sides of the above equation by $\boldsymbol{M}^{-1}$ and then by $\boldsymbol{K}^{-1}(-\boldsymbol{K}\boldsymbol{E}^{\mathrm{T}}\boldsymbol{L}^{-1}\boldsymbol{E})$, we get

$$\lambda \widetilde{\boldsymbol{M}}\boldsymbol{\tau} = \boldsymbol{E}^{\mathrm{T}}\boldsymbol{L}^{-1}\boldsymbol{E}\widetilde{\boldsymbol{K}}\boldsymbol{E}^{\mathrm{T}}\boldsymbol{L}^{-1}\boldsymbol{E}\boldsymbol{\tau} = \widetilde{\boldsymbol{A}}\boldsymbol{\tau}, \tag{4.3.3}$$

Table 4.1 Relationship of mode shapes and boundary supports between original and conjugate beams

	Original beam	Conjugate beam
Mode shape	Bending moment	Displacement
	Shear force	Angle of rotation
	Displacement	Bending moment
	Angle of rotation	Shear force
Boundary support	Clamped	Free
	Pinned	Pinned
	Free	Clamped
	Sliding	Sliding

where

$$\widetilde{\boldsymbol{M}} = \boldsymbol{K}^{-1} = \mathrm{diag}(\tilde{m}_0, \tilde{m}_1, \ldots, \tilde{m}_n) \text{ and } \widetilde{\boldsymbol{K}} = \boldsymbol{M}^{-1} = \mathrm{diag}(\tilde{k}_0, \tilde{k}_1, \ldots, \tilde{k}_n).$$

Since Eq. (4.3.3) is of the same form as the modal equation (4.1.27) when $\phi_0 = 0$ and $\phi_{n+1} = 0$, we may view it as the modal equation of a system consisting of springs, masses, and rigid bars with the following parameters:

$$l_r\ (r = 1, 2, \ldots, n), \quad \tilde{m}_r = k_r^{-1}\ (r = 0, 1, \ldots, n), \quad \text{and} \quad \tilde{k}_r = m_r^{-1}\ (r = 0, 1, \ldots, n).$$

The system has a "displacement" vector $\boldsymbol{\tau} = \boldsymbol{K}\boldsymbol{w}$. Thus, we may refer to such a system as the conjugate of the original finite difference model of the beam.

Per the definition of a conjugate beam and Eq. (4.3.1), the mode shape of the shear force associated with the original beam corresponds to the mode shape of the angle of rotation associated with the conjugate beam. In addition, after both sides of Eq. (4.3.2) are multiplied by $\boldsymbol{M}^{-1}$, the right-hand side of the resulting equation may be viewed as the "bending moment" of the conjugate beam, and thus the displacement mode shape of the original beam is related to the mode shape of the bending moment associated with the conjugate beam. Finally, we see from the first and last equations in (4.3.1) that the mode shape of the angle of rotation associated the original beam corresponds to the mode shape of the shear force associated with the conjugate beam.

Based on the above observation, we may establish boundary conditions of the conjugate beam from known end supports of the original beam. Specifically, the clamped end, where both the displacement and angle of rotation are zero, of the original beam corresponds to the free end, where both the bending moment and shear force are zero, of the conjugate beam; the free end of the original beam associates with the clamped end of the conjugate beam; the pinned end, where both the displacement and bending moment are zero, of the original beam relates to the pinned end of the conjugate beam; and the sliding end, where both the angle of rotation and shear force are zero, matches the sliding end of the conjugate beam.

The relationship of mode shapes and boundary constraints between the original and conjugate beams are summarized in Table 4.1. This table also reveals the following two important facts:

(1) Since Eq. (4.3.3) is derived from Eq. (4.1.27) through simple arithmetic operations, the original and conjugate beams must share a common set of nonzero frequencies.
(2) The conjugate of any statically determinate beam studied in Sect. 4.2 must also be statically determinate. Furthermore, a statically indeterminate beam given in Sect. 4.2 must correspond to a conjugate beam with some rigid-body motion and conversely, a beam allowing rigid-body motion must be related to a conjugate beam that is statically indeterminate.

Notice that the two matrices, $\widetilde{\boldsymbol{A}}$ in Eq. (4.3.3) and $\underline{\boldsymbol{A}}$ in Eq. (4.1.28), are of identical structure. Based on this observation, the discussion in Sect. 4.2 regarding sign-oscillation of the matrix $\underline{\boldsymbol{A}}$ and its truncated submatrices, as well as the aforementioned two important facts, we may obtain a theorem that parallels Theorem 4.1.

Theorem 4.2 *If natural frequencies of a beam in the pinned–pinned, clamped–free, or pinned–sliding configuration have been sorted in an ascending order, then mode shapes of the bending moment* $\boldsymbol{\tau}^{(i)}$ *and relative rotation* $\boldsymbol{w}^{(i)}$*, which correspond to the natural frequency* f_i*, each have exactly* $i-1$ *sign reversals, namely,*

$$S_{\boldsymbol{\tau}^{(i)}} = S_{\boldsymbol{w}^{(i)}} = i - 1, \quad i = 1, 2, \ldots, N,$$

where $N = n - 1$ *when the beam is subject to pinned–pinned constraints and* $N = n$ *otherwise. In addition, nodes of two successive mode shapes* $\boldsymbol{\tau}^{(i)}$ *and* $\boldsymbol{\tau}^{(i+1)}$ *interlace, and so are those of* $\boldsymbol{w}^{(i)}$ *and* $\boldsymbol{w}^{(i+1)}$.

Theorem 4.3 *Nonzero natural frequencies of a beam in the free–pinned, free–free, or free–sliding configuration are all distinct. If they are arranged in an ascending order, as*

$$f_1^* < f_2^* < \cdots < f_N^*,$$

then the mode shape of the bending moment ${\boldsymbol{\tau}_*}^{(i)}$*, which corresponds to the natural frequency* f_i^**, has exactly* $i - 1$ *sign reversals. Here,* $N = n$ *when the beam is in the free–sliding configuration and* $N = n - 1$ *otherwise; while* $i = 1, 2, \ldots, N$.

It should be pointed out that the free–pinned and free–sliding beams each have one natural frequency of zero while an unconstrained beam has two natural frequencies of zero. When zero frequencies are taken into account, we obtain the following results for these three types of beams, with regard to the sequencing of natural frequencies as well as the counting of sign reversals associated with the mode shape of the bending moment, $\boldsymbol{\tau}^{(i)}$, which corresponds to the natural frequency $f_i = \sqrt{\lambda_i}/2\pi$:

(1) *Free–pinned beam* Natural frequencies may be sorted from small to large as $0 = f_1 < f_2 < \cdots < f_n$. In addition, the number of sign reversals associated with the mode shape of the bending moment $\boldsymbol{\tau}^{(i)} = \boldsymbol{\tau}_*^{(i-1)}$, which corresponds to the frequency $f_i = f_{i-1}^*$, is

$$S_{\tau^{(i)}} = i - 2, \quad i = 2, 3, \ldots, n. \tag{4.3.4}$$

(2) *Free–free beam* Natural frequencies can be arranged in the order $0 = f_1 = f_2 < f_3 < \cdots < f_{n+1}$. Additionally, the number of sign reversals associated with the mode shape of the bending moment $\boldsymbol{\tau}^{(i)} = \boldsymbol{\tau}_*^{(i-2)}$, which corresponds to the frequency $f_i = f_{i-2}^*$, is

$$S_{\tau^{(i)}} = i - 3, \quad i = 3, 4, \ldots, n + 1. \tag{4.3.5}$$

(3) *Free–sliding beam* Natural frequencies may be put in the order $0 = f_1 < f_2 < \cdots < f_{n+1}$. Furthermore, the number of sign reversals associated with the mode shape of the bending moment $\boldsymbol{\tau}^{(i)} = \boldsymbol{\tau}_*^{(i-1)}$, which corresponds to the frequency $f_i = f_{i-1}^*$, is

$$S_{\tau^{(i)}} = i - 2, \quad i = 2, 3, \ldots, n + 1. \tag{4.3.6}$$

Finally, let us look at the beam with sliding supports at both ends. Rewrite the first half of the modal equation of natural vibration (4.3.2) as

$$\lambda\boldsymbol{\theta} = \boldsymbol{L}^{-1}\boldsymbol{E}\boldsymbol{M}^{-1}\boldsymbol{E}^{\mathrm{T}}\boldsymbol{L}^{-1}\boldsymbol{E}\boldsymbol{K}\boldsymbol{E}^{\mathrm{T}}\boldsymbol{\theta}, \tag{4.3.7}$$

$$\lambda\boldsymbol{\phi} = \boldsymbol{L}^{-1}\boldsymbol{E}\boldsymbol{K}\boldsymbol{E}^{\mathrm{T}}\boldsymbol{L}^{-1}\boldsymbol{E}\boldsymbol{M}^{-1}\boldsymbol{E}^{\mathrm{T}}\boldsymbol{\phi}. \tag{4.3.8}$$

Similar to Sect. 4.2, sign-reverse matrices $(\boldsymbol{E}\boldsymbol{M}^{-1}\boldsymbol{E}^{\mathrm{T}})^*$ and $(\boldsymbol{E}\boldsymbol{K}\boldsymbol{E}^{\mathrm{T}})^*$ of matrices $\boldsymbol{E}\boldsymbol{M}^{-1}\boldsymbol{E}^{\mathrm{T}}$ and $\boldsymbol{E}\boldsymbol{K}\boldsymbol{E}^{\mathrm{T}}$, respectively, are both totally nonnegative and their subdiagonal elements are all positive. In addition, $\det(\boldsymbol{E}\boldsymbol{K}\boldsymbol{E}^{\mathrm{T}})$ and $\det(\boldsymbol{E}\boldsymbol{M}^{-1}\boldsymbol{E}^{\mathrm{T}})$ are positive when $k_0 + k_n \neq 0$ and $m_0^{-1} + m_n^{-1} \neq 0$. Therefore, $\boldsymbol{E}\boldsymbol{K}\boldsymbol{E}^{\mathrm{T}}$ 和 $\boldsymbol{E}\boldsymbol{M}^{-1}\boldsymbol{E}^{\mathrm{T}}$ are both sign-oscillatory matrices. By further taking into account the fact that $\boldsymbol{L}^{-1}$ is a positive definite and diagonal matrix, we may conclude that the coefficient matrices in Eqs. (4.3.7) and (4.3.8) are both sign-oscillatory. Thus, a direct application of the theory of oscillatory matrices yields the following result:

(4) *Sliding-Sliding beam* Nonzero natural frequencies of a sliding–sliding beam are all distinct. They can be sorted in the order $0 = f_1 < f_2 < \cdots < f_{n+1}$. Furthermore, both mode shapes of the angle of rotation and shear force associated with a nonzero natural frequency have fixed numbers of sign reversals, namely,

$$S_{\boldsymbol{\theta}_{\mathrm{ss}}^{(i)}} = S_{\boldsymbol{\phi}_{\mathrm{ss}}^{(i)}} = i - 2, \quad i = 2, 3, \ldots, n + 1. \tag{4.3.9}$$

The content of this section and Sect. 4.4 is largely taken from Ref. [11].

4.4 Numbers of Sign Reversals of Various Mode Shapes Associated with the Finite Difference Systems of Beams

4.4.1 Prerequisite Propositions

In previous sections, we primarily studied qualitative properties of natural frequencies and displacement mode shapes of a beam subject to various boundary constraints. When the support conditions allow for rigid-body motion, our discussion on mode shapes has instead focused on the bending moment, or the angle of rotation and the shear force. To complete the study on qualitative properties of beams, we need to introduce a few propositions for later use.

Set

$$\boldsymbol{y} = -\boldsymbol{L}^{-1}\boldsymbol{E}\boldsymbol{x} \quad \text{and} \quad \boldsymbol{z} = \boldsymbol{K}\boldsymbol{E}^{\mathrm{T}}\boldsymbol{y}, \tag{4.4.1}$$

where $\boldsymbol{x} = (x_0, x_1, \ldots, x_n)^{\mathrm{T}}$, $\boldsymbol{y} = (y_1, y_2, \ldots, y_n)^{\mathrm{T}}$, $\boldsymbol{z} = (z_0, z_1, \ldots, z_n)^{\mathrm{T}}$, $\boldsymbol{K} = \mathrm{diag}(k_0, k_1, \ldots, k_n)$, $\boldsymbol{L} = \mathrm{diag}(l_1, l_2, \ldots, l_n)$, and the matrix $\boldsymbol{E}$ is defined by Eq. (4.1.23). Then, we may establish the proposition below:

Proposition 4.2 *If the minimum number of sign reversals of the vector* $\boldsymbol{x}$ *is* $S_x^- = j$, *then numbers of sign reversals of vectors* $\boldsymbol{y}$ *and* $\boldsymbol{z}$ *satisfy the following inequalities:*

$$S_y^- \geq j + 1 - H(x_0) - H(x_n) \text{ and } S_z^- \geq j + H(k_0) + H(k_n) - H(x_0) - H(x_n),$$

where $H(t)$ *denotes the influence function of the boundary condition and*

$$H(t) = \begin{cases} 0, & t = 0, \\ 1, & t \neq 0. \end{cases}$$

Proof According to the definition of the number of sign reversals of a vector introduced in Sect. 2.3, when $S_x^- = j$, there must exist a set of indices $(0 \leq) r_1 < r_2 < \cdots < r_{j+1} (\leq n)$ such that

$$x_{r_k} \cdot x_{r_{k+1}} < 0, \quad k = 1, 2, \ldots, j.$$

Furthermore, per the definition of the vector $\boldsymbol{y}$,

$$y_s = (x_s - x_{s-1})/l_s, \quad s = 1, 2, \ldots, n.$$

Thus, there must be an index s_i satisfying $r_{i-1} < s_i \leq r_i$ such that $y_{s_i} \cdot x_{r_i} > 0$, where $i = 2, 3, \ldots, j + 1$. Combining these results, we get

$$y_{s_i} \cdot y_{s_{i+1}} < 0, \quad i = 2, 3, \ldots, j.$$

This shows: $S_{\boldsymbol{y}}^{-} \geq j-1$ when both x_0 and x_n are different from zero.

Now, let us consider the influence of boundary conditions. When $x_0 = 0$, there must be $r_1 > 0$ and $x_{r_1} \neq 0$. Thus, there is also an index s_1 satisfying $0 < s_1 \leq r_1$ such that $y_{s_1} \cdot x_{r_1} > 0$, which leads to $y_{s_1} \cdot y_{s_2} < 0$. It implies that $S_{\boldsymbol{y}}^{-}$ increases by 1 when $x_0 = 0$. By the same token, when $x_n = 0$, $S_{\boldsymbol{y}}^{-}$ increases by 1, too. This completes the proof of the first inequality.

Next, we try to prove the second inequality involving $S_{\boldsymbol{z}}^{-}$. Notice that the relation between vectors $\boldsymbol{z}$ and $\boldsymbol{y}$ is similar to that between vectors $\boldsymbol{y}$ and $\boldsymbol{x}$. In addition, both y_{s_2} (or y_{s_1}) and $y_{s_{n-1}}$ (or y_{s_n}) are nonzero. Therefore, by applying the same argument, we may get

$$S_{\boldsymbol{z}}^{-} \geq j - H(x_0) - H(x_n),$$

when k_0 and k_n are both zero. If $k_0 \neq 0$, let us assume y_t to be the first nonzero component of the vector $\boldsymbol{y}$. Apparently, $0 < t \leq s_2$ (or s_1). Thus, $z_t = k_{t-1}(y_t - y_{t-1})$ is of the same sign as y_t. Consequently, whether t is equal to or smaller than s_2 (or s_1), the minimum number of sign reversals $S_{\boldsymbol{z}}^{-}$ of the vector $\boldsymbol{z}$ always increases by 1 at least. We can also draw similar conclusion for the case where $k_n \neq 0$. This completes the proof of the second inequality. ■

Proposition 4.3 *If the maximum number of sign reversals of the vector* $\boldsymbol{z}$ *satisfies the equation* $S_{\boldsymbol{z}}^{+} = j$, *then* $S_{\boldsymbol{y}}^{+} \leq j-1$ *and* $S_{\boldsymbol{x}}^{+} \leq j$.

Proof We will proceed with proof by contradiction. Suppose $S_{\boldsymbol{y}}^{+} \geq j$. Then, from the definition for the maximum number of sign reversals, there exists at least one set of indices $(1 =)s_1 < s_2 < \cdots < s_{j+1} (\leq n)$ such that

$$y_{s_i} \cdot y_{s_{i+1}} \leq 0, \quad i = 1, 2, \ldots, j.$$

Similar to Proposition 4.2, there is a set of indices $\{t_i\}$ satisfying $s_i < t_i \leq s_{i+1}$ such that

$$z_{t_i} \cdot z_{t_{i+1}} \leq 0, \quad i = 1, 2, \ldots, j.$$

Without loss of generality, let us assume $y_1 \geq 0$. Then, $z_{t_1} \leq 0$. However, since $t_1 > 0$, we get $z_0 z_{t_1} \leq 0$, and this implies $S_{\boldsymbol{z}}^{+} > j$. Similarly, if $S_{\boldsymbol{x}}^{+} > j$, we ought have $S_{\boldsymbol{y}}^{+} \geq j$, too. These results are in conflict with the given condition. Therefore, Proposition 4.3 has to be valid. ■

Proposition 4.4 *If the number of sign reversals of the vector* $\boldsymbol{y}$ *is* $S_{\boldsymbol{y}} = j$, *then* $S_{\boldsymbol{x}}^{+} \leq j+1$ *while* $S_{\boldsymbol{z}}^{-} \geq j - 1 + H(k_0) + H(k_n)$.

In fact, $S_{\boldsymbol{y}} = j$ implies $S_{\boldsymbol{y}}^{+} = S_{\boldsymbol{y}}^{-} = j$. Thus, Proposition 4.3 yields the first inequality, while Proposition 4.2 leads to the second one.

4.4.2 Numbers of Sign Reversals of Mode Shapes of the Displacement, Angle of Rotation, Bending Moment, and Shear Force Associated with a Beam Subject to Arbitrary End Constraints

With the previous preparation, we are now ready to discuss qualitative properties of mode shapes of the displacement, angle of rotation, bending moment (or relative rotation), and shear force.

1. *Statically Determinate Beams*

For a beam in the pinned–pinned configuration, Theorems 4.1 and 4.2 yield

$$S_{\boldsymbol{u}^{(i)}} = S_{\boldsymbol{\tau}^{(i)}} = i - 1, \quad i = 1, 2, \ldots, n - 1. \tag{4.4.2}$$

This is a necessary condition for the displacement mode shape of the beam. Here, $\boldsymbol{\tau}^{(i)} = (\tau_{1i}, \tau_{2i}, \ldots, \tau_{n-1,i})^{\mathrm{T}}$ is the mode shape of the bending moment associated with $\boldsymbol{u}^{(i)}$, and its components are determined by Eq. (4.3.1).

Notice that the relationship between $\boldsymbol{\phi}^{(i)}$ and $\boldsymbol{\tau}^{(i)}$ is similar to that between $\boldsymbol{\theta}^{(i)}$ and $\boldsymbol{u}^{(i)}$ (see Eq. (4.3.1)). In addition, for a beam pinned on both ends, $u_{0i} = u_{ni} = 0$ and $\tau_{0i} = \tau_{ni} = 0$. Thus, starting from Eq. (4.4.2) and applying Proposition 4.2, we obtain

$$S^-_{\boldsymbol{\theta}^{(i)}} \geq i \text{ and } S^-_{\boldsymbol{\phi}^{(i)}} \geq i.$$

Moreover, for the expanded vector of the bending moment $\underline{\boldsymbol{\tau}}^{(i)} = (0, \tau_{1i}, \tau_{2i}, \ldots, \tau_{n-1,i}, 0)^{\mathrm{T}}$, $S^+_{\underline{\boldsymbol{\tau}}^{(i)}} = i+1$. Thus, we get from Proposition 4.3, $S^+_{\boldsymbol{\theta}^{(i)}} \leq i$.

Per the modal equation of natural vibration

$$\lambda \boldsymbol{u} = \boldsymbol{M}^{-1}\boldsymbol{E}^{\mathrm{T}}\boldsymbol{\phi},$$

$\boldsymbol{\phi}^{(i)}$ is related to the expanded vector $\underline{\boldsymbol{u}}^{(i)} = (0, u_{1i}, u_{2i}, \ldots, u_{n-1,i}, 0)^{\mathrm{T}}$ in a way similar to how $\boldsymbol{y}$ and $\boldsymbol{z}$ are related in Eq. (4.4.1). Additionally, there is the equality $S^+_{\underline{\boldsymbol{u}}^{(i)}} = i + 1$. Applying Proposition 4.3 one more time, we get $S^+_{\boldsymbol{\phi}^{(i)}} \leq i$. Thus, we may obtain numbers of sign reversals of mode shapes associated with the angle of rotation and shear force of the pinned–pinned beam as follows:

$$S_{\boldsymbol{\theta}^{(i)}} = S_{\boldsymbol{\phi}^{(i)}} = i, \quad i = 1, 2, \ldots, n - 1. \tag{4.4.3}$$

By the same token, we can find numbers of sign reversals of various mode shapes associated with pinned–sliding and clamped–free beams:

$$S_{\boldsymbol{u}^{(i)}} = S_{\boldsymbol{\theta}^{(i)}} = S_{\boldsymbol{\tau}^{(i)}} = S_{\boldsymbol{\phi}^{(i)}} = i - 1, \quad i = 1, 2, \ldots, n, \tag{4.4.4}$$

where $\boldsymbol{\tau}_{\text{ps}}^{(i)} = (\tau_{1i}, \tau_{2i}, \ldots, \tau_{ni})^{\text{T}}$ and $\boldsymbol{\tau}_{\text{cf}}^{(i)} = (\tau_{0i}, \tau_{1i}, \ldots, \tau_{n-1,i})^{\text{T}}$.

2. *Statically Indeterminate Beams As Well As Those With Rigid-Body Motion*

Let us first determine the modal qualitative properties of a beam in the clamped–pinned configuration. To this end, consider the following vector that includes displacements at both end points

$$\underline{\boldsymbol{u}}^{(i)} = (0, u_{1i}, u_{2i}, \ldots, u_{n-1,i}, 0)^{\text{T}}.$$

Per Theorem 4.1,

$$S_{\boldsymbol{u}^{(i)}} = i - 1, \quad i = 1, 2, \ldots, n-1.$$

Thus, $S^-_{\underline{\boldsymbol{u}}^{(i)}} = i - 1$. By taking into account of Proposition 4.2 and also noticing that $u_{0i} = u_{ni} = 0$ and $k_n = 0$, we get

$$S^-_{\boldsymbol{\theta}^{(i)}} \geq i \text{ and } S^-_{\boldsymbol{\tau}^{(i)}} \geq i, \quad i = 1, 2, \ldots, n-1.$$

Furthermore, the modal equation of natural vibration

$$\lambda_i \underline{\boldsymbol{u}}^{(i)} = \boldsymbol{M}^{-1}\boldsymbol{E}^{\text{T}}\boldsymbol{L}^{-1}\boldsymbol{E}\boldsymbol{\tau}^{(i)} = \boldsymbol{M}^{-1}\boldsymbol{E}^{\text{T}}\boldsymbol{\phi}^{(i)},$$

Proposition 4.3, as well as the conditions $S^+_{\underline{\boldsymbol{u}}^{(i)}} = i+1$ and $k_n^* = 0$ lead to the following inequalities:

$$S^+_{\boldsymbol{\phi}^{(i)}} \leq i, \, S^+_{\underline{\boldsymbol{\tau}}^{(i)}} \leq i+1, \text{ and } S^+_{\boldsymbol{\tau}^{(i)}} \leq i,$$

where $\boldsymbol{\tau}_{\text{cp}}^{(i)} = (\tau_{0i}, \tau_{1i}, \ldots, \tau_{n-1,i})^{\text{T}}$ and $\underline{\boldsymbol{\tau}}_{\text{cp}}^{(i)} = (\tau_{0i}, \tau_{1i}, \ldots, \tau_{n-1,i}, 0)^{\text{T}}$. By comparing the two inequalities above that are related to $\boldsymbol{\tau}^{(i)}$, we find

$$S_{\boldsymbol{\tau}^{(i)}} = i, \quad i = 1, 2, \ldots, n-1. \tag{4.4.5}$$

Starting from $S^+_{\underline{\boldsymbol{\tau}}^{(i)}} = i + 1$ and $S^-_{\boldsymbol{\tau}^{(i)}} = i$ and then applying Prepositions 4.2 and 4.3 one more time, we get

$$S^+_{\boldsymbol{\theta}^{(i)}} \leq i \text{ and } S^-_{\boldsymbol{\phi}^{(i)}} \geq i.$$

Thus, we have in the end

$$S_{\boldsymbol{\theta}^{(i)}} = S_{\boldsymbol{\phi}^{(i)}} = i, \quad i = 1, 2, \ldots, n-1. \tag{4.4.6}$$

This completes the task of determining numbers of sign reversals of various mode shapes associated with the clamped-pinned beam.

Since a free–pinned beam is the conjugate of a clamped–pinned beam, we may obtain numbers of sign reversals of the mode shape corresponding to a nonzero natural frequency f_i for the beam in the free–pinned configuration, based on the relationship between mode shapes of the original and conjugate beams introduced in Sect. 4.3. Specifically,

$$S_{\boldsymbol{u}^{(i)}} = S_{\boldsymbol{\theta}^{(i)}} = S_{\boldsymbol{\phi}^{(i)}} = i - 1, \quad S_{\boldsymbol{\tau}^{(i)}} = i - 2, \quad i = 2, 3, \ldots, n. \tag{4.4.7}$$

By using similar arguments, we may derive the following equations for numbers of sign reversals of mode shapes associated with beams in clamped–clamped and free–free configurations, respectively:

$$S_{\boldsymbol{u}_{\mathrm{cc}}^{(i)}} = i - 1, \quad S_{\boldsymbol{\tau}_{\mathrm{cc}}^{(i)}} = i + 1, \quad S_{\boldsymbol{\theta}_{\mathrm{cc}}^{(i)}} = S_{\boldsymbol{\phi}_{\mathrm{cc}}^{(i)}} = i, \quad i = 1, 2, \ldots, n - 1. \tag{4.4.8}$$

$$S_{\boldsymbol{u}_{\mathrm{ff}}^{(i)}} = i - 1, \quad S_{\boldsymbol{\tau}_{\mathrm{ff}}^{(i)}} = i - 3, \quad S_{\boldsymbol{\theta}_{\mathrm{ff}}^{(i)}} = S_{\boldsymbol{\phi}_{\mathrm{ff}}^{(i)}} = i - 2, \quad i = 3, 4, \ldots, n + 1. \tag{4.4.9}$$

Moreover, numbers of sign reversals of mode shapes associated with beams in clamped–sliding and free–sliding configurations are, respectively,

$$S_{\boldsymbol{u}_{\mathrm{cs}}^{(i)}} = S_{\boldsymbol{\theta}_{\mathrm{cs}}^{(i)}} = S_{\boldsymbol{\phi}_{\mathrm{cs}}^{(i)}} = i - 1, \quad S_{\boldsymbol{\tau}_{\mathrm{cs}}^{(i)}} = i, \quad i = 1, 2, \ldots, n; \tag{4.4.10}$$

$$S_{\boldsymbol{u}_{\mathrm{fs}}^{(i)}} = i - 1, \quad S_{\boldsymbol{\theta}_{\mathrm{fs}}^{(i)}} = S_{\boldsymbol{\tau}_{\mathrm{fs}}^{(i)}} = S_{\boldsymbol{\phi}_{\mathrm{fs}}^{(i)}} = i - 2, \quad i = 2, 3, \ldots, n + 1. \tag{4.4.11}$$

We would like to remind readers that $\boldsymbol{u}^{(i)}$, $\boldsymbol{\theta}^{(i)}$, $\boldsymbol{\tau}^{(i)}$, and $\boldsymbol{\phi}^{(i)}$ in Eqs. (4.4.6)–(4.4.11) are mode shapes corresponding to the same natural frequency f_i, and they are all calculated from Eq. (4.3.1). Although both $\boldsymbol{\theta}$ and $\boldsymbol{\phi}$ have n components, the number of components in $\boldsymbol{\tau}$ varies depending upon the type of boundary supports. Specifically,

$$\begin{aligned}
\boldsymbol{\tau}_{\mathrm{cc}} &= (\tau_0, \tau_1, \ldots, \tau_n)^{\mathrm{T}}, \quad \boldsymbol{\tau}_{\mathrm{cp}} = (\tau_0, \tau_1, \ldots, \tau_{n-1})^{\mathrm{T}}, \\
\boldsymbol{\tau}_{\mathrm{fp}} &= (\tau_1, \tau_2, \ldots, \tau_{n-1})^{\mathrm{T}}, \quad \boldsymbol{\tau}_{\mathrm{cs}} = (\tau_0, \tau_1, \ldots, \tau_n)^{\mathrm{T}}, \\
\boldsymbol{\tau}_{\mathrm{ff}} &= (\tau_1, \tau_2, \ldots, \tau_{n-1})^{\mathrm{T}}, \quad \boldsymbol{\tau}_{\mathrm{fs}} = (\tau_1, \tau_2, \ldots, \tau_n)^{\mathrm{T}}.
\end{aligned}$$

3. *A Beam in the Sliding–Sliding Configuration*

When the beam is sliding at both ends, the numbers of sign reversals of mode shapes associated with the angle of rotation and shear force have been given in Eq. (4.3.9) of Sect. 4.3. Specifically,

$$S_{\boldsymbol{\theta}_{\mathrm{ss}}^{(i)}} = S_{\boldsymbol{\phi}_{\mathrm{ss}}^{(i)}} = i - 2, \quad i = 2, 3, \ldots, n + 1.$$

According to Proposition 4.4, we have

$$S^{+}_{\boldsymbol{u}_{\mathrm{ss}}^{(i)}} \le i - 1, \quad S^{-}_{\boldsymbol{\tau}_{\mathrm{ss}}^{(i)}} \ge i - 1, \quad S^{+}_{\boldsymbol{\tau}_{\mathrm{ss}}^{(i)}} \le i - 1, \quad i = 2, 3, \ldots, n + 1.$$

Similarly, we get $S^{-}_{\boldsymbol{u}_{\mathrm{ss}}^{(i)}} \ge i - 1$ from the modal equation of natural vibration

Table 4.2 Numbers of sign reversals of mode shapes of the displacement, angle of rotation, bending moment, and shear force associated with the discrete system of a beam

Type of end support	Support parameters				Numbers of sign reversals			
	h_1	β_1	h_2	β_2	$S_{u^{(i)}}$	$S_{\theta^{(i)}}$	$S_{\tau^{(i)}}$	$S_{\phi^{(i)}}$
Clamped-free	∞	∞	0	0	$i-1$	$i-1$	$i-1$	$i-1$
Clamped-sliding	∞	∞	0	∞	$i-1$	$i-1$	i	$i-1$
Clamped-pinned	∞	∞	∞	0	$i-1$	i	i	i
Clamped-clamped	∞	∞	∞	∞	$i-1$	i	$i+1$	i
Pinned-pinned	∞	0	∞	0	$i-1$	i	$i-1$	i
Pinned-sliding	∞	0	0	∞	$i-1$	$i-1$	$i-1$	$i-1$
Free-pinned	0	0	∞	0	$i-1$	$i-1$	$i-2$*	$i-1$
Free-sliding	0	0	0	∞	$i-1$	$i-2$*	$i-2$*	$i-2$*
Free-free	0	0	0	0	$i-1$	$i-2$*	$i-3$[a]	$i-2$*
Sliding-sliding	0	∞	0	∞	$i-1$	$i-2$*	$i-1$	$i-2$*

*For a beam in the free-pinned configuration, $S_{\tau^{(1)}} = 0$; for a beam in the free-sliding configuration, $S_{\theta^{(1)}} = S_{\tau^{(1)}} = S_{\phi^{(1)}} = 0$; and for beams in free-free and sliding-sliding configuration, $S_{\theta^{(1)}} = S_{\phi^{(1)}} = 0$

[a]For an unconstrained (free-free) beam, $S_{\tau^{(1)}} = S_{\tau^{(2)}} = 0$

$$\lambda_i \boldsymbol{u}^{(i)} = \boldsymbol{M}^{-1}\boldsymbol{E}^{\mathrm{T}}\boldsymbol{\phi}^{(i)}$$

as well as Proposition 4.4. Thus,

$$S_{\boldsymbol{u}_{\mathrm{ss}}^{(i)}} = S_{\boldsymbol{\tau}_{\mathrm{ss}}^{(i)}} = i - 1, \quad i = 2, 3, \ldots, n + 1. \tag{4.4.12}$$

Table 4.2 summarizes results on numbers of sign reversals, i.e., sign-reversal counts, of different mode shapes associated with the finite difference model of a beam subject to various boundary constraints. It should be emphasized that $S_{\tau^{(i)}} = S_{\boldsymbol{w}^{(i)}}$ remains true, regardless of the type of constraints applied.

At the end of this section, let us prove that Property (3) in Theorem 4.1 continues to be valid, even when the discrete system of a beam has rigid-body motion. But we will have to proceed by considering two separate cases below. Also see Ref. [18].

Case 1 This case includes beams in the free-pinned, free–free, and free–sliding configurations. All of these systems have some rigid-body motion. In addition, "Mode shapes of the displacement" $\boldsymbol{\tau}_*^{(i)}$ $(i = 1, 2, \ldots, N)$ of their respective conjugate beams are all eigenvectors of sign-oscillatory matrices. Thus, Property (3) in Theorem 4.1 is applicable here. In other words, for any set of real numbers c_i $(i = p, p+1, \ldots, q)$ that do not vanish simultaneously, the number of sign reversals of the vector

$$\boldsymbol{\tau} = c_p\boldsymbol{\tau}_*^{(p)} + c_{p+1}\boldsymbol{\tau}_*^{(p+1)} + \cdots + c_q\boldsymbol{\tau}_*^{(q)}, \quad 1 \le p \le q \le N \tag{4.4.13}$$

lies in between $p - 1$ and $q - 1$, i.e.,

$$p - 1 \leq S_{\tau}^{-} \leq S_{\tau}^{+} \leq q - 1.$$

For the discrete system of a beam in the free–pinned configuration, there are two things deserving some attention: First, "the mode shape of the displacement" of the conjugate system is $\boldsymbol{\tau}_*^{(i)}$ and the related natural frequency f_i^* corresponds to the nonzero natural frequency f_{i+1} of the original system, where $i = 1, 2, \ldots, n - 1$. Thus,

$$\boldsymbol{\tau}_*^{(i)} = -\boldsymbol{K}\boldsymbol{E}^{\mathrm{T}}\boldsymbol{L}^{-1}\boldsymbol{E}\boldsymbol{u}^{(i+1)}. \tag{4.4.14}$$

Second, it is known that $\boldsymbol{\tau}_*^{(i)}$ has only $n - 1$ components. In order to apply Proposition 4.3, where vectors $\boldsymbol{x}$ and $\boldsymbol{z}$ both consist of $n + 1$ components, we should add two zero components to each of the vectors $\boldsymbol{\tau}$ and $\boldsymbol{\tau}_*^{(i)}$ $(i = p, p + 1, \ldots, q)$ in Eq. (4.4.13), one at the beginning of the vector under consideration while the other at its end, namely,

$$\begin{aligned} \underline{\boldsymbol{\tau}} &= (0, \tau_1, \ldots, \tau_{n-1}, 0)^{\mathrm{T}}, \\ \underline{\boldsymbol{\tau}}_*^{(i)} &= (0, \tau_{*1i}, \ldots, \tau_{*n-1,i}, 0)^{\mathrm{T}}, \quad (i = p, p + 1, \ldots, q). \end{aligned}$$

Furthermore, one zero component should be appended to the beginning of each displacement mode shape $\boldsymbol{u}^{(i)}$ $(i = p + 1, p + 2, \ldots, q + 1)$ of the original system as well as to the vector $\boldsymbol{u}$, which is a linear combination of the mode shape vectors, namely,

$$\begin{aligned} \underline{\boldsymbol{u}} &= (0, u_1, u_2, \ldots, u_n)^{\mathrm{T}}, \\ \underline{\boldsymbol{u}}^{(i)} &= (0, u_{1i}, u_{2i}, \ldots, u_{ni})^{\mathrm{T}}, \quad (i = p + 1, p + 2, \ldots, q + 1). \end{aligned}$$

Thus, $S_{\underline{\tau}}^{+} \leq q + 1$. By comparing Eqs. (4.3.1) and (4.4.1), it is clear that we can apply Proposition 4.3 to $\underline{\boldsymbol{\tau}}$ and the respective displacement vector $\underline{\boldsymbol{u}}$. This yields $S_{\underline{u}}^{+} \leq q + 1$. By further comparing components of $\boldsymbol{u} = (u_1, u_2, \ldots, u_n)^{\mathrm{T}}$ and $\underline{\boldsymbol{u}}$, we obtain

$$S_{\boldsymbol{u}}^{+} \leq q.$$

Now consider the rewritten modal equations of the discrete system of the pinned–free beam,

$$\lambda \boldsymbol{M}\underline{\boldsymbol{u}} = -\boldsymbol{E}^{\mathrm{T}}\boldsymbol{L}^{-1}\boldsymbol{E}\underline{\boldsymbol{\tau}},$$

Notice that locations of $\underline{\boldsymbol{\tau}}$ and $\underline{\boldsymbol{u}}$ in the above equation correspond, respectively, to those of $\boldsymbol{x}$ and $\boldsymbol{z}$ in Eq. (4.4.1). It also holds true that $u_0 = 0$ and $u_n \neq 0$. Additionally, the existence of the first and last zero components in a vector does not affect the minimum number of sign reversals. Thus, $S_{\underline{\tau}}^- \geq p - 1$. Then, by applying Proposition 4.2 to $\underline{\tau}$ and the corresponding $\underline{u}$, we see

$$S_u^- = S_{\underline{u}}^- \geq S_{\underline{\tau}}^- + H(u_0) + H(u_n) - H(\tau_0) - H(\tau_n) \geq p - 1 + 0 + 1 - 0 - 0 = p.$$

After combining these results, we have

$$p \leq S_u^- \leq S_u^+ \leq q. \tag{4.4.15}$$

From Eq. (4.4.14), we then get the vector $\boldsymbol{u}$ that is related to the vector $\boldsymbol{\tau}$ in Eq. (4.4.13):

$$\boldsymbol{u} = c_p \boldsymbol{u}^{(p+1)} + c_{p+1} \boldsymbol{u}^{(p+2)} + \cdots + c_q \boldsymbol{u}^{(q+1)}, \quad 1 \leq p \leq q \leq N. \tag{4.4.16}$$

Hence, Eq. (4.4.15) is what needs to be proved.

As for the discrete system of a beam unconstrained at both ends, there are also two things that we should pay attention to: First, "the displacement mode shape" of its conjugate system is $\boldsymbol{\tau}_*^{(i)}$, while its related natural frequency f_i^* corresponds to the nonzero natural frequency f_{i+2} of the original system, where $i = 1, 2, \ldots, n-1$. Thus,

$$\boldsymbol{\tau}_*^{(i)} = -\boldsymbol{K}\boldsymbol{E}^{\mathrm{T}}\boldsymbol{L}^{-1}\boldsymbol{E}\boldsymbol{u}^{(i+2)}. \tag{4.4.17}$$

Second, it is known that the mode shape of the moment $\boldsymbol{\tau}_*^{(i)}$ has only $n-1$ components, similar to the free-pinned beam we have just discussed. Consequently, we should also expand $\boldsymbol{\tau}$ and $\boldsymbol{\tau}_*^{(i)}$ in Eq. (4.4.13) to

$$\underline{\boldsymbol{\tau}} = (0, \tau_1, \ldots, \tau_{n-1}, 0)^{\mathrm{T}},$$
$$\underline{\boldsymbol{\tau}}_*^{(i)} = (0, \tau_{*1i}, \ldots, \tau_{*n-1,i}, 0)^{\mathrm{T}}, \quad i = p, p+1, \ldots, q.$$

Thus, $S_{\underline{\tau}}^+ \leq q + 1$. By comparing Eqs. (4.3.1) and (4.4.1), we see that Proposition 4.3 may be applied to $\underline{\boldsymbol{\tau}}$ and the related $\boldsymbol{u}$. This leads to

$$S_u^+ \leq q + 1.$$

Next, let us look at the modified modal equation of the discrete system associated with this beam in the free–free configuration

$$\lambda \boldsymbol{M}\boldsymbol{u} = -\boldsymbol{E}^{\mathrm{T}}\boldsymbol{L}^{-1}\boldsymbol{E}\underline{\boldsymbol{\tau}}.$$

Again, notice that the locations of $\underline{\boldsymbol{\tau}}$ and $\boldsymbol{u}$ in the above equation correspond to, respectively, those of $\boldsymbol{x}$ and z in Eq. (4.4.1). We also have $u_0 \neq 0$ and $u_n \neq 0$. Furthermore, the existence of zero components at the beginning and end of a vector has no effect on its minimum number of sign reversals. By applying Proposition 4.2 to $\underline{\boldsymbol{\tau}}$ and the corresponding $\boldsymbol{u}$, we get

$$S_{\boldsymbol{u}}^{-} \geq S_{\underline{\boldsymbol{\tau}}}^{-} + H(u_0) + H(u_n) - H(\tau_0) - H(\tau_n) \geq p - 1 + 1 + 1 - 0 - 0 = p + 1.$$

Combination of the above results yields

$$p + 1 \leq S_{\boldsymbol{u}}^{-} \leq S_{\boldsymbol{u}}^{+} \leq q + 1. \tag{4.4.18}$$

The vector $\boldsymbol{u}$ that corresponds to the vector $\boldsymbol{\tau}$ may be obtained from Eq. (4.4.17). Specifically,

$$\boldsymbol{u} = c_p \boldsymbol{u}^{(p+2)} + c_{p+1} \boldsymbol{u}^{(p+3)} + \cdots + c_q \boldsymbol{u}^{(q+2)}, \quad 1 \leq p \leq q \leq N. \tag{4.4.19}$$

Thus, Eq. (4.4.18) is indeed what needs to be proved.

A similar proof can be given for the discrete system of a beam in the free–sliding configuration. Details are left out here.

Case 2 Now we turn to the other case in the effort to show that Property (3) in Theorem 4.1 is applicable to beams with rigid-body motion. This case only includes the discrete system of a beam in the sliding-sliding configuration.

Let us refer to systems represented by Eqs. (4.3.7) and (4.3.8) as transformed systems associated with the discrete model of the sliding-sliding beam. Then, "displacement mode shapes" $\boldsymbol{\theta}_{\text{tr}}^{(i)}$ and $\boldsymbol{\phi}_{\text{tr}}^{(i)}$ $(i = 1, 2, \ldots, n)$ of the two transformed systems are all eigenvectors of some sign oscillatory matrices, where the subscript tr indicates association with the transformed system. Therefore, Property (3) in Theorem 4.1 should be applicable to these vectors. In other words, for any set of real numbers c_i $(i = p, p+1, \ldots, q)$ that are not zero simultaneously, counts of sign-reversals of vectors

$$\boldsymbol{\theta} = c_p \boldsymbol{\theta}_{\text{tr}}^{(p)} + c_{p+1} \boldsymbol{\theta}_{\text{tr}}^{(p+1)} + \cdots + c_q \boldsymbol{\theta}_{\text{tr}}^{(q)}, \quad 1 \leq p \leq q \leq n \tag{4.4.20}$$

and

$$\boldsymbol{\phi} = c_p \boldsymbol{\phi}_{\text{tr}}^{(p)} + c_{p+1} \boldsymbol{\phi}_{\text{tr}}^{(p+1)} + \cdots + c_q \boldsymbol{\phi}_{\text{tr}}^{(q)}, \quad 1 \leq p \leq q \leq n \tag{4.4.21}$$

are between $p - 1$ and $q - 1$, i.e.,

$$p - 1 \leq S_{\boldsymbol{\theta}}^{-} \leq S_{\boldsymbol{\theta}}^{+} \leq q - 1 \text{ and } p - 1 \leq S_{\boldsymbol{\phi}}^{-} \leq S_{\boldsymbol{\phi}}^{+} \leq q - 1. \tag{4.4.22}$$

After comparing Eqs. (4.3.1) and (4.4.1), we can see that the relationship between $\boldsymbol{\theta}$ and $\boldsymbol{u}$ parallels that of $\boldsymbol{y}$ and $\boldsymbol{x}$ in Proposition 4.4. By applying Proposition 4.4 to vectors $\boldsymbol{\theta}$ and $\boldsymbol{u}$, we then get $S_{\boldsymbol{u}}^{+} \leq q$.

Next, comparing Eq. (4.4.1) with the modified modal equation of the discrete system of the sliding-sliding beam

$$\lambda \boldsymbol{u} = \boldsymbol{M}^{-1}\boldsymbol{E}^{\mathrm{T}}\boldsymbol{\phi},$$

we find that the relationship between $\boldsymbol{\phi}$ and $\boldsymbol{u}$ is equivalent to that between $\boldsymbol{y}$ and $\boldsymbol{z}$ in Proposition 4.4. Thus, we obtain

$$S_{\boldsymbol{u}}^{-} \geq S_{\boldsymbol{\phi}}^{-} - 1 + H(u_0) + H(u_n) \geq (p-1) - 1 + 1 + 1 = p,$$

after applying Proposition 4.4 to vectors $\boldsymbol{\phi}$ and $\boldsymbol{u}$.

Finally, by combining the above results, we arrive at the result:

$$p \leq S_{\boldsymbol{u}}^{-} \leq S_{\boldsymbol{u}}^{+} \leq q. \tag{4.4.23}$$

The vector $\boldsymbol{u}$ corresponding to both $\boldsymbol{\theta}$ in Eq. (4.4.20) and $\boldsymbol{\phi}$ in Eq. (4.4.21) is

$$\boldsymbol{u} = c_p \boldsymbol{u}^{(p+1)} + c_{p+1}\boldsymbol{u}^{(p+2)} + \cdots + c_q \boldsymbol{u}^{(q+1)}, \quad 1 \leq p \leq q \leq N.$$

Thus, Eq. (4.4.23) is indeed the one that needs to be proved.

4.5 Construction of the Finite Difference System of a Beam Using Modes

4.5.1 *Finite Difference System of a Beam Constructed with One Mode*

Given a vector $\boldsymbol{u}$ as well as a set of positive numbers ω and l_i $(i = 1, 2, \ldots, n)$, it is possible to calculate the vector $\boldsymbol{w}$ in Eq. (4.3.1). Now let us prove the following theorem:

Theorem 4.4 *Under a given set of boundary conditions, in order for* $\boldsymbol{u}$ *and* ω *to be the displacement mode shape and the corresponding natural angular frequency, respectively, of a system consisting of springs, masses, and rigid bars as well as for* l_i $(i = 1, 2, \ldots, n)$ *to represent lengths of the rigid bars, the necessary and sufficient conditions are: Numbers of sign reversals of both* $\boldsymbol{u}$ *and* $\boldsymbol{w}$ *are fixed. In addition, they satisfy various conditions specified in Sect. 4.4.*

Proof The necessary part of the conditions has been proved in Sect. 4.4. We will attempt to establish the sufficient portion in this section by actually constructing a system.

As an example, let us look at a beam pinned at both ends. In this case, the vector $\boldsymbol{u} = (u_1, u_2, \ldots, u_{n-1})^{\mathrm{T}}$ is known, while $\boldsymbol{w} = (w_1, w_2, \ldots, w_{n-1})^{\mathrm{T}}$ can be calculated. We are going to prove if the two vectors satisfy the condition $S_{\boldsymbol{u}} = S_{\boldsymbol{w}} = i - 1$ then there exists a spring–mass–rigid-bar system of the beam in the pinned–pinned configuration such that $\boldsymbol{u}$ and $\boldsymbol{w}$ are the i-th mode shapes of the displacement and relative rotation, respectively. To help with this effort, we look at the geometrical interpretation of the condition

$$S_{\boldsymbol{u}} = S_{\boldsymbol{w}} = i - 1, \quad 1 \leq i \leq n - 1.$$

Consider the following piecewise linear function:

$$u_r(\xi - L_{r-1})/l_r + u_{r-1}(L_r - \xi)/l_r, \quad r = 1, 2, \ldots, n, \tag{4.5.1}$$

where

$$L_0 = 0, \quad L_r = \sum_{p=1}^{r} l_p, \quad L_{r-1} \leq \xi \leq L_r, \text{ and } u_0 = 0 = u_n.$$

This, in fact, represents the curve formed by rigid bars of the system shown in Fig. 4.1b, where, in our particular case, the system is pinned at both ends. Since $S_{\boldsymbol{u}} = i - 1$, the curve (4.5.1) has $i - 1$ nodes. Thus, it is possible to divide the curve into i segments, such that within each segment the function defined by Eq. (4.5.1) does not vanish and has a fixed sign while in between two adjacent segments the function is of opposite signs. Consequently, there is one extremum $u_{t_j} \neq 0\,(j = 1, 2, \ldots, i)$ in each segment that shares a common sign. Here, the subscript t_j is the sequential number of the mass, whose displacement reaches an extremum. When $u_{t_j} > 0$, the extremum is a local maximum; while when $u_{t_j} < 0$, the extremum is a local minimum. Moreover, since $S_{\boldsymbol{w}} = i - 1$, it is impossible to have $u_{t_j-1} = u_{t_j} = u_{t_j+1}$. Thus,

$$w_{t_j} \cdot u_{t_j} < 0, \quad j = 1, 2, \ldots, i.$$

This implies: There is one and only one sign reversal in the subsequence consisting of components of $\boldsymbol{w}$, i.e., $(w_{t_j}, w_{t_j+1}, \ldots, w_{t_{j+1}})$, where $1 \leq j \leq i - 1$; while there is no sign reversal in each of the two subsequences $(w_1, w_2, \ldots, w_{t_1})$ and $(w_{t_i}, w_{t_i+1}, \ldots, w_{n-1})$. Therefore, it is possible to divide the nonzero components of $\boldsymbol{w}$ into i segments

$$w_{\alpha_j}, w_{\alpha_j+1}, \ldots, w_{\beta_j}, \quad j = 1, 2, \ldots, i, \ \alpha_1 = 1, \text{ and } \beta_i = n - 1$$

such that the following inequalities hold within the j-th segment $(1 \leq j \leq i)$:

$$w_{\alpha_j-1} \cdot w_{t_j} \leq 0, \quad w_{\beta_j+1} \cdot w_{t_j} \leq 0,$$

$$w_r \cdot w_{t_j} > 0, \quad r = \alpha_j, \alpha_j + 1, \ldots, t_j, \ldots, \beta_j.$$

For any j $(1 \leq j \leq i)$, let us look at the following system of equations:

$$\begin{cases} -\left(\frac{1}{l_{\alpha_j}} + \frac{1}{l_{\alpha_j+1}}\right)\tau_{\alpha_j} + \frac{\tau_{\alpha_j+1}}{l_{\alpha_j+1}} = \lambda m_{\alpha_j} \cdot u_{\alpha_j} - \frac{\tau_{\alpha_j-1}}{l_{\alpha_j}}, \\ \frac{\tau_{r-1}}{l_r} - \left(\frac{1}{l_r} + \frac{1}{l_{r+1}}\right)\tau_r + \frac{\tau_{r+1}}{l_{r+1}} = \lambda m_r \cdot u_r, \quad r = \alpha_j + 1, \alpha_j + 2, \ldots, \beta_j - 1, \\ \frac{\tau_{\beta_j-1}}{l_{\beta_j}} - \left(\frac{1}{l_{\beta_j}} + \frac{1}{l_{\beta_j+1}}\right)\tau_{\beta_j} = \lambda m_{\beta_j} \cdot u_{\beta_j} - \frac{\tau_{\beta_j+1}}{l_{\beta_j+1}}, \end{cases} \tag{4.5.2}$$

where

$$\lambda = \omega^2 \quad \text{and} \quad \tau_{\alpha_1-1} = 0 = \tau_{\beta_i+1}, \text{ i.e., } \tau_0 = 0 = \tau_n.$$

To construct the pinned–pinned beam, we need to consider four different cases. These are explained in details in the next a few paragraphs.

Case 1 If $u_{\alpha_j} \neq 0$ or $w_{\alpha_j-1} = 0$ and, additionally, $u_{\beta_j} \neq 0$ or $w_{\beta_j+1} = 0$, then set

$$m_r = \begin{cases} \left(-w_{t_j}\rho_j + \delta_{\alpha_j r}\frac{\tau_{\alpha_j-1}}{l_{\alpha_j}} + \delta_{\beta_j r}\frac{\tau_{\beta_j+1}}{l_{\beta_j+1}}\right) \cdot \frac{1}{\lambda u_r}, & w_r u_r < 0, \\ 1, & u_r = 0, \\ \left(\varepsilon_j w_{t_j}\rho_j + \delta_{\alpha_j r}\frac{\tau_{\alpha_j-1}}{l_{\alpha_j}} + \delta_{\beta_j r}\frac{\tau_{\beta_j+1}}{l_{\beta_j+1}}\right) \cdot \frac{1}{\lambda u_r}, & w_r u_r > 0, \end{cases} \tag{4.5.3}$$

where the parameters are defined as:

$$\rho_j \text{ and } \varepsilon_j > 0, \quad \delta_{\alpha\beta} = \begin{cases} 1, & \alpha = \beta, \\ 0, & \alpha \neq \beta, \end{cases} \quad \text{and } r = \alpha_j, \alpha_j + 1, \ldots, \beta_j.$$

After assigning a proper value to ε_j, plugging Eq. (4.5.3) into Eq. (4.5.2), and solving the resulting system of equations, we obtain

$$\tau_r = \xi_r \rho_j w_{t_j}, \quad r = \alpha_j, \alpha_j + 1, \ldots, \beta_j. \tag{4.5.4}$$

Since the coefficient matrix in Eq. (4.5.2) is a Jacobi matrix, it can be shown that $\xi_r > 0$. Thus, it is clear from Eq. (4.3.1) that $\{k_r\}_{\alpha_j}^{\beta_j} = \{\tau_r/w_r\}_{\alpha_j}^{\beta_j}$ are all positive numbers. Moreover, we know from the way they are constructed that $\{m_r\}_{\alpha_j+1}^{\beta_j-1}$ are also positive numbers. Now let us look at m_{α_j}. When $w_{\alpha_j-1} = 0$ or $w_{\alpha_j}u_{\alpha_j} < 0$, m_{α_j} obviously remains a positive number. If

$$w_{\alpha_j-1} \neq 0 \quad \text{and} \quad w_{\alpha_j} \cdot u_{\alpha_j} > 0,$$

in which case

$$\beta_{j-1} = \alpha_j - 1 \quad \text{and} \quad w_{\beta_{j-1}} \cdot u_{\beta_{j-1}} < 0,$$

then

$$m_{\alpha_j} = \left(\varepsilon_j w_{t_j} \rho_j + \frac{\tau_{\alpha_j - 1}}{l_{\alpha_j}}\right) \cdot \frac{1}{\lambda u_{\alpha_j}} = \left(\varepsilon_j w_{t_j} \rho_j + \frac{\xi_{\beta_{j-1}} \rho_{j-1} w_{t_{j-1}}}{l_{\alpha_j}}\right) \cdot \frac{1}{\lambda u_{\alpha_j}}.$$

Thus, we can have $m_{\alpha_j} > 0$ by setting ρ_{j-1}/ρ_j to be sufficiently small, while $m_{\beta_{j-1}} > 0$ always holds true regardless of what positive value is assigned to ρ_{j-1}/ρ_j. Similarly, it can be shown that m_{β_j} calculated from Eq. (4.5.3) is either positive on its own or can be made positive by choosing an appropriate value for ρ_j/ρ_{j+1}.

Case 2 If $u_{\alpha_j} = 0$ and $u_{\beta_j} = 0$, then we must have

$$u_r w_r < 0 \quad \text{and} \quad r = \alpha_j + 1, \alpha_j + 2, \ldots, \beta_j - 1.$$

In this case, make ρ_j/ρ_{j-1} and ρ_j/ρ_{j+1} sufficiently large, and set

$$m_{\alpha_j} = m_{\beta_j} = 1, \quad m_r = -w_{t_j} \rho_j / \lambda u_r, \quad \text{and} \quad r = \alpha_j + 1, \alpha_j + 2, \ldots, \beta_j - 1.$$

Next, solve the system of equations (4.5.2). We again get Eq. (4.5.4) and $\xi_r > 0$. Thus, $\{k_r\}_{\alpha_j}^{\beta_j} = \{\tau_r/w_r\}_{\alpha_j}^{\beta_j}$ are guaranteed to be all positive.

Case 3 If $u_{\alpha_j} = 0$ and $w_{\alpha_j - 1} \neq 0$ and additionally, $u_{\beta_j} \neq 0$ or $w_{\beta_j + 1} = 0$, then we need to make a few changes to the proof presented above. Specifically, the equation of motion for m_{α_j} should be taken out of the system of equations (4.5.2) and rewritten as

$$\left(\frac{1}{l_{\alpha_j}} + \frac{1}{l_{\alpha_j + 1}}\right) \tau_{\alpha_j} = \frac{\tau_{\alpha_j - 1}}{l_{\alpha_j}} + \frac{\tau_{\alpha_j + 1}}{l_{\alpha_j + 1}}. \tag{4.5.5}$$

Notice that we must have here

$$\beta_{j-1} = \alpha_j - 1, \quad w_{\alpha_j - 1} \cdot u_{\alpha_j - 1} < 0, \text{ and } w_{\alpha_j + 1} \cdot u_{\alpha_j + 1} < 0.$$

Set

$$m_r = \begin{cases} \left(-w_{t_j} \rho_j + \delta_{\alpha_j + 1, r} \dfrac{\tau_{\alpha_j}}{l_{\alpha_j + 1}} + \delta_{\beta_j r} \dfrac{\tau_{\beta_j + 1}}{l_{\beta_j + 1}}\right) \cdot \dfrac{1}{\lambda u_r}, & w_{t_j} u_r < 0, \\ 1, & u_r = 0, \\ \left(\varepsilon_j w_{t_j} \rho_j + \delta_{\beta_j r} \dfrac{\tau_{\beta_j + 1}}{l_{\beta_j + 1}}\right) \cdot \dfrac{1}{\lambda u_r}, & w_{t_j} u_r > 0, \end{cases} \tag{4.5.6}$$

where $r = \alpha_j, \alpha_j + 1, \ldots, \beta_j$. As before, it may be shown that

$$\{k_r\}_{\alpha_j+1}^{\beta_j}, \quad \{m_r\}_{\alpha_j+2}^{\beta_j}, \text{ and } m_{\alpha_j}$$

can all be made positive by either properly selecting the positive number ε_j or appropriately adjusting the ratio ρ_j/ρ_{j+1}. As for m_{α_j+1}, Eqs. (4.5.5) and (4.5.6) yield

$$m_{\alpha_j+1} = \left[-w_{t_j}\rho_j + \frac{l_{\alpha_j}}{l_{\alpha_j} + l_{\alpha_j+1}}\left(\frac{\tau_{\alpha_j+1}}{l_{\alpha_j+1}} + \frac{\tau_{\alpha_j-1}}{l_{\alpha_j}}\right)\right] \cdot \frac{1}{\lambda u_{\alpha_j+1}}. \tag{4.5.7}$$

To make τ_{α_j} and m_{α_j+1} both positive, the value of ρ_{j-1}/ρ_j needs to satisfy the following inequality, per Eqs. (4.5.5) and (4.5.7):

$$-\frac{w_{t_j}}{w_{t_{j-1}}} \cdot \frac{l_{\alpha_j}}{\xi_{\alpha_j-1}}\left(\frac{\xi_{\alpha_j+1}}{l_{\alpha_j+1}} - \frac{l_{\alpha_j} + l_{\alpha_j+1}}{l_{\alpha_j}}\right) < \frac{\rho_{j-1}}{\rho_j} < -\frac{w_{t_j}}{w_{t_{j-1}}} \cdot \frac{l_{\alpha_j}}{\xi_{\alpha_j-1}} \cdot \frac{\xi_{\alpha_j+1}}{l_{\alpha_j+1}}. \tag{4.5.8}$$

Apparently, there always exists a ratio ρ_{j-1}/ρ_j that meets the requirement above. Since $w_{\beta_{j-1}}u_{\beta_{j-1}} < 0$, we have $m_{\beta_{j-1}} > 0$, without the need to change the value of ρ_{j-1}/ρ_j. Other situations can be discussed in a similar fashion.

Case 4 We should take care of the scenario where certain component w_s of the Vector $\boldsymbol{w}$ is zero, In this situation, there must exist an index $j\,(1 \le j \le i-2)$ such that $s-1 = \beta_j$ and $s+1 = \alpha_{j+1}$. As the coefficient of the equation $\tau_s = k_s w_s$, k_s can be assigned any positive value. The previous discussion does not involve the equation of motion associated with the s-th mass. Thus, in order to satisfy the equation here, we need to set

$$m_s = \begin{cases} (\tau_{s-1}/l_s + \tau_{s+1}/l_{s+1})/(\lambda u_s), & u_s \ne 0, \\ 1, & u_s = 0. \end{cases} \tag{4.5.9}$$

Because

$$\tau_{s-1} = \xi_{\beta_j} w_{t_j}\rho_j \quad \text{and} \quad \tau_{s+1} = \xi_{\alpha_{j+1}} w_{t_{j+1}}\rho_{j+1},$$

we can make

$$(\tau_{s-1}/l_s + \tau_{s+1}/l_{s+1})/u_s > 0, \quad \text{when } u_s \ne 0$$

or

$$\tau_{s-1}/l_s + \tau_{s+1}/l_{s+1} = 0, \quad \text{when } u_s = 0$$

by adjusting the ratio ρ_j/ρ_{j+1}. This will guarantee $m_s > 0$. Moreover, m_{β_j} and $m_{\alpha_{j+1}}$ are positive without the need to change the value of ρ_j/ρ_{j+1}.

Thus, we have demonstrated that the condition $S_u = S_w$ ensures the existence of a pinned–pinned beam with positive structural parameters. This process can be easily generalized to cover beams subject to other types of boundary constraints. Due to space limitation, we have to leave out the additional details here.

As of now, we have completely proved that the requirements on numbers of sign reversals S_u and $S_w = S_\tau$ in Sect. 4.4 are sufficient conditions for a vector $\boldsymbol{u}$ to be one of the displacement modes of the finite difference system of a beam. ■

The content of this subsection is mainly taken from Refs. [8, 11].

4.5.2 *Finite Difference System of a Beam Constructed with Two Modes*

Similar to the spring–mass system, when the step sizes l_r $(r = 1, 2, \ldots, n)$ are known, all parameters of the finite difference system of a beam can be determined by two modes. Below, we will use beams in clamped–free and pinned–pinned configurations as examples to demonstrate the process to derive the result.

Let us begin with the discrete system of a clamped–free beam. When two modes $(\omega_i, \boldsymbol{u}^{(i)})$ and $(\omega_j, \boldsymbol{u}^{(j)})$ are given and parameters l_r $(r = 1, 2, \ldots, n)$ are known, we can calculate the respective mode shapes of the relative rotation $\boldsymbol{w}^{(i)}$ and $\boldsymbol{w}^{(j)}$ using Eq. (4.3.1). Then, we get from modal equations of the finite difference system associated with the beam in the clamped–free configuration:

$$\begin{cases} \omega_i^2 m_r u_{ri} = k_{r-1}\dfrac{w_{r-1,i}}{l_r} - k_r w_{ri}\left(\dfrac{1}{l_r} + \dfrac{1}{l_{r+1}}\right) + k_{r+1}\dfrac{w_{r+1,i}}{l_{r+1}}, \\ \omega_j^2 m_r u_{rj} = k_{r-1}\dfrac{w_{r-1,j}}{l_r} - k_r w_{rj}\left(\dfrac{1}{l_r} + \dfrac{1}{l_{r+1}}\right) + k_{r+1}\dfrac{w_{r+1,j}}{l_{r+1}}, \\ \qquad r = 1, 2, \ldots, n-1 \text{ and } k_n = 0, \end{cases} \tag{4.5.10}$$

$$\omega_i^2 m_n u_{ni} = k_{n-1}\frac{w_{n-1,i}}{l_n}, \quad \omega_j^2 m_n u_{nj} = k_{n-1}\frac{w_{n-1,j}}{l_n}. \tag{4.5.11}$$

Without loss of generality, suppose the bending stiffness s_{n-1} at the $(n-1)$-th grid point is known and thus the parameter k_{n-1} can be calculated. Thus, Eq. (4.5.11) yields

$$m_n = k_{n-1} w_{n-1,i} / \omega_i^2 l_n u_{ni}.$$

Set

$$\begin{cases} a_r = (\omega_i^2 u_{ri} w_{r-1,j} - \omega_j^2 u_{rj} w_{r-1,i})/l_r, \quad r = 1, 2, \ldots, n-1, \\ b_r = (\omega_i^2 u_{ri} w_{rj} - \omega_j^2 u_{rj} w_{ri})(l_r + l_{r+1})/l_r l_{r+1}, \quad r = 1, 2, \ldots, n-1, \\ c_r = (\omega_i^2 u_{ri} w_{r+1,j} - \omega_j^2 u_{rj} w_{r+1,i})/l_{r+1}, \quad r = 1, 2, \ldots, n-2, \\ e_r = (w_{r-1,i} w_{rj} - w_{ri} w_{r-1,j})(l_r + l_{r+1})/l_r l_{r+1}, \quad r = 1, 2, \ldots, n-1, \\ g_r = (w_{r-1,i} w_{r+1,j} - w_{r+1,i} w_{r-1,j})/l_{r+1}, \quad r = 1, 2, \ldots, n-2. \end{cases} \tag{4.5.12}$$

Then, we obtain by solving Eq. (4.5.10):

$$\begin{cases} k_{r-1} = \dfrac{b_r k_r - c_r k_{r+1}}{a_r}, \quad r = n-1, n-2, \ldots, 1 \text{ and } k_n = 0, \\ m_r = \dfrac{e_r k_r - g_r k_{r+1}}{a_r l_r}, \quad r = n-1, n-2, \ldots, 1 \text{ and } k_n = 0, \end{cases} \tag{4.5.13}$$

The sufficient conditions that ensure all physical parameters calculated to be positive are:

(1) $a_r \neq 0$ $(r = 1, 2, \ldots, n-1)$.
(2) $\omega_i^2 u_{ni}/w_{ni} = \omega_j^2 u_{nj}/w_{nj} > 0$ $(i \neq j)$.
(3) $\det \boldsymbol{A}^{(r)} > 0$ and $\det \boldsymbol{B}^{(r)} > 0$. Here, $r = 1, 2, \ldots, n-1$; both $\boldsymbol{A}^{(r)}$ and $\boldsymbol{B}^{(r)}$ are tridiagonal matrices of order $n-r$; elements along the main diagonal of $\boldsymbol{A}^{(r)}$ are of the form $\{b_i/a_i\}_r^{n-1}$, while elements on the left and right subdiagonals are of the forms $\{c_i/a_i\}_r^{n-2}$ and $\{-1\}$, respectively; and the matrix $\boldsymbol{B}^{(r)}$ can be obtained from $\boldsymbol{A}^{(r)}$ by replacing the numerator b_r of the element at the first row and the first column with e_r and also substituting the numerator c_r of the element at the second row and the first column with g_r.

After $\{k_r\}_0^{n-1}$ and $\{m_r\}_1^n$ are calculated, the sectional bending stiffness and linear density at each of the discretization points can be computed from Eq. (4.1.4).

As for the finite difference system of a beam pinned at both ends, the process to construct it from two modes is almost identical to that used for the beam in the clamped–free configuration. There are only two differences: First, the system of modal equations associated with a pinned–pinned beam does not contain Eq. (4.5.11), and thus there is no need to compute the parameter m_n. Second, because $k_0 = 0$ for a beam pinned at both ends, the first set of equations in (4.5.13) only includes those with $r = n-1, n-2, \ldots, 3, 2$ and, in addition, $\det \boldsymbol{A}^{(1)} = 0$ needs to be satisfied.

Thus, we may reach the following conclusion:

Theorem 4.5 *Suppose the step sizes* l_r $(r = 1, 2, \ldots, n)$ *are known. Then within a constant factor, there exists a unique finite difference system of the beam in the clamped–free or pinned–pinned configuration that has two specified modes* $(\omega_i, \boldsymbol{u}^{(i)})$ *and* $(\omega_j, \boldsymbol{u}^{(j)})$.

This shows that among the n modes of the finite difference system modeling a beam, only two and, in fact, any two are independent.

The content of this subsection is largely taken from Refs. [9, 10].

4.6 Interlacement of Natural Frequencies of the Finite Difference System Modeling a Beam Subject to Various Boundary Constraints

In this section, we study the interlacing properties of frequencies of the finite difference system modeling a beam with one end to be under the same boundary support while the other end to be subject to different constraints.

4.6.1 Two Propositions

The following two propositions are needed in later discussion.

Proposition 4.5 *Consider a set of real numbers $c_i > 0\,(i = 1, 2, \dots, n)$ and some points $x_1 < x_2 < \cdots < x_n$. Then, the function*

$$f(x) = \sum_{i=1}^{n} \frac{c_i}{x_i - x}$$

has one and only one real root s_k in the interval $(x_k,\ x_{k+1})$, where $k = 1, 2, \dots, n-1$. Consequently, the function has a total of $n-1$ real roots $\{s_r\}_1^{n-1}$ in the interval (x_1, x_n).

Proof Notice that

$$\lim_{x\to x_k^+} f(x) = -\infty \text{ and } \lim_{x\to x_k^-} f(x) = +\infty.$$

Thus, we only need to prove $f(x)$ is monotonically increasing in each interval (x_k, x_{k+1}). But this is obviously true, because we always have

$$f'(x) = \sum_{i=1}^{n} \frac{c_i}{(x_i - x)^2} > 0.$$

■

Proposition 4.6 *For the function $f(x)$ in Proposition 4.5, the n roots $\{\eta_r\}_1^n$ of the equation*

$$f(x) + q = 0$$

satisfy

$$0 < x_1 < \eta_1 < s_1 < x_2 < \cdots < x_{n-1} < \eta_{n-1} < s_{n-1} < x_n < \eta_n \tag{4.6.1}$$

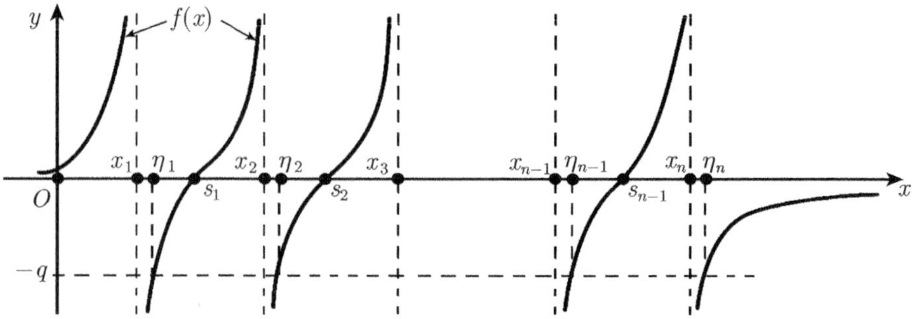

Fig. 4.3 Relationship between roots of equations $f(x) = 0$ and $f(x) + q = 0\,(q > 0)$

if $q > 0$*; while the inequality*

$$\eta_1 < x_1 < s_1 < \eta_2 < x_2 < \cdots < x_{n-1} < s_{n-1} < \eta_n < x_n \tag{4.6.2}$$

holds true if $q < 0$*. In addition,* $\eta_1 > 0$ *only if* $-q > f(0)$.

Proof In fact, we know two things from the proof of Proposition 4.5: First, $f(x)$ is also monotonically increasing in $(-\infty, x_1)$ and (x_n, ∞). Second, the plot of $y = f(x) + q$ can be obtained simply by shifting that of $y = f(x)$ upward or downward along the y-axis (see Fig. 4.3). ■

4.6.2 *Interlacement of Natural Frequencies of the Finite Difference System Modeling a Beam Clamped at One End and Constrained in Various Ways at the Other End*

Consider the finite difference system modeling a beam that is always clamped at the left end but subject to different constraints at the right end: free, sliding, pinned, anti-resonant, or clamped. Let $\{\omega_i\}_1^n$, $\{\sigma_i\}_1^n$, $\{\mu_i\}_1^{n-1}$, $\{\nu_i\}_1^{n-1}$, and $\{\eta_i\}_1^{n-1}$ represent natural angular frequencies of the same finite difference system in clamped–free, clamped–sliding, clamped–pinned, clamped–anti-resonant, and clamped–clamped configurations, respectively. With the help of the above two propositions as well as another proposition, it has been proved in Refs. [2, 3] that these frequencies have the following interlacing relationship:

$$(\eta_{i-1}, \omega_i) < \sigma_i < \nu_i < \mu_i < (\eta_i, \omega_{i+1}) < \sigma_{i+1}, \quad i = 1, 2, \ldots, n-1. \tag{4.6.3}$$

Details can be found in Sect. 8.4 of Ref. [3].

4.6.3 Interlacement of Natural Frequencies of the Finite Difference System Modeling a Beam Pinned at One End and Subject to Different Support Conditions at the Other End

We now study interlacing properties of frequencies associated with the finite difference system modeling a beam pinned at the left end and constrained in three different ways at the right end: anti-resonant, pinned, or clamped.

Let us begin with the finite difference system of the beam pinned on both sides, i.e., a simply supported beam. Denote its eigenvector matrix that has been normalized with respect to the mass matrix $\boldsymbol{M}_{\mathrm{pp}}$ as $\boldsymbol{U} = (u_{ji})$. The governing equation for the forced vibration is

$$\omega^2 \boldsymbol{M}_{\mathrm{pp}} \boldsymbol{u} = \boldsymbol{A}_{\mathrm{pp}} \boldsymbol{u} + \tau_n \boldsymbol{e}^{(n-1)} / l_n, \tag{4.6.4}$$

where $\boldsymbol{M}_{\mathrm{pp}}$ and $\boldsymbol{A}_{\mathrm{pp}}$ are, respectively, the mass and stiffness matrices of the finite difference system modeling the pinned–pinned beam, while $\tau = \tau_n \sin \omega t$ represents a concentrated harmonic couple acting on the right end of the beam.

Next, by applying the method of modal superposition, we can express the displacement vector as

$$\boldsymbol{u} = \sum_{i=1}^{n} \alpha_i \boldsymbol{u}^{(i)}. \tag{4.6.5}$$

Substitute the above expression into Eq. (4.6.4), left multiply both sides of the resulting equation by $(\boldsymbol{u}^{(k)})^{\mathrm{T}}$ $(k = 1, 2, \ldots, n-1)$, and take into account the orthonormal properties of mode shapes $\boldsymbol{u}^{(i)}$ $(i = 1, 2, \ldots, n-1)$ of the finite difference system modeling the simply supported beam. After these steps, we get

$$\omega^2 \alpha_i = \omega_i^2 \alpha_i - \tau_n \theta_{ni}, \quad i = 1, 2, \ldots, n-1.$$

Solving for α_i $(i = 1, 2, \ldots, n-1)$ and inserting them back into Eq. (4.6.5), we obtain

$$\boldsymbol{u} = \sum_{i=1}^{n-1} \boldsymbol{u}^{(i)} \frac{\theta_{ni} \tau_n}{\omega_i^2 - \omega^2}. \tag{4.6.6}$$

In the derivation above, we have used the condition $\theta_{ni} = -u_{n-1,i}/l_n$. Furthermore, we can derive

$$\theta_n = \sum_{i=1}^{n-1} \frac{\theta_{ni}^2 \tau_n}{\omega_i^2 - \omega^2}, \tag{4.6.7}$$

$$\phi_n = \sum_{i=1}^{n-1} \frac{\theta_{ni}\phi_{ni}\tau_n}{\omega_i^2 - \omega^2} - \frac{\tau_n}{l_n}. \tag{4.6.8}$$

When the beam is in the pinned–anti-resonant configuration, the boundary conditions at the right end are $u_n = 0$ and $\phi_{n+1} = 0$. From these and the equilibrium equation at the right boundary, we should also have $\phi_n = 0$. Thus, Eq. (4.6.8) leads to the following frequency equation for the finite difference system of the pinned–anti-resonant beam:

$$\sum_{i=1}^{n-1} \frac{\theta_{ni}\phi_{ni}}{\omega_i^2 - \omega^2} = \frac{1}{l_n}. \tag{4.6.9}$$

It is known that natural angular frequencies $\{\omega_i\}_1^{n-1}$ of the finite difference system modeling the simply supported beam are all distinct. In addition, it can be shown that $\theta_{ni}\phi_{ni} > 0 (i = 1, 2, \ldots, n-1)$. Thus, Proposition 4.6 shows that natural angular frequencies $\{\xi_i\}_1^{n-1}$ of the finite difference system of the pinned–anti-resonant beam satisfy the following inequalities:

$$\omega_i < \xi_{i+1} < \omega_{i+1}, \quad i = 1, 2, \ldots, n-2. \tag{4.6.10}$$

As for ξ_1, because the inequality

$$\sum_{i=1}^{n-1} \frac{\theta_{ni}\phi_{ni}}{\omega_i^2} < \frac{1}{l_n}$$

holds true and the function on the left-hand side of Eq. (4.6.9) is monotonic in the interval $(0,\ \omega_1)$, we conclude that $0 < \xi_1 < \omega_1$.

Now let us turn to the finite difference system of the beam in the pinned–clamped configuration. Again, denote its natural angular frequencies as $\{\mu_i\}_1^{n-1}$. Since the boundary constraints at the right end are $u_n = 0$ and $\theta_{n+1} = 0$, which lead to $\tau_n = -k_n\theta_n$, Eq. (4.6.7) yields the following frequency equation:

$$\sum_{i=1}^{n-1} \frac{\theta_{ni}^2}{\omega_i^2 - \omega^2} = -\frac{1}{k_n}. \tag{4.6.11}$$

Based on Proposition 4.6 as well as the monotonicity of the left-hand side of Eq. (4.6.11) on the interval $(\omega_{n-1}, +\infty)$, we derive

$$\omega_i < \mu_i < \omega_{i+1} < \mu_{i+1}, \quad i = 1, 2, \ldots, n-2. \tag{4.6.12}$$

In order to determine the interlacing properties of $\{\xi_i\}_1^{n-1}$ and $\{\mu_i\}_1^{n-1}$, i.e., the relationship between ξ_{i+1} and $\mu_i (i = 1, 2, \ldots, n-2)$, we may use the finite difference system of the pinned–anti-resonant beam as the basis, apply the same process employed in the earlier discussion, introduce some relatively complicated transformation, and eventually prove $\mu_i < \xi_{i+1}$ $(i = 1, 2, \ldots, n-2)$. Thus, we have finally established the following relationship among natural angular frequencies of the finite difference system modeling a beam pinned at the left end but subject to anti-resonant, pinned, or clamped constraint at the right end:

$$0 < \xi_1 < \omega_1 < \mu_1 < \xi_2 < \omega_2 < \cdots < \omega_{n-2} < \mu_{n-2} < \xi_{n-1} < \omega_{n-1} < \mu_{n-1}. \tag{4.6.13}$$

The content of this subsection is mainly taken from Ref. [12].

4.6.4 Construction of the Finite Difference System of a Beam Using Three Sets of Spectra

In Sects. 4.6.2 and 4.6.3, we discussed interlacing properties of angular frequencies associated the finite difference system modeling a beam subject to various boundary constraints. Based on these results, Gladwell [3, 4], He et al. [6, 7], and Wang et al. [12] studied the inverse problem in vibration, where finite difference systems of beams are reconstructed from three sets of spectra corresponding to different types of boundary constraints. Here is a quick summary of the work that has been completed:

(1) Given natural angular frequencies $\{\omega_i\}_1^n$ of the finite difference system modeling a clamped-free beam as well as any two among the three sets of natural angular frequencies $\{\sigma_i\}_1^n$, $\{\mu_i\}_1^{n-1}$, and $\{\nu_i\}_1^{n-1}$ associated with the finite difference system of the same beam in the clamped–sliding, clamped–pinned, and clamped–anti-resonant configurations, respectively, the finite difference system of this beam can be constructed with the block Lanczos algorithm. See Refs. [3, 7].
(2) Given three sets of natural angular frequencies $\{\xi_i, \ \omega_i, \ \mu_i\}_1^{n-1}$ associated with the finite difference system of the same beam in the pinned–anti-resonant, pinned–pinned, and pinned–clamped configurations, respectively, the finite difference system of the beam can again be built with the block Lanczos algorithm. See Ref. [12].

4.7 Oscillatory Properties of Finite Element Systems of a Beam

Due to the critical role that finite element systems of beams play in engineering, it is natural for us to pay particular attention to their oscillatory properties in static deformation and vibration. See Ref. [5]. These subjects will be our focus in this section.

In Chap. 3, we proved that several discrete systems of the string and bar have oscillatory properties in static deformation and vibration. This was accomplished by demonstrating the respective stiffness matrices to be positive definite standard Jacobi matrices. In earlier sections of this chapter, we showed the stiffness matrix associated with the finite difference system of a single-span beam to be sign-oscillatory, and thus proved such a system has oscillatory properties in vibration.

For the finite element systems of a beam to be discussed in this section, their stiffness matrices are of the seven-diagonal type. It is difficult to apply the previously employed methods to such systems. Thus, Zheng, Chen, and Wang introduced a simple and elegant approach to determine whether some relatively complex systems possess oscillatory properties in static deformation and vibration. See Refs. [19–22].

4.7.1 The Preservation Theorem of Oscillatory Properties

First, let us establish a fundamental theorem that shows how oscillatory properties in static deformation and vibration can be preserved from the continuous system of a one-dimensional single-span structure to its discrete counterpart.

Theorem 4.6 *If*

(1) *System A is a one-dimensional, single-span, well-constrained, and discrete system;*
(2) *there exists another System B that can be either discrete or continuous but needs to have oscillatory properties in static deformation and in case of the system being discrete, it consists of the same set of discretization nodes as System A does; and*
(3) *when an arbitrary set of forces $\{F\}$ is simultaneously applied to the discretization nodes of System A and the corresponding points of System B, the two systems share identical displacement at each discretization node of A and its respective point of B;*

then the discrete system A should also have oscillatory properties in static deformation. Furthermore, if a lumped mass matrix is used in System A, it should have oscillatory properties in vibration as well.

Proof Suppose the discrete system A in the theorem is subject to an arbitrary set of k nodal forces $\{F\}$, where $k \leq$ the number of degrees of freedom of the system A. If the corresponding system B is also discrete, then the number of sign reversals of its u-line does not exceed $k-1$, and the u-line between any two adjacent discretization nodes is a straight line segment. If B is a continuous system, then the number of sign reversals of its displacement function u does not exceed $k-1$, and u is a linear or higher-order function in between two points corresponding to the two adjacent discretization nodes of System A. Since the u-line of the discrete system A between two adjacent discretization nodes is a straight line segment, its number of sign reversals definitely should not exceed $k - 1$. Furthermore, when only one nodal force is applied to System A, the displacements at all nodes are obviously in the direction of this force. Therefore, System A ought to have oscillatory properties in static deformation.

Based on the final conclusion of Sect. 2.10.1, the flexibility matrix of System A is an oscillatory matrix. Furthermore, if a lumped mass matrix is adopted in System A, then it has oscillatory properties in vibration. ■

When the discrete system A happens to be derived from the continuous system B, we get a special case of Theorem 4.6, where the theorem shows how the oscillatory properties of the one-dimensional continuous system are preserved and passed on to the respective discrete system. This result is of great significance in both theory and application, and is thus warranted to be included in a separate theorem.

Theorem 4.7 *If*

(1) *System B is a one-dimensional, single-span, well-constrained, and continuous system, and has oscillatory properties in static deformation;*
(2) *System A is an outcome of discretizing System B; and*
(3) *the same concentrated force, simultaneously applied to any discretization node of System A and the corresponding point of System B, results in an identical displacement at every discretization node of System A and the corresponding point of System B;*

then the discrete system A should have oscillatory properties in static deformation. Furthermore, if a lumped mass matrix is used in System A, it has oscillatory properties in vibration, just like System B.

Proof Apply the same set of concentrated forces to any k discretization nodes of System A as well as its corresponding points of System B, where $k \leq$ number of degrees of freedom of System A. Because Systems B and A are both linear, the displacement at every node of System A should be identical to that at the corresponding point of System B. Thus, Theorem 4.6 leads to the conclusions of this theorem. ■

Theorem 4.7 may be referred to as the preservation theorem that preserves oscillatory properties in static deformation and vibration from the continuous system to its discrete counterpart.

It will be proved in Chap. 6 that the continuous and well-constrained system of a one-dimensional single-span beam has oscillatory properties in static deformation. Thus, it is easy to conclude, based on Theorem 4.7, that discrete systems of the following two types of beams have oscillatory properties in static deformation and vibration:

(1) the system consisting of a massless elastic beam and some masses, as shown in Fig. 4.4;
(2) the finite element system of a beam consisting of several uniform beam segments and with Hermite polynomials as the shape functions for the displacement, as shown in Fig. 4.5.

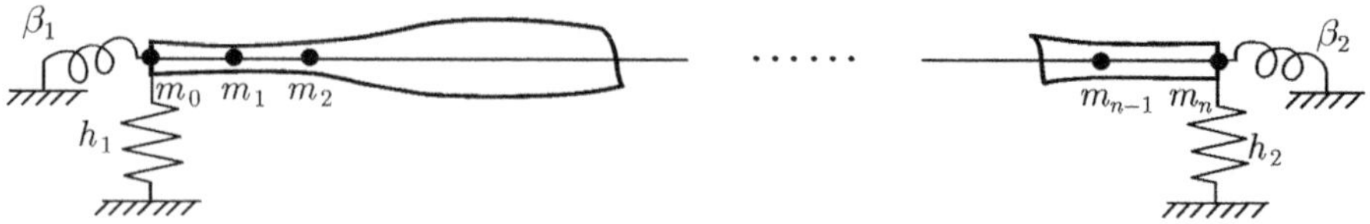

Fig. 4.4 An illustration of the system consisting of a massless elastic beam and some masses

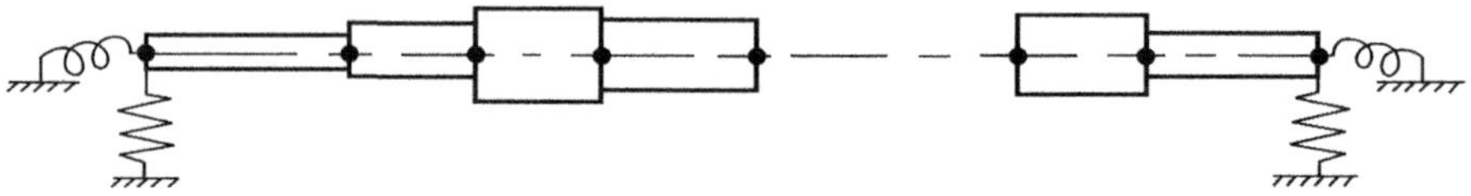

Fig. 4.5 An illustration of the system consisting of some masses and several massless beam segments with uniform cross sections

4.7.2 *Oscillatory Properties of the Finite Element Beam Model Constructed with the Force Method*

Among finite element software packages for engineering application, the Hellinger–Reissner principle is often used to construct element stiffness matrices of finite element systems modeling bars, beams, and other one-dimensional structures. This approach is often called the force or flexibility method. The finite element system of a beam built this way is referred to as the HR-FE beam in short. By comparing this discrete system with its continuous counterpart, we notice an important advantage of it. Specifically, if the same set of concentrated forces is applied to nodes of the discrete system and their corresponding points in the continuous system, nodal displacements of discrete system calculated with the finite element methods are identical to their counterparts of the continuous system obtained analytically. Thus, according to Theorem 4.7, an HR-FE beam with any stiffness distribution should have oscillatory properties. These will be elaborated in the discussion below.

Now let us first derive the element stiffness matrix of an HR-FE beam. If there are only concentrated forces on element nodes, the bending moment diagram of the beam within an element is a straight line, and they form a continuous broken line along the entire beam. Thus, the bending moment within an element can be expressed as a linear interpolation of bending moments $(\tau_i, \tau_j)^{\mathrm{T}}$ at the two end points of the element:

$$\tau = \boldsymbol{N}^F(\xi)\begin{pmatrix} \tau_i \\ \tau_j \end{pmatrix},$$

where $\boldsymbol{N}^F(\xi) = (1 - \xi, \xi)$. We may then derive the following equilibrium equation at the element level by applying the Hellinger–Reissner principal

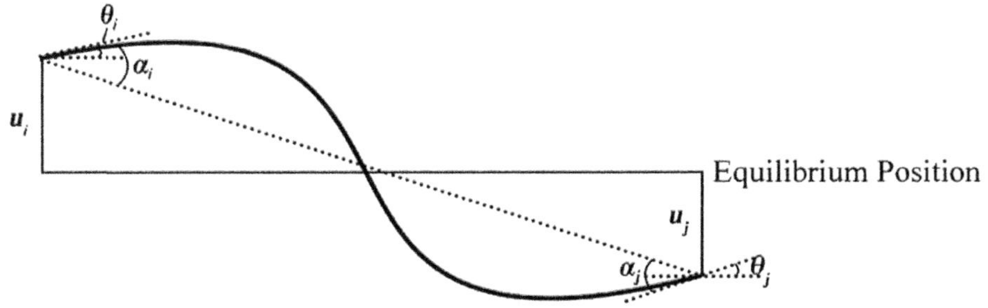

Fig. 4.6 Definition of the relative rotation

$$\begin{pmatrix} \alpha_i \\ \alpha_j \end{pmatrix} = l \int_0^1 \frac{(\boldsymbol{N}^F(\xi))^{\mathrm{T}} \boldsymbol{N}^F(\xi)}{\overline{EJ}(\xi)} \mathrm{d}\xi = \overline{\boldsymbol{F}}^e_{HR} \begin{pmatrix} \tau_i \\ \tau_j \end{pmatrix}, \tag{4.7.1}$$

where $(\alpha_i, \alpha_j)^{\mathrm{T}}$ is the relative rotation at the two ends of the element, as shown in Fig. 4.6, $\overline{EJ}$ is the bending stiffness within the element, the matrix $\overline{\boldsymbol{F}}^e_{\mathrm{HR}}$ may be interpreted as the flexibility matrix of a simply supported beam while its inverse is the corresponding stiffness matrix, and the relative rotation is related to the deflection and rotation at element nodes by

$$\begin{pmatrix} \alpha_i \\ \alpha_j \end{pmatrix} = \boldsymbol{S}\boldsymbol{u} = \begin{pmatrix} \frac{1}{l} & 1 & -\frac{1}{l} & 0 \\ \frac{1}{l} & 0 & -\frac{1}{l} & 1 \end{pmatrix} \begin{pmatrix} u_i \\ \theta_i \\ u_j \\ \theta_j \end{pmatrix}. \tag{4.7.2}$$

The element stiffness matrix and another form of the element equilibrium equation are, respectively,

$$\boldsymbol{K}^e_{HR} = \boldsymbol{S}^{\mathrm{T}} (\overline{\boldsymbol{F}}^e_{HR})^{-1} \boldsymbol{S} \tag{4.7.3}$$

and

$$(\phi_i, \tau_i, \phi_j, \tau_j)^{\mathrm{T}} = \boldsymbol{K}^e_{HR} (u_i, \theta_i, u_j, \theta_j)^{\mathrm{T}}, \tag{4.7.4}$$

where ϕ_i and ϕ_j are shear forces acting at the two ends of the element. The global stiffness matrix and equilibrium equation of the HR-FE beam can be obtained by assembling all of their counterparts at the element level, which are of the forms shown in Eqs. (4.7.3) and (4.7.4).

Next, we look at the continuous system of the beam that is related to the HR-FE beam. When one set of concentrated forces is applied to the continuous system at points that are nodes of the HR-FE beam, the bending moment within the beam segment that is an element of the HR-FE beam should be a linear function. Therefore, the continuous system and its discrete counterpart, i.e., the HR-FE beam, have identical internal forces. In addition, the two systems share the same displacements at points that are nodes of the HR-FE beam. Since the beam has oscillatory properties, we get the following result from Theorem 4.7:

Theorem 4.8 *Consider a single-span and well-constrained beam. The distribution of its bending stiffness may be arbitrary. Then, the HR-FE beam of this structure possesses oscillatory properties in static deformation. Furthermore, if the lumped mass matrix is used, the HR-FE beam also has oscillatory properties in vibration.*

4.7.3 Oscillatory Properties of the Finite Element Beam Model Constructed with the Displacement Method and Hermite Shape Functions

A PE-FE beam is an abbreviation for the finite element system of the beam established by using the principle of minimum potential energy, the two-node elements, and the third-degree Hermite polynomials as shape functions. For the beam whose section parameters are piecewise constant functions as well as its PE-FE counterpart, their displacements are identical piecewise third-degree polynomials, when external forces are applied only to the finite element nodes and their corresponding points of the continuous system modeling the beam. Thus, both systems should have oscillatory properties. In the following discussion, we will consider the more general situation, where the section parameters within an element are not constants.

We begin by deriving the element stiffness matrix of a PE-FE beam. From the principle of minimum potential energy, the element equilibrium equation, in terms of the relative rotation and bending moment at element nodes, assumes the following form:

$$\bar{\boldsymbol{K}}^{e}_{\mathrm{PE}}\begin{pmatrix}\alpha_i\\ \alpha_j\end{pmatrix}=\int_0^1\frac{EJ(\xi)}{l^3}\left(\frac{\mathrm{d}^2\boldsymbol{N}^d}{\mathrm{d}\xi^2}\right)^{\mathrm{T}}\frac{\mathrm{d}^2\boldsymbol{N}^d}{\mathrm{d}\xi^2}\mathrm{d}\xi\begin{pmatrix}\alpha_i\\ \alpha_j\end{pmatrix}=\begin{pmatrix}\tau_i\\ \tau_j\end{pmatrix},\tag{4.7.5}$$

where $\boldsymbol{N}^d(\xi)=(l(\xi-2\xi^2+\xi^3), l(\xi^3-\xi^2))$ and the element stiffness matrix is

$$\boldsymbol{K}^{e}_{\mathrm{PE}}=\boldsymbol{S}^{\mathrm{T}}\bar{\boldsymbol{K}}^{e}_{\mathrm{PE}}\boldsymbol{S}.\tag{4.7.6}$$

After rewriting the element equilibrium equation in terms of displacements, angles of rotation, shear forces, and bending moments at element nodes, we get

$$(\phi_i,\tau_i,\phi_j,\tau_j)^{\mathrm{T}}=\boldsymbol{K}^{e}_{\mathrm{PE}}(u_i,\theta_i,u_j,\theta_j)^{\mathrm{T}}.\tag{4.7.7}$$

The global stiffness matrix and equilibrium equation of the PE-FE beam can be obtained by assembling, respectively, the element stiffness matrices and equilibrium equations given in Eqs. (4.7.6) and (4.7.7).

Based on Theorems 4.6 and 4.8, for a given PE-FE beam, if we can construct an HR-FE beam such that the two beams share identical nodal displacements and angles of rotation under any set of nodal forces, then this PE-FE beam should have oscillatory properties in static deformation. Such an HR-FE beam can be obtained by requiring all of its element flexibility matrices be equal to their counterparts of the PE-FE beam. Specifically, suppose the bending stiffness of two corresponding elements in the PE-FE and HR-FE beams are EJ and $\overline{EJ}$, respectively. By comparing Eqs. (4.7.1) and (4.7.5), we conclude that the following condition must be satisfied:

$$\boldsymbol{F}^e_{HR} = (\overline{\boldsymbol{K}}^e_{PE})^{-1}, \tag{4.7.8}$$

where $\overline{\boldsymbol{K}}^e_{PE}$ and $\boldsymbol{F}^e_{HR}$ are given in Eqs. (4.7.5) and (4.7.1), respectively.

Notice that we have to skip the complicated proof and only present the final conclusion here: When the bending stiffness $EJ(\xi)$ of each element in the PE-FE beam satisfies the condition,

$$\int_0^1 EJ(9\xi^2 - 9\xi + 2)\mathrm{d}\xi > 0, \tag{4.7.9}$$

it is possible to construct an HR-FE beam with the bending stiffness of the respective element to be $\overline{EJ}(\xi)$. Thus, we obtain the following theorem:

Theorem 4.9 *Consider the PE-FE beam model of a one-dimensional, single-span, and well-constrained beam. If the bending stiffness of each of its element satisfies* Eq. (4.7.9), *then this beam has oscillatory properties in static deformation. Furthermore, if a lumped mass matrix is used with the PE-FE beam, then it has oscillatory properties in vibration.*

The weight function $w = 9\xi^2 - 9\xi + 2$ in the integral in Eq. (4.7.9) is illustrated in Fig. 4.7.

For the majority of beams in engineering, we can ensure that the bending stiffness of their finite element systems satisfies Eq. (4.7.9) by properly partitioning the beams. Thus, the resulting PE-FE beams should have oscillatory properties in static deformation. However, Eq. (4.7.9) is not a necessary condition. Consequently, there are finite element systems that have oscillatory properties in static deformation but may not satisfy Eq. (4.7.9). Also, there are beams with a special form of bending stiffness distribution and involving inappropriate finite element partition. We found in these cases, Eq. (4.7.9) is not satisfied, and the respective PE-FE beams do not have oscillatory properties. A real example was presented in Refs. [20, 22].

In addition to what have been covered above, Zijun Zheng has studied oscillatory properties of higher order PE-FE beams and the Timoshenko beam in Ref. [22]. Some significant results have been obtained.

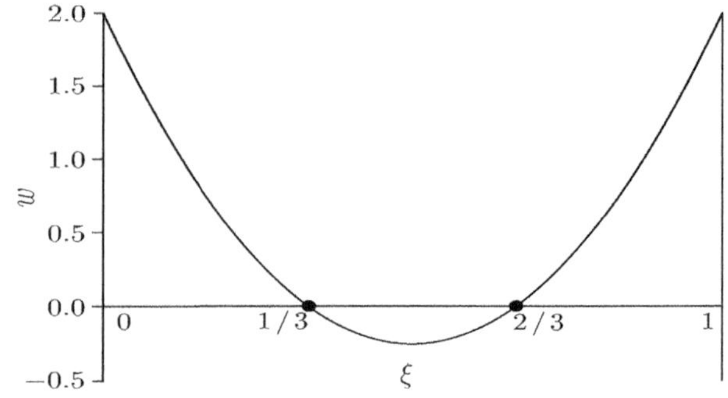

Fig. 4.7 The weight function in the integral in Eq. (4.7.9)

4.8 Modal Qualitative Properties of Discrete Systems of Multi-Span Beams

4.8.1 The Finite Difference Model of a Multi-Span Beam Subject to Arbitrary Boundary Supports at Two Ends

Consider a beam with length l, linear density $\rho(x)$, and bending stiffness $EJ(x)$. Suppose it vibrates laterally under the general form of boundary constraints given by Eqs. (4.1.2) and (4.1.3). In Sect. 4.1, we discretized this beam into an arbitrarily supported single-span model shown in Fig. 4.1b, using the second-order central finite difference scheme. In the figure, parameters m_r and k_r $(r = 0, 1, \ldots, n)$ represent, respectively, the mass of the r-th particle and the stiffness of the rotational springs controling relative rotation between the two rigid bars at the r-th particle; l_r $(r = 1, 2, \ldots, n)$ is the length of the r-th massless rigid bar; u_r $(r = 0, 1, \ldots, n)$ denotes the lateral displacement of the r-th mass; and x_r $(r = 0, 1, \ldots, n)$ designates the coordinate of the r-th mass.

Now suppose p pinned supports are applied to points x_{c_k} $(k = 1, 2, \ldots, p)$, which is a subset of the discretization nodes $\{x_i\}_0^n$. Thus,

$$u(x_{c_k}) = 0, \quad k = 1, 2, \ldots, p. \tag{4.8.1}$$

In addition, the following relationship exists among the points:

$$0 = x_0 < x_1 < \cdots < x_{r_1-1} < x_{r_1}(= x_{c_1}) < x_{r_1+1} < \cdots < x_{r_p}(= x_{c_p}) < x_{r_p+1} < \cdots < x_n = l.$$

The corresponding physical model is shown in Fig. 4.8, where the parameters are interpreted in the same way as before. In the following discussion, we refer to the model in Fig. 4.1b, which does not have the internal supports, as System S; and call the model in Fig. 4.8, which is constrained by pinned mid-supports, System S^* corresponding to S.

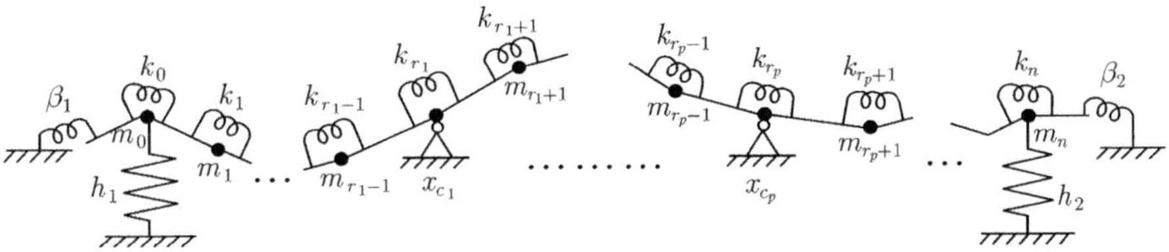

Fig. 4.8 The finite difference model of an arbitrarily supported multi-span beam

4.8.2 *Stiffness Matrix of the Multi-Span Beam System S**

Set $\lambda = \omega^2$ and $u_r = u(x_r)$. It has been shown in Sect. 4.1 that the modal equations of the aforementioned system S are of the form

$$\begin{cases} \lambda m_0 u_0 = a_0 u_0 - b_1 u_1 + c_1 u_2 + h_1 u_0, \\ \lambda m_1 u_1 = -b_1 u_0 + a_1 u_1 - b_2 u_2 + c_2 u_3, \\ \lambda m_r u_r = c_{r-1} u_{r-2} - b_r u_{r-1} + a_r u_r - b_{r+1} u_{r+1} + c_{r+1} u_{r+2}, \quad r = 2, 3, \ldots, n-2, \\ \lambda m_{n-1} u_{n-1} = c_{n-2} u_{n-3} - b_{n-1} u_{n-2} + a_{n-1} u_{n-1} - b_n u_n, \\ \lambda m_n u_n = c_{n-1} u_{n-2} - b_n u_{n-1} + a_n u_n + h_2 u_n, \end{cases} \tag{4.8.2}$$

where

$$\begin{cases} a_0 = \dfrac{k_0^* + k_1}{l_1^2}, \quad a_1 = \dfrac{k_0^*}{l_1^2} + \left(\dfrac{1}{l_1} + \dfrac{1}{l_2}\right)^2 k_1 + \dfrac{k_2}{l_2^2}, \\ a_r = \dfrac{k_{r-1}}{l_r^2} + \left(\dfrac{1}{l_r} + \dfrac{1}{l_{r+1}}\right)^2 k_r + \dfrac{k_{r+1}}{l_{r+1}^2}, \quad r = 2, 3, \ldots, n-2, \\ a_{n-1} = \dfrac{k_{n-2}}{l_{n-1}^2} + \left(\dfrac{1}{l_{n-1}} + \dfrac{1}{l_n}\right)^2 k_{n-1} + \dfrac{k_n^*}{l_n^2}, \quad a_n = \dfrac{k_n^* + k_{n-1}}{l_n^2}, \end{cases} \tag{4.8.3}$$

$$\begin{cases} b_1 = \dfrac{k_0^*}{l_1^2} + \dfrac{k_1}{l_1}\left(\dfrac{1}{l_1} + \dfrac{1}{l_2}\right), \\ b_r = \dfrac{k_{r-1}}{l_r}\left(\dfrac{1}{l_{r-1}} + \dfrac{1}{l_r}\right) + \dfrac{k_r}{l_r}\left(\dfrac{1}{l_r} + \dfrac{1}{l_{r+1}}\right), \quad r = 2, 3, \ldots, n-1, \\ b_n = \dfrac{k_{n-1}}{l_n}\left(\dfrac{1}{l_{n-1}} + \dfrac{1}{l_n}\right) + \dfrac{k_n^*}{l_n^2}, \end{cases} \tag{4.8.4}$$

$$c_r = \frac{k_r}{l_r l_{r+1}}, \quad r = 1, 2, \ldots, n-1, \tag{4.8.5}$$

and both k_0^* and k_n^* are defined in Eq. (4.1.12).

Apparently, differences between equations of System S^* in Fig. 4.8 and System S in Fig. 4.1b are attributed only to those equations related to masses immediately adjacent to the mid-supports; hence we may derive the governing equations of System S^* from those of System S. In fact, according to Fig. 4.8, the masses at x_{r_k} $(k = 1, 2, \ldots, p)$ are still and thus, the respective equations of motion are degenerated to equilibrium equations. As a result, they drop out of the system of modal equations. Furthermore, the modal components at x_{r_k} $(k = 1, 2, \ldots, p)$ should be replaced with 0 in all other equations. After these steps, we may get the following system of modal equations for System S^*

$$\begin{cases} \lambda m_0 u_0 = a_0 u_0 - b_1 u_1 + c_1 u_2 + h_1 u_0, \\ \lambda m_1 u_1 = -b_1 u_0 + a_1 u_1 - b_2 u_2 + c_2 u_3, \\ \lambda m_r u_r = c_{r-1} u_{r-2} - b_r u_{r-1} + a_r u_r - b_{r+1} u_{r+1} + c_{r+1} u_{r+2} \\ \quad (r = 2, 3, \ldots, r_1 - 3; r_1 + 3, \ldots, r_2 - 3; \ldots; r_p + 3, \ldots, n-2), \\ \lambda m_{r_k-2} u_{r_k-2} = c_{r_k-3} u_{r_k-4} - b_{r_k-2} u_{r_k-3} + a_{r_k-2} u_{r_k-2} - b_{r_k-1} u_{r_k-1}, \\ \lambda m_{r_k-1} u_{r_k-1} = c_{r_k-2} u_{r_k-3} - b_{r_k-1} u_{r_k-2} + a_{r_k-1} u_{r_k-1} + c_{r_k} u_{r_k+1}, \\ \lambda m_{r_k+1} u_{r_k+1} = c_{r_k} u_{r_k-1} + a_{r_k+1} u_{r_k+1} - b_{r_k+2} u_{r_k+2} + c_{r_k+2} u_{r_k+3}, \\ \lambda m_{r_k+2} u_{r_k+2} = -b_{r_k+2} u_{r_k+1} + a_{r_k+2} u_{r_k+2} - b_{r_k+3} u_{r_k+3} + c_{r_k+3} u_{r_k+4} \\ \quad (k = 1, 2, \ldots, p), \\ \lambda m_{n-1} u_{n-1} = c_{n-2} u_{n-3} - b_{n-1} u_{n-2} + a_{n-1} u_{n-1} - b_n u_n, \\ \lambda m_n u_n = c_{n-1} u_{n-2} - b_n u_{n-1} + a_n u_n + h_2 u_n. \end{cases} \tag{4.8.6}$$

Rewrite the equations in the vector form as

$$\lambda \boldsymbol{M} \boldsymbol{u} = \boldsymbol{A}_p \boldsymbol{u}. \tag{4.8.7}$$

Here,

$$\boldsymbol{u} = (u_0, u_1, \ldots, u_{r_1-1}, u_{r_1+1}, u_{r_1+2}, \ldots, u_{r_p-1}, u_{r_p+1}, u_{r_p+2}, \ldots, u_n)^{\mathrm{T}}$$

denotes the mode shape vector of System S^*;

$$\boldsymbol{M} = \mathrm{diag}(m_0, m_1, \ldots, m_{r_1-1}, m_{r_1+1}, \ldots, m_{r_p-1}, m_{r_p+1}, \ldots, m_n)$$

is the mass matrix; and the stiffness matrix is of the block tridiagonal form:

$$
\boldsymbol{A}_p = \begin{bmatrix}
\boldsymbol{A}_{11} & \boldsymbol{A}_{12} & \boldsymbol{0} & \cdots & \boldsymbol{0} & \boldsymbol{0} & \boldsymbol{0} \\
\boldsymbol{A}_{21} & \boldsymbol{A}_{22} & \boldsymbol{A}_{23} & \cdots & \boldsymbol{0} & \boldsymbol{0} & \boldsymbol{0} \\
\boldsymbol{0} & \boldsymbol{A}_{32} & \boldsymbol{A}_{33} & \cdots & \boldsymbol{0} & \boldsymbol{0} & \boldsymbol{0} \\
\vdots & \vdots & \vdots & & \vdots & \vdots & \vdots \\
\boldsymbol{0} & \boldsymbol{0} & \boldsymbol{0} & \cdots & \boldsymbol{A}_{p-1,p-1} & \boldsymbol{A}_{p-1,p} & \boldsymbol{0} \\
\boldsymbol{0} & \boldsymbol{0} & \boldsymbol{0} & \cdots & \boldsymbol{A}_{p,p-1} & \boldsymbol{A}_{pp} & \boldsymbol{A}_{p,p+1} \\
\boldsymbol{0} & \boldsymbol{0} & \boldsymbol{0} & \cdots & \boldsymbol{0} & \boldsymbol{A}_{p+1,p} & \boldsymbol{A}_{p+1,p+1}
\end{bmatrix}, \tag{4.8.8}
$$

where **0** denotes a submatrix whose elements are all zeros,

$$
\boldsymbol{A}_{ii} = \begin{bmatrix}
a_{r_{i-1}+1} + h_1\delta_{i1} & -b_{r_{i-1}+2} & c_{r_{i-1}+2} & & & & & \\
-b_{r_{i-1}+2} & a_{r_{i-1}+2} & -b_{r_{i-1}+3} & c_{r_{i-1}+3} & & 0 & & \\
c_{r_{i-1}+2} & -b_{r_{i-1}+3} & a_{r_{i-1}+3} & -b_{r_{i-1}+4} & c_{r_{i-1}+4} & & & \\
& \ddots & \ddots & \ddots & \ddots & \ddots & & \\
& & c_{r_i-4} & -b_{r_i-3} & a_{r_i-3} & -b_{r_i-2} & c_{r_i-2} & \\
& 0 & & c_{r_i-3} & -b_{r_i-2} & a_{r_i-2} & -b_{r_i-1} & \\
& & & & c_{r_i-2} & -b_{r_i-1} & a_{r_i-1} + h_2\delta_{i,p+1} &
\end{bmatrix},
$$

$$
i = 1, 2, \ldots, p+1, r_0 = -1, r_{p+1} = n+1,
$$

and

$$
\boldsymbol{A}_{i,i+1} = \boldsymbol{A}_{i+1,i}^{\mathrm{T}} = \begin{bmatrix}
0 & 0 & \cdots & 0 & 0 \\
\vdots & \vdots & & \vdots & \vdots \\
0 & 0 & \cdots & 0 & 0 \\
c_{r_i} & 0 & \cdots & 0 & 0
\end{bmatrix}_{t_i \times n_i}, \quad i = 1, 2, \ldots, p,
$$

with

$$
t_1 = r_1, \quad t_i = r_i - r_{i-1} - 1 = n_{i-1}, \quad i = 2, 3, \ldots, p,
$$

$$
n_p = n - r_p.
$$

4.8.3 *Sign Oscillation of the Stiffness Matrix of a Two-Span Beam*

For ease of explanation, we first discuss the sign oscillation property of the stiffness matrix for the case involving only one mid-support, i.e., $p = 1$, and then generalize the result to other situations involving additional mid-supports in the next subsection.

Denote the two-span beam with one mid-support as S_1^*. By setting $p = 1$ in Eq. (4.8.8), we get the stiffness matrix of System S_1^*:

$$\boldsymbol{A}_1 = \begin{bmatrix} \boldsymbol{A}_{11} & \boldsymbol{A}_{12} \\ \boldsymbol{A}_{21} & \boldsymbol{A}_{22} \end{bmatrix}.$$

Obviously, the matrix $\boldsymbol{A}_1$ is not a sign-oscillatory matrix, because its element c_{r_1} is positive. Here, c_{r_1} is the element located at the r_1-th row and $(r_1 + 1)$-th column as well as at the $(r_1 + 1)$-th row and r_1-th column, i.e., the bottom left corner of $\boldsymbol{A}_{12}$ and top right corner of $\boldsymbol{A}_{21}$. A positive c_{r_1} violates the requirement of a sign-oscillatory matrix, which stipulates that all subdiagonal elements must be negative. To overcome this difficulty, we introduce the vector

$$\widetilde{\boldsymbol{u}} = (u_0, u_1, \ldots, u_{r_1-1}, -u_{r_1+1}, -u_{r_1+2}, \ldots, -u_n)^{\mathrm{T}}, \tag{4.8.9}$$

and transform the modal equation (4.8.7) to

$$\lambda \boldsymbol{M}\widetilde{\boldsymbol{u}} = \widetilde{\boldsymbol{A}}_1\widetilde{\boldsymbol{u}}. \tag{4.8.10}$$

Let us denote the transformed system as $\widetilde{S}_1$. Apparently, its stiffness matrix $\widetilde{\boldsymbol{A}}_1$ is of the same form as the stiffness matrix $\boldsymbol{A}_1$ of the original system. To get $\widetilde{\boldsymbol{A}}_1$ from $\boldsymbol{A}_1$, we only need to replace the element c_{r_1} at the r_1-th row and $(r_1 + 1)$-th column as well as at the $(r_1 + 1)$-th row and r_1-th column of the latter matrix with $-c_{r_1}$. According to the criteria for the sign-oscillatory matrix (see Sect. 2.4.2), we still need to demonstrate that the following two conditions are also met:

Condition 1 The sign-reverse matrix $\widetilde{\boldsymbol{A}}_1^* = ((-1)^{i+j}a_{ij})$ corresponding to $\widetilde{\boldsymbol{A}}_1 = (a_{ij})$ is totally nonnegative.

This result is self-evident. In fact, it has been established in Sect. 4.2 that the sign-reverse matrix $\boldsymbol{C}^* = ((-1)^{i+j}c_{ij})_{(n+1)\times(n+1)}$ corresponding to the coefficient matrix $\boldsymbol{C} = (c_{ij})_{(n+1)\times(n+1)}$ of Eq. (4.8.2) is totally nonnegative. Since $\widetilde{\boldsymbol{A}}_1^*$ is a submatrix of $\boldsymbol{C}^*$, obtained by deleting its $(r_1 + 1)$-th row and $(r_1 + 1)$-th column, all minors of $\widetilde{\boldsymbol{A}}_1^*$ are also minors of $\boldsymbol{C}^*$. Thus, $\widetilde{\boldsymbol{A}}_1^*$ has to be a totally nonnegative matrix.

Condition 2 $\widetilde{\boldsymbol{A}}_1$ is nonsingular, i.e., $\left|\widetilde{\boldsymbol{A}}_1\right| = |\boldsymbol{A}_1| > 0$.

In fact, by applying the Laplace expansion theorem to $|\boldsymbol{A}_1|$, which expresses a determinant in term of its minors, we get

$$
\begin{aligned}
|\boldsymbol{A}_1| &= A_1\begin{pmatrix} 1\ 2\cdots r_1 \\ 1\ 2\cdots r_1 \end{pmatrix} A_1\begin{pmatrix} r_1+1\ r_1+2\cdots n \\ r_1+1\ r_1+2\cdots n \end{pmatrix} \\
&\quad - A_1\begin{pmatrix} 1\ 2\cdots r_1-1\ \ r_1 \\ 1\ 2\cdots r_1-1\ r_1+1 \end{pmatrix} A_1\begin{pmatrix} r_1+1\ r_1+2\cdots n \\ r_1\ \ r_1+2\cdots n \end{pmatrix} \\
&= \left[A_1\begin{pmatrix} 1\ 2\cdots r_1 \\ 1\ 2\cdots r_1 \end{pmatrix} - A_1\begin{pmatrix} 1\ 2\cdots r_1-1\ \ r_1 \\ 1\ 2\cdots r_1-1\ r_1+1 \end{pmatrix}\right] A_1\begin{pmatrix} r_1+1\ r_1+2\cdots n \\ r_1+1\ r_1+2\cdots n \end{pmatrix} \\
&\quad + A_1\begin{pmatrix} 1\ 2\cdots r_1-1\ \ r_1 \\ 1\ 2\cdots r_1-1\ r_1+1 \end{pmatrix}\left[A_1\begin{pmatrix} r_1+1\ r_1+2\cdots n \\ r_1+1\ r_1+2\cdots n \end{pmatrix} - A_1\begin{pmatrix} r_1+1\ r_1+2\cdots n \\ r_1\ \ r_1+2\cdots n \end{pmatrix}\right].
\end{aligned}
$$

Notice that the two minors

$$
A_1\begin{pmatrix} 1\ 2\cdots r_1 \\ 1\ 2\cdots r_1 \end{pmatrix} \text{ and } A_1\begin{pmatrix} 1\ 2\cdots r_1-1\ \ r_1 \\ 1\ 2\cdots r_1-1\ r_1+1 \end{pmatrix}
$$

differ only at the last column. In addition, there is only one nonzero element c_{r_1} in the last column and at the bottom right corner of the latter minor. Let us introduce a determinant $|\boldsymbol{A}_{1c}|$ to represent the difference between the above two minors. It can be written in the following form, based on the rule to split a determinant into the sum of two determinants,

$$
|\boldsymbol{A}_{1c}| = A_1\begin{pmatrix} 1\ 2\cdots r_1 \\ 1\ 2\cdots r_1 \end{pmatrix} - A_1\begin{pmatrix} 1\ 2\cdots r_1-1\ \ r_1 \\ 1\ 2\cdots r_1-1\ r_1+1 \end{pmatrix}
$$

$$
= \begin{vmatrix}
a_0^* & -b_1 & c_1 & & & & \\
-b_1 & a_1 & -b_2 & c_2 & & 0 & \\
c_1 & -b_2 & a_2 & -b_3 & c_3 & & \\
& \ddots & \ddots & \ddots & \ddots & \ddots & \\
& & c_{r_1-4} & -b_{r_1-3} & a_{r_1-3} & -b_{r_1-2} & c_{r_1-2} \\
& 0 & & c_{r_1-3} & -b_{r_1-2} & a_{r_1-2} & -b_{r_1-1} \\
& & & & c_{r_1-2} & -b_{r_1-1} & a_{r_1-1}-c_{r_1}
\end{vmatrix},
$$

where $a_0^* = a_0 + h_1$. By the same token, we may introduce a determinant $|\boldsymbol{B}_{1c}|$, and express it in the following form:

$$
|\boldsymbol{B}_{1c}| = A_1\begin{pmatrix} r_1+1\ r_1+2\cdots n \\ r_1+1\ r_1+2\cdots n \end{pmatrix} - A_1\begin{pmatrix} r_1+1\ r_1+2\cdots n \\ r_1\ \ r_1+2\cdots n \end{pmatrix}
$$

$$
= \begin{bmatrix}
a_{r_1+1}-c_{r_1} & -b_{r_1+2} & c_{r_1+2} & & & & & 0 \\
-b_{r_1+2} & a_{r_1+2} & -b_{r_1+3} & c_{r_1+3} & & & & \\
c_{r_1+2} & -b_{r_1+3} & a_{r_1+3} & -b_{r_1+4} & c_{r_1+4} & & & \\
 & \ddots & \ddots & \ddots & \ddots & \ddots & & \\
 & & c_{n-3} & -b_{n-2} & a_{n-2} & -b_{n-1} & c_{n-1} & \\
 & 0 & & c_{n-2} & -b_{n-1} & a_{n-1} & -b_n & \\
 & & & & c_{n-1} & -b_n & a_n^* &
\end{bmatrix},
$$

where $a_n^* = a_n + h_2$.

Notice that

$$
a_{r_1-1} - c_{r_1} = \frac{k_{r_1-2}}{l_{r_1-1}^2} + \left(\frac{1}{l_{r_1-1}} + \frac{1}{l_{r_1}}\right)^2 k_{r_1-1} + \frac{k_{r_1}}{l_{r_1}^2} - \frac{k_{r_1}}{l_{r_1} l_{r_1+1}}
$$

and

$$
a_{r_1+1} - c_{r_1} = \frac{k_{r_1}}{l_{r_1+1}^2} + \left(\frac{1}{l_{r_1+1}} + \frac{1}{l_{r_1+2}}\right)^2 k_{r_1+1} + \frac{k_{r_1+2}}{l_{r_1+2}^2} - \frac{k_{r_1}}{l_{r_1} l_{r_1+1}}.
$$

Thus, when $l_{r_1} = l_{r_1+1}$, $\boldsymbol{A}_{1c}$ represents the stiffness matrix of a beam arbitrarily supported on the left but pinned on the right, while $\boldsymbol{B}_{1c}$ is the stiffness matrix of a beam pinned on the left but arbitrarily constrained on the right. Consequently, as long as one of the two ends of the original beam is clamped, pinned, or sliding, one of the two determinants $|\boldsymbol{A}_{1c}|$ and $|\boldsymbol{B}_{1c}|$ must be greater than zero. Furthermore, we should have

$$
A_1\begin{pmatrix} 1\ 2 \cdots r_1-1 & r_1 \\ 1\ 2 \cdots r_1-1 & r_1+1 \end{pmatrix} = c_{r_1} A_1\begin{pmatrix} 1\ 2 \cdots r_1-1 \\ 1\ 2 \cdots r_1-1 \end{pmatrix} > 0
$$

and

$$
A_1\begin{pmatrix} r_1+1\ r_1+2 \cdots n \\ r_1+1\ r_1+2 \cdots n \end{pmatrix} > 0,
$$

because they both correspond to the stiffness matrices of certain single-span beams with one end clamped. Since the condition $l_{r_1} = l_{r_1+1}$ can always be satisfied in our finite difference model, we conclude

$$
|\boldsymbol{A}_1| = |\boldsymbol{A}_{1c}| A_1\begin{pmatrix} r_1+1\ r_1+2 \cdots n \\ r_1+1\ r_1+2 \cdots n \end{pmatrix} + A_1\begin{pmatrix} 1\ 2 \cdots r_1-1 & r_1 \\ 1\ 2 \cdots r_1-1 & r_1+1 \end{pmatrix} |\boldsymbol{B}_{1c}| > 0.
$$

In summary, we have shown that if the same grid spacing is used on both sides of the mid-support and the original beam is, at least, clamped, pinned, or sliding at one end, then the transformed matrix $\widetilde{\boldsymbol{A}}_1$, which corresponds to the stiffness matrix of the finite difference model of the beam with one mid-support, is sign-oscillatory.

4.8.4 Sign Oscillation of the Stiffness Matrix Associated with the Discrete System of a Multi-Span Beam Subject to Arbitrary End Constraints

Based on results derived in the previous subsection, sign oscillation of the stiffness matrix associated with the finite difference system of a multi-span beam is likely to be affected by the boundary supports at the two ends of the beam. We will start our discussion by considering a three-span beam, i.e., a beam with two mid-supports. Let us denote such a system as S_2^*. According to Sect. 4.8.2, the stiffness matrix of a three-span beam is of the form

$$\boldsymbol{A}_2 = \begin{bmatrix} \boldsymbol{A}_{11} & \boldsymbol{A}_{12} & \boldsymbol{0} \\ \boldsymbol{A}_{21} & \boldsymbol{A}_{22} & \boldsymbol{A}_{23} \\ \boldsymbol{0} & \boldsymbol{A}_{32} & \boldsymbol{A}_{33} \end{bmatrix}. \tag{4.8.11}$$

Following the approach of the previous subsection, introduce the vector

$$\tilde{\boldsymbol{u}} = (u_0, u_1, \ldots, u_{r_1-1}, -u_{r_1+1}, -u_{r_1+2}, \ldots, -u_{r_2-1}, u_{r_2+1}, u_{r_2+2}, \ldots, u_n)^{\mathrm{T}}. \tag{4.8.12}$$

Then, the modal equation in (4.8.7) is converted to

$$\lambda \boldsymbol{M}\tilde{\boldsymbol{u}} = \widetilde{\boldsymbol{A}}_2\tilde{\boldsymbol{u}}. \tag{4.8.13}$$

Here, the stiffness matrix $\widetilde{\boldsymbol{A}}_2$ of the transformed system is very much like the stiffness matrix $\boldsymbol{A}_2$ of the original system. The only differences between them are that c_{r_1} at rows r_1 and $r_1 + 1$ in matrix $\boldsymbol{A}_2$ (i.e., the bottom left corner of $\boldsymbol{A}_{12}$ and the top right corner of $\boldsymbol{A}_{21}$) is replaced with $-c_{r_1}$ in $\widetilde{\boldsymbol{A}}_2$, while c_{r_2} at rows $r_2 - 1$ and r_2 in $\boldsymbol{A}_2$ (i.e., the bottom left corner of $\boldsymbol{A}_{23}$ and top right corner of $\boldsymbol{A}_{32}$) is substituted with $-c_{r_2}$ in $\widetilde{\boldsymbol{A}}_2$. The sign-reverse matrix $\widetilde{\boldsymbol{A}}_2^*$ corresponding to $\widetilde{\boldsymbol{A}}_2$ remains to be a submatrix of the sign-reverse matrix $\boldsymbol{C}^*$. In fact, Matrix $\widetilde{\boldsymbol{A}}_2^*$ is obtained from $\boldsymbol{C}^*$ by deleting its $(r_1 + 1)$-th row and $(r_1 + 1)$-th column as well as its $(r_2 + 1)$-th row and $(r_2 + 1)$-th column. Hence, $\widetilde{\boldsymbol{A}}_2^*$ is also a totally nonnegative matrix.

Again, similar to Sect. 4.8.3, we have

$$\begin{aligned}\left|\widetilde{\boldsymbol{A}}_2\right| = |\boldsymbol{A}_2| &= A_2\begin{pmatrix}1\,2\cdots r_1\\1\,2\cdots r_1\end{pmatrix}A_2\begin{pmatrix}r_1+1\;r_1+2\cdots n-1\\r_1+1\;r_1+2\cdots n-1\end{pmatrix}\\&\quad-A_2\begin{pmatrix}1\,2\cdots r-1_1\quad r_1\\1\,2\cdots r_1-1\;r_1+1\end{pmatrix}A_2\begin{pmatrix}r_1+1\;r_1+2\cdots n-1\\r_1\quad r_1+2\cdots n-1\end{pmatrix}\\&=|\boldsymbol{A}_{2c}|A_2\begin{pmatrix}r_1+1\;r_1+2\cdots n-1\\r_1+1\;r_1+2\cdots n-1\end{pmatrix}+A_2\begin{pmatrix}1\,2\cdots r_1-1\quad r_1\\1\,2\cdots r_1-1\;r_1+1\end{pmatrix}|\boldsymbol{B}_{2c}|,\end{aligned}$$

where

$$|\boldsymbol{A}_{2c}| = A_2\begin{pmatrix}1\,2\cdots r_1\\1\,2\cdots r_1\end{pmatrix}-A_2\begin{pmatrix}1\,2\cdots r_1-1\quad r_1\\1\,2\cdots r_1-1\;r_1+1\end{pmatrix}\geq 0$$

and

$$|\boldsymbol{B}_{2c}| = A_2\begin{pmatrix}r_1+1\;r_1+2\cdots n-1\\r_1+1\;r_1+2\cdots n-1\end{pmatrix}-A_2\begin{pmatrix}r_1+1\;r_1+2\cdots n-1\\r_1\quad r_1+2\cdots n-1\end{pmatrix}>0.$$

The reasons are: when $l_{r_1} = l_{r_1+1}$, Matrix $\boldsymbol{A}_{2c}$ represents the stiffness matrix of a single-span beam arbitrarily supported (could be free) on the left side and pinned on the right side, while $\boldsymbol{B}_{2c}$ happens to be the stiffness matrix of a two-span beam pinned on the left, arbitrarily constrained on the right, and hinged at the r_2-th mass. In addition, we get, after applying the same argument used in Sect. 4.8.3,

$$A_2\begin{pmatrix}1\,2\cdots r_1-1\quad r_1\\1\,2\cdots r_1-1\;r_1+1\end{pmatrix}=c_{r_1}A_2\begin{pmatrix}1\,2\cdots r_1-1\\1\,2\cdots r_1-1\end{pmatrix}>0.$$

Thus, as long as $l_{r_1} = l_{r_1+1}$ and $l_{r_2} = l_{r_2+1}$, it always occurs that

$$\left|\widetilde{\boldsymbol{A}}_2\right| = |\boldsymbol{A}_2| > 0.$$

This completes the proof that $\widetilde{\boldsymbol{A}}_2$ is a sign-oscillatory matrix.

The above discussion on two- and three-span beams can be generalized to cover multi-span beams arbitrarily supported at the two ends. Denote the beam system with $p\,(p > 2)$ mid-supports as S_p^*. The displacement component of the corresponding transformed system is

$$\tilde{u}_r = \varepsilon_k u_r. \tag{4.8.14}$$

Here, $r = r_{k-1}+1, r_{k-1}+2, \ldots, r_k-1; k = 1, 2, \ldots, p+1; r_0 = -1$ and $r_{p+1}-1 = n$; and ε_k is defined by

$$\varepsilon_k = (-1)^{k-1}, \quad k = 1, 2, \ldots, p + 1.$$

Then, let us replace the vector

$$\boldsymbol{u} = (u_0, u_1, \ldots, u_{r_1-1}, u_{r_1+1}, u_{r_1+2}, \ldots, u_{r_2-1}, u_{r_2+1}, \ldots, u_{r_p-1}, u_{r_p+1}, u_{r_p+2}, \ldots, u_n)^{\mathrm{T}}$$

with

$$\tilde{\boldsymbol{u}} = (\tilde{u}_0, \tilde{u}_1, \ldots, \tilde{u}_{r_1-1}, \tilde{u}_{r_1+1}, \tilde{u}_{r_1+2}, \ldots, \tilde{u}_{r_2-1}, \tilde{u}_{r_2+1}, \ldots, \tilde{u}_{r_p-1}, \tilde{u}_{r_p+1}, \tilde{u}_{r_p+2}, \ldots, \tilde{u}_n)^{\mathrm{T}}.$$

The resulting mathematical model is referred to as $\widetilde{S}_p$. It can be shown through mathematical induction that the stiffness matrix $\widetilde{\boldsymbol{A}}_p$ related to System $\widetilde{S}_p$ is sign-oscillatory.

In fact, we have proved the base case of $\widetilde{\boldsymbol{A}}_1$ in Sect. 4.8.3. In the inductive step, assume the stiffness matrix $\widetilde{\boldsymbol{A}}_{p-1}$ of System $\widetilde{S}_{p-1}$ to be sign-oscillatory. Next, we can essentially copy the proof word by word for the case of $\widetilde{\boldsymbol{A}}_2$ being sign-oscillatory, after replacing $\boldsymbol{A}_2$ and $\widetilde{\boldsymbol{A}}_2$ with $\boldsymbol{A}_p$ and $\widetilde{\boldsymbol{A}}_p$, respectively. This process leads to the results: (1) $\widetilde{\boldsymbol{A}}_p$ is totally nonnegative, and (2) $|\widetilde{\boldsymbol{A}}_p| = |\boldsymbol{A}_p| > 0$. Thus, we may conclude that $\widetilde{\boldsymbol{A}}_p$ is indeed a sign-oscillatory matrix.

4.8.5 Modal Qualitative Properties of the Discrete System Associated with a Multi-Span Beam

Since $\widetilde{\boldsymbol{A}}_p$ is sign-oscillatory, we find, by applying the theory of oscillatory matrices (see Refs. [1, 3]), that eigenvalues and eigenvectors of System $\widetilde{S}_p$ have the following qualitative properties:

(1) Its eigenvalues $\lambda_i = \omega_i^2$ are all positive and distinct. In addition, they can be arranged from small to big as follows

$$(0 <)\lambda_1 < \lambda_2 < \cdots < \lambda_N,$$

where N is the number of degrees of freedom of the system $\widetilde{S}_p$.

(2) The $\tilde{u}^{(i)}$-line corresponding to the eigenvector $\tilde{\boldsymbol{u}}^{(i)}$ of System $\widetilde{S}_p$ has exactly $(i-1)$ nodes (simple zeroes). In other words, there are precisely $(i-1)$ sign reversals in the component sequence of the eigenvector $\tilde{\boldsymbol{u}}^{(i)}$ of System $\widetilde{S}_p$.

(3) Nodes of $\tilde{u}^{(i)}$- and $\tilde{u}^{(i+1)}$-lines corresponding to eigenvectors $\tilde{\boldsymbol{u}}^{(i)}$ and $\tilde{\boldsymbol{u}}^{(i+1)}$, respectively, interlace, where $i = 2, 3, \ldots, N-1$.

From Eq. (4.8.14), the displacement components u_i of the system S_p^* and $\tilde{u}_i$ of the transformed system $\widetilde{S}_p$ are identical at odd-numbered spans, while they are equal in value but opposite in sign at even-numbered spans. Thus, the aforementioned three properties of System $\widetilde{S}_p$ can be transferred to the multi-span beam system S_p^*. Namely, S_p^* should have the following qualitative properties:

(1) Set

$$\boldsymbol{G}_p = \mathrm{diag}\left(\underbrace{1, 1, \ldots, 1}_{r_1 \text{ items}}, \underbrace{-1, -1, \ldots, -1}_{r_2 - r_1 - 1 \text{ items}}, \underbrace{1, 1, \ldots, 1}_{r_3 - r_2 - 1 \text{ items}}, \ldots, \underbrace{(-1)^p, (-1)^p, \ldots, (-1)^p}_{n - r_p \text{ items}}\right).$$

Since $\boldsymbol{G}_p^{-1} = \boldsymbol{G}_p$, $\widetilde{\boldsymbol{A}}_p = \boldsymbol{G}_p^{-1}\boldsymbol{A}_p\boldsymbol{G}_p$, i.e., matrices $\widetilde{\boldsymbol{A}}_p$ and $\boldsymbol{A}_p$ are similar, which implies that the two share identical eigenvalues. Therefore, the $n - p + 1$ natural frequencies of the multi-span beam are all distinct.

(2) The first mode shape of a multi-span beam has exactly p nodes, where p is the number of internal supports. If we refer to the situation where there is no mass of S_p^* between a node and an internal support as the node overlapping with the support, then each node of the first mode shape overlaps with its respective internal support.

(3) The i-th ($i \geq 2$) mode shape $\boldsymbol{u}^{(i)}$ of a multi-span beam has exactly $i - 1 + p - 2s$ nodes, where s is the number of times nodes of the $\tilde{u}^{(i)}$-line overlap with the internal supports and $s \leq \min(i-1, p)$.
In fact, we know the $\tilde{u}^{(i)}$-line has $i - 1$ nodes. If there are s nodes overlapping with the internal supports, then the $u^{(i)}$-line of the multi-span beam loses s nodes but picks up $p - s$ nodes at the remaining $p - s$ internal supports. Therefore, the node count of the i-th ($i \geq 2$) mode shape of the multi-span beam has to be

$$(i - 1) - s + (p - s) = i - 1 + p - 2s.$$

(4) When nodes of the $\tilde{u}^{(i)}$- and $\tilde{u}^{(i+1)}$-lines do not overlap with any of the internal supports, where $i = 2, 3, \ldots, N-1$, nodes of the two u-lines of the multi-span beam, excluding those at the internal supports, should interlace. However, when one node of the $\tilde{u}^{(i)}$- or $\tilde{u}^{(i+1)}$-line overlaps with an internal support, the multi-span beam may no longer behave in a way similar to that described in Properties (3) of System $\widetilde{S}_p$, i.e., nodes of the $u^{(i)}$- and $u^{(i+1)}$-lines ($i = 2, 3, \ldots, N-1$) associated with the multi-span beam might not interlace.

The content of this section is mainly derived from Refs. [14] and [16].

4.9 Modal Qualitative Properties of Discrete Systems of Overhang Beams

Actually, we have demonstrated in the previous section the sign oscillation characteristics of stiffness matrices $\widetilde{\boldsymbol{A}}_1$ and $\widetilde{\boldsymbol{A}}_2$ associated with the transformed systems $\widetilde{S}_1$ and $\widetilde{S}_2$ of the two-span beam with one overhang in Fig. 4.9 and the three-span beam with two overhangs in Fig. 4.10, respectively. Starting from this result and also applying the theory of oscillatory matrices, we may derive the following properties for eigenvalues and eigenvectors of Systems $\widetilde{S}_1$ and $\widetilde{S}_2$:

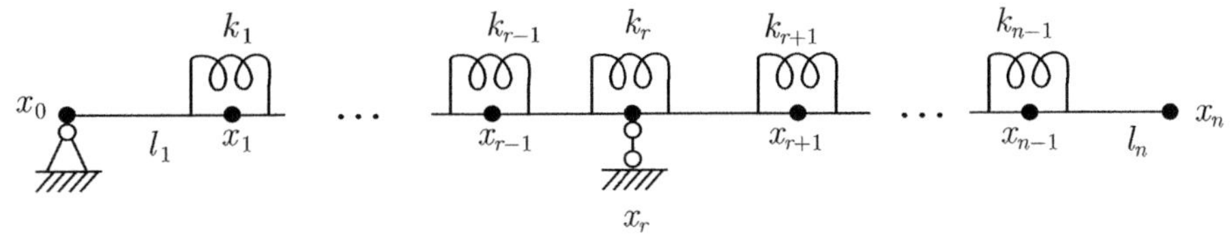

Fig. 4.9 The model of a two-span overhang beam consists of springs, masses, and rigid bars

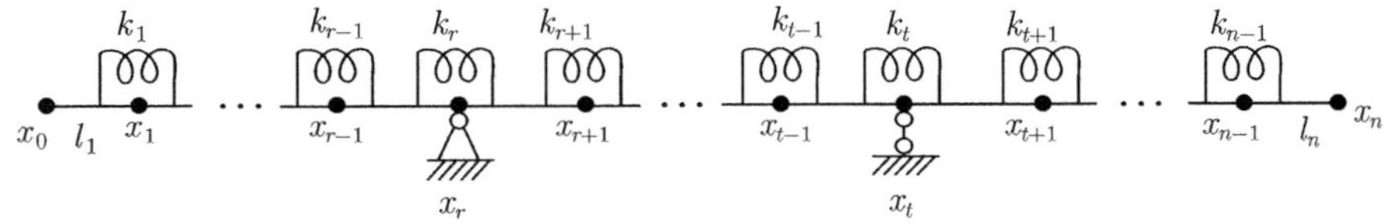

Fig. 4.10 The model of a three-span overhang beam, composed of springs, masses, and rigid bars

(1) Eigenvalues $\lambda = \omega^2$ of Systems $\widetilde{S}_1$ and $\widetilde{S}_2$ are positive and distinct. In addition, the set of eigenvalues of each system $\{\lambda_i\}_1^{n-1}$ may be sorted in an ascending order as

$$(0 <)\lambda_1 < \lambda_2 < \cdots < \lambda_{n-1}.$$

(2) The $\tilde{u}^{(i)}$-line corresponding to the i-th eigenvector $\tilde{\boldsymbol{u}}^{(i)}$ of $\widetilde{S}_1$ and $\widetilde{S}_2$ has exactly $i-1$ nodes (simple zeros). In other words, the sequence of components associated with the eigenvector $\tilde{\boldsymbol{u}}^{(i)}$ has precisely $i-1$ sign reversals.
(3) Nodes of the $\tilde{u}^{(i)}$- and $\tilde{u}^{(i+1)}$-lines corresponding to eigenvectors $\tilde{\boldsymbol{u}}^{(i)}$ and $\tilde{\boldsymbol{u}}^{(i+1)}$, respectively, must interlace, where $i = 2, 3, \cdots, n-2$.

Similar to those arbitrarily supported multi-span beams, the displacement component u_i of Systems S_1^* and S_2^* is also equal to its counterpart $\tilde{u}_i$ of Systems $\widetilde{S}_1$ and $\widetilde{S}_2$ at an odd-numbered span, while the two are of identical value but opposite signs at an even-numbered span. Consequently, the aforementioned properties of Systems $\widetilde{S}_1$ and $\widetilde{S}_2$ can be transferred, respectively, to systems S_1^* and S_2^*. Specifically, the two- and three-span overhang beams have the following properties:

Property 1 *Since $\widetilde{\boldsymbol{A}}_p = \boldsymbol{G}_p^{-1}\boldsymbol{A}_p\boldsymbol{G}_p$, i.e., $\widetilde{\boldsymbol{A}}_p$ is similar to $\boldsymbol{A}_p$, the two share the same set of eigenvalues, where $p = 1$ and 2. Hence, the $n-1$ natural frequencies of a two- or three-span overhang beam are all distinct.*

Property 2 *The first mode shape of an overhang beam has exactly $p\,(p = 1, 2)$ nodes, and these nodes overlap with the internal supports.*

Property 3 *The i-th $(i \geq 2)$ mode shape of an overhang beam has exactly $i-1+p-2s$ nodes, where $p = 1$ or 2 and s is the number of times that nodes of the $\tilde{u}^{(i)}$-line overlap with internal supports. For a two-span overhang beam, $s \leq 1$; while for a three-span overhang beam, $s \leq 2$.*

Property 4 *Set $i = 2, 3, \ldots, n-2$. When none of the nodes of $\tilde{u}^{(i)}$- and $\tilde{u}^{(i+1)}$-lines overlaps with any of the internal supports, nodes of the two u-lines of the overhang beam, excluding those at the internal supports, shall interlace. However, when at least one of the nodes of $\tilde{u}^{(i)}$- and $\tilde{u}^{(i+1)}$-lines overlaps with an internal support, the overhang beam may no longer possess the aforementioned Property (3) of Systems $\widetilde{S}_p$, where $p = 1, 2$. In other words, nodes of $u^{(i)}$- and $u^{(i+1)}$-lines might not interlace.*

Furthermore, this type of beams has four other important qualitative properties, i.e., Properties 5–8. Since their proof is more involved, they will be presented in Sects. 4.9.2 and 4.9.3 instead.

4.9.1 Conjugate Structures of Overhang Beams

In order to further study modal qualitative properties of the angle of rotation, bending moment, and shear force, we need to introduce the conjugate structure of an overhang beam following the approach of Sect. 4.3. Let us first consider a two-span overhang beam. The system of modal equations of the discrete model is given below:

$$\begin{cases} \lambda m_1 u_1 = a_1 u_1 - b_2 u_2 + c_2 u_3, \\ \lambda m_1 u_1 = -b_2 u_1 + a_2 u_2 - b_3 u_3 + c_3 u_4, \\ \lambda m_k u_k = c_{k-1} u_{k-2} - b_k u_{k-1} + a_k u_k - b_{k+1} u_{k+1} + c_{k+1} u_{k+2}, \\ \quad (k = 3, 4, \ldots, r-3), \\ \lambda m_{r-2} u_{r-2} = c_{r-3} u_{r-4} - b_{r-2} u_{r-3} + a_{r-2} u_{r-2} - b_{r-1} u_{r-1}, \\ \lambda m_{r-1} u_{r-1} = c_{r-2} u_{r-3} - b_{r-1} u_{r-2} + a_{r-1} u_{r-1} + c_r u_{r+1}, \\ \lambda m_{r+1} u_{r+1} = c_r u_{r-1} + a_{r+1} u_{r+1} - b_{r+2} u_{r+2} + c_{r+2} u_{r+3}, \\ \lambda m_{r+2} u_{r+2} = -b_{r+2} u_{r+1} + a_{r+2} u_{r+2} - b_{r+3} u_{r+3} + c_{r+3} u_{r+4}, \\ \lambda m_k u_k = c_{k-1} u_{k-2} - b_k u_{k-1} + a_k u_k - b_{k+1} u_{k+1} + c_{k+1} u_{k+2}, \\ \quad (k = r+3, r+4, \ldots, n-2), \\ \lambda m_{n-1} u_{n-1} = c_{n-2} u_{n-3} - b_{n-1} u_{n-2} + a_{n-1} u_{n-1} - b_n u_n, \\ \lambda m_n u_n = c_{n-1} u_{n-2} - b_n u_{n-1} + a_n u_n. \end{cases} \tag{4.9.1}$$

Set

$$\boldsymbol{M} = \mathrm{diag}(m_1, m_2, \ldots, m_r, \ldots, m_n), \quad \boldsymbol{K} = \mathrm{diag}(k_1, k_2, \ldots, k_r, \ldots, k_{n-1}),$$
$$\overline{\boldsymbol{u}} = (u_1, u_2, \ldots, u_{r-1}, 0, u_{r+1}, u_{r+2}, \ldots, u_n)^{\mathrm{T}}, \text{ and } \boldsymbol{L} = \mathrm{diag}(l_1, l_2, \ldots, l_n).$$

The mode shape of the bending moment of the two-span beam may be expressed as

$$\boldsymbol{\tau} = (\tau_1, \tau_2, \ldots, \tau_{n-1})^{\mathrm{T}} = \boldsymbol{K}\boldsymbol{E}_{n-1}\boldsymbol{L}^{-1}\widetilde{\boldsymbol{E}}_n^{\mathrm{T}}\overline{\boldsymbol{u}}, \tag{4.9.2}$$

where matrices $\boldsymbol{E}_{n-1}$ and $\widetilde{\boldsymbol{E}}_n$ are still of the forms defined in (4.1.23), with the exception that orders of $\boldsymbol{E}_{n-1}$ and $\widetilde{\boldsymbol{E}}_n$ are now $(n-1)\times n$ and $n\times n$, respectively. Thus, Eq. (4.9.1) can be rewritten as

$$\lambda \boldsymbol{M}\bar{\boldsymbol{u}} = \widetilde{\boldsymbol{E}}_n \boldsymbol{L}^{-1}\boldsymbol{E}_{n-1}^{\mathrm{T}}\boldsymbol{K}\boldsymbol{E}_{n-1}\boldsymbol{L}^{-1}\widetilde{\boldsymbol{E}}_n^{\mathrm{T}}\bar{\boldsymbol{u}} + R_r\boldsymbol{e}^{(r)} = \widetilde{\boldsymbol{E}}_n\boldsymbol{L}^{-1}\boldsymbol{E}_{n-1}^{\mathrm{T}}\boldsymbol{\tau} + R_r\boldsymbol{e}^{(r)}. \quad (4.9.3)$$

Here, we have included the following equilibrium equation for the r-th mass at the mid-support in Eq. (4.9.3), which has been left out in the modal equation (4.9.1) of the two-span overhang beam:

$$0 = \frac{1}{m_r}\left(\frac{\tau_r - \tau_{r-1}}{l_r} - \frac{\tau_{r+1} - \tau_r}{l_{r+1}}\right) + \frac{R_r}{m_r}. \quad (4.9.4)$$

Additionally, in Eq. (4.9.3), R_r is the reaction force from the mid-support, while $\boldsymbol{e}^{(r)}$ stands for the n-dimensional unit column vector whose r-th element is 1 and all other elements are zeroes. If we let m_r approaches infinity, then $R_r/m_r \to 0$. Thus, Eq. (4.9.3) can be further reformatted into

$$\lambda \boldsymbol{E}_{n-1}\boldsymbol{L}^{-1}\widetilde{\boldsymbol{E}}_n^{\mathrm{T}}\bar{\boldsymbol{u}} = \boldsymbol{E}_{n-1}\boldsymbol{L}^{-1}\widetilde{\boldsymbol{E}}_n^{\mathrm{T}}\boldsymbol{M}^{-1}\widetilde{\boldsymbol{E}}_n\boldsymbol{L}^{-1}\boldsymbol{E}_{n-1}^{\mathrm{T}}\boldsymbol{\tau}$$

or

$$\lambda \boldsymbol{K}^{-1}\boldsymbol{\tau} = \boldsymbol{E}_{n-1}\boldsymbol{L}^{-1}\widetilde{\boldsymbol{E}}_n^{\mathrm{T}}\boldsymbol{M}^{-1}\widetilde{\boldsymbol{E}}_n\boldsymbol{L}^{-1}\boldsymbol{E}_{n-1}^{\mathrm{T}}\boldsymbol{\tau} = \boldsymbol{A}_{1\tau}\boldsymbol{\tau}. \quad (4.9.5)$$

The above expression also represents the modal equation of a "beam" with $n-1$ degrees of freedom. We refer to the respective system characterized by Eq. (4.9.5) as the conjugate of the original two-span overhang beam. By taking into account boundary conditions of the two-span overhang beam as well as the equilibrium Eq. (4.9.4), we get

$$\tau_0 = 0,\quad \tau_0'' = 0,\quad \tau_n = 0,\quad \tau_n' = 0, \text{ and } \tau_r''/m_r = 0.$$

Therefore, the conjugate system of the two-span overhang beam is a two-span continuous beam pinned at the left end, clamped at the right end, and hinged at $x_r = c$, as shown in Fig. 4.11a. Its "displacement" mode shape is $\boldsymbol{\tau} = (\tau_1, \tau_2, \ldots, \tau_{n-1})^{\mathrm{T}}$, while its parameters are

$$\{\bar{m}_i\}_1^{n-1} = \{k_i^{-1}\}_1^{n-1},\quad \bar{k}_i = m_i^{-1}(i = 1, 2, \ldots, n; i \neq r),\quad \{l_i\}_1^n, \text{ and } \bar{k}_r = m_r^{-1} = 0.$$

Similarly, we may find the conjugate system of a three-span overhang beam to be a three-span beam clamped at both ends and hinged at x_r and x_t, as shown in Fig. 4.11b. Again, the "displacement" of the system is $\boldsymbol{\tau} = (\tau_1, \tau_2, \ldots, \tau_{n-1})^{\mathrm{T}}$, while its parameters include

$$\{\bar{m}_i\}_1^{n-1} = \{k_i^{-1}\}_1^{n-1},\quad \bar{k}_i = m_i^{-1}(i = 0, 1, \ldots, n; i \neq r, t),\quad \{l_i\}_1^n,\ \bar{k}_r = 0, \text{ and } \bar{k}_t = 0.$$

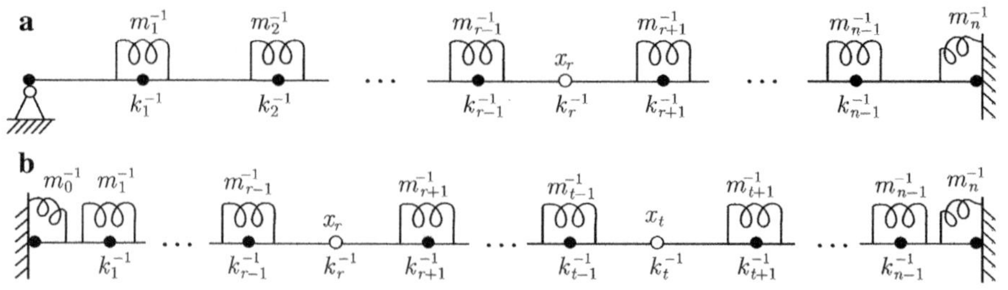

Fig. 4.11 Conjugate systems of **a** two-span and **b** three-span overhang beams

The above derivation also reveals that a conjugate system shares the same set frequencies with the original overhang beam.

4.9.2 Basic Qualitative Properties of the Conjugate Systems of Overhang Beams

We will first study the sign-oscillation property of stiffness matrices associated with conjugate systems of two- and three-span overhang beams. Based on this work, modal qualitative properties of the bending moment in the overhang beams can be established.

1. ***Sign-Oscillation of the Stiffness Matrix Associated with the Conjugate System Of a Two-Span Overhang Beam***

Set

$$\overline{\boldsymbol{A}}_{11}=\begin{bmatrix} \bar{a}_1 & -\bar{b}_2 & \bar{c}_2 & & & & & \\ -\bar{b}_2 & \bar{a}_2 & -\bar{b}_3 & \bar{c}_3 & & & 0 & \\ \bar{c}_2 & -\bar{b}_3 & \bar{a}_3 & -\bar{b}_4 & \bar{c}_4 & & & \\ & \ddots & \ddots & \ddots & \ddots & \ddots & & \\ & & \bar{c}_{r-3} & -\bar{b}_{r-2} & \bar{a}_{r-2} & -\bar{b}_{r-1} & \bar{c}_{r-1} \\ & 0 & & \bar{c}_{r-2} & -\bar{b}_{r-1} & \bar{a}'_{r-1} & -\bar{b}'_r \\ & & & & \bar{c}_{r-1} & -\bar{b}'_r & \bar{a}'_r \end{bmatrix},$$

$$\overline{\boldsymbol{A}}_{12} = \overline{\boldsymbol{A}}_{21}^{\mathrm{T}} = \begin{bmatrix} 0 & 0 & 0 & \cdots & 0 \\ \vdots & \vdots & \vdots & & \vdots \\ 0 & 0 & 0 & \cdots & 0 \\ -\overline{b}'_{r+1} & \overline{c}_{r+1} & 0 & \cdots & 0 \end{bmatrix}_{r\times(n-r-1)},$$

$$\widetilde{\boldsymbol{A}}_{22} = \begin{bmatrix} \overline{a}'_r & -\overline{b}'_{r+1} & \overline{c}_{r+1} & & & & & \\ -\overline{b}'_{r+1} & \overline{a}'_{r+1} & -\overline{b}_{r+2} & \overline{c}_{r+2} & & 0 & & \\ \overline{c}_{r+1} & -\overline{b}_{r+2} & \overline{a}_{r+2} & -\overline{b}_{r+3} & \overline{c}_{r+3} & & & \\ & \ddots & \ddots & \ddots & \ddots & \ddots & & \\ & & \overline{c}_{n-4} & -\overline{b}_{n-3} & \overline{a}_{n-3} & -\overline{b}_{n-2} & \overline{c}_{n-2} \\ & 0 & & \overline{c}_{n-3} & -\overline{b}_{n-2} & \overline{a}_{n-2} & -\overline{b}_{n-1} \\ & & & & \overline{c}_{n-2} & -\overline{b}_{n-1} & \overline{a}_{n-1} \end{bmatrix},$$

$$\widetilde{\boldsymbol{A}}_{12} = \widetilde{\boldsymbol{A}}_{21}^{\mathrm{T}} = \begin{bmatrix} 0 & 0 & \cdots & 0 \\ \vdots & \vdots & & \vdots \\ 0 & 0 & \cdots & 0 \\ \overline{c}_{r-1} & 0 & \cdots & 0 \\ -\overline{b}'_r & 0 & \cdots & 0 \end{bmatrix}_{(r-1)\times(n-r)}.$$

In the matrices above, parameters $\overline{a}_i$ ($i = 1, 2, \ldots, r-2, r+2, \ldots n-2$), $\overline{b}_i$ ($i = 2, 3, \ldots, r-1, r+2, \ldots, n-1$), and $\overline{c}_i$ ($i = 2, 3, \ldots, n-2$) are of forms almost identical to their respective counterparts given in Eqs. (4.8.3)–(4.8.5). The only differences between the two sets of parameters are that $\overline{k}_i$ should replace k_i in Eqs. (4.8.3)–(4.8.5), and

$$\overline{a}'_{r-1} = \frac{\overline{k}_{r-2}}{l_{r-1}^2} + \left(\frac{1}{l_{r-1}} + \frac{1}{l_r}\right)^2 \overline{k}_{r-1}, \quad \overline{a}'_r = \frac{\overline{k}_{r-1}}{l_r^2} + \frac{\overline{k}_{r+1}}{l_{r+1}^2},$$

$$\overline{a}'_{r+1} = \left(\frac{1}{l_{r+1}} + \frac{1}{l_{r+2}}\right)^2 \overline{k}_{r+1} + \frac{\overline{k}_{r+2}}{l_{r+2}^2}, \quad \overline{a}_{n-1} = \frac{\overline{k}_{n-2}}{l_{n-1}^2} + \left(\frac{1}{l_{n-1}} + \frac{1}{l_n}\right)^2 \overline{k}_{n-1} + \frac{\overline{k}_n}{l_n^2}.$$

$$\overline{b}'_r = \left(\frac{1}{l_r} + \frac{1}{l_{r+1}}\right)\frac{\overline{k}_{r-1}}{l_r}, \quad \overline{b}'_{r+1} = \left(\frac{1}{l_{r+1}} + \frac{1}{l_{r+2}}\right)\frac{\overline{k}_{r+1}}{l_{r+1}}.$$

Thus, we get the following stiffness matrix of the discrete model associated with the conjugate system of the two-span overhang beam derived in the previous subsection:

$$\boldsymbol{A}_{1\tau} = \begin{bmatrix} \overline{\boldsymbol{A}}_{11} & \overline{\boldsymbol{A}}_{12} \\ \overline{\boldsymbol{A}}_{21} & \overline{\boldsymbol{A}}_{22} \end{bmatrix} = \begin{bmatrix} \widetilde{\boldsymbol{A}}_{11} & \widetilde{\boldsymbol{A}}_{12} \\ \widetilde{\boldsymbol{A}}_{21} & \widetilde{\boldsymbol{A}}_{22} \end{bmatrix}, \tag{4.9.6}$$

where $\widetilde{\boldsymbol{A}}_{11}$ is obtained by deleting the last row and column of $\overline{\boldsymbol{A}}_{11}$ while $\overline{\boldsymbol{A}}_{22}$ is derived from $\widetilde{\boldsymbol{A}}_{22}$ by removing its first row and column.

As can be seen in Eq. (4.9.6), subdiagonal elements of the stiffness matrix $\boldsymbol{A}_{1\tau}$ are all negative, and its sign-reverse matrix $\boldsymbol{A}_{1\tau}^*$ is totally nonnegative. Additionally, the r-th row of the determinant of $\boldsymbol{A}_{1\tau}$ can be decomposed into

$$\begin{aligned} &\left(0 \ \cdots \ 0 \ \overline{c}_{r-1} \ -\overline{b}'_r \ \overline{a}'_r \ -\overline{b}'_{r+1} \ \overline{c}_{r+1} \ 0 \ \cdots \ 0\right) \\ &= \left(0 \ \cdots \ 0 \ \overline{c}_{r-1} \ -\overline{b}'_r \ \overline{a}'_{r1} \quad 0 \quad\ 0 \ \ 0 \ \cdots \ 0\right) \\ &+ \left(0 \ \cdots \ 0 \ \ 0 \quad 0 \ \ \overline{a}'_{r2} \ -\overline{b}'_{r+1} \ \overline{c}_{r+1} \ 0 \ \cdots \ 0\right), \end{aligned}$$

where

$$\overline{a}'_{r1} = \frac{\overline{k}_{r-1}}{l_r^2}, \quad \overline{a}'_{r2} = \frac{\overline{k}_{r+1}}{l_{r+1}^2}, \quad \text{and} \quad \overline{a}'_{r1} + \overline{a}'_{r2} = \overline{a}'_r.$$

Thus, the determinant $\det \boldsymbol{A}_{1\tau}$ can be expanded into the sum of two determinants by decomposing its r-th row, i.e.,

$$\det \boldsymbol{A}_{1\tau} = \det \begin{bmatrix} \hat{\boldsymbol{A}}_{11} & \boldsymbol{0} \\ \overline{\boldsymbol{A}}_{21} & \overline{\boldsymbol{A}}_{22} \end{bmatrix} + \det \begin{bmatrix} \widetilde{\boldsymbol{A}}_{11} & \widetilde{\boldsymbol{A}}_{12} \\ \boldsymbol{0} & \hat{\boldsymbol{A}}_{22} \end{bmatrix}, \tag{4.9.7}$$

where the only difference between $\hat{\boldsymbol{A}}_{11}$ and $\overline{\boldsymbol{A}}_{11}$ is that the element $\overline{a}'_r$ at the bottom right corner of $\overline{\boldsymbol{A}}_{11}$ is replaced with $\overline{a}'_{r1}$ in $\hat{\boldsymbol{A}}_{11}$, the only difference between $\hat{\boldsymbol{A}}_{22}$ and $\widetilde{\boldsymbol{A}}_{22}$ is that the element $\overline{a}'_r$ at the top left corner of $\widetilde{\boldsymbol{A}}_{22}$ is replaced with $\overline{a}'_{r2}$ in $\hat{\boldsymbol{A}}_{22}$, and $\boldsymbol{0}$ denotes a submatrix whose elements are all zeros.

Notice that the submatrix $\hat{\boldsymbol{A}}_{11}$ at the top left corner of the first determinant in Eq. (4.9.7) happens to be the stiffness matrix of a beam in the pinned–free configuration. Thus, the corresponding determinant should vanish, which yields

$$\det \begin{bmatrix} \hat{\boldsymbol{A}}_{11} & \boldsymbol{0} \\ \overline{\boldsymbol{A}}_{21} & \overline{\boldsymbol{A}}_{22} \end{bmatrix} = 0.$$

Furthermore, the submatrix $\widetilde{A}_{11}$ at the top left corner of the second determinant in Eq. (4.9.7) is the stiffness matrix of a beam in the pinned–pinned configuration, while the submatrix $\hat{\boldsymbol{A}}_{22}$ at the bottom right corner is the stiffness matrix of a beam in the free–clamped configuration. Thus, their respective determinants should be both positive, and this lead to

$$\det\begin{bmatrix} \widetilde{\boldsymbol{A}}_{11} & \widetilde{\boldsymbol{A}}_{12} \\ \boldsymbol{0} & \hat{\boldsymbol{A}}_{22} \end{bmatrix} > 0.$$

As a result,

$$\det \boldsymbol{A}_{1\tau} > 0.$$

Per the criteria for an oscillatory matrix (see Refs. [1, 3]), the stiffness matrix of the conjugate system of a two-span overhang beam is sign-oscillatory.

2. ***Sign-Oscillation of the Stiffness Matrix Associated with the Conjugate System of a Three-Span Overhang Beam***

The stiffness matrix $\boldsymbol{A}_{2\tau}$ of the discrete conjugate system associated with a three-span overhang beam was derived in the previous subsection. Similar to the case of the two-span overhang beam, it has negative subdiagonal elements, and its sign-reverse matrix $\boldsymbol{A}_{2\tau}^*$ is totally nonnegative. In addition, its determinant $|\boldsymbol{A}_{2\tau}|$ can be expanded into the sum of two determinants by decomposing its r-th row:

$$\det \boldsymbol{A}_{2\tau} = \det\begin{bmatrix} \hat{\boldsymbol{B}}_{11} & \boldsymbol{0} \\ \overline{\boldsymbol{B}}_{21} & \overline{\boldsymbol{B}}_{22} \end{bmatrix} + \det\begin{bmatrix} \widetilde{\boldsymbol{B}}_{11} & \widetilde{\boldsymbol{B}}_{12} \\ \boldsymbol{0} & \hat{\boldsymbol{B}}_{22} \end{bmatrix}.$$

Here, $\hat{\boldsymbol{B}}_{11}$ can be obtained by taking the first r rows and the first r columns of $\boldsymbol{A}_{2\tau}$ and then replacing the element $\overline{a}'_r = \overline{a}'_{r1} + \overline{a}'_{r2}$, at the bottom right corner of the resulting submatrix, with $\overline{a}'_{r1}$. It happens to be the stiffness matrix of a beam in the clamped–free configuration, and thus the determinant of $\hat{\boldsymbol{B}}_{11}$ ought to be positive. Next, $\overline{\boldsymbol{B}}_{22}$ consists of the last $n-r-1$ rows and the last $n-r-1$ columns of $\boldsymbol{A}_{2\tau}$. It is the stiffness matrix of the conjugate system associated with a two-span overhang beam, pinned at the left end, clamped at the right end, and pinned somewhere in the middle. Thus, the determinant of $\overline{\boldsymbol{B}}_{22}$ should also be positive. Finally, $\hat{\boldsymbol{B}}_{22}$ in the above equation is constructed by taking the last $n-r$ rows and the last $n-r$ columns of $\boldsymbol{A}_{2\tau}$ and then replacing the element $\overline{a}'_r$ at the top left corner of the resulting submatrix with $\overline{a}'_{r2}$. It is none other than the stiffness matrix of a beam unconstrained at the left end, clamped at the right end, and pinned somewhere in the middle. Thus, the determinant of $\hat{\boldsymbol{B}}_{22}$ must be zero. Based on the above observations, we conclude

$$\det \boldsymbol{A}_{2\tau} > 0.$$

According to the criteria for an oscillatory matrix, the stiffness matrix of the conjugate system associated with a three-span overhang beam is sign-oscillatory.

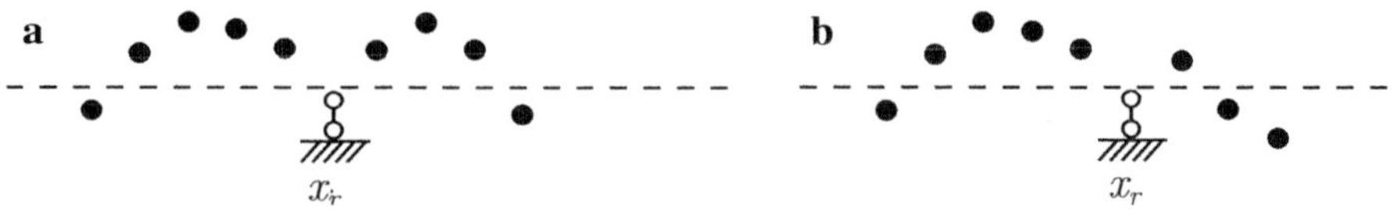

Fig. 4.12 Two possible patterns of the $\tilde{u}^{(i)}$-line when one of its nodes overlaps with an internal support

3. *Qualitative Properties Of Mode Shapes Associated with the Bending Moment of an Overhang Beam*

Denote the i-th natural angular frequency as ω_i and the respective mode shape of the bending moment as $\boldsymbol{\tau}^{(i)}$, where $i = 1, 2, \ldots, n-1$. It is known from the previous discussion in this subsection that stiffness matrices of the conjugate systems associated with both two- and three-span overhang beams are sign-oscillatory. Furthermore, their mass matrices are both positive definite and diagonal. Thus, according to the theory of oscillatory matrices (see Ref. [1, 3]), mode shapes of the bending moment have the following qualitative properties:

Property 5 $\boldsymbol{\tau}^{(i)}$ *has exactly* $i - 1$ *nodes, where* $i = 1, 2, \ldots, n - 1$.

Property 6 *Nodes of* $\tau^{(i)}$*- and* $\tau^{(i+1)}$*-lines interlace, where* $i = 2, 3, \ldots, n - 2$.

Notice that Properties 1–4 has been introduced at the beginning of Sect. 4.9.

4.9.3 *Other Qualitative Properties of Overhang Beams*

Some additional qualitative properties of overhang beams are discussed in this subsection.

1. *Numbers of Nodes and Zeros of the Displacement Mode Shapes*

As before, the displacement mode shape $\overline{\boldsymbol{u}}^{(i)}$ of an overhang beam includes the zero displacement(s) at mid-support(s). We would like to point out an important fact: When a node of the $\tilde{u}^{(i)}$-line overlaps with the mid-support at x_r, this point must be a null anti-node of $\overline{\boldsymbol{u}}^{(i)}$. In fact, based on the rule that a displacement mode shape has a sign reversal when it passes through an internal support, the two masses adjacent to the support must move to the same side of the equilibrium position of the beam, as shown in Fig. 4.12. Thus, although the i-th mode shape $\boldsymbol{u}^{(i)}$ has only $i - 1 + p - 2s$ nodes, the corresponding $\overline{u}^{(i)}$-line has $i - 1 + p$ zeroes, where $p = 1,\ 2$ and a null anti-node is counted as two distinct zeros.

2. Extrema of the Displacement Mode Shape

Suppose u_t is an internal local maximum among components of $\boldsymbol{u}^{(i)}$, i.e.,

$$u_t \geq u_{t-1} \quad \text{and} \quad u_t \geq u_{t+1}.$$

These two inequalities cannot become equalities simultaneously. Otherwise, according to the definition of $\boldsymbol{\tau}$, we would have

$$\tau_t = k_t\left(\frac{u_t - u_{t-1}}{l_t} - \frac{u_{t+1} - u_t}{l_{t+1}}\right) = 0, \quad \tau_{t-1} \leq 0, \text{ and } \tau_{t+1} \leq 0,$$

which contradicts the fact that $\boldsymbol{\tau}^{(i)}$ has a fixed number of sign reversals. Therefore, whenever u_t is an internal local maximum, $\tau_t < 0$. By the same token, when u_t is an internal local minimum, $\tau_t > 0$.

3. Qualitative Properties of the Mode Shape of the Angle of Rotation

Define the angle of rotation as

$$\boldsymbol{\theta} = \boldsymbol{L}^{-1}\widetilde{\boldsymbol{E}}_n^{\mathrm{T}}\overline{\boldsymbol{u}}$$

for a two-span overhang beam, and

$$\boldsymbol{\theta} = -\boldsymbol{L}^{-1}\boldsymbol{E}\,\overline{\boldsymbol{v}}, \quad \overline{\boldsymbol{v}} = (u_0, u_1, \ldots, u_{r-1}, 0, u_{r+1}, \ldots, u_{t-1}, 0, u_{t+1}, \ldots, u_n)^{\mathrm{T}}$$

for a three-span overhang beam, where $\boldsymbol{E}$ is given in Eq. (4.1.23) and $\widetilde{\boldsymbol{E}}_n$ is the n-th order square matrix of the form similar to $\widetilde{\boldsymbol{E}}$, also defined in Eq. (4.1.23). Then, we have the result below:

Property 7 *Consider the i-th mode shape of the angle of rotation $\boldsymbol{\theta}^{(i)}$ associated with an overhang beam. The sequence consisting of its components has exactly i sign reversals.*

In fact, the i-th displacement mode shape $\boldsymbol{u}^{(i)}$ of the overhang beam has precisely $i-1+p-2s$ ($p = 1, 2$) nodes. These nodes devide the component sequence of $\boldsymbol{u}^{(i)}$ into $i+p-2s$ segments, within each of which component members share a common sign. Excluding the p segments containing the unconstrained end(s), there is at least one extremum within each of these common-sign segments. Hence, there are, at a minimum, $i - 2s$ internal extrema in the component sequence of $\boldsymbol{u}^{(i)}$. If none of the nodes of the $\tilde{u}^{(i)}$-line associated with the overhang beam overlaps with a support, then $s = 0$. In contrast, if one of its nodes overlaps with a support, there will be three extrema of the $\tilde{u}^{(i)}$-line in between the two nodes of $\boldsymbol{u}^{(i)}$ that are immediately adjacent to this mid-support, as shown in Fig. 4.12. In other words, there are two additional internal extrema corresponding to each node of the $\tilde{u}^{(i)}$-line that overlaps with a support. Notice there can be, at most, s nodes that overlap with the mid-supports. Thus, on the one hand, we must be able to find at least i internal extrema of the $\bar{u}^{(i)}$-line. On the other hand, there are at most i internal extrema of the $\bar{u}^{(i)}$ line, due to the fact that the mode shape of the bending moment $\boldsymbol{\tau}^{(i)}$ associated with the overhang beam has exactly $i-1$ nodes (or, equivalently, the component sequence of $\boldsymbol{\tau}^{(i)}$ has precisely $i-1$ sign reversals) and also based on the result given in Part 2 of this subsection. These observations show that there are i and only i internal extrema in the component sequence of the $\bar{u}^{(i)}$-line. Because the sign of $\boldsymbol{\theta}^{(i)}$ changes once on the two sides of each of the extrema, we may derive Property 7.

The above discussion also reveals: Excluding the portions immediately adjacent to a mid-support on both sides, each of the common-sign segments of $\boldsymbol{u}^{(i)}$ has only one extremum that is a local maximum if elements in the segment are positive or a local minimum if the elements are negative. It should be pointed out that if, in the definition of $\boldsymbol{\theta}$, zero components in $\overline{\boldsymbol{u}}$ and $\overline{\boldsymbol{v}}$ are removed, i.e., $\overline{\boldsymbol{u}}$ and $\overline{\boldsymbol{v}}$ are reduced to $\boldsymbol{u}$ and $\boldsymbol{v}$, respectively, then the new vector $\boldsymbol{\theta}$ obtained this way may not have a fixed number of sign reversals.

4. A Qualitative Property of the Mode Shape of the Shear Force

Define the mode shape of the shear force associated with an overhang beam as

$$\boldsymbol{\phi} = \boldsymbol{L}^{-1}\boldsymbol{E}_{n-1}^{\mathrm{T}}\boldsymbol{\tau}.$$

Then, based on the node count of $\boldsymbol{\tau}^{(i)}$, the number of zeroes of $\overline{\boldsymbol{u}}^{(i)}$ or $\overline{\boldsymbol{v}}^{(i)}$, and Eq. (4.9.3), we obtain the following result, apply applying an argument similar to that of Part 3:

Property 8 *The component sequence of the i-th mode shape of the shear force* $\boldsymbol{\phi}^{(i)}$ *has exactly i sign reversals, where* $i = 1, 2, \ldots, n-1$.

The content of this section is derived primarily from Refs. [15] and [17].

References

1. Gantmacher FP, Krein MG (1961) Oscillation matrices and kernels and small vibrations of mechanical systems. U. S. Atomic Energy Commission, Washington
2. Gladwell GML (1985) Qualitative properties of vibrating systems. Proc Roy Soc Lond A 401:299–315
3. Gladwell GML (2004) Inverse problems in vibration, 2nd edn. Springer, Dordrecht (1986, 1st edn, Martinus Nijhoff Publishers, Dordrecht)
4. Gladwell GML, England AH, Wang DJ (1987) Examples of reconstruction of an Euler-Bernoulli beam from spectral data. J Sound Vibr 19(1):81–94
5. Gladwell GML (1991) Qualitative properties of finite-element models II: the Euler-Bernoulli beam. Q J Mech Appl Math 44(2):267–284
6. He BC, Wang DJ, Low KH (1989) Inverse problem for a vibrating beam using a finite difference model. In: Proceedings international conference on noise & vibration, Singapore, B, pp 92–104
7. He BC, Wang DJ, Wang QS (1989) Inverse problem for the finite difference model of Euler beam in vibration. J Vibr Eng 2(2):1–9 (in Chinese)

8. He BC, Wang DJ, Wang QS (1991) The single mode inverse problem of Euler-Bernoulli beam. Acta Mech Solida Sin 12(1):85–89 (in Chinese)
9. Wang DJ, He BC, et al (1990) Inverse problems of the finite difference model of a vibrating Euler beam. In: Proceedings international conference on vibration problems in engineering, June 1990, China, vol 1(1), pp 21–26. International Academic Publishers, Singapore
10. Wang DJ, He BC, Wang QS (1990) On the construction of the Euler-Bernoulli beam via two sets of modes and the corresponding frequencies. Acta Mech Sin 22(4):479–483 (in Chinese)
11. Wang QS, He BC, Wang DJ (1990) Some qualitative properties of frequencies and modes of Euler beams. J Vibr Eng 3(4):58–66 (in Chinese)
12. Wang QS, Wang DJ, He BC (1991) Construction of the difference discrete system for the simple-support beam from the spectra data. Eng Mech 8(4):10–19 (in Chinese)
13. Wang QS, Wang DJ (2006) Difference discrete system of the Euler beam with arbitrary supports and sign oscillatory property of stiffness matrices. Appl Math & Mech 27(3):393–398
14. Wang QS, Wu L, Wang DJ (2009) Some qualitative properties of frequency spectrum and modes of difference discrete system of multi-bearing beam. Chin J Theor Appl Mech 41(6):61–68 (in Chinese)
15. Wang QS, Wang DJ, Wu L et al (2009) Sign-oscillating property of stiffness matrix of difference discrete system and qualitative property of beam with overhang. J Vibr Shock 28(6):113–117 (in Chinese)
16. Wang QS, Wang DJ, Wu L et al (2011) Qualitative properties of frequencies and modes of vibrating multi-span beam. Q J Mech Appl Math 64(1):75–86
17. Wang QS, Zhang LH, Wang DJ (2012) Some qualitative properties of modes of discrete system of beam with overhang. Chin J Theor Appl Mech 44(6):1071–1074 (in Chinese)
18. Wang QS, Wang DJ (2014) Supplementary proof of some oscillation property for discrete systems of rod and beam having rigid modes. Chin Q Mech 35(2):262–269 (in Chinese)
19. Zheng ZJ, Chen P, Wang DJ (2012) Oscillation property of modes for FE models of bars and beams. J Vibr Shock 31(20):79–83 (in Chinese)
20. Zheng ZJ, Chen P, Wang DJ (2013) Oscillation property of the vibrations for finite element models of Euler beam. Q J Mech Appl Math 66(4):587–608
21. Zheng ZJ, Chen P, Wang DJ (2014) A unified proof to oscillation property of discrete beam models. Appl Math Mech 35(5):621–636
22. Zheng ZJ (2014) The qualitative vibrational property and modal inverse problems of rods and Euler beams [D]. Department of Mechanics and Engineering Science, College of Engineering, Peking University (in Chinese)

Chapter 5
Qualitative Properties of Vibration and Static Deformation of the Sturm–Liouville System

Chapters 3 and 4 were devoted to qualitative properties in vibration and static deformation of discrete systems, while the focus of this chapter is shifted to similar topics related to Sturm–Liouville systems. We will primarily discuss the qualitative properties of a bar with distributed parameters and in longitudinal vibration.

5.1 Natural Vibration of the Sturm–Liouville System

The unified form of the differential equation for the natural angular frequency ω and mode shape $u(x)$ of a bar is of the form:

$$[p(x)u'(x)]' + \omega^2 \rho(x)u(x) = 0, \quad 0 < x < l. \tag{5.1.1}$$

Here, if the bar is in longitudinal or axial vibration, $\rho(x)$ is the linear density of the bar and

$$p(x) = E(x)A(x),$$

where $A(x)$ and $E(x)$ represent the section area of the bar and the elastic modulus of the material, respectively; while, if the bar is in torsional vibration,

$$p(x) = G(x)J_t(x) \quad \text{and} \quad \rho(x) = I_t(x),$$

where $G(x)$, $J_t(x)$, and $I_t(x)$ are all positive functions of x and denote, respectively, the shear modulus of the bar, the polarmoment of inertia of the section, and the moment of inertia of the section. In addition, $u(x)$ in Eq. (5.1.1) designates either the amplitude of the displacement or angle of rotation at the location x.

D. Wang et al., *Qualitative Theory in Structural Mechanics*,
https://doi.org/10.1007/978-981-13-1376-9_5

The boundary conditions at two ends of the bar can be collectively expressed as

$$p(0)u'(0) - hu(0) = 0 = p(l)u'(l) + Hu(l), \tag{5.1.2}$$

where l is the length of the bar while non-negative parameters h and H represent elastic constants of springs at the left and right boundaries, respectively. As h or H assumes the value of 0, the respective end of the bar is unconstrained or free; while as h or $H \to \infty$, the corresponding end of the bar becomes fixed.

Now let us introduce the Sturm–Liouville operator

$$Lu = -[p(x)u'(x)]' + q(x)u(x), \tag{5.1.3}$$

where $q(x) \geq 0$ and $p(x)$ is either a continuous or a piecewise continuous positive function on the interval $[0, l]$. The operator captures the behavior of a physical system known as the Sturm–Liouville system. The modal equation governing the natural vibration of such a system is of the form:

$$Lu = \lambda\rho(x)u(x), \tag{5.1.4}$$

where $\lambda = \omega^2$, ω is the natural angular frequency, and $\rho(x)$ denotes the linear density of the inertia. The boundary conditions are given by Eq. (5.1.2). If $h \geq 0$, $H \geq 0$, $h + H > 0$, and $q(x) \geq 0$ for $x \in [0, l]$, then the system is called a well-constrained Sturm–Liouville system, which is also referred to as a positive Sturm–Liouville system in mathematics. It is easy to see that the transverse or lateral vibration of a string and the longitudinal vibration of a bar with or without an elastic foundation are all examples of the Sturm–Liouville system. Furthermore, when the string or bar is fixed or elastically constrained at one end at the minimum, it is a well-constrained Sturm–Liouville system.

Set

$$(u, v) = \int_0^l u(x)v(x)\mathrm{d}x. \tag{5.1.5}$$

We are all familiar with the following properties of a well-constrained Sturm–Liouville system, the proof of which is left out:

Theorem 5.1 *If $u(x)$ and $v(x)$ are functions defined on the interval $[0, l]$ and satisfy boundary conditions given by Eq. (5.1.2), then the Sturm–Liouville operator L is self-adjoint, i.e.,*

$$(u, Lv) = (v, Lu). \tag{5.1.6}$$

Equation (5.1.6) is also known as Green's identity (see Ref. [2]).

Starting from Theorem 5.1, we may derive some additional properties for eigenvalues and eigenfunctions of a Sturm–Liouville system:

(1) The eigenvalue λ of a Sturm–Liouville system is real and distinct, while the corresponding eigenfunction is a real function.
(2) Eigenvalues $\{\lambda_i\}$ of a well-constrained Sturm–Liouville system are all positive.
(3) Eigenfunctions of a Sturm–Liouville system corresponding to different eigenvalues are orthogonal with respect to the weight $\rho(x)$, namely,

$$\int_0^l \rho(x)u_k(x)u_r(x)\mathrm{d}x = 0, \quad k \neq r, \tag{5.1.7}$$

where $u_r(x)$ is the eigenfunction corresponding to the eigenvalue λ_r.

5.2 Green's Function of the Sturm–Liouville System

To obtain additional properties of the Sturm–Liouville system, we will study the Green's function from different perspectives in this section, such as alternative ways to introduce the Green's function, its expression under some specific boundary conditions, the connection between the Green's function and the integral equations, etc. By the way, the Green's function is also known as the influence function.

5.2.1 Green's Function of a Well-Constrained Sturm–Liouville System

The Green's function $G(x, s)$ of a well-constrained Sturm–Liouville system is determined by the following equations:

$$L_x G(x, s) = \delta(x - s), \quad x \text{ and } s \in I, \tag{5.2.1}$$

$$p(0)G_x(0, s) - hG(0, s) = 0 = p(l)G_x(l, s) + HG(l, s), \tag{5.2.2}$$

where Operator L is defined in Eq. (5.1.3), L_x indicates that the operator L only contains derivatives with respect to the variable x, $\delta(x - s)$ is the δ function, and I has been introduced in Chap. 2 to represent the set of all movable points in the interval $[0, l]$. The physical meanings of the above two equations are:

(1) Excluding the point $x = s$, the Green's function satisfies the homogeneous equation $L_x G(x, s) = 0$ everywhere in I.
(2) At $x = s$, the derivative of the Green's function satisfies the condition

$$G_x(s - 0, s) - G_x(s + 0, s) = \frac{1}{p(s)}. \tag{5.2.3}$$

(3) At the two end points, the Green's function must satisfy boundary conditions of the original Sturm–Liouville system, which are given in Eq. (5.1.2).

We may construct the Green's function of a Sturm–Liouville system by following the process outlined in the next two paragraphs. Also see Ref. [2].

Suppose $\varphi(x)$ and $\psi(x)$ are, respectively, non-trivial solutions of following equations:

$$L\varphi = 0 \quad (0 < x < l) \quad \text{and} \quad p(0)\varphi'(0) - h\varphi(0) = 0; \tag{5.2.4}$$

$$L\psi = 0 \quad (0 < x < l) \quad \text{and} \quad p(l)\psi'(l) + H\psi(l) = 0. \tag{5.2.5}$$

Then, the Green's function of the operator L can be expressed as

$$G(x,s) = \begin{cases} \varphi(x)\psi(s), & x \le s, \\ \varphi(s)\psi(x), & x > s, \end{cases} \quad x \text{ and } s \in I. \tag{5.2.6}$$

Apparently, the Green's function obtained this way satisfies the aforementioned conditions (1) and (3). As for Condition (2), we have

$$G_x(s-0,s) - G_x(s+0,s) = \varphi'(s)\psi(s) - \varphi(s)\psi'(s).$$

Equations (5.2.4) and (5.2.5), which are satisfied by $\varphi(x)$ and $\psi(x)$, respectively, further lead to

$$(L\varphi)\psi - (L\psi)\varphi = -\psi(p\varphi')' + \varphi(p\psi')' = -[p(\varphi'\psi - \varphi\psi')]' = 0.$$

This implies that the following equation must hold for any s satisfying $0 < s < l$:

$$p(s)[\varphi'(s)\psi(s) - \varphi(s)\psi'(s)] = \text{constant}. \tag{5.2.7}$$

The constant cannot be zero. Otherwise, we could find $\varphi(s)$ and $\psi(s)$ to be proportional, and thus zero would be an eigenvalue of the Sturm–Liouville system. However, this contradicts the condition that the system is well-constrained. Therefore, by properly choosing the functions $\varphi(s)$ and $\psi(s)$, it is possible to make this constant be 1.

It is obvious that the Green's function given by Eq. (5.2.6) is symmetric. This result is captured in the next theorem.

Theorem 5.2 *The Green's function defined by Eqs. (5.2.1) and (5.2.2) is symmetric, i.e.,*

$$G(x,s) = G(s,x), \quad x \text{ and } s \in I.$$

Due to space limitation, we skip the proof here.

The physical interpretation of this theorem is well known. Specifically, a unit force applied to one point s on a Sturm–Liouville system causes a displacement at another point x on the system. This displacement is equal to that obtained at the point s if a unit force is applied to the point x.

5.2.2 Green's Functions of Bars Elastically Constrained at One End and Either Elastically Supported or Free at the Other End

As examples, let us look at Green's functions of a bar elastically supported at both ends as well as a bar subject to the elastic constraint at one end but letting free at the other end. In both cases, we assume there is no elastic foundation.

Example 1 In the absence of an elastic foundation, we have $q(x) = 0$. Then, Eqs. (5.2.4) and (5.2.5) yield

$$\varphi(x) = C\left(\int_0^x \frac{dt}{p(t)} + \frac{1}{h}\right) \quad \text{and} \quad \psi(x) = D\left(\int_x^l \frac{dt}{p(t)} + \frac{1}{H}\right).$$

Thus, the Green's function of the bar elastically supported at both ends becomes

$$G(x,s) = \begin{cases} k\left(\int_0^x \frac{dt}{p(t)} + \frac{1}{h}\right)\left(\int_s^l \frac{dt}{p(t)} + \frac{1}{H}\right), & x \le s, \\ k\left(\int_0^s \frac{dt}{p(t)} + \frac{1}{h}\right)\left(\int_x^l \frac{dt}{p(t)} + \frac{1}{H}\right), & x > s. \end{cases}$$

Here, the parameter $k = CD$, and its value can be calculated from the condition $p(s)[\varphi'(s)\psi(s) - \varphi(s)\psi'(s)] = 1$ as below:

$$k = \frac{1}{\frac{1}{h} + \int_0^l \frac{dt}{p(t)} + \frac{1}{H}}.$$

In the end, we get the Green's function of the form:

$$G(x,s) = \frac{\left(\int_0^\alpha \frac{dt}{p(t)} + \frac{1}{h}\right)\left(\int_\beta^l \frac{dt}{p(t)} + \frac{1}{H}\right)}{\frac{1}{h} + \int_0^l \frac{dt}{p(t)} + \frac{1}{H}}, \tag{5.2.8}$$

where

$$\alpha = \min(x, s), \quad \beta = \max(x, s).$$

In particular, when the bar is fixed at both ends, we get

$$G(x, s) = \int_0^{\alpha} \frac{\mathrm{d}t}{p(t)} \cdot \int_{\beta}^{l} \frac{\mathrm{d}t}{p(t)} \bigg/ \int_0^{l} \frac{\mathrm{d}t}{p(t)}. \tag{5.2.9}$$

Example 2 We now turn to the case of a bar elastically supported on the left but unconstrained on the right. Again, there is no elastic foundation. As before, we have

$$\varphi(x) = C\left(\int_0^{x} \frac{\mathrm{d}t}{p(t)} + \frac{1}{h}\right), \quad \psi(x) = D.$$

Again, just like in Example 1, set $k = CD$, and Eq. (5.2.7) leads to $k = 1$. Then, we obtain the Green's function,

$$\mathrm{G}(x, s) = \int_0^{\alpha} \frac{\mathrm{d}t}{p(t)} + \frac{1}{h}. \tag{5.2.10}$$

5.2.3 Derivation of the Green's Function from Flexibility Coefficients

An alternative way to derive the Green's function is to let the flexibility coefficients of the respective discrete system approach their limits. In the absence of an elastic foundation, the following stiffness matrix $\boldsymbol{K}$ of an arbitrarily supported bar was given in Sect. 3.1.1:

$$\boldsymbol{K} = \begin{bmatrix} k_1 + h & -k_1 & 0 & \cdots & 0 & 0 & 0 \\ -k_1 & k_1 + k_2 & -k_2 & \cdots & 0 & 0 & 0 \\ \vdots & \vdots & \vdots & & \vdots & \vdots & \vdots \\ 0 & 0 & 0 & \cdots & -k_{n-1} & k_{n-1} + k_n & -k_n \\ 0 & 0 & 0 & \cdots & 0 & -k_n & k_n + H \end{bmatrix}. \tag{5.2.11}$$

Here, $h = 0$, $h \to \infty$, and $0 < h < +\infty$ correspond to free, fixed, and elastic boundary conditions at the left end, respectively; H, at the right end, is interpreted in a similar way; and

$$k_r = \frac{p_r + p_{r-1}}{2l_r}, \quad r = 1, 2, \ldots, n,$$

where $p_r = p(x_r)$ $(r = 0, 1, \ldots, n)$ and $l_r = x_r - x_{r-1} = \Delta x_r$ $(r = 1, 2, \ldots, n)$.

Next, we calculate elements in the flexibility matrix $\boldsymbol{R} = (r_{ij})$ of the discretized bar, i.e., those flexibility coefficients. To this end, set $k_0 = h$, $k_{n+1} = H$, and

$$\Delta_{st} = K\begin{pmatrix} s \; s+1 \cdots t \\ s \; s+1 \cdots t \end{pmatrix} = \begin{vmatrix} k_{s-1} + k_s & -k_s & 0 & \cdots & 0 & 0 \\ -k_s & k_s + k_{s+1} & -k_{s+1} & \cdots & 0 & 0 \\ \vdots & \vdots & \vdots & & \vdots & \vdots \\ 0 & 0 & 0 & \cdots & -k_{t-1} & k_{t-1} + k_t \end{vmatrix},$$

$$s = 1, 2, \ldots, n+1; \; t = s, s+1, \ldots, n+1.$$

After using the property of expanding one determinant as the sum of two other determiants and also performing some simplification, we may get

$$\Delta_{st} = k_{s-1}\Delta_{s+1,t} + k_s k_{s+1} \cdots k_{t-1} k_t.$$

Starting from here and further applying the method of mathematical induction, we obtain

$$\Delta_{st} = k_{s-1}k_s \cdots k_{t-1}k_t \sum_{\gamma=s-1}^{t} k_\gamma^{-1}.$$

In particular, the above result yields

$$\det \boldsymbol{K} = \Delta_{1,n+1} = k_0 k_1 \cdots k_n k_{n+1} \sum_{\gamma=0}^{n+1} k_\gamma^{-1}.$$

Moreover, the cofactor of the element a_{ij} in Matrix $\boldsymbol{K} = (a_{ij})_{(n+1)\times(n+1)}$ is defined by

$$K_{ij} = (-1)^{i+j} K\begin{pmatrix} 1\;2 \cdots i-1 \;\; i+1 \cdots n+1 \\ 1\;2 \cdots j-1 \;\; j+1 \cdots n+1 \end{pmatrix}.$$

Since $\boldsymbol{K}$ is a tridiagonal matrix, $a_{ij} = 0$ when $|i - j| > 1$, and $a_{i,i+1} = a_{i+1,i} = -k_i$ when $|i - j| = 1$. Therefore, if $i \leq j$,

$$K_{ij} = K\begin{pmatrix} 1\;2 \cdots i-1 \\ 1\;2 \cdots i-1 \end{pmatrix} \cdot k_i \cdots k_{j-1} \cdot K\begin{pmatrix} j+1 \;\; j+2 \cdots n+1 \\ j+1 \;\; j+2 \cdots n+1 \end{pmatrix}$$

$$= k_0 k_1 \cdots k_n k_{n+1} \sum_{\alpha=0}^{i-1} k_\alpha^{-1} \cdot \sum_{\beta=j}^{n+1} k_\beta^{-1};$$

while if $i > j$,

$$K_{ij} = K\begin{pmatrix} 1\ 2 \cdots j-1 \\ 1\ 2 \cdots j-1 \end{pmatrix} \cdot k_j \cdots k_{i-1} \cdot K\begin{pmatrix} i+1\ i+2 \cdots n+1 \\ i+1\ i+2 \cdots n+1 \end{pmatrix}$$

$$= k_0 k_1 \cdots k_n k_{n+1} \sum_{\alpha=0}^{j-1} k_\alpha^{-1} \cdot \sum_{\beta=i}^{n+1} k_\beta^{-1}.$$

Because $\boldsymbol{R} = \boldsymbol{K}^{-1}$, we get, in the end,

$$r_{ij} = \frac{K_{ij}}{\det K} = \begin{cases} \left(\sum_{\alpha=0}^{i-1} k_\alpha^{-1} \cdot \sum_{\beta=j}^{n+1} k_\beta^{-1}\right) \Big/ \sum_{\gamma=0}^{n+1} k_\gamma^{-1}, & i \leqslant j, \\ \left(\sum_{\alpha=0}^{j-1} k_\alpha^{-1} \cdot \sum_{\beta=i}^{n+1} k_\beta^{-1}\right) \Big/ \sum_{\gamma=0}^{n+1} k_\gamma^{-1}, & i > j. \end{cases} \tag{5.2.12}$$

Equation (5.2.12) is applicable to bars under any boundary conditions that lead to well-constrained systems. As a special case, when there is a free end, the above equation is greatly simplified. For example, if the bar is in the fixed-free configuration, then $k_0 = h \to \infty$, $k_{n+1} = H = 0$, and its flexibility coefficients are reduced to

$$r_{ij} = \begin{cases} \sum_{\alpha=1}^{i-1} k_\alpha^{-1}, & i \le j, \\ \sum_{\alpha=1}^{j-1} k_\alpha^{-1}, & i > j. \end{cases} \tag{5.2.13}$$

Now we will make some further observations. Consider any fixed points x and s in the interval $[0, l]$. There must exist indices i and j in the finite difference discretization such that the following conditions are satisfied:

(1) $x_{i-1} \le x \le x_i$ and $x_{j-1} \le s \le x_j$ $(1 \le i, j \le n)$.
(2) If $x < s$, then $i < j$; while if $x > s$, then $i > j$.
(3) As the finite difference grid is gradually refined, i.e.,

$$n \to \infty \quad \text{and} \quad \delta = \max \Delta x_r \to 0, \quad r = 1, 2, \ldots, n,$$

$i, j, n-i$, and $n-j$ all approach infinity.

In addition, for a bar in engineering application, its $p(x)$ must be either a continuous or piecewise continuous function with, at most, finite number of Type I discontinuities. As a result, $1/p(x)$ is integrable, and the following relationship holds:

$$\frac{2}{p_r + p_{r+1}} = \frac{1 + b_r}{p(\xi_r)}, \quad x_r \le \xi_r \le x_{r+1},$$

where $b_r = 0$ if $p(x)$ is continuous in (x_r, x_{r+1}), or b_r is bounded if $p(x)$ has discontinuities inside (x_r, x_{r+1}).

Thus, when the bar is fixed on the left, we get the following equation in the limit:

$$\lim_{n\to\infty,\,\delta\to 0} \sum_{r=1}^{i-1} k_r^{-1} = \lim_{n\to\infty,\,\delta\to 0} \left[\sum_{r=1}^{i-1} \frac{\Delta x_{r+1}}{p(\xi_r)} + \sum_{r=1}^{i-1} \frac{b_r \Delta x_{r+1}}{p(\xi_r)} \right] = \int_0^x \frac{dt}{p(t)}. \tag{5.2.14}$$

Here, since there is only a finite number of non-zero b_r's, the second sum in the expression between two equal signs in Eq. (5.2.14) approaches zero in the limit. In a more general situation, when the bar is elastically constrained on the left, $k_0 = h$, and thus,

$$\lim_{n\to\infty,\,\delta\to 0} \sum_{r=0}^{i-1} k_r^{-1} = \int_0^x \frac{dt}{p(t)} + \frac{1}{h}. \tag{5.2.15}$$

Since $h \to \infty$ leads to a fixed end, Eq. (5.2.14) can be viewed as a special case of Eq. (5.2.15). By the same token, we can calculate all parts of Eq. (5.2.12) by assuming both ends of the bar to be elastically constrained:

$$\lim_{n\to\infty,\,\delta\to 0} \sum_{r=0}^{j-1} k_r^{-1} = \int_0^s \frac{dt}{p(t)} + \frac{1}{h}, \quad \lim_{n\to\infty,\,\delta\to 0} \sum_{r=j}^{n+1} k_r^{-1} = \int_s^l \frac{dt}{p(t)} + \frac{1}{H},$$

$$\lim_{n\to\infty,\,\delta\to 0} \sum_{r=i}^{n+1} k_r^{-1} = \int_x^l \frac{dt}{p(t)} + \frac{1}{H}, \quad \lim_{n\to\infty,\,\delta\to 0} \sum_{\gamma=0}^{n+1} k_\gamma^{-1} = \frac{1}{h} + \int_0^l \frac{dt}{p(t)} + \frac{1}{H}.$$

Comparing Eq. (5.2.12) with Eqs. (5.2.8) and (5.2.9), we see easily that

$$\lim_{n\to\infty,\,\delta\to 0} r_{ij} = G(x, s),$$

when the bar is either elastically supported or fixed at the boundary. As for the case involving a free end, for instance, a bar in the fixed-free configuration, its flexibility coefficient r_{ij} is calculated from Eq. (5.2.13). Apparently, as $n \to \infty$ and $\max \Delta x_r \to 0\ (r = 1, 2, \ldots, n)$, r_{ij} becomes Eq. (5.2.10) in the limit as $h \to \infty$. Thus, regardless of the type of boundary conditions involved, the flexibility coefficients of a bar approach the corresponding Green's function in the limit.

The content of Sects. 5.2.2 and 5.2.3 is taken primarily from Ref. [4].

5.2.4 Green's Function and Integral Equation

The greatest value of the Green's function is that it helps convert the equations of motion (5.1.4) and (5.1.2) of a well-constrained Sturm–Liouville system into an integral equation of the form:

$$u(x) = \lambda \int_0^l G(x, s)\rho(s)u(s)\mathrm{d}s. \tag{5.2.16}$$

In fact, if the Green's function satisfies Eqs. (5.2.1) and (5.2.2), it can be shown, by applying D'Alembert's principle of inertial forces as well as the principle of supposition, that Eq. (5.2.16) is obviously a solution to the governing equations of a well-constrained Sturm–Liouville system. Conversely, if Eq. (5.2.16) is satisfied, we have

$$Lu = \lambda \int_0^l L_x G(x, s)\rho(s)u(s)\mathrm{d}s = \lambda \int_0^l \delta(x - s)\rho(s)u(s)\mathrm{d}s = \lambda\rho(x)u(x),$$

$$p(0)u'(0) - hu(0) = \lambda \int_0^l [p(0)G_x(0, s) - hG(0, s)]\rho(s)u(s)\mathrm{d}s = 0,$$

$$p(l)u'(l) + Hu(l) = \lambda \int_0^l [p(l)G_x(l, s) + HG(l, s)]\rho(s)u(s)\mathrm{d}s = 0.$$

This shows that the solution $\{\lambda,\ u(x, \lambda)\}$ to Eq. (5.2.16) must also satisfy Eqs. (5.1.4) and (5.1.2).

We often set

$$\tilde{u}(x) = \sqrt{\rho(x)}u(x) \quad \text{and} \quad K(x, s) = G(x, s)\sqrt{\rho(x)\rho(s)}, \tag{5.2.17}$$

and transform Eq. (5.2.16) into the following symmetric form:

$$\tilde{u}(x) = \lambda \int_0^l K(x, s)\tilde{u}(s)\mathrm{d}s, \quad x \in I. \tag{5.2.18}$$

Functions $G(x, s)$ and $K(x, s)$ are referred to as kernels of the integral Eqs. (5.2.16) and (5.2.18), respectively.

5.3 Oscillatory Properties in Vibration and Static Deformation of the Sturm–Liouville System

5.3.1 Oscillatory Properties in Vibration of the Sturm–Liouville System

The general properties of the integral equation with a symmetric kernel were introduced in Sect. 2.7. However, $G(x, s)$ and $K(x, s)$, defined in Eqs. (5.2.6) and (5.2.17), respectively, are not ordinary symmetric kernels, because eigenvalues of Eq. (5.2.16) are not only real but also positive and distinct. This indicates that the kernels $G(x, s)$ and $K(x, s)$ possess some special constructs. In fact, it will become evident later that they belong to the group of oscillatory kernels defined in Sect. 2.7. This result can be established in two ways outlined in the next two paragraphs.

With regard to the first approach, recall that we have demonstrated in Chap. 3 the sign-oscillation characteristics of stiffness matrices associated with both finite difference and finite element models of a bar subject to either fixed-free or fixed–fixed constraints. This leads to the conclusion that the corresponding flexibility matrices have to be oscillatory. In addition, it was proved in Sect. 5.2.3 that flexibility coefficients of a bar under various constraints become the respective Green's functions in the limit. Thus, according to Theorem 2.29 in Sect. 2.11, we conclude that the Green's function of a bar, in either the fixed-free or fixed–fixed configuration, must be an oscillatory kernel (see Ref. [5]). However, this approach has a shortcoming that it excludes bars resting on elastic foundations and thus cannot deal with oscillatory properties of a Sturm–Liouville system in the most general form.

In the second approach, we take advantage of the relationship between oscillatory kernels and oscillatory matrices to establish oscillatory properties of the Green's function associated with a well-constrained Sturm–Liouville system. Before elaborating further on this approach, let us introduce the following definition:

Definition 5.1 *If a continuous function $f(x)$ satisfies the condition $f(x) \geq 0$ (or ≤ 0) everywhere in $I \subset [0, l]$, then the function $f(x)$ has a fixed sign on the interval I. Furthermore, if $f(x) > 0$ (or < 0) at any point in I, then the function $f(x)$ has a strictly fixed sign on I.*

In next a few paragraphs, we will prove the Green's function of a well-constrained Sturm–Liouville system to be an oscillatory kernel. Let us begin by introducing a theorem (see Refs. [1, 2]):

Theorem 5.3 *For a well-constrained Sturm–Liouville system, functions $\varphi(x)$ and $\psi(x)$, determined by Eqs. (5.2.4) and (5.2.5), respectively, have the following properties:*

(1) $\varphi(x)$ and $\psi(x)$ have strictly fixed sign on the interval I.
(2) $\varphi(x)/\psi(x)$ is monotonically increasing on the interval I.
(3) $\varphi(x)\psi(x) > 0$ for any $x \in I$. Consequently, we may assume, without loss of generality,

$$\varphi(x) > 0 \quad and \quad \psi(x) > 0, \quad x \in I.$$

Proof We first apply the method of proof by contradiction to the function $\varphi(x)$. Suppose there exists a point $x_0 \in I$ such that $\varphi(x_0) = 0$. Then,

$$0 = \int_0^{x_0} \varphi L\varphi \mathrm{d}x = -p(x)\varphi(x)\varphi'(x)\big|_0^{x_0} + \int_0^{x_0} [p(\varphi')^2 + q\varphi^2]\mathrm{d}x$$

$$= h\varphi^2(0) + \int_0^{x_0} [p(\varphi')^2 + q\varphi^2]\mathrm{d}x.$$

If $\varphi'(x)$ is not identically zero; then the left-hand side of the above equation vanishes, but the right-hand side is positive, regardless of the value of the parameter h. This is a contradiction, which implies that $\varphi(x)$ cannot be equal to zero anywhere in the interval I. Moreover, since the function $\varphi(x)$ is continuous on I, it cannot change sign. As for the case of $\varphi'(x) \equiv 0$, we have $\varphi(x) =$ constant, and the conclusion remains correct obviously.

Very similarly, we may prove that $\psi(x)$ has no sign reversal in I, too.

Next, it is known from Eq. (5.2.7) that

$$p(x)[\varphi'(x)\psi(x) - \varphi(x)\psi'(x)] = 1, \quad x \in I. \tag{5.3.1}$$

Rewrite the above equation as

$$\frac{\mathrm{d}}{\mathrm{d}x}\left[\frac{\varphi(x)}{\psi(x)}\right] = \frac{1}{p(x)\psi^2(x)} > 0, \quad x \in I. \tag{5.3.2}$$

Thus, Property (2) is valid.

Finally, Property (3), which needs to be proved, is equivalent to

$$\frac{\varphi(x)}{\psi(x)} > 0, \quad x \in I. \tag{5.3.3}$$

Taking into consideration of Property (2), we only need to show $\varphi(0)/\psi(0) \geq 0$ in order to prove Eq. (5.3.3).

When $h \to \infty$, which corresponds to the bar being fixed on the left side, $\varphi(0) = 0$, and the above inequality is apparently satisfied. As a result, we only need to consider the case of $\varphi(0) > 0$. Let us again follow the process of proof by contradiction by assuming $\psi(0) < 0$. Then, we get from Property (1) as well as the boundary condition for $\psi(x)$ at the right end,

$$\psi(x) < 0\,(x \in I) \quad \text{and} \quad \psi'(l) > 0.$$

Moreover, the first equation in (5.2.5) yields

$$[p(x)\psi'(x)]' = q(x)\psi(x) \leq 0, \quad x \in I,$$

which shows $p(x)\psi'(x)$ to be monotonically decreasing on I. Thus, we must have

$$p(0)\psi'(0) > 0.$$

However, this would invalidate Eq. (5.3.1), because its left-hand side now becomes negative at $x = 0$. This contradiction helps establish the validity of Property (3). ■

We may get the following corollary directly from Property (3) of Theorem 5.3:

Corollary *Kernels $G(x, s)$ and $K(x, s)$, defined by Eqs. (5.2.6) and (5.2.17), respectively, satisfy inequalities $G(x, s) > 0$ and $K(x, s) > 0$, where $x \in I$ and $s \in I$.*

By starting from Theorem 5.3, it is also not difficult to obtain the theorem below:

Theorem 5.4 *For a well-constrained Sturm–Liouville system, $G(x, s)$ and $K(x, s)$, defined by Eqs. (5.2.6) and (5.2.17), respectively, are both oscillatory kernels.*

Proof Consider an arbitrary set of points x_i $(i = 1, 2, \cdots, n)$, satisfying the conditions: $0 \leq x_1 < x_2 < \cdots < x_n \leq l$. Suppose at least one of them is an interior point. Let us look at the matrix $\boldsymbol{L} = (G(x_i, x_j))$, where

$$G(x_i, x_j) = \begin{cases} \varphi_i \psi_j, & i \leq j \\ \varphi_j \psi_i, & i > j \end{cases}, \quad i \text{ and } j = 1, 2, \cdots, n,$$

$$\varphi_i = \varphi(x_i) \text{ and } \psi_i = \psi(x_i), \quad i = 1,\ 2,\ \cdots,\ n.$$

Here, $\varphi(x)$ and $\psi(x)$ represent solutions to Eqs. (5.2.4) and (5.2.5), respectively. Apparently, $\boldsymbol{L}$ is none other than the matrix we have studied as an example of oscillatory matrices in Sect. 2.4. According to Theorem 5.3 above, for a well-constrained Sturm–Liouville system, φ_i and ψ_i $(i = 1,\ 2,\ \cdots,\ n)$ are of the same sign, and satisfy inequalities below:

$$\frac{\varphi_1}{\psi_1} < \frac{\varphi_2}{\psi_2} < \cdots < \frac{\varphi_n}{\psi_n}.$$

Consequently, $\boldsymbol{L}$ is an oscillatory matrix. Further application of Theorem 2.18, which is for identifying an oscillatory kernel, leads to the conclusion that $G(x, s)$ is indeed an oscillatory kernel. It is then clear from the corollary of Theorem 2.18 that $K(x, s)$ is also an oscillatory kernel. ■

Last, based on the previous discussion as well as Theorem 2.23, which summarizes properties of an integral equation with an oscillatory kernel, we may immediately obtain the next corollary.

Corollary *Consider a well-constrained Sturm-Liouville system. It has the following oscillatory properties in vibration:*

(1) *Natural frequencies of a well-constrained Sturm–Liouville system are all distinct, i.e.,*

$$0 < f_1 < f_2 < f_3 < \cdots.$$

(2) *The family of mode shapes $u_i(x)\,(i = 1, 2, \ldots)$ constitutes a sequence of Markov functions on the interval $[0, l]$. Thus, they possess all properties summarized in Theorem 2.22, namely:*

   (a) *The first mode shape $u_1(x)$ has no zero in the interval I.*
   (b) *The i-th mode shape $u_i(x)$ has $i - 1$ nodes and no other zero in the interval I. Consequently, $u_i(x)$ has $i - 1$ sign reversals in the interval $(0, l)$.*
   (c) *Consider the following function defined on the interval I:*

$$u(x) = \sum_{i=k}^{m} c_i u_i(x), \quad 1 \le k \le m \text{ and } \sum_{i=k}^{m} c_i^2 > 0.$$

   Its number of nodes is no less than $k - 1$, while its count of zeroes is no greater than $m - 1$. Specifically, if $u(x)$ has $m - 1$ different zeroes, then all of them have to be nodes.
   (d) *Nodes of successive mode shapes $u_i(x)$ and $u_{i+1}(x)$ $(i = 2,\ 3,\ \cdots)$ interlace.*

5.3.2 Oscillatory Properties in Static Deformation of the Sturm–Liouville System

Now that it has been shown the Green's function of a well-constrained Sturm–Liouville system to be oscillatory, we know from the discussion in Sect. 2.10.1 that the well-constrained continuous system of a bar or string must have oscillatory properties in static deformation, A and B. Specifically,

(1) if a concentrated force is applied to a movable point on a well-constrained Sturm–Liouville system, the deflection at every movable point of the system is nonzero and in the direction of the force applied;
(2) when a set of n concentrated forces are applied to some movable points on the well-constrained Sturm–Liouville system, its deflection $u(x)$ has at most $n - 1$ sign reversals, i.e., $S_u \le n - 1$, where n is a positive integer.

5.4 Number of Independent Modes of a Bar and Additional Properties of Its Modal Shapes

5.4.1 *Further Properties of Modal Shapes of Well-Constrained Bars*

It was pointed out earlier that the modal equation of a bar and the associated boundary conditions are:

$$[p(x)u'(x)]' + \lambda\rho(x)u(x) = 0, \tag{5.4.1}$$

$$p(0)u'(0) - hu(0) = 0, \quad p(l)u'(l) + Hu(l) = 0. \tag{5.4.2}$$

Additionally, the oscillatory properties in vibration of the well-constrained system of a bar were presented in Sect. 5.3. However, we still need to introduce the following lemma, prior to discussing further properties of mode shapes of well-constrained bars.

Lemma 5.1 *Suppose that $u(x)$ is a differentiable function defined on the interval* $[0, l]$, *and satisfies Eq. (5.4.1) and boundary conditions in Eq. (5.4.2). If $u(x)$ has j nodes in the interval* $(0, l)$, *then the node count of $u'(x)$ in* $(0, l)$ *is*

$$S_{u'} = j + 1 - \Delta(u'(0)) - \Delta(u'(l)),$$

where

$$\Delta(t) = \begin{cases} 1, & t = 0, \\ 0, & t \neq 0. \end{cases}$$

Proof Let $\xi_1, \xi_2, \ldots,$ and ξ_j represent nodes of $u(x)$. They divide the interval $[0, l]$ into $j + 1$ subintervals: $[0, \xi_1]$, $[\xi_1, \xi_2], \ldots,$ and $[\xi_j, l]$. Per Rolle's Theorem, the function $u'(x)$ has at least one zero in each subinterval (ξ_r, ξ_{r+1}), where $r = 1, 2, \ldots, j - 1$. As for the subinterval $(0, \xi_1)$, more elaborate analysis is needed. When $h = 0$ or ∞, it is obvious that $u'(x)$ has at least one zero in the subinterval. If $0 < h < \infty$, then there must exist an $s \in (0, \xi_1)$ such that $u'(0)u'(s) \leq 0$. Otherwise, $u(x)$ must be strictly increasing or decreasing in $(0, \xi_1)$, which would contradict the condition $u(\xi_1) = 0$. Therefore, the continuity of $u'(x)$ ensures that there is at least one zero of $u'(x)$ in $(0, \xi_1)$. We can reach a similar conclusion regarding the interval (ξ_j, l).

Next, assume there are two or more zeros of $u'(x)$ in a subinterval $[\xi_r, \xi_{r+1}]$ $(r = 0, 1, \ldots, j;\ \xi_0 = 0,\ \xi_{j+1} = l)$ and denote them as $c, d, \ldots$. After integrating the left-hand side of Eq. (5.4.1) with respect to x over the subinterval interval $[c, d] \subset [\xi_r, \xi_{r+1}]$, we get

$$\int_c^d \{[p(x)u'(x)]' + \lambda\rho(x)u(x)\}\mathrm{d}x = \lambda \int_c^d \rho(x)u(x)\mathrm{d}x.$$

Since there is no sign reversal for $u(x)$ in $[c, d]$, the right-hand side of the above expression cannot be zero, but it is apparently untrue. This shows that there can be one and only one zero of $u'(x)$ in the subinterval (ξ_r, ξ_{r+1}). We know, from the process to prove Rolle's Theorem, that the zero x_r of $u'(x)$ must lie within the open sub-interval (ξ_r, ξ_{r+1}) and furthermore, it has to be a node, where $r = 1, 2, \ldots, j-1$. However, the leading and ending subintervals $[0, \xi_1]$ and $[\xi_j, l]$ may be exceptions to this rule. In summary, we have

$$0 \le x_1 < \xi_1 < x_2 < \xi_2 < \cdots < x_j < \xi_j < x_{j+1} \le l, \tag{5.4.3}$$

where $u'(x_r) = 0\,(r = 1, 2, \ldots, j+1)$, and the equality in (5.4.3) holds only when $h = 0$ or $H = 0$. Notice that when $h = 0$ and $H = 0$, $x_1 = 0$ and $x_{j+1} = l$ may not be nodes, respectively. This concludes the proof of Lemma 5.1. ■

Based on Eqs. (5.4.1) and (5.4.2) as well as Lemma 5.1, we can make the following three remarks:

First, it can be seen from boundary conditions in Eq. (5.4.2) that displacements and strains at the two ends of the bar satisfy inequalities

$$u_i(0)u_i'(0) \ge 0 \quad \text{and} \quad u_i(l)u_i'(l) \le 0, \tag{5.4.4}$$

where equalities in (5.4.4) may hold only when h and H either assume the value of 0 or approach ∞. Moreover, if $u_i(x)$ and $u_k(x)$ represent two different mode shapes of the bar, Eq. (5.4.2) yields

$$u_i(0)u_k'(0) - u_k(0)u_i'(0) = 0 = u_i(l)u_k'(l) - u_k(l)u_i'(l), \tag{5.4.5}$$

where $i, k = 1, 2, 3, \cdots$ and $i \ne k$.

Second, the modified modal equation of the bar is

$$p(x)u_i''(x) + p'(x)u_i'(x) + \lambda_i\rho(x)u_i(x) = 0, \quad i = 1, 2, \ldots, \tag{5.4.6}$$

and the condition $u_i'(x_r) = 0$ leads to

$$u_i(x_r)u_i''(x_r) < 0, \quad i = 1, 2, \ldots \quad \text{and} \quad r = 1, 2, \ldots, i. \tag{5.4.7}$$

Third, notice that the mode shape $u_i(x)$ of the well-constrained system of a bar has and only has $i - 1$ $(i = 1, 2, \ldots)$ nodes. In addition, $h = 0$ and $H = 0$ are equivalent to $u'(0) = 0$ and $u'(l) = 0$, respectively. Thus, by applying Lemma 5.1 and Eq. (5.4.3), we may establish a new theorem.

Theorem 5.5 *Following properties are associated with the well-constrained system of a bar:*

(1) The node count of its strain mode shape $u_i'(x)$ can determined by

$$S_{u_i'} = i - \Delta(h) - \Delta(H), \quad i = 2, 3, \ldots. \tag{5.4.8}$$

(2) Let ξ_r $(r = 1, 2, \ldots, i-1)$ and x_r $(r = 1, 2, \ldots, i - \Delta(h) - \Delta(H))$ denote, respectively, nodes of the i-th mode shapes of the displacement and strain, $u_i(x)$ and $u_i'(x)$. Then, the two sets of nodes interlace, as shown in Eq. (5.4.3).

Notice that the case of $u_1(x)$ is not included in the above discussion. Since it has only one subinterval, which is $[0, l]$, Eq. (5.4.8) is obviously applicable to $u_1'(x)$, too. Based on results obtained this far, we get another theorem.

Theorem 5.6 *The necessary conditions for a function $\varphi(x)$ to be the mode shape of a well-constrained bar are: $\varphi(x)$ and its derivative have fixed numbers of sign reversals in the interval I, and satisfy both Eq. (5.4.4) and the condition*

$$S_{\varphi'} = S_{\varphi} + 1 - \Delta(\varphi'(0)) - \Delta(\varphi'(l)). \tag{5.4.9}$$

This set of conditions is referred to as Condition A in the subsequent discussion. If $\varphi(x)$ is the i-th mode shape of the bar, then it is further required that $S_\varphi = i - 1$ $(i = 1, 2, \cdots)$.

Next, we will show that Condition A, introduced above, is not only necessary but also sufficient for the mode shape of a bar. To this end, we first introduce a lemma.

Lemma 5.2 *Suppose that the function $\varphi(x)$ has the continuous first derivative on the interval $[0, l]$. Moreover, it is assumed that $\varphi(x)$ satisfies Condition A on the same interval. Then, nodes $\{\xi_r\}$ of $\varphi(x)$ interlace with nodes $\{x_r\}$ of $\varphi'(x)$, as shown in Eq. (5.4.3). Furthermore, we have*

$$\varphi(x_r)\varphi''(x_r) < 0, r = 1, 2, \ldots, S_{\varphi'}.$$

Proof Since $\varphi(x)$ has a fixed number of sign reversals in I, which may be represented by $S_\varphi = i - 1$, we can find points $\{\xi_r\}_1^{i-1}$ such that $\varphi(\xi_r) = 0$ and

$$0 < \xi_1 < \xi_2 < \cdots < \xi_{i-1} < l.$$

We also know from Eq. (5.4.9) that $\varphi'(x)$ changes sign exactly once in each of the subintervals (ξ_r, ξ_{r+1}) $(r = 1, 2, \ldots, i-2)$. In comparison, it may either retain its sign or have precisely one sign reversal in $(0, \xi_1)$, depending upon whether $\varphi'(0)$ is zero; while $\varphi'(x)$ behave similarly in (ξ_{i-1}, l), depending on whether $\varphi'(l)$ vanishes. Thus, according to Rolle's Theorem, $\varphi'(x)$ must have exactly one zero that has to be a node in each of the subintervals mentioned above, i.e., Equation (5.4.3) is satisfied.

Furthermore, if $\varphi(x) > 0$ on a subinterval, then $\varphi(x)$ has to reach the local maximum at the point where $\varphi'(x)$ changes sign; while if $\varphi(x) < 0$ on a subinterval, then $\varphi(x)$ must achieve the local minimum at the place where the sign of $\varphi'(x)$ is reversed. Otherwise, there will be contradiction to Eq. (5.4.9). Hence, $\varphi(x_r)\varphi''(x_r) < 0$ must also hold true. ■

The argument given above also shows that $u_i(x)$ has to be either monotonically ascending or monotonically descending in between two adjacent nodes of $u_i'(x)$. Thus, we may further get the following result.

Corollary *If x_r represents a node of $u_i'(x)$, then*

$$u_i(x_r)u_i'(x) > 0 \text{ when } x_{r-1} < x < x_r$$

and

$$u_i(x_r)u_i'(x) < 0 \text{ when } x_r < x < x_{r+1},$$

where $i = 1, 2, \ldots$ and $r = 1, 2, \ldots, i$.

Theorem 5.7 *Condition* A *in Theorem 5.6 is sufficient for the function $\varphi(x)$ to be one mode shape of a well-constrained bar.*

Proof Let us follow the approach of a constructive proof. Given a positive number λ as well as a function $\varphi(x)$, which has a continuous first derivative and a piecewise continuous second derivative and, at the same time, satisfies Condition A, we should be able to build a bar with positive parameters $p(x)$ and $\rho(x)$ such that $\varphi(x)$ is one of its mode shapes in longitudinal vibration while $\sqrt{\lambda}$ is the corresponding angular frequency.

In fact, when $\varphi(x)$ satisfies Condition A, Eq. (5.4.4) yields

$$h = \frac{p(0)\varphi'(0)}{\varphi(0)} \geq 0 \quad \text{and} \quad H = -\frac{p(l)\varphi'(l)}{\varphi(l)} \geq 0. \tag{5.4.10}$$

It is further known from Lemma 5.2 that Eq. (5.4.3) holds true and $\varphi(x_r)\varphi''(x_r) < 0$. Additionally, Eq. (5.4.1) leads to

$$p(x) = \frac{1}{\varphi'(x)}\left[p(0)\varphi'(0) - \lambda\int_0^x \rho(s)\varphi(s)\mathrm{d}s\right]. \tag{5.4.11}$$

Set $S_\varphi = i - 1$. We will proceed by considering three cases.

Case 1 $x_1 > 0$ and $x_i < l$. We may find a positive function $\rho(x)$ such that

$$\lambda\int_0^{x_r} \rho(s)\varphi(s)\mathrm{d}s = B, \quad r = 1, 2, \ldots, i \text{ and } \varphi'(x_r) = 0, \tag{5.4.12}$$

where B is a constant having the same sign as $\varphi(0+0)$. Let us define

$$p(x)=\begin{cases}\dfrac{1}{\varphi'(x)}\left[B-\lambda\int\limits_0^x\rho(s)\varphi(s)\mathrm{d}s\right], & x_r<x<x_{r+1} \text{ and } r=0,1,\ldots,i \text{ with } x_0=0 \text{ and } x_{i+1}=l;\\ -\lambda\rho(x_r)\dfrac{\varphi(x_r)}{\varphi''(x_r)}, & x=x_r \text{ and } r=1,2,\ldots,i.\end{cases} \tag{5.4.13}$$

$p(x)$ calculated above and $\rho(x)$ to be further determined, together with the specified h and H, constitute a set of parameters associated with the bar we are looking for.

Now the question is: How are we going to select $\rho(x)$ such that $p(x)$ defined by Eq. (5.4.13) remains positive all the time? In fact, we have much leeway in the selection of $\rho(x)$. For instance, we may choose

$$\rho(x)=\begin{cases}a_r & x_r\le x\le\xi_r,\\ a_r+\dfrac{a_{r+1}-a_r}{c_r}(x-\xi_r), & \xi_r\le x\le\xi_r+c_r,\\ a_{r+1} & \xi_r+c_r\le x\le x_{r+1},\end{cases}$$

$$r=1,2,\ldots,i-1, \tag{5.4.14}$$

where

$$c_r=(x_{r+1}-\xi_r)/10,\quad r=1,2,\ldots,i-1. \tag{5.4.15}$$

As for $\rho(x)$ in the intervals $[0,x_1]$ and $[x_i,l]$, it may be assigned to constant values a_1 and a_i, respectively. Thus, after selecting a positive value for a_1, a_{r+1} can be uniquely determined by the equation:

$$\int\limits_{x_r}^{x_{r+1}}\rho(s)\varphi(s)\mathrm{d}s=0,\quad r=1,2,\ldots,i-1.$$

Apparently, $\rho(x)$ defined by Eq. (5.4.14) is positive. Based on the above condition and Eq. (5.4.12), the expression

$$B-\lambda\int\limits_0^x\rho(s)\varphi(s)\mathrm{d}s=\lambda\int\limits_x^{x_{r+1}}\rho(s)\varphi(s)\mathrm{d}s$$
$$x_r<x<x_{r+1};\quad r=0,1,\ldots,i-1;\quad x_0=0,\ x_{i+1}=l$$

must be of the same sign as $\varphi(x_{r+1})$. We then know, by applying the corollary of Lemma 5.2 to this case, that the two functions

$$B - \lambda \int_0^x \rho(s)\varphi(s)\mathrm{d}s \quad \text{and} \quad \varphi'(x),$$

have to share a common sign in the interval (x_r, x_{r+1}) $(r = 0, 1, \ldots, i-1)$. Therefore, $p(x)$ determined by Eq. (5.4.13) must be greater than zero inside the interval (x_r, x_{r+1}) and at its end points, where $r = 0, 1, \ldots, i-1$.

As for the value of $p(x)$ in the interval (x_i, l), we have

$$B - \lambda \int_0^x \rho(s)\varphi(s)\mathrm{d}s = -\lambda \int_{x_i}^x \rho(s)\varphi(s)\mathrm{d}s, \quad x_i < x < l.$$

Obviously, the above expression and, consequently, $p(x)$ are of signs opposite to $\varphi(x_i)$. Thus, they must be of the same sign as $\varphi'(x)$. As a result, $p(x)$ determined by Eq. (5.4.13) must also be positive in this interval.

Case 2 $x_1 = 0$ and $x_i < l$. Obviously, the above discussion is completely applicable to this case. But we need to limit $r = 2, 3, \ldots, i$ in Eq. (5.4.12) as well as in the second expression in Eq. (5.4.13). Furthermore, when assigning a value to r in all other places, we need to make sure that the subinterval $[0, x_1]$ is excluded.

Case 3 $x_1 > 0$ and $x_i = l$. The previous argument remains valid in this case. But, we have to limit $r = 1, 2, \ldots, i-1$ in Eq. (5.4.12) as well as in the second expression in Eq. (5.4.13). Additionally, when assigning a value to r in all other places, we need to ensure the subinterval $[x_i, l]$ to be excluded.

These results show that as long as $\varphi(x)$ satisfies Condition A, it will be one of the mode shapes of a well-constrained bar, namely, Condition A is sufficient. ■

To help a reader better understand the discussion above, let us look at an example.

Consider the function $\varphi(x) = \cos x + \sin x$, which satisfies Condition A related to the mode shape of a bar in the interval $[0, 5\pi/4]$:

$$S_\varphi = S_{\varphi'} = 1, \quad h = \frac{p(0)\varphi'(0)}{\varphi(0)} = p(0) > 0, \quad \text{and} \quad \varphi'(5\pi/4) = 0.$$

If $\rho(x)$ is taken to be a positive constant, then Eq. (5.4.13) yields $p = \lambda\rho$, which should also be a positive constant. Thus, when both $p(x)$ and $\rho(x)$ are constants, $\omega = \sqrt{p/\rho}$ is the second natural angular frequency of a bar with the length $l = 5\pi/4$, elastically supported on the left ($h = p(0)$), and unconstrained on the right. The function $\cos x + \sin x$ is the corresponding mode shape. The result is identical to the one obtained by directly solving the differential equation. Notice that this function continues to satisfy Condition A in the expanded interval $[0, l]$, where $5\pi/4 \leq l \leq 3\pi/2$, if the boundary constraint at the right end is changed from the unconstrained condition to an elastic support. Consequently, it can still serve as the mode shape of a bar. However, if the interval $[0, l]$ is not properly chosen, for instance, if $l \in (3\pi/4, 5\pi/4)$, then the same function $\varphi(x)$ will violate Eq. (5.4.10), and thus cannot be a mode shape of any bar.

In summary, we have established the necessary and sufficient conditions for the mode shape of a well-constrained bar. Apparently, the discussion should cover, as special cases, bars in both fixed-free and fixed–fixed configurations, which are commonplace in engineering applications.

5.4.2 Compatibility Conditions of Modes and the Number of Independent Modes

Similar to related work on the discrete system, we may also study the compatibility of two modes associated with the continuous system of a bar.

Let $\varphi(x)$ and $\psi(x)$ denote two different mode shapes of a bar; while have λ and μ represent, respectively, squares of the corresponding natural angular frequencies. Since $\varphi(x)$ and $\psi(x)$ satisfy the following equations:

$$\begin{cases} [p(x)\varphi'(x)]' + \lambda\rho(x)\varphi(x) = 0, \\ [p(x)\psi'(x)]' + \mu\rho(x)\psi(x) = 0, \end{cases} \tag{5.4.16}$$

or equivalently,

$$\begin{cases} \varphi''(x) + \varphi'(x)\dfrac{p'(x)}{p(x)} + \lambda\varphi(x)\dfrac{\rho(x)}{p(x)} = 0, \\[2ex] \psi''(x) + \psi'(x)\dfrac{p'(x)}{p(x)} + \mu\psi(x)\dfrac{\rho(x)}{p(x)} = 0, \end{cases}$$

we get,

$$\frac{\rho(x)}{p(x)} = \frac{z(x)}{f(x)} \quad \text{and} \quad p(x) = C\exp\left\{-\int_0^x \frac{g(s)}{f(s)}\mathrm{d}s\right\}, \tag{5.4.17}$$

where

$$\begin{aligned} f(x) &= \mu\varphi'(x)\psi(x) - \lambda\varphi(x)\psi'(x), \\ g(x) &= \mu\varphi''(x)\psi(x) - \lambda\varphi(x)\psi''(x), \\ z(x) &= \varphi''(x)\psi'(x) - \varphi'(x)\psi''(x). \end{aligned}$$

Thus, in addition to Condition A and Eq. (5.4.5), $\varphi(x)$ and $\psi(x)$ have to satisfy another set of constraints: At any given point x in $[0, l]$, $z(x)$ and $f(x)$ must either be zero simultaneously or share the same sign. If $f(x) = 0$ at $x = s$, then $g(x)/f(x)$

should have a finite limit while $z(x)/f(x)$ ought to approach a positive limit, as $x \to s$. This set of conditions, together with Eq. (5.4.5), are referred to as the compatibility conditions for two different modes of a bar.

The above discussion leads to the following theorem:

Theorem 5.8 *Consider two sets of modes* $(\omega_i = \sqrt{\lambda},\ \varphi(x))$ *and* $(\omega_j = \sqrt{\mu},\ \psi(x))$. *If they satisfy the necessary conditions (Condition A) of a mode shape as well as the aforementioned compatibility requirements, then we may construct a bar with the density function* $\rho(x)$ *and stiffness function* $p(x)$ *defined in Eq. (5.4.17). All other modes of this bar can be derived afterwards.*

Due to the importance of this result, we would like to offer an alternative interpretation of it in the theorem below.

Theorem 5.9 *Among the infinite sets of displacement modes associated with the continuous system of a bar,* $(\omega_i,\ u_i(x))\ (i = 1, 2, \ldots)$, *there are only two, and arbitrarily two, displacement modes* $(\omega_{i_1},\ u_{i_1}(x))$ *and* $(\omega_{i_2},\ u_{i_2}(x))\ (i_1 \neq i_2)$ *that are independent.*

The implication of the above theorem is: If two bars share two sets of modes, including both natural frequencies and related mode shapes, then the two systems must be identical. It also implies that in order to design a bar to have certain modal properties, we may specify, at most, two modes.

It needs to be pointed out that there is only one independent mode for a string. In fact, the natural angular frequency ω and mode shape $\varphi(x)$ of the string satisfy the following equation and boundary conditions:

$$\begin{cases} T\varphi''(x) + \omega^2\rho(x)\varphi(x) = 0, \\ \varphi(0) = \varphi(l) = 0, \end{cases} \tag{5.4.18}$$

where T is a constant representing the tension force in the string. After reformatting Eq. (5.4.18), we get

$$\rho(x) = -T\varphi''(x)/\omega^2\varphi(x). \tag{5.4.19}$$

Suppose a set of modal data $(\omega, \varphi(x))$ satisfies the conditions: $\varphi''(x)$ and $\varphi(x)$ are of opposite signs in the interval $(0,\ l)$; or if it appears that $\varphi(x) = 0$ at $x = \xi$, then $\varphi''(\xi) = 0$ while the limit of $\varphi''(x)/\varphi(x)$ as $x \to \xi$ exists and is of negative value. Under one of these conditions, we may compute $\rho(x)$ from Eq. (5.4.19). Thus, we conclude: When the tension of the string T is specified and the density function $\rho(x)$ is the only unknown, it needs only one mode to determine $\rho(x)$. This leads to the theorem below:

Theorem 5.10 *Although a string has infinite sets of modes, there is only one that is independent and all others can be derived from this mode.*

For two strings of equal length and under the same tension T, if they share one set of the natural frequency and mode shape, then the two strings are identical and all of their other modes match.

The content of Sects. 5.4.1 and 5.4.2 comes primarily from Ref. [3].

5.4.3 Modal Qualitative Properties of a Bar Unconstrained at Both Ends

It is necessary to remind the reader that the previous arguments are not suitable for bars unconstrained at both ends. In order to study qualitative properties of this type of bars, we need to follow the approach used when dealing with the corresponding discrete systems. Set

$$\sigma(x) = p(x)u'(x), \tag{5.4.20}$$

and rewrite Eq. (5.4.1) as

$$[\rho^{-1}(x)\sigma'(x)]' + \lambda p^{-1}(x)\sigma(x) = 0. \tag{5.4.21}$$

Equation (5.4.21) is still the modal equation of a "bar". Let us refer to this "bar" with section parameters

$$p^*(x) = \rho^{-1}(x) \quad \text{and} \quad \rho^*(x) = p^{-1}(x)$$

as the conjugate of the original bar. Suppose the original bar is in the free-free configuration. Under the transformation (5.4.20), the boundary conditions assume the following form:

$$\sigma(0) = 0 = \sigma(l), \tag{5.4.22}$$

namely, the corresponding conjugate bar is fixed at both ends.

Thus, it is not difficult to identify the following modal qualitative properties of a bar unconstrained at both ends:

(1) The non-zero natural frequencies of a bar in the free–free configuration are not only positive but also distinct. They can be arranged in an ascending order as

$$0 = f_1 < f_2 < f_3 < \cdots.$$

(2) Consider the conjugate of a bar unconstrained at both ends. Denote its natural frequencies and related mode shapes as $(f_i^*, \sigma_i^*(x))$ $(i = 1, 2, \ldots)$, where the superscript "*" indicates the related variable to be associated with the nonzero natural frequencies of a bar in the free-free configuration. The corresponding

Table 5.1 Node counts of displacement and strain mode shapes associated with the continuous system of a bar

Constraints			S_{u_i}	$S_{u'_i}$
Type	h	H		
Fixed–fixed	∞	∞	$i-1$	i
Fixed–free	∞	0	$i-1$	$i-1$
Free–free	0	0	$i-1$	$i-2^*$

*For a bar unconstrained at both ends, $S_{u'_1}=0$ when $i=1$

mode pairs of the original unconstrained bar are $(f_{i+1}, \sigma_{i+1}(x))\,(i=1,2,\ldots)$. Notice that $f_i^*=f_{i+1}$. Furthermore, since the mode shape $\sigma_i^*(x)$ has $i-1$ sign reversals in the interval $[0,l]$, the respective mode shape $\sigma_{i+1}(x)$ and thus $u'_{i+1}(x)$ should have $i-1$ sign reversals in the same interval, where $i=1,2,\ldots$. In other words, $\sigma_i(x)$ and thus $u'_i(x)$ of the bar unconstrained at both ends should have $i-2$ sign reversals in $[0,l]$, where $i=2,3,\ldots$.

(3) Eqs. (5.4.1) and (5.4.20) yield

$$\sigma'_i(x)=-\lambda_i\rho(x)u_i(x).$$

This shows that $\sigma'_i(x)$ and $u_i(x)$ should have identical qualitative properties. Based on Lemma 5.1 and taking into account $\sigma_i(0)=0=\sigma_i(l)$, we see that if $\sigma_i(x)$ has $i-2$ sign reversals in $[0,l]$ then $\sigma'_i(x)$ has precisely $i-1$ sign reversals within the same interval, where $i=2,3,\ldots$. Thus, we may conclude that $u_i(x)$ should also have $i-1$ sign reversals in $[0,l]$, where $i=1,2,\ldots$. As for $u_1(x)$, which corresponds to $f_1=0$, it can obviously be set to a constant value, and thus the number of sign reversals of $u_1(x)$ possess the same Property (3) stated above. However, the number of sign reversals of $u'_1(x)$ is zero in this case. Thus, Property (4) to be introduced next will not cover $u_1(x)$.

(4) The necessary and sufficient conditions for the function $u_i(x)$ to be the i-th mode shape of a bar in the free–free configuration is

$$S_{u_i}=S_{u'_i}+1=i-1,\quad i=2,3,\ldots. \tag{5.4.23}$$

The sufficient part of this result can be proved by following the same process used to prove Theorem 5.7. The details are left out here.

(5) All of the discussion in Subsect. 5.4.2 is applicable to a bar unconstrained at both ends, after its rigid-body modes are excluded.

Based on the above discussion, we can tabulate node counts (numbers of sign reversals) of mode shapes of the displacement and strain associated with the continuous systems of a bar subject to three different boundary constraints. See Table 5.1.

At the conclusion of this subsection, let us prove that the oscillatory property in vibration, (2)(c), established at the end of Subsect. 5.3.1 for a well-constrained

Sturm–Liouville system, should also hold true for a bar unconstrained at both ends. This result is captured in detail in the next theorem (see Ref. [6]).

Theorem 5.11 *Consider the following function defined on the interval I and expressed as a linear combination of a few mode shapes* $u_i(x)$ $(i = k, k+1, \ldots, m$ and $1 \leq k \leq m)$ *of a bar unconstrained at both ends:*

$$u(x) = \sum_{i=k}^{m} c_i u_i(x), \quad 1 \leq k \leq m \text{ and } \sum_{i=k}^{m} c_i^2 > 0.$$

Its node count is no less than $k - 1$*, while its number of zeroes is no more than* $m - 1$.

Proof Let us look at the linear combination of some mode shapes $\sigma_i^*(x)$ $(i = k, k+1, \ldots, m$ and $1 \leq k \leq m)$ associated with the conjugate of the bar unconstrained at both ends:

$$\sigma(x) = \sum_{i=k}^{m} c_i \sigma_i^*(x), \quad 1 \leq k \leq m \text{ and } \sum_{i=k}^{m} c_i^2 > 0.$$

$\sigma(x)$ is the "displacement" of the conjugate bar, which is obviously a well-constrained Sturm–Liouville system. Based on the oscillatory property in vibration, (2)(c), of such system, $\sigma(x)$ should have no less than $k - 1$ nodes and no more than $m - 1$ zeroes in the interval $[0, l]$.

Now, let us consider the linear combination of some mode shapes $u_i'(x)$ $(i = k+1, k+2, \ldots, m+1$ and $1 \leq k \leq m)$ associated with the bar unconstrained at both ends:

$$u'(x) = \sum_{i=k}^{m} c_i u_{i+1}'(x), \quad 1 \leq k \leq m \text{ and } \sum_{i=k}^{m} c_i^2 > 0. \tag{5.4.24}$$

By taking into account Eq. (5.4.20), the result of $\sigma(x)$ can be transferred to $u'(x)$. Thus, $u'(x)$ has no less than $k - 1$ nodes and no more than $m + 1$ zeroes in the interval $[0, l]$. Here, the zeroes of $u'(x)$ at the two ends of $[0, l]$ have been taken into account.

With the above preparation, we are now ready to prove the theorem. First, let us establish the second part of the theorem by using the method of proof by contradiction. Specifically, we assume $u(x)$ associated with Eq. (5.4.24) to have more than m zeroes. By applying Lemma 5.1 and noticing the conditions $h = 0 = H$, we find the zero count of $u'(x)$ in Eq. (5.4.24) to be more than $m+1$. This leads to a contradiction. Thus, the number of zeroes of $u(x)$ associated with Eq. (5.4.24) should be no more than m.

Next, we prove the first part of the theorem. To this end, let us look at the modified modal equation of the bar unconstrained at both ends

$$\sigma'(x) = -\lambda \rho(x) u(x).$$

It shows that $u(x)$ and $\sigma'(x)$ share identical nodes. Start from the result that the node count of $\sigma(x)$ is no less than $k-1$, take into account the fact that both $\sigma'(0)$ and $\sigma'(l)$ are nonzero, and apply Lemma 5.1 to $\sigma(x)$ and $\sigma'(x)$. We may then conclude that the number of nodes of $\sigma'(x)$ is no less than k. Consequently, the number of nodes of $u(x)$ associated with Eq. (5.4.24) is no less than k. Notice that $u(x)$ corresponding to Eq. (5.4.24) is of the following form:

$$u(x) = \sum_{i=k}^{m} c_i u_{i+1}(x), \quad 1 \le k \le m \quad \text{and} \quad \sum_{i=k}^{m} c_i^2 > 0.$$

This completes the proof. ■

5.4.4 *Interlacement of Nodes Associated with Various Types of Mode Shapes of a Bar*

Recall that when proving nodes of two successive functions of a Markov function sequence to be interlacing in Theorem 2.22, we only used Properties (2) and (3) of the Markov function sequence. In the context of a Sturm–Liouville system, these two properties correspond to oscillatory properties in vibration, (2)(b) and (2)(c), in the corollary of Theorem 5.4. Since these properties are valid for a well-constrained bar, nodes of successive mode shapes of its displacement should interlace. We then derive similar results for nodes of mode shapes of its displacement and strain associated with the same frequency in Theorem 5.5.

In Sect. 5.4.3, we showed that a bar unconstrained at both ends also possesses oscillatory properties in vibration, (2)(b) and (2)(c), described in the corollary of Theorem 5.4. Because of this and also taking into account of the relationship between mode shapes of the bar and its conjugate, we may further reach conclusions as follows:

(1) Nodes of two successive displacement mode shapes $u_i(x)$ and $u_{i+1}(x)$ ($i = 2, 3, \ldots$) of a bar unconstrained at both ends should interlace.
(2) Nodes of two successive strain mode shapes $u_i'(x)$ and $u_{i+1}'(x)$ ($i = 2, 3, \ldots$) of an arbitrarily supported bar interlace.
(3) Consider a bar unconstrained at both ends. Nodes of its displacement and strain mode shapes $u_i(x)$ and $u_i'(x)$ corresponding to the same frequency should interlace, where $i = 2, 3, \ldots$.

Notice that Property (3) above can be established by using Rolle's Theorem and rules related to sign reversals, similar to the derivation of Theorem 5.5.

Based on discussions in this section, we may sketch mode shapes of a bar in Fig. 5.1.

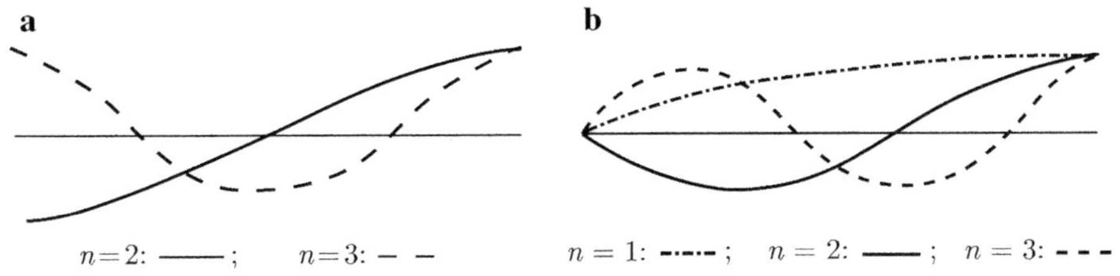

Fig. 5.1 **a** Second and third mode shapes of a bar unconstrained at both ends; **b** first, second, and third mode shapes of a bar in the fixed-free configuration

5.5 Interlacement of Natural Frequencies of a Bar Subject to Various Boundary Constraints

The primary objective of this section is to derive interlacing properties of natural frequencies of a bar subject to various boundary constraints. We proceed by comparing results of the bar subject to fixed-free constraint with those under other types of constraints separately, and then combining them to reach the final conclusion.

5.5.1 Interlacement of Natural Frequencies of a Bar in Fixed–Fixed and Fixed-Free Configurations

Suppose there is a distributed load of $F(x)\sin\omega t$ acting on the entire span of a bar in the fixed-free configuration. Solution to the forced vibration problem is of the form $u(x,t) = u(x)\sin\omega t$, where $u(x)$ satisfies the equation

$$[p(x)u'(x)]' + \omega^2\rho(x)u(x) = F(x), \quad 0 < x \le l. \tag{5.5.1}$$

Also, assume $\{\omega_i^2, u_i(x)\}$ $(i = 1, 2, \ldots)$ to be the set of normalized modal solutions to the homogeneous equation corresponding to (5.5.1). Obviously,

$$\int_0^l \rho(x)u_i^2(x)\mathrm{d}x = 1, \quad i = 1, 2, \ldots.$$

After applying the method of superposition, we may express the forced response as the sum of mode shape series, similar to the discrete system. Specifically,

$$u(x) = \sum_{i=1}^{\infty} c_i u_i(x). \tag{5.5.2}$$

Substituting this expression into Eq. (5.5.1) and taking into account the orthogonality and normality of mode shapes, we get

$$c_i = \frac{1}{\omega^2 - \omega_i^2} \int_0^l F(x) u_i(x) \mathrm{d}x, \quad i = 1, 2, \ldots.$$

If the external force is a concentrated load and acts right at the free end of the bar, namely,

$$F(x) = F_0 \delta(x - l),$$

then coefficients in the expansion (5.5.2) are given by

$$c_i = \frac{F_0 u_i(l)}{\omega^2 - \omega_i^2}, \quad i = 1, 2, \ldots.$$

Thus, the corresponding forced response becomes

$$u(x) = F_0 \sum_{i=1}^{\infty} \frac{u_i(l)}{\omega^2 - \omega_i^2} u_i(x). \tag{5.5.3}$$

In particular, we have

$$u(l) = F_0 \sum_{i=1}^{\infty} \frac{u_i^2(l)}{\omega^2 - \omega_i^2}. \tag{5.5.4}$$

When $u(l) \equiv 0$, the right end of the bar is fixed. Equation (5.5.4) is changed to

$$\sum_{i=1}^{\infty} \frac{u_i^2(l)}{\omega^2 - \omega_i^2} = 0, \tag{5.5.5}$$

which is none other than the frequency equation of a bar fixed at both ends. In other words, under the action of a harmonic external load, the frequency that reduces the amplitude at the free end to zero is one of the natural frequencies of the original bar subject to a fixed constraint at the right end. Furthermore, if the forced response function $u(x, \omega)$ has $i - 1$ nodes in the interval $(0, l)$, then ω is the i-th angular frequency of the bar fixed at both ends, where $i = 1, 2, \ldots$. Nevertheless, the forced response function itself is not a mode shape of the bar fixed at both ends. This is because the forced response function $u(x, \omega)$ always satisfies the condition

$$u'(l) = F_0 \sum_{i=1}^{\infty} \frac{u_i(l) u_i'(l)}{\omega^2 - \omega_i^2} = 0.$$

Notice that $u_i(l) \neq 0$ all the time for a bar in the fixed-free configuration. By applying Proposition 4.5 to Eq. (5.5.5), we get

$$\omega_1^{\mathrm{cf}} < \omega_1^{\mathrm{cc}} < \omega_2^{\mathrm{cf}} < \omega_2^{\mathrm{cc}} < \cdots,$$

or equivalently,

$$f_1^{\mathrm{cf}} < f_1^{\mathrm{cc}} < f_2^{\mathrm{cf}} < f_2^{\mathrm{cc}} < \cdots, \tag{5.5.6}$$

where all superscripts follow the same convention introduced in Chap. 3. Thus, natural frequencies of fixed–fixed and fixed-free bars interlace.

5.5.2 *Interlacement of Natural Frequencies of a Bar in Free–Free and Fixed-Free Configurations*

By following the thought process of Sect. 5.5.1, and by applying the concept of conjugate bars, we may obtain interlacing properties of natural frequencies associated with a bar unconstrained at both ends and the same bar subject to the fixed-free constraint.

Consider a bar with non-uniform cross sections. It is first put in the free–free configuration, and then in the fixed-free configuration. Suppose natural frequencies corresponding to these two cases are, respectively,

$$0 = f_1^{\mathrm{ff}} < f_2^{\mathrm{ff}} < \cdots \quad \text{and} \quad 0 < f_1^{\mathrm{cf}} < f_2^{\mathrm{cf}} < \cdots.$$

It is known from our earlier discussion on conjugate bars in Sect. 5.4.3 that the conjugate system of a bar unconstrained at both ends is one fixed at both ends, while a bar subject to the fixed-free constraints corresponds to a conjugate "bar" in the free–fixed configuration. Apparently, these two conjugate systems can still be viewed as the same bar with different boundary supports at the left end. Let us denote their natural frequencies as, respectively,

$$0 < \tilde{f}_1^{\mathrm{cc}} < \tilde{f}_2^{\mathrm{cc}} < \cdots \quad \text{and} \quad 0 < \tilde{f}_1^{\mathrm{fc}} < \tilde{f}_2^{\mathrm{fc}} < \cdots.$$

Applying the method identical to that used in the previous subsection, we can prove

$$0 < \tilde{f}_1^{\mathrm{fc}} < \tilde{f}_1^{\mathrm{cc}} < \tilde{f}_2^{\mathrm{fc}} < \tilde{f}_2^{\mathrm{cc}} < \cdots.$$

Next, take into account the relationship between natural frequencies of original and conjugate systems, specially,

$$\tilde{f}_i^{\mathrm{fc}} = f_i^{\mathrm{cf}}, \quad \tilde{f}_i^{\mathrm{cc}} = f_{i+1}^{\mathrm{ff}}, \quad i = 1, 2, \ldots.$$

Thus, we may conclude that natural frequencies of the bar in free–free and fixed-free configurations interlace, i.e.,

$$0 = f_1^{\mathrm{ff}} < f_1^{\mathrm{cf}} < f_2^{\mathrm{ff}} < f_2^{\mathrm{cf}} < \cdots. \tag{5.5.7}$$

By combining Eqs. (5.5.6) and (5.5.7), we get

$$f_i^{\mathrm{ff}} < f_i^{\mathrm{cf}} < (f_i^{\mathrm{cc}}, f_{i+1}^{\mathrm{ff}}) < f_{i+1}^{\mathrm{cf}}, \quad i = 1, 2, \cdots. \tag{5.5.8}$$

Here, the parenthesis in Eq. (5.5.8) indicates uncertain relationship between frequencies of free–free and fixed–fixed bars. In fact, when all parameters of the bar are constants, frequencies associated with the two types of boundary supports, one unconstrained at both ends while the other fixed at both ends, happen to satisfy the condition below:

$$f_i^{\mathrm{cc}} = f_{i+1}^{\mathrm{ff}}, \quad i = 1, 2, \ldots.$$

5.6 Comparison Between Discrete and Continuous Systems

We studied qualitative properties of vibration associated with the continuous systems of bars in this chapter. Comparing these results with their counterparts of the discrete systems in Chap. 3, we find the two share many commonalities, but there are also some important differences between them.

With regard to the oscillatory properties of modes, discrete and continuous systems are in complete agreement. This provides some assurance regarding the validity of finite difference and finite element models of bars. In other words, properties obtained from the discrete systems of vibrating bars are indeed some reasonable approximation of those associated with the related continuous systems.

However, one detail deserves some attention. With the discrete model, an elastic support may be viewed as a special case of the fixed constraint. They differ only by one degree of freedom. As a result, the necessary and sufficient conditions of the mode shape associated with a discrete system do not involve anything related to elastic constraints. In contrast, the mode shape of a continuous system must first satisfy boundary conditions defined in Eq. (5.4.4), where the less-than and greater-than operators are related to the elastic supports.

References

1. Gantmacher FP, Krein MG (1961) Oscillation matrices and kernels and small vibrations of mechanical systems. U. S. Atomic Energy Commission, Washington
2. Gladwell GML (2004) Inverse Problems in Vibration, 2nd edn. Springer, Dordrecht (1986, 1st edn, Martinus Nijhoff Publishers, Dordrecht)

3. Wang QS, Wang DJ (1994) An inverse mode problem for continuous second-order systems. In: Proceedings of international conference on vibration engineering, ICVE'94, Beijing. International Academic Publishers, Singapore, pp 167–170
4. Wang QS, Wang DJ (1996) The flexibility matrices and its limits of difference discrete system for a rod and beam. Mech Pr 18(5):43–47 (in Chinese)
5. Wang QS, Wang DJ (1997) United proof for qualitative properties of discrete and continuous systems of vibrating rod and beam. Acta Mech Sin 29(1):99–102 (in Chinese)
6. Wang QS, Wang DJ (2014) The supplementary proof of some oscillation property for continuous systems of rod and beam having rigid modes. J AQTC (Nat Sci Ed) 20(1):1–5 (in Chinese)

Chapter 6
Qualitative Properties of Vibration and Static Deformation Associated with Continuous Systems of Beams

In this chapter, we prove the Green's function of a well-constrained beam to be an oscillatory kernel by verifying that the beam has the oscillatory properties in static deformation. Based on this result, a number of modal qualitative properties can be derived for the continuous system of a single-span and well-constrained beam. These qualitative properties are also extended to a single-span and under-constrained beam, by taking advantage of the concept of a conjugate beam. Furthermore, several other modal qualitative properties, such as the number of independent modes and the frequency dependence on beam parameters, are introduced. At the end of the chapter, we discuss the modal qualitative properties of overhang beams.

6.1 Differential Equation and Boundary Constraints of a Vibrating Beam

Consider a slender straight beam of length l and with geometric and physical parameters as functions of the section location x. See Fig. 6.1. Its natural frequency f and mode shape $u(x)$ satisfy the following equation:

$$[EJ(x)u''(x)]'' = \lambda\rho(x)u(x), \quad 0 < x < l, \tag{6.1.1}$$

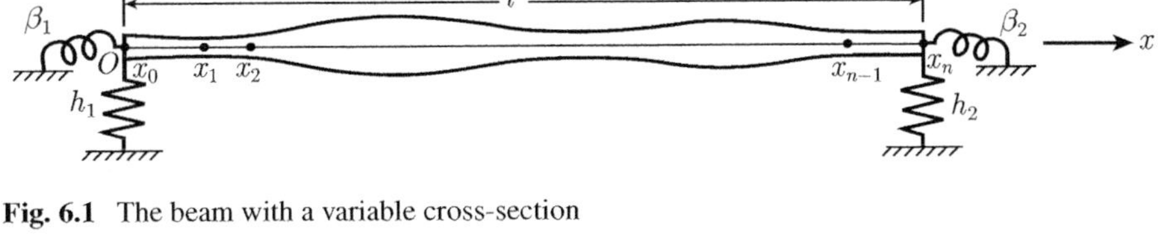

Fig. 6.1 The beam with a variable cross-section

D. Wang et al., *Qualitative Theory in Structural Mechanics*,
https://doi.org/10.1007/978-981-13-1376-9_6

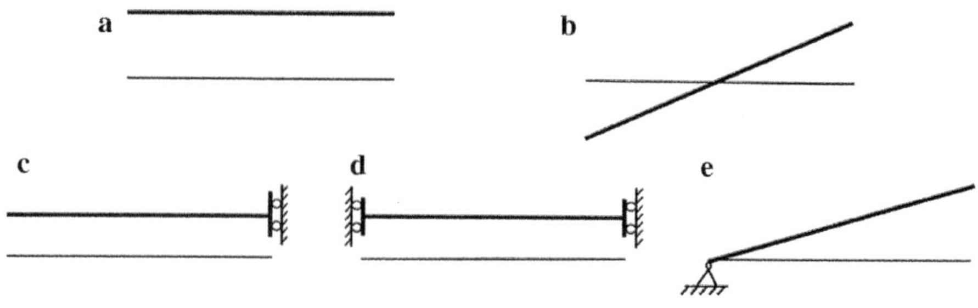

Fig. 6.2 Various rigid-body mode shapes of a beam

where $\lambda = (2\pi f)^2$, $\rho(x)$ is the linear mass density of the material, and $EJ(x)$ is the bending stiffness or flexural rigidity of the beam. Obviously, $EJ(x) > 0$ and $\rho(x) > 0$. It is assumed in the subsequent discussion that second derivatives of both $EJ(x)$ and $\rho(x)$ always exist. Listed below are beam support conditions at the ends that are commonly seen in engineering applications:

$$\begin{aligned} &\text{free } u'' = 0 \text{ and } (EJu'')' = 0, \quad \text{pinned } u = 0 \text{ and } u'' = 0, \\ &\text{sliding } u' = 0 \text{ and } (EJu'')' = 0, \quad \text{clamped } u = 0 \text{ and } u' = 0. \end{aligned} \tag{6.1.2}$$

Certain combinations of the above four types of boundary constraints at the two ends of the beam allow for rigid-body motions. Figure 6.2 illustrates these combinations and their possible ways of free movement. The related mode shapes all correspond to $\lambda = 0$, and are known as rigid-body mode shapes. They can be obtained by solving Eq. (6.1.1) under some form of boundary conditions given in Eq. (6.1.2). Notice that a beam unconstrained at both ends has two kinds of rigid-body mode shapes, depicted in Figs. 6.2a and 6.2b. Although the mode shape in Fig. 6.2a may not be directly orthogonal to that in Fig, 6.2b with respect to the weight function $\rho(x)$, certain combination of the two can always be found such that it is orthogonal to the mode shape in Fig. 6.2a with respect to the weight function $\rho(x)$.

A unified representation of the four types of boundary conditions of the beam is

$$\left[EJ(x)u''(x)\right]'_{x=0} + h_1 u(0) = 0 = \left[EJ(x)u''(x)\right]'_{x=l} - h_2 u(l), \tag{6.1.3}$$

$$EJ(0)u''(0) - \beta_1 u'(0) = 0 = EJ(l)u''(l) + \beta_2 u'(l). \tag{6.1.4}$$

Physically, they represent extensional and rotational spring constraints applied to the ends of the beam. Here, h_1 and h_2 designate the stiffness of the extension springs, while β_1 and β_2 denote the stiffness of the rotation springs. They are all non-negative constants. We can recover boundary conditions in (6.1.2) by setting:

$$\begin{aligned} &\text{free } h = 0 \text{ and } \beta = 0, \quad \text{sliding } h = 0 \text{ and } \beta \to \infty, \\ &\text{pinned } h \to \infty \text{ and } \beta = 0, \quad \text{clamped } h \to \infty \text{ and } \beta \to \infty. \end{aligned}$$

Similar to its discrete counterpart, if

$$h_1 + h_2 > 0 \quad \text{and} \quad \beta_1 + \beta_2 > 0 \tag{6.1.5}$$

or

$$h_1 \cdot h_2 > 0, \tag{6.1.6}$$

then the continuous system governed by Eqs. (6.1.1), (6.1.3), and (6.1.4) are well-constrained. See Refs. [3, 5]. Since the spring stiffness can never be negative, Condition (6.1.5) implies that at least one of h_1 and h_2 must be greater than zero while at least one of β_1 and β_2 has to be positive. In addition, Condition (6.1.6) indicates that it is all right for β_1 and β_2 to vanish simultaneously but both h_1 and h_2 have to be positive in this case. It will be demonstrated later that when either Inequality (6.1.5) or Eq. (6.1.6) is satisfied none of the rigid-body mode shapes shown in Fig. 6.2 can appear. Now, we present a few well-known properties associated with the modal equation of a vibrating beam.

Theorem 6.1 *The differential operator of a beam $B \equiv [EJ(x)u''(x)]''$ is symmetric under boundary conditions (6.1.3) and (6.1.4), namely,*

$$(Bu, v) = (Bv, u). \tag{6.1.7}$$

Proof In fact,

$$\begin{aligned}(Bu, v) - (Bv, u) &= \int_0^l \left\{\left[EJ(x)u''(x)\right]''v(x) - [EJ(x)v''(x)]''u(x)\right\}\mathrm{d}x \\ &= \{[EJ(x)u''(x)]'v(x) - [EJ(x)v''(x)]'u(x)\}_0^l \\ &\quad - EJ(x)[u''(x)v'(x) - v''(x)u'(x)]_0^l.\end{aligned}$$

Under boundary conditions (6.1.3) and (6.1.4), every term on the right-hand side of the second equality in the above equation vanishes. Thus, we get Eq. (6.1.7). ■

Theorem 6.2 *Under boundary conditions (6.1.3) and (6.1.4), eigenvalues of the modal equation (6.1.1) are real and non-negative. The necessary and sufficient condition for these eigenvalues to be positive is that the system is well-constrained.*

Proof Suppose λ and $u(x)$ to be the eigenvalue and the corresponding eigenfunction, respectively, of Eq. (6.1.1) subject to boundary conditions (6.1.3) and (6.1.4), i.e.,

$$\mathrm{B}u(x) = \lambda\rho(x)u(x).$$

Since both $EJ(x)$ and $\rho(x)$ are real functions, we get

$$\lambda(\rho u, \bar{u}) = (Bu, \bar{u}) = (B\bar{u}, u) = \overline{(Bu, \bar{u})} = \bar{\lambda}\overline{(\rho u, \bar{u})} = \bar{\lambda}(\rho u, \bar{u}).$$

This confirms λ to be a real number.

Moreover,

$$\begin{aligned}\lambda(\rho u, u) &= (Bu, u) \\ &= \{[EJ(x)u''(x)]'u(x) - EJ(x)u''(x)u'(x)\}_0^l + \int_0^l EJ(x)[u''(x)]^2\mathrm{d}x \\ &= h_1u^2(0) + h_2u^2(l) + \beta_1u'^2(0) + \beta_2u'^2(l) + \int_0^l EJ(x)[u''(x)]^2\mathrm{d}x.\end{aligned}$$

This shows that as long as $u(x)$ is not identically zero, we obviously have $\lambda \geq 0$.

The above equation further reveals that λ is zero, i.e., the right-hand side of the above equation vanishes, if and only if $u''(x) \equiv 0$, i.e., $u(x) = cx + d$, and

$$h_1d^2 = h_2(cl + d)^2 = \beta_1c^2 = \beta_2c^2 = 0.$$

These conditions lead to the following possible scenarios: (1) If $c = 0$, then $h_1 = h_2 = 0$, corresponding to the case where the beam is only constrained by rotational springs at two ends and its rigid-body motion involves just translation; (2) if $d = 0$, then $h_2 = 0$ and $\beta_1 = \beta_2 = 0$, leading to the situation where there is an extension spring at the left end of the beam and the rigid-body motion is restricted to a rotation about the point O, namely, the left end; (3) if $c \neq 0$ and $d \neq 0$, then $h_1 = h_2 = 0$ and $\beta_1 = \beta_2 = 0$, representing an unconstrained beam with both translational and rotational rigid-body motions. In summary, as long as the system is well-constrained, we must have $\lambda > 0$. If the condition is not satisfied, there must exist a zero eigenvalue. The above discussion is also applicable to situations where some or all of h_r and β_r $(r = 1,\ 2)$ are of infinite value. ■

Theorem 6.3 *Under boundary conditions (6.1.3) and (6.1.4), eigenfunctions of the modal equation corresponding to different eigenvalues are orthogonal to each other with respect to the weight function* $\rho(x)$.

The proof of this theorem is left out here.

6.2 Green's Function of a Beam

As is the case with a bar, the Green's function also plays a critical role in the study of qualitative properties of a beam. Thus, in this section we are going to introduce basic properties of the Green's function, derive its general form, and obtain the specific expressions of Green's functions for two common types of beams, i.e., beams in the clamped-free and pinned-pinned configurations. See Refs. [4, 7].

The Green's function of a beam is the solution of the following special form of the boundary value problem:

$$\begin{cases} B_x G(x,s) = \delta(x-s), \quad 0 < x < l, \quad 0 < s < l, \\ [EJ(x)G_{xx}(x,s)]_x|_{x=0} + h_1 G(0,s) = 0 = [EJ(x)G_{xx}(x,s)]_x|_{x=l} - h_2 G(l,s), \\ EJ(0)G_{xx}(0,s) - \beta_1 G_x(0,s) = 0 = EJ(l)G_{xx}(l,s) + \beta_2 G_x(l,s), \end{cases} \tag{6.2.1}$$

where h_r and β_r $(r = 1, 2)$ are all non-negative constants and satisfy either Eq. (6.1.5) or Eq. (6.1.6), while B_x indicates that differentiation in the operator B is with respect to the variable x. This definition shows that the Green's function of a beam $G(x, s)$ is, in fact, the static displacement of the beam caused by a unit concentrated force applied to the point s of the beam. Furthermore, the Green's function has the following properties:

(1) For any $x \in [0, l]$, $G(x, s)$ is continuous and satisfies boundary conditions (6.1.3) and (6.1.4).
(2) Excluding the point $x = s$, the fourth derivative of $G(x, s)$ with respect to $x \in [0, l]$ exists. But at $x = s$, its third derivative has a discontinuity. Specifically,

$$[EJ(x)G_{xx}(x,s)]_x\Big|_{x=s-0}^{x=s+0} = 1. \tag{6.2.2}$$

(3) $B_x G(x, s) = 0$ for any $x \in [0, l]$ and $x \neq s$.

Based on the above properties, the Green's function of a well-constrained beam system may be constructed following steps presented in the next a few paragraph.

First, by integrating the first equation in (6.2.1) twice with respect to the variable x, we obtain

$$EJ(x)G''(x,s) = \begin{cases} C_1 x + C_2, & x < s, \\ (1 + C_1)x + C_3, & x > s. \end{cases}$$

Dividing both sides of the above equation by $EJ(x)$ and also integrating the resulting equation twice with respect to x, we get

$$G(x,s) = \begin{cases} \int_0^x \mathrm{d}z \int_0^z \dfrac{C_1 t + C_2}{EJ(t)} \mathrm{d}t + C_4 x + C_6, & x < s, \\ \int_0^x \mathrm{d}z \int_0^z \dfrac{(C_1 + 1)t + C_3}{EJ(t)} \mathrm{d}t + C_5 x + C_7, & x > s. \end{cases}$$

By imposing continuity conditions on the displacement, angle of rotation, and bending moment at $x = s$, we have

$$C_3 = C_2 - s, \quad C_5 = C_4 - \int_0^s \frac{t-s}{EJ(t)}\mathrm{d}t,$$

$$C_7 = C_6 - \int_0^s \mathrm{d}z \int_0^z \frac{t-s}{EJ(t)}\mathrm{d}t + s\int_0^s \frac{t-s}{EJ(t)}\mathrm{d}t.$$

Thus,

$$G(x,s) = \begin{cases} \int_0^x \mathrm{d}z \int_0^z \frac{C_1 t + C_2}{EJ(t)}\mathrm{d}t + C_4 x + C_6, & x < s, \\ \int_0^x \mathrm{d}z \int_0^z \frac{C_1 t + C_2}{EJ(t)}\mathrm{d}t + C_4 x + C_6 + \int_s^x \mathrm{d}z \int_s^z \frac{t-s}{EJ(t)}\mathrm{d}t, & x > s. \end{cases}$$

Next, remove the intermediate variable z through integration by parts. This leads to

$$G(x,s) = \begin{cases} \int_0^x \frac{(x-t)(C_1 t + C_2)}{EJ(t)}\mathrm{d}t + C_4 x + C_6, & x < s, \\ \int_0^x \frac{(x-t)(C_1 t + C_2)}{EJ(t)}\mathrm{d}t + C_4 x + C_6 + \int_s^x \frac{(x-t)(t-s)}{EJ(t)}\mathrm{d}t, & x > s. \end{cases} \tag{6.2.3}$$

Finally, we can determine values of the integration constants by applying specific boundary conditions, and thus obtain Green's functions of a beam subject to various boundary conditions that produce well-constrained systems.

Specifically, for a beam in the clamped-free configuration, it is easy to find

$$C_4 = C_6 = 0, \quad C_1 = -1, \quad C_2 = s.$$

Thus, the Green's function of the clamped-free beam is

$$G^{\mathrm{cf}}(x,s) = \int_0^{\min(x,s)} \frac{(x-t)(s-t)}{EJ(t)}\mathrm{d}t. \tag{6.2.4}$$

As another example, for a beam pinned at both ends, its integration constants satisfy

$$C_2 = C_6 = 0, \quad (C_1 + 1)l - s = 0,$$

$$\int_0^l \frac{(l-t)C_1 t}{EJ(t)}\mathrm{d}t + C_4 l + \int_s^l \frac{(l-t)(t-s)}{EJ(t)}\mathrm{d}t = 0.$$

Hence, the Green's function of the beam pinned at two ends is

$$G^{\mathrm{pp}}(x,s) = \begin{cases} \dfrac{l-s}{l}\displaystyle\int_0^x \frac{t(t-x)}{EJ(t)}\mathrm{d}t - \frac{x}{l}\int_0^s \frac{(l-t)(s-t)}{EJ(t)}\mathrm{d}t + \frac{xs}{l^2}\int_0^l \frac{(l-t)^2}{EJ(t)}\mathrm{d}t, & x < s, \\ \dfrac{l-x}{l}\displaystyle\int_0^s \frac{t(t-s)}{EJ(t)}\mathrm{d}t - \frac{s}{l}\int_0^x \frac{(l-t)(x-t)}{EJ(t)}\mathrm{d}t + \frac{xs}{l^2}\int_0^l \frac{(l-t)^2}{EJ(t)}\mathrm{d}t, & x > s. \end{cases} \tag{6.2.5}$$

Similarly, we can derive Green's functions of the beam subject to other types of boundary conditions: pinned-sliding, clamped-sliding, clamped-pinned, and clamped-clamped. Details are skipped here. Notice that in the above discussion, indices c, f, and p indicate the respective end points being clamped, free, and pinned, respectively.

An alternative way to derive the Green's function of a beam is to start from flexibility coefficients of the corresponding discrete system and then find their limits. We will demonstrate this method with beams in the clamped-free and pinned-pinned configurations. See Ref. [4].

Let us first look at the beam subject to the clamped-free constraints. The stiffness matrix $\boldsymbol{A}$ of the finite difference system associated with this beam has been derived in Chap. 4, which is of the following form:

$$\boldsymbol{A} = \tilde{\boldsymbol{E}}_n \boldsymbol{L}^{-1} \tilde{\boldsymbol{E}}_n \boldsymbol{K}_{\mathrm{cf}} \tilde{\boldsymbol{E}}_n^{\mathrm{T}} \boldsymbol{L}^{-1} \tilde{\boldsymbol{E}}_n^{\mathrm{T}},$$

where $\boldsymbol{L} = \mathrm{diag}(l_1, l_2, \ldots, l_n)$, $\boldsymbol{K}_{\mathrm{cf}} = \mathrm{diag}(k_0, k_1, \ldots, k_{n-1})$, and the square matrix $\tilde{\boldsymbol{E}}_n$ of the n-th–order and its inverse $\boldsymbol{F}$ are, respectively

$$\tilde{\boldsymbol{E}}_n = \begin{bmatrix} 1 & -1 & & & \\ & 1 & -1 & & 0 \\ & & \ddots & \ddots & \\ & 0 & & 1 & -1 \\ & & & & 1 \end{bmatrix}_{n\times n} \quad \text{and} \quad \boldsymbol{F} = \begin{bmatrix} 1 & 1 & \cdots & 1 \\ & 1 & \cdots & 1 \\ & & \ddots & \vdots \\ 0 & & & 1 \end{bmatrix}_{n\times n}.$$

Thus, the corresponding flexibility matrix $\boldsymbol{R}$ is

$$\boldsymbol{R} = \boldsymbol{F}^{\mathrm{T}} \boldsymbol{L} \boldsymbol{F}^{\mathrm{T}} \boldsymbol{K}_{\mathrm{cf}}^{-1} \boldsymbol{F} \boldsymbol{L} \boldsymbol{F}, \tag{6.2.6}$$

or, in terms of the flexibility coefficients,

$$r_{ij} = \begin{cases} \sum_{\alpha=1}^{i}\left(\sum_{p=\alpha}^{i} l_p\right)k_{\alpha-1}^{-1}\left(\sum_{q=\alpha}^{j} l_q\right), & i \le j, \\ \sum_{\alpha=1}^{j}\left(\sum_{p=\alpha}^{i} l_p\right)k_{\alpha-1}^{-1}\left(\sum_{q=\alpha}^{j} l_q\right), & i > j, \end{cases} \tag{6.2.7}$$

where, per Eq. (4.1.4),

$$k_\alpha^{-1} = \frac{1}{2}\left[\frac{\Delta x_\alpha}{EJ(x_\alpha)} + \frac{\Delta x_{\alpha+1}}{EJ(x_\alpha)}\right], \quad \alpha = 1, 2, \ldots, n-1, \tag{6.2.8}$$

in the context of our specific finite difference scheme.

As with the discussion of bars, we can always find indices i and j in the discretization process such that

$$x_{i-1} \le x \le x_i \text{ and } x_{j-1} \le s \le x_j$$

for any x and s. Thus,

$$\sum_{p=\alpha}^{i} l_p = x - x_{\alpha-1} + b_i \Delta x_i \text{ and } \sum_{q=\alpha}^{j} l_q = s - x_{\alpha-1} + b_j' \Delta x_j,$$

where b_i and b_j' are both positive numbers less than 1. When the number of finite difference points increases to infinity and all step sizes $\{\Delta x_\alpha\}$ approach zero uniformly, Eq. (6.2.7) becomes

$$\begin{aligned} \lim_{n\to\infty,\delta\to 0} r_{ij} &= \lim_{n\to\infty,\delta\to 0} \sum_{\alpha=1}^{\min(i,j)} (x - x_{\alpha-1} + b_i\Delta x_i)\frac{\Delta x_\alpha + \Delta x_{\alpha+1}}{2EJ(x_\alpha)}\left(s - x_{\alpha-1} + b_j'\Delta x_j\right) \\ &= \int_0^{\min(x,s)} \frac{(x-t)(s-t)}{EJ(t)}\mathrm{d}t, \quad 0 \le x, s \le l, \end{aligned}$$

where $\delta = \max \Delta x_\alpha$ and $\alpha = 1, 2, \ldots, n$. Notice that the term on the right-hand side of the second equality above is none other than the Green's function of a beam in the clamped-free configuration, presented in Eq. (6.2.4).

Next, let us consider the beam pinned at both ends. The stiffness matrix $\boldsymbol{A}$ of a beam subject to the pinned-pinned constraints has been established in Chap. 4, which is of the form:

$$\boldsymbol{A} = \boldsymbol{E}_{n-1}\boldsymbol{L}^{-1}\boldsymbol{E}_{n-1}^{\mathrm{T}}\boldsymbol{K}_{\mathrm{pp}}\boldsymbol{E}_{n-1}\boldsymbol{L}^{-1}\boldsymbol{E}_{n-1}^{\mathrm{T}}.$$

Here, $\boldsymbol{K}_{\text{pp}} = \text{diag}(k_1, k_2, \ldots, k_{n-1})$; while $\boldsymbol{E}_{n-1}$ resembles $\boldsymbol{E}$ in Eq. (4.1.23), and the two differ only in that they are of orders $(n-1) \times n$ and $n \times (n+1)$, respectively. Notice that elements of the matrix $\boldsymbol{E}_{n-1}\boldsymbol{L}^{-1}\boldsymbol{E}_{n-1}^{\text{T}}$ happen to be of the same form as those given in Eq. (5.2.12) after k_i is replaced with l_i^{-1}. Thus, if we set

$$(\boldsymbol{E}_{n-1}\boldsymbol{L}^{-1}\boldsymbol{E}_{n-1}^{\text{T}})^{-1} = \{b_{ij}\}_1^{n-1},$$

then

$$b_{ij} = \begin{cases} \sum_{p=1}^{i-1} l_p \cdot \sum_{q=j}^{n} l_q \Big/ \sum_{t=1}^{n} l_t = \dfrac{x_{i-1}(l - x_{j-1})}{l}, & i \le j, \\ \sum_{p=1}^{j-1} l_p \cdot \sum_{q=i}^{n} l_q \Big/ \sum_{t=1}^{n} l_t = \dfrac{x_{j-1}(l - x_{i-1})}{l}, & i > j. \end{cases}$$

The flexibility matrix of the beam in the pinned-pinned configuration is given by

$$\boldsymbol{R} = (r_{ij})_{(n-1)\times(n-1)} = (\boldsymbol{E}_{n-1}\boldsymbol{L}^{-1}\boldsymbol{E}_{n-1}^{\text{T}})^{-1}\boldsymbol{K}_{\text{pp}}^{-1}(\boldsymbol{E}_{n-1}\boldsymbol{L}^{-1}\boldsymbol{E}_{n-1}^{\text{T}})^{-1},$$

and the corresponding flexibility coefficients are

$$\begin{aligned} r_{ij} &= \sum_{\alpha=1}^{n-1} k_\alpha^{-1} b_{i\alpha} b_{\alpha j} \\ &= \begin{cases} \sum_{\alpha=1}^{i-1} k_\alpha^{-1} b_{i\alpha} b_{\alpha j} + \sum_{\alpha=i}^{j-1} k_\alpha^{-1} b_{i\alpha} b_{\alpha j} + \sum_{\alpha=j}^{n-1} k_\alpha^{-1} b_{i\alpha} b_{\alpha j}, & i \le j, \\ \sum_{\alpha=1}^{j-1} k_\alpha^{-1} b_{i\alpha} b_{\alpha j} + \sum_{\alpha=j}^{i-1} k_\alpha^{-1} b_{i\alpha} b_{\alpha j} + \sum_{\alpha=i}^{n-1} k_\alpha^{-1} b_{i\alpha} b_{\alpha j}, & i > j. \end{cases} \end{aligned}$$

Substituting k_α^{-1} in Eq. (6.2.8) as well as the expression of b_{ij} into the above equation, we get

$$r_{ij} = \begin{cases} \sum_{\alpha=1}^{i-1} \dfrac{x_\alpha(l-x)(\Delta x_\alpha + \Delta x_{\alpha+1})x_\alpha(l-s)}{2EJ(x_\alpha)l^2} + \sum_{\alpha=i}^{j-1} \dfrac{x(l-x_\alpha)(\Delta x_\alpha + \Delta x_{\alpha+1})x_\alpha(l-s)}{2EJ(x_\alpha)l^2} \\ \quad + \sum_{\alpha=j}^{n-1} \dfrac{x(l-x_\alpha)(\Delta x_\alpha + \Delta x_{\alpha+1})s(l-x_\alpha)}{2EJ(x_\alpha)l^2}, \quad i \le j, \\ \sum_{\alpha=1}^{j-1} \dfrac{x_\alpha(l-s)(\Delta x_\alpha + \Delta x_{\alpha+1})x_\alpha(l-x)}{2EJ(x_\alpha)l^2} + \sum_{\alpha=j}^{i-1} \dfrac{s(l-x_\alpha)(\Delta x_\alpha + \Delta x_{\alpha+1})x_\alpha(l-x)}{2EJ(x_\alpha)l^2} \\ \quad + \sum_{\alpha=i}^{n-1} \dfrac{x(l-x_\alpha)(\Delta x_\alpha + \Delta x_{\alpha+1})s(l-x_\alpha)}{2EJ(x_\alpha)l^2}, \quad i > j, \end{cases} \tag{6.2.9}$$

where terms of second-order or higher have been ignored in the above equation. As $n \to \infty$ and all step sizes of the finite difference discretization $\Delta x_\alpha \to 0$, indices $i, j, n-i, n-j$, and $j-i$ approach to ∞ simultaneously. Thus,

$$\lim_{n\to\infty,\delta\to 0} r_{ij} = \begin{cases} \int_0^x \frac{(l-x)(l-s)}{l^2}\cdot\frac{t^2}{EJ}\mathrm{d}t + \int_x^s \frac{x(l-s)}{l^2}\cdot\frac{t(l-t)}{EJ}\mathrm{d}t + \int_s^l \frac{xs}{l^2}\cdot\frac{(l-t)^2}{EJ}\mathrm{d}t, & x \le s, \\ \int_0^s \frac{(l-x)(l-s)}{l^2}\cdot\frac{t^2}{EJ}\mathrm{d}t + \int_s^x \frac{s(l-x)}{l^2}\cdot\frac{t(l-t)}{EJ}\mathrm{d}t + \int_x^l \frac{xs}{l^2}\cdot\frac{(l-t)^2}{EJ}\mathrm{d}t, & x > s. \end{cases}$$

It is easy to verify that this result is identical to Eq. (6.2.5).

By the same token, we may study flexibility coefficients r_{ij} of discretized beams subject to clamped-sliding, clamped-pinned, clamped-clamped, and pinned-sliding constraints. It can be verified that in all cases they approach Green's functions of their respective continuous systems in the limit. See Ref. [7].

The Green's function of a beam, as with its counterpart of a bar, has two important properties. They are summarized in the next two theorems.

Theorem 6.4 *The Green's function of a beam is symmetric, i.e.,* $G(x, s) = G(s, x)$.

Readers can easily justify the symmetry of the Green's function by applying the principle of virtual work. Due to space limitation, we skip the proof here.

Theorem 6.5 *Function*

$$u(x) = \int_0^l G(x, s) f(s) ds \tag{6.2.10}$$

must be the solution to the following equation

$$Bu = f(x), \tag{6.2.11}$$

subject to boundary conditions defined in Eqs. (6.1.3) and (6.1.4). Conversely, the solution to Eq. (6.2.11) under boundary conditions (6.1.3) and (6.1.4) can always be expressed in the form given in Eq. (6.2.10).

Proof In fact, by applying the differential operator of the beam to both sides of Eq. (6.2.10) and also taking into account the definition (6.2.1) of the Green's function, we will get Eq. (6.2.11).

Conversely, notice that

$$(Bu, G) = \int_0^l G(x, s) f(x) \mathrm{d}x,$$

and

$$(u, B_x G) = \int_0^l u(x)\delta(x - s)\mathrm{d}x = u(s).$$

Since the operator of an Euler–Bernoulli beam is symmetric, we obtain

$$u(s) = \int_0^l G(x, s)f(x)\mathrm{d}x.$$

Furthermore, because the Green's function is symmetric, we know the above equation is equivalent to Eq. (6.2.10). ■

We may, based on Theorem 6.5, rewrite the modal equation for the natural angular frequency $\omega = \sqrt{\lambda}$ and the related mode shape $u(x)$ into the integral equation form,

$$u(x) = \lambda \int_0^l G(x, s)\rho(s)u(s)\mathrm{d}s. \tag{6.2.12}$$

As a matter of fact, Eq. (6.2.12) is the reason for us to introduce the Green's function in the first place. Moreover, Eq. (6.2.12) can be transformed into a symmetric form by setting

$$\tilde{u}(x) = \sqrt{\rho(x)}u(x), \quad K(x, s) = \sqrt{\rho(x)\rho(s)}G(x, s). \tag{6.2.13}$$

Then, Eq. (6.2.12) becomes

$$\tilde{u}(x) = \lambda \int_0^l K(x, s)\tilde{u}(s)\mathrm{d}s, \quad 0 < x < l. \tag{6.2.14}$$

Functions $G(x, s)$ and $K(x, s)$ are called kernels of the integral equations (6.2.12) and (6.2.14), respectively. In the subsequent discussion, we will first prove $K(x, s)$ to be an oscillatory kernel and then derive oscillatory properties of beams subject to various constraints.

6.3 Oscillatory Properties in Static Deformation and Vibration of a Well-Constrained Beam

6.3.1 Oscillatory Properties in Static Deformation of an Arbitrarily Supported But Well-Constrained Beam

As is the case with a bar, there are also two ways to prove the Green's function of a well-constrained beam system to be an oscillatory kernel.

Let us begin with the first method. On the one hand, it has been pointed out in Chap. 4 that, as long as an arbitrarily supported beam produces a well-constrained system, the stiffness matrix of its finite difference model is sign oscillatory while the corresponding flexibility matrix is oscillatory. On the other hand, it has been shown in Sect. 6.2 that the flexibility coefficients of a beam subject to various boundary constraints approach the respective Green's function in the limit. Thus, based on Theorem 2.29, Green's functions of the six types of well-constrained beams, with boundary conditions defined in (6.1.2), belong to the group of oscillatory kernels.

In the second method, we may directly prove the Green's function of a well-constrained beam system to be an oscillatory kernel by taking advantage of oscillatory properties in static deformation. There is no need to actually derive the Green's function with this approach. Let us now introduce a related theorem.

Theorem 6.6 *Set*

$$u(x) = G(x, s), \quad 0 \le x,\ s \le l,$$

where $G(x, s)$ is the Green's function of the well-constrained system of a beam, i.e., it is the solution to Eq. (6.2.1). If h_r and β_r $(r = 1,\ 2)$ are all of finite and positive value, then

$$\tau'(x) = [EJ(x)u''(x)]' = \begin{cases} -c, & 0 \le x < s, \\ 1 - c, & s < x \le l, \end{cases} \tag{6.3.1}$$

where $0 < c < 1$.

Proof According to Eq. (6.2.2) as well as Property (3) of the Green's function given immediately below the equation, there must be a constant c such that Eq. (6.3.1) is satisfied. If h_r and β_r $(r = 1,\ 2)$ are all of positive and finite value, we know in Statics that it is impossible to have either $c = 0$ or $c = 1$. As a result, it only needs to prove $0 < c < 1$ here.

Let us take a closer look at $\tau(x)$. We can show it cannot be always positive in the interval $(0, l)$. In fact, if otherwise, there would be only three possibilities for the boundary conditions of $\tau(x)$: (1) $\tau(0) \ge 0$ and $\tau(l) > 0$. Then, the third equation in (6.2.1), which is satisfied by $G(x, s)$, would yield $u'(0) \ge 0$ and $u'(l) < 0$. (2) $\tau(0) > 0$ and $\tau(l) = 0$. As a result, the third equation in (6.2.1) would lead to

$u'(0) > 0$ and $u'(l) = 0$. (3) $\tau(0) = 0$ and $\tau(l) = 0$. In this case, the third equation in (6.2.1) would produce $u'(0) = 0$ and $u'(l) = 0$, where $u'(x) \equiv 0$ obviously should not occur. But all these possible situations conflict with the condition that $u'(x)$ is monotonically increasing in the interval $(0, l)$.

By the same token, it can be shown that the function $\tau(x)$ cannot stay negative constantly, too.

Now let us assume $c < 0$. Then, $\tau'(x)$ is always positive on the interval $[0, l]$, and thus $\tau(x)$ is monotonically ascending on the same interval. Since all other possibilities have been excluded, the only thing that can happen is $\tau(0) < 0$ and $\tau(l) > 0$. The third equation in (6.2.1) yields $u'(0) < 0$ and $u'(l) < 0$. Notice that $\tau(x)$ is a piecewise linear function and it has only one zero. Consequently, $u'(x)$ has only one local minimum inside $[0, l]$. The above discussion implies $u'(x) < 0$ for any $x \in [0, l]$, and thus $u(x)$ is monotonically decreasing on the interval. However, since $\tau'(0) > 0$ and $\tau'(l) > 0$, the second equation in (6.2.1) suggests $u(0) < 0$ and $u(l) > 0$, which contradicts the earlier conclusion that $u(x)$ is descending on $[0, l]$.

As for the case of $c > 1$, $\tau'(x)$ is invariably negative in $[0, l]$, and thus $\tau(x)$ decreases monotonically on this interval. Proceeding with an argument similar to that used in the previous paragraph, we conclude that $u'(x) > 0$, where $x \in [0, l]$. However, this contradicts the results of $u(0) > 0$ and $u(l) < 0$, obtained from the second equation in (6.2.1).

Thus, we have proved $0 < c < 1$. ∎

A similar discussion can be made for situations where h_r and β_r are either 0 or ∞. We may reach conclusions analogous to Theorem 6.6, but have to modify the range of the constant c to be $0 \leq c \leq 1$ in the end result. For instance, $c = 1$ for a beam in the clamped-free configuration, while $c = 0$ for a beam subject to the free-clamped constraints.

Corollary $\tau(x)$ *must have at least one sign reversal in the interval* $[0, l]$.

Proof We have shown in the proof of Theorem 6.6 that $\tau(x)$ can stay neither positive nor negative all the time. Hence, sign reversal must occur with $\tau(x)$ in the interval $[0, l]$. ∎

Theorem 6.7 *Under boundary conditions given in Eq. (6.2.1), the Green's function of the well-constrained system of a beam satisfies the inequality*

$$G(x, s) > 0, \quad x \text{ and } s \in I.$$

Proof In fact, when all of h_r and β_r $(r = 1, 2)$ are of finite positive value, we may, per Theorem 6.6, create plots of $\tau'(x)$, as shown in Fig. 6.3a. Thus, there are only three possibilities for the function $\tau(x)$, and they are depicted in Fig. 6.3b. Based on the third equation in (6.2.1), $u'(0)$ and $\tau(0)$ are of the same sign, while $u'(l)$ and $\tau(l)$ have opposite signs, By applying rules for determining the monotonic behavior and extremum of a function, we conclude that the shape of $u'(x)$ can only be one of the three forms shown in Fig. 6.3c. Finally, based on the second equation in (6.2.1),

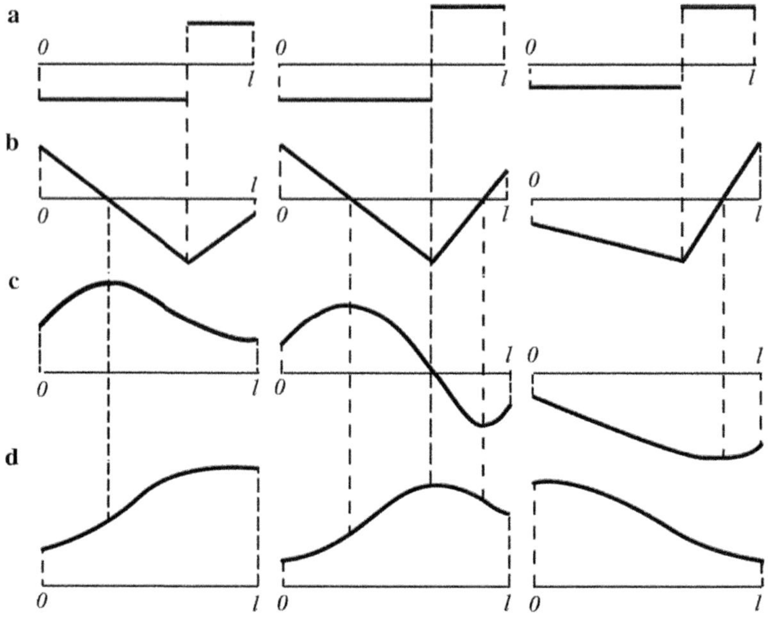

Fig. 6.3 Three possible shapes for functions: **a** $\tau'(x)$, **b** $\tau(x)$, **c** $u'(x)$, and **d** $u(x)$

$u(0)$ and $\tau'(0)$ are of opposite signs, while $u(l)$ and $\tau'(l)$ are of the same sign. This leads to the results $u(0) > 0$ and $u(l) > 0$. Therefore, $u(x)$ has to assume one of the three shapes sketched in Fig. 6.3d. These graphs have clearly demonstrated that $u(x) = G(x, s) > 0$ (x and $s \in I$) anyhow, i.e., Theorem 6.7 is valid.

Now let us turn to the situation where h_r and β_r may independently assume the value of either 0 or ∞ but the system remains well-constrained. They correspond to the six types of well-constrained beams discussed in Sect. 6.2, i.e., those subject to clamped-free, clamped-sliding, clamped-pinned, clamped-clamped, pinned-sliding, and pinned-pinned constraints. The deflection of a beam under the action of a unit concentrated load can be derived from the theory of Mechanics of Materials. It is clear that Theorem 6.7 is also valid for these beams. ■

Theorem 6.7 indicates that as long as an arbitrarily supported beam leads to a well-constrained system, the beam always has the oscillatory property in static deformation, A .

Next, we are going to verify that if an arbitrarily supported beam results in a well-constrained system, then the beam also has the oscillatory property in static deformation, B. To this end, let us first introduce a lemma.

Lemma 6.1 *Suppose that $\varphi'(x)$ is continuous on $[a, b]$, its nodes are represented by ξ_i $(i = 1, 2, \ldots, n)$ where $\xi_1 < \xi_2 < \cdots < \xi_n$, and there is no other zero of $\varphi'(x)$ in the interval. Then, there exists at most one zero of $\varphi(x)$ in each of the* *subintervals* (a, ξ_1), (ξ_1, ξ_2), ..., *and* (ξ_n, b). *Thus,* $\varphi(x)$ *has at most* $n+1$ *zeroes in* (a, b). *Furthermore, if* $\varphi(a)\varphi'(a) > 0$, $\varphi(x)$ *has no zero in subintervals* (a, ξ_1)*; while if* $\varphi(b)\varphi'(b) < 0$, $\varphi(x)$ *has no zero in the subinterval* (ξ_n, b). *In other words, when any one of these conditions is met, the zero count of* $\varphi(x)$ *in* (a, b) *is reduced by 1.*

Proof Let us prove the first half of the lemma using the method of proof by contradiction. Suppose $\varphi(x)$ has two zeroes c and d in a subinterval (ξ_{r-1}, ξ_r). Then, according to Rolle's Theorem, there exists a point ξ such that $\varphi'(\xi) = 0$. But this contradicts the condition of the lemma.

As for the second half of the lemma, when $\varphi(a)\varphi'(a) > 0$, the continuity of $\varphi'(x)$ stipulates that $\varphi(a)\varphi'(x) > 0$ in the subinterval (a, ξ_1). Otherwise, it would contradict the condition that ξ_1 is the first node of $\varphi'(x)$. Furthermore, per the mean value theorem, there exists a point $x_0 \in (a, x)$ such that

$$\varphi(a)\varphi(x) = \varphi(a)[\varphi(a) + \varphi'(x_0)(x - a)] > 0$$

for any $x \in (a, \xi_1)$. This indicates that $\varphi(x)$ does not have a zero in the subinterval (a, ξ_1).

By the same token, when $\varphi(b)\varphi'(b) < 0$, there exists a point $x_1 \in (x, b)$ such that

$$\varphi(b)\varphi(x) = \varphi(b)[\varphi(b) - \varphi'(x_1)(b - x)] > 0$$

for any $x \in (\xi_n, b]$, i.e., $\varphi(x)$ does not have a zero in the subinterval (ξ_n, b). ■

Note It is not difficult to see that the second part of the above lemma remains valid even when the conditions $\varphi(a)\varphi'(a) > 0$ and $\varphi(b)\varphi'(b) < 0$ are replaced by $\varphi(a) = 0$ and $\varphi(b) = 0$, respectively.

For the ease of future application, we further generalize Lemma 6.1 to the following form:

Lemma 6.2 *Assume that* $\varphi'(x)$ *is piecewise continuous on the interval* $[a, b]$, $\xi_1 < \xi_2 < \cdots < \xi_n$ *represent a complete set of points where the function experiences sign reversal, and* $\varphi'(x)$ *does not have any other zero in the interval* $[a, b]$. *Under these conditions, the conclusions of Lemma 6.1 remain valid.*

Theorem 6.8 *Suppose a beam is subject to* n *concentrated forces of magnitude* F_i $(i = 1, 2, \ldots, n)$ *at locations* s_i $(i = 1, 2, \ldots n)$*, respectively, where* $s_1 < s_2 < \cdots < s_n$, $0 \le s_1$, *and* $s_n \le l$. *Let* $u(x)$ *represent the displacement of a well-constrained system of the beam. Then, the number of sign reversals of* $u(x)$ *does not exceed* $n - 1$.

Proof Let us first assume $0 < s_1$ and $s_n < l$. Under the conditions of the theorem, the displacement of the beam can be expressed as

$$u(x) = \sum_{i=1}^{n} F_i G(x, s_i). \tag{6.3.2}$$

Set $\tau(x) = EJ(x)u''(x)$, as before. We know from earlier discussion:

$$\tau'(x) = c_i, \quad s_i < x < s_{i+1}, \; i = 0, 1, \ldots, n, \; s_0 = 0, \; s_{n+1} = l, \tag{6.3.3}$$

where

$$c_i = c_0 + \sum_{r=1}^{i} F_r, \quad i = 1, 2, \ldots, n. \tag{6.3.4}$$

Since $\tau'(x)$ satisfies all conditions of Lemma 6.2, the number of sign reversals of $\tau(x)$ (or the equivalent $u''(x)$) does not exceed $n+1$ on $[0, l]$. Moreover, Lemma 6.1 shows that numbers of sign reversals of $u'(x)$ and $u(x)$ are no greater than $n+2$ and $n+3$, respectively.

Next, let us look at the function sequence:

$$\tau'(x), \; M(x), \; u'(x), \; \text{and} \; u(x).$$

When both β_1 and β_2 are of finite value, boundary conditions in Eq. (6.1.4) yield

$$u'(0)u''(0) > 0 \quad \text{and} \quad u'(l)u''(l) < 0. \tag{6.3.5}$$

Thus, the number of sign reversals of $u'(x)$ should not be more than n. Correspondingly, the number of sign reversals of $u(x)$ should not exceed $n+1$. Furthermore, when both h_1 and h_2 are also of finite value, boundary conditions in Eq. (6.1.3) lead to

$$\tau'(0)u(0) < 0 \quad \text{and} \quad \tau'(l)u(l) > 0. \tag{6.3.6}$$

Consequently, there are only two possibilities at $x = 0$: Either $\tau'(0)\tau(0) > 0$ or $\tau'(0)\tau(0) < 0$. In the first case, numbers of sign reversals of $\tau(x)$, $u'(x)$, and $u(x)$ are all reduced by 1; while in the latter case, Eqs. (6.3.5) and (6.3.6) yield $u'(0)u(0) > 0$ and the number of sign reversals of $u(x)$ is also decreased by 1. By the same token, there are only two scenarios at $x = l$: Either $\tau'(l)\tau(l) < 0$ or $\tau'(l)\tau(l) > 0$. In the first case, numbers of sign reversals of $\tau(x)$, $u'(x)$, and $u(x)$ all go down by 1; while in the second case, we will have $u'(l)u(l) < 0$, and thus the number of sign reversals of $u(x)$ also decreases by 1. In summary, when all of h_r and β_r are of positive finite value, the number of sign reversals of $u(x)$ is no larger than $n-1$.

With regard to proofs for other special cases, we will just point out a few facts below without going into details.

(1) The influence of parameter pairs (h_1, β_1) and (h_2, β_2) on $u''(x)$, $u'(x)$, and $u(x)$ is confined, respectively, to the first and last subintervals where the respective function has no sign change. Thus, it is not difficult to understand how boundary parameters affect sign reversals of these functions. Take the left end of the beam as an example. When β_1 becomes ∞, $u'(0) = 0$. The note to Lemma 6.1

shows that the number of sign reversals of $u'(x)$ is reduced by one, and so is the sign reversal count of $u(x)$. In addition, there are only three possibilities at the left boundary: First, if $\tau'(0)\tau(0) > 0$, the number of sign reversals of $u(x)$ is further reduced by one; second, if $\tau'(0)\tau(0) < 0$, then $u''(0)u(0) > 0$, $u(x)$ either starts positive and then monotonically increases or begins negative and then monotonically decreases in the first subinterval where $u'(x)$ retains its sign, and the number of sign reversals of $u(x)$ is still reduced by one; and third, if $\tau'(0) = 0$, then $\tau(x)$ is constant in the first subinterval where $u'(x)$ experience no sign interchange and thus the number of sign reversals of $\tau(x)$ also goes down by 1. As a result, the conclusion of Theorem 6.8 continue to be correct. As another example, when $\beta_1 \to \infty$ and $h_1 \to \infty$, $u(0) = u'(0) = 0$. This case can be discussed in a way almost identical to the previous one. The only difference we have to deal with is: $u(x)$ now starts from zero and then either monotonically increases or monotonically decreases.

(2) When $s_1 = 0$ and/or $s_n = l$, the number of segments of the piecewise function $\tau'(x)$ is reduced. But, obviously, this should have little impact on our proof. ■

6.3.2 *Oscillatory Properties in Vibration of an Arbitrarily Supported But Well-Constrained Beam*

Based on discussion in the previous subsection, we may derive the theorem below.

Theorem 6.9 *Consider an arbitrarily supported but well-constrained beam. Its Green's function is a symmetric oscillatory kernel.*

Proof In fact, Theorems 6.7 and 6.8 show, respectively, that the Green's function of a well-constrained beam system satisfy Eqs. (2.7.4) and (2.7.5) in the definition of an oscillatory kernel introduced in Chap. 2. In addition, the discussion on the strain energy indicates that the Green's function of such a system also satisfies Eq. (2.7.6) in the definition of an oscillatory kernel. Consequently, the Green's function of an arbitrarily supported but well-constrained beam is a symmetric oscillatory kernel.

Starting from Theorems 6.9 and 2.23, we may further derive some important results summarized in the next corollary.

Corollary *Consider an arbitrarily supported beam that leads to a well-constrained system. It has the following oscillatory properties in vibration:*

(1) Its natural frequencies are all distinct, i.e.,

$$0 < f_1 < f_2 < f_3 < \cdots.$$

(2) The corresponding mode shapes $u_i(x)$ ($i = 1, 2, \ldots$) constitutes a Markov function sequence on the interval $[0, l]$. As a result, they possess all properties described in Theorem 2.22, namely,

(a) *$u_1(x)$ has no zero in the point set $I \subset [0, l]$.*

(b) *There are $i-1$ nodes and no other zero of $u_i(x)$ in the point set $I \subset [0, l]$, where $i = 2, 3, \ldots$. Consequently, the function $u_i(x)$ has $i-1$ sign reversals in the interval $[0, l]$.*

(c) *The function*

$$u(x) = \sum_{i=p}^{q} c_i u_i(x), \quad \sum_{i=p}^{q} c_i^2 > 0, \quad 1 \leq p \leq q,$$

has at least $p-1$ nodes but no more than $q-1$ zeroes in the point set $I \subset [0, l]$. In particular, if $u(x)$ has $q-1$ different zeroes in the interval, then all of them are nodes.

(d) *Nodes of two successive mode shapes $u_i(x)$ and $u_{i+1}(x)$ interlace, where $i = 2, 3, \ldots$.*

The content of this section is primarily taken from Refs. [1,3]

6.4 Oscillatory Properties in Vibration of an Under-Constrained Beam

6.4.1 Modal Qualitative Properties of the Angle of Rotation, Bending Moment (or Curvature), and Shear Force Associated with Well-Constrained Beams

Only basic qualitative properties of the well-constrained beam system, i.e., the so-called oscillatory properties in vibration, have been introduced in the previous section. This classical theory will be extended here. However, we will not follow the approach of Sect. 4.3 to achieve this goal, which involves the conjugate beam. Instead, we are going to use the lemma to be introduced below, base our discussion on the general form of boundary conditions (6.1.3) and (6.1.4) that produce a well-constrained beam system, and directly determine requirements on sign-reversal counts of the shear force, bending moment (or curvature), and angle of rotation. Starting from this result, we will derive the necessary conditions of mode shapes for six types of well-constrained beams subject to a combination of boundary conditions listed in Eq. (6.1.2).

Lemma 6.3 *Suppose $\varphi(x)$ is differentiable on the interval $I \subset [0, l]$. Furthermore, it has n nodes and no other zero. Then, the number of sign reversals of $\varphi'(x)$ is no less than $n-1$ inside I. In addition, under each of the conditions $\varphi(0)\varphi'(0) > 0$ and $\varphi(l)\varphi'(l) < 0$, the count of sign reversals of $\varphi'(x)$ is increased by 1 at the minimum.*

Proof Let ξ_r $(r = 1, 2, \ldots, n)$ represent those n nodes of $\varphi(x)$ in the interval I and

$$0 < \xi_1 < \xi_2 < \cdots < \xi_n < l.$$

These points divide the interval I into $n+1$ subintervals $[\xi_r, \xi_{r+1}]$ ($r = 0, 1, \ldots, n$ and $\xi_0 = 0, \xi_{n+1} = l$). Per Rolle's Theorem, $\varphi'(x)$ has at least one zero in each of the subintervals (ξ_r, ξ_{r+1}) ($r = 1, 2, \ldots, n-1$), and hence

$$S_{\varphi'} \geq n - 1.$$

If $\varphi(0)\varphi'(0) > 0$, the continuous function $\varphi(x)$ can behave only in one of the following two ways in $[0, \xi_1]$: Either it starts from a positive value, monotonically increases to the local maximum, and then goes down to zero; or it begins with a negative value, monotonically decreases to the local minimum, and then rises up to zero. In either case, $\varphi(x)$ should have at least one extremum in $[0, \xi_1]$, namely, the number of sign reversals of $\varphi'(x)$ must increase by at least one in the interval I. By the same token, the condition $\varphi(l)\varphi'(l) < 0$ should also allow the number of sign reversals of $\varphi'(x)$ to be raised by one, at a minimum, in the interval I. ■

It is easy to see that the second half of the lemma remains valid even when the conditions $\varphi(0)\varphi'(0) > 0$ and $\varphi(l)\varphi'(l) < 0$ are replaced by $\varphi(0) = 0$ and $\varphi(l) = 0$, respectively.

Theorem 6.10 *Suppose $\{\lambda_i,\ u_i(x)\}$ is an eigenpair of the well-constrained system of a beam, where $i = 1, 2, \ldots$. Set*

$$\tau_i(x) = EJ(x)u_i''(x), \quad i = 1, 2, \ldots \tag{6.4.1}$$

Then, numbers of sign reversals of $\tau_i'(x)$, $\tau_i(x)$, and $u_i'(x)$ ($i = 1, 2, \ldots$) in $[0, l]$ satisfy the following inequalities:

$$i - 2 + \Delta\left(h_1^{-1}\right) + \Delta\left(h_2^{-1}\right) \leq S^-_{\tau_i'} \leq S^+_{\tau_i'} \leq i - \Delta(h_1) - \Delta(h_2), \tag{6.4.2}$$

$$\begin{aligned} i - 1 - \Delta(\beta_1) - \Delta(\beta_2) + \Delta(h_1^{-1}) + \Delta(h_2^{-1}) &\leq S^-_{\tau_i} \\ \leq S^+_{\tau_i} \leq i + 1 - \Delta(h_1) - \Delta(h_2) - \Delta(\beta_1) - \Delta(\beta_2),& \end{aligned} \tag{6.4.3}$$

$$i - 2 + \Delta(h_1^{-1}) + \Delta(h_2^{-1}) \leq S^-_{u_i'} \leq S^+_{u_i'} \leq i - \Delta(h_1) - \Delta(h_2), \tag{6.4.4}$$

where

$$\Delta(t) = \begin{cases} 1, & t = 0, \\ 0, & t \neq 0. \end{cases}$$

Proof It has been shown in the previous section that the number of sign reversals of $u_i(x)$ is $S_{u_i} = i - 1$. This, together with Lemma 6.3, lead to the inequality

$$S^-_{u_i'} \geq i - 2.$$

Furthermore, as $h_1 \to \infty$ and $h_2 \to \infty$, boundary conditions in Eq. (6.1.3) yield $u_i(0) = 0$ and $u_i(l) = 0$, respectively. Thus, we have proved the leftmost inequality in Eq. (6.4.4).

When $\beta_1 \neq 0$ and $\beta_2 \neq 0$, boundary conditions in Eq. (6.1.4) lead to $u_i'(0)u_i''(0) > 0$ and $u_i'(l)u_i''(l) < 0$, respectively. Similarly, we get, from Lemma 6.3 and the leftmost inequality in Eq. (6.4.4),

$$S_{\tau_i}^- \geq i - 1 + \Delta(h_1^{-1}) + \Delta(h_2^{-1}).$$

When $\beta_1 = 0$ and $\beta_2 = 0$, we have $\tau_i(0) = 0$ and $\tau_i(l) = 0$, respectively, from boundary conditions in Eq. (6.1.4). Since the minimum number of sign reversals of a function is not affected by its zero at an end point, this helps establish the first inequality in Eq. (6.4.3).

By taking advantage of Lemma 6.3 as well as the first inequality in Eq. (6.4.3), we obtain

$$S_{\tau_i'}^- \geq i - 2 + \Delta(h_1^{-1}) + \Delta(h_2^{-1}) - \Delta(\beta_1) - \Delta(\beta_2). \tag{6.4.5}$$

If $\beta_1 \neq 0$ and $\beta_2 \neq 0$, then $\Delta(\beta_1) = 0$ and $\Delta(\beta_2) = 0$, respectively; while if $\beta_1 = 0$ and $\beta_2 = 0$, then $\tau_i(0) = 0$ and $\tau_i(l) = 0$, respectively. In the latter case, each of the conditions $\beta_r = 0\,(r = 1, 2)$ helps increase the count of sign reversals of $\tau_i'(x)$ by at least one. Thus, regardless of whether the values of $\beta_r\,(r = 1, 2)$ to be zero, the last two terms on the right-hand side of Eq. (6.4.5) always vanish. This produces the leftmost inequality in Eq. (6.4.2).

Next, it can be seen from the modal equation of a beam, namely,

$$\lambda\rho(x)u(x) = [EJ(x)u''(x)]'' = [\tau'(x)]',$$

that the derivative of the mode shape of the shear force $\phi_i(x) = \tau_i'(x)$, i.e., $\phi_i'(x)$, has the same set of nodes and an identical number of sign reversals as $u_i(x)$. After applying Lemma 6.1 to $\phi_i'(x)$, and also noticing that when $h_1 = 0$ and $h_2 = 0$ boundary conditions in Eq. (6.1.3) lead to $\phi_i(0) = 0$ and $\phi_i(l) = 0$, respectively, we may derive the rightmost inequality in (6.4.2) regarding the maximum number of sign reversals of $\tau_i'(x)$.

Apply Lemma 6.1 to $\tau_i'(x)$, and also take into account the fact that boundary conditions in Eq. (6.1.4) yield $\tau_i(0) = 0$ and $\tau_i(l) = 0$, respectively, when $\beta_1 = 0$ and $\beta_2 = 0$. We will then get the last inequality in Eq. (6.4.3) regarding the maximum number of sign reversals of $\tau_i(x)$.

By applying Lemma 6.1 to $\tau_i(x) = EJu''(x)$, we find

$$S_{u_i'}^+ \leq i + 2 - \Delta(h_1^{-1}) - \Delta(h_2^{-1}) - \Delta(\beta_1) - \Delta(\beta_2).$$

Similar to the discussion on Eq. (6.4.5), we may examine how values of β_1 and β_2 affect the inequality above. Specifically, if $\beta_1 \neq 0$ and $\beta_2 \neq 0$, then we have, respectively, $\Delta(\beta_1) = 0$ and $\Delta(\beta_2) = 0$, and furthermore, $u_i'(0)u_i''(0) > 0$

Table 6.1 Numbers of sign reversals of mode shapes associated with the displacement, angle of rotation, bending moment, and shear force of continuous systems of beams

No.	Boundary Conditions	h_1	β_1	h_2	β_2	S_{u_i}	$S_{u_i'}$	S_{τ_i}	S_{ϕ_i}
1	Clamped-Free	∞	∞	0	0	$i-1$	$i-1$	$i-1$	$i-1$
2	Clamped-Sliding	∞	∞	0	∞	$i-1$	$i-1$	i	$i-1$
3	Clamped-Pinned	∞	∞	∞	0	$i-1$	i	i	i
4	Clamped-Clamped	∞	∞	∞	∞	$i-1$	i	$i+1$	i
5	Pinned-Pinned	∞	0	∞	0	$i-1$	i	$i-1$	i
6	Pinned-Sliding	∞	0	0	∞	$i-1$	$i-1$	$i-1$	$i-1$
7	Free-Pinned	0	0	∞	0	$i-1$	$i-1$	$i-2^*$	$i-1$
8	Free-Sliding	0	0	0	∞	$i-1$	$i-2^*$	$i-2^*$	$i-2^*$
9	Free-Free	0	0	0	0	$i-1$	$i-2^*$	$i-3^{\mathrm{a}}$	$i-2^*$
10	Sliding-Sliding	0	∞	0	∞	$i-1$	$i-2^*$	$i-1$	$i-2^*$

$^*S_{\tau_1} = 0$ for a beam in the free-pinned configuration, $S_{u_1'} = S_{\tau_1} = S_{\phi_1} = 0$ for a beam in the free-sliding configuration, and $S_{u_1'} = S_{\phi_1} = 0$ for a beam in the free-free or sliding-sliding configuration

$^{\mathrm{a}}S_{\tau_1} = S_{\tau_2} = 0$ for a beam in the free-free configuration

and $u_i'(l)u_i''(l) < 0$ based on boundary conditions in Eq. (6.1.4). Thus, the number of sign reversals of $u_i'(x)$ will decrease by 2 accordingly. In a different scenario, if $\beta_1 = 0$ and $\beta_2 = 0$, we have $\Delta(\beta_1) = 1$ and $\Delta(\beta_2) = 1$, respectively. Thus, we see that the rightmost inequality in Eq. (6.4.4) should be satisfied in all situations. ■

Based on this theorem, we may immediately conclude that mode shapes of the displacement, angle of rotation, bending moment, and shear force associated with six types of well-constrained beams have fixed numbers of sign reversals. The results are detailed in the first six rows of Table 6.1. However, it is regrettable that when h_r and β_r $(r = 1, 2)$ are all of finite values, we are not able to determine completely from Eqs. (6.4.2)–(6.4.4) the numbers of sign reversals of mode shapes associated with the angle of rotation, bending moment, and shear force. In addition, we would like to remind the reader that, per Eq. (6.4.1), mode shapes of the bending moment $\tau_i(x)$ and curvature $u_i''(x)$ share the same number of sign reversals or, equivalently, the same number of nodes.

Let ξ_r $(r = 1, 2, \ldots, i-1)$ denote nodes of $u_i(x)$ inside $(0, l)$. Some important corollaries can be established based on Theorem 6.10.

Corollary 1 *If a beam is unconstrained at the right end, i.e., $h_2 = 0 = \beta_2$, then*

$$u_i(l)u_i'(x) > 0, \quad \xi_{i-1} \leq x \leq l \text{ and } i = 1, 2, \ldots. \tag{6.4.6}$$

In fact, we may look at the angle of rotation $u_i'(x)$ of this beam when it is subject to the same constraint at the left end but under free, pinned, or clamped support at the right end. According to Table 6.1, the sign-reversal count corresponding to an unconstrained right end is one less than its counterpart associated with an either clamped or pinned right end. This implies that due to the existence of a free end on the right side, $u_i'(x)$ cannot have any sign reversal within the last segment where the function $u_i(x)$ retains its sign. Hence, Eq. (6.4.6) must be valid.

Apparently, some additional results may be obtained below:

(1) If the beam is unconstrained at the left end, then the following inequalities hold true, which are similar to those in Eq. (6.4.6),

$$u_i(0)u_i'(x) < 0, \quad 0 \le x \le \xi_1 \text{ and } i = 1, 2, \ldots. \tag{6.4.7}$$

(2) If the beam is constrained with a sliding support at the left end, rather than being left unconstrained, then there exist inequalities similar but not identical to those in Eq. (6.4.6). Specifically,

$$u_i(0)u_i'(x) < 0, \quad 0 < x \le \xi_1 \text{ and } i = 1, 2, \ldots; \tag{6.4.8}$$

while if the beam slides at the right end, the following inequalities, which correspond to those in Eq. (6.4.7), are satisfied,

$$u_i(l)u_i'(x) > 0, \quad \xi_{i-1} \le x < l \text{ and } i = 1, 2, \ldots. \tag{6.4.9}$$

Corollary 2 *There is one and only one extremum x_r in the subinterval $(\xi_r, \ \xi_{r+1})$, where $r = 1, 2, \ldots, i-2$. In addition, we have*

$$u_i(x_r)u_i''(x_r) < 0, \quad i = 1, 2, \ldots. \tag{6.4.10}$$

In fact, assume that the continuous third derivative of the function $u_i(x)$ exists. If $u_i(x)$ has more than one extremum in $(\xi_r, \ \xi_{r+1})$, then there must be at least three extrema in the subinterval. But this would result in $\left|S_{u_i'} - S_{u_i}\right| > 2$, which contradicts Eq. (6.4.4). Thus, there can be only one local maximum in a subinterval where $u_i(x)$ remains positive, while there can be only one local minimum in a subinterval where $u_i(x)$ stays negative. Eq. (6.4.10) is, therefore, valid.

Finally, Eqs. (6.4.3) and (6.4.4) obviously lead to the corollary below.

Corollary 3 *Compared with a free end, the existence of a clamped end produces an additional sign reversal of $u_i''(x)$ inside $(0, \xi_1)$ or (ξ_{i-1}, l), while each of the clamped and pinned ends also causes the number of sign reversals of $u_i'(x)$ to increase by one inside $(0, \xi_1)$ or (ξ_{i-1}, l).*

The content of this subsection is taken primarily from Ref. [6].

6.4.2 Oscillatory Properties in Vibration of Under-Constrained Beams

1. Conjugate Systems of Under-Constrained Beams

As of now we have been dealing with well-constrained systems of beams. However, it is common to encounter situations involving beams with some rigid-body motion in engineering applications. To study oscillatory properties of this type of beams, we need to introduce the concept of conjugate beams, just like what we have done with the discrete systems. See Ref. [6].

Set

$$\tau(x) = EJ(x)u''(x). \tag{6.4.11}$$

Then, Eq. (6.1.1) can be rewritten as

$$[(EJ(x))^*\tau''(x)]'' = \lambda\rho^*(x)\tau(x). \tag{6.4.12}$$

This may be viewed as the modal equation of a "beam" with parameters

$$[EJ(x)]^* = \rho^{-1}(x) \text{ and } \rho^*(x) = [EJ(x)]^{-1}$$

on the interval $[0, l]$. We refer to this "beam" as the conjugate of the original beam, and call $\tau(x)$ the "displacement" of the conjugate beam.

It is clear from Transform (6.4.11) that a free end of the original beam corresponds to a clamped end of the conjugate beam. By taking into account the modified form of the modal equation associated with the original beam, namely,

$$\tau''(x) = \lambda\rho(x)u(x) \text{ and } [\rho^{-1}(x)\tau''(x)]' = \lambda u'(x),$$

we may find the general relationship between original and conjugate beams in regard to boundary conditions and mode shapes. These are summarized in Table 6.2.

In the remainder of this subsection, we will present oscillatory properties in vibration of beams having rigid-body motion.

2. Oscillatory Properties of Three Under-Constrained Beams Having Well-Constrained Conjugates

Based on the relationship shown in Table 6.2, we may obtain the following oscillatory properties for three types of beams with rigid-body motion:

(1) The conjugates of beams in free-pinned and free-sliding configurations are, respectively, beams in the clamped-pinned and clamped-sliding configurations. Thus, their non-zero natural frequencies are all distinct, i.e.,

$$0 = f_1 < f_2 < \cdots < f_n < \cdots.$$

Table 6.2 Relationship between original and conjugate beams regarding boundary conditions and mode shapes

	Original beams	Conjugate beams
Support condition at the boundary	Free	Clamped
	Sliding	Sliding
	Pinned	Pinned
	Clamped	Free
Mode shape	Displacement	Bending moment
	Angle of rotation	Shear force
	Bending moment	Displacement
	Shear force	Angle of rotation

The mode shape of the bending moment (the "displacement" of the conjugate beam) corresponding to f_i has precisely $i-2$ sign reversals in the interval $[0, l]$, where $i = 2, 3, \ldots$. Consequently, all mode shapes of the displacement, angle of rotation, shear force, etc. have fixed numbers of sign reversals. The detailed results are listed in Rows 7 and 8 of Table 6.1.

(2) The conjugate of a beam unconstrained at both ends is the beam clamped at both ends. Thus, its non-zero natural frequencies are also distinct, namely,

$$0 = f_1 = f_2 < f_3 < \cdots < f_n < \cdots.$$

The mode shape of the bending moment (the "displacement" the conjugate beam) corresponding to f_i has $i - 3$ sign reversals in the interval $[0, l]$, where $i = 3, 4, \ldots$. The numbers of sign reversals associated with mode shapes of the displacement, angle of rotation, shear force, etc. are listed in Row 9 of Table 6.1.

3. *Oscillatory Properties of a Beam in Sliding-Sliding Configuration*

For a beam in the sliding-sliding configuration, its conjugate is still a beam subject to the sliding constraint at both ends. Thus, the concept of conjugate beam is not helpful in this situation. In order to derive the qualitative properties, we need to rewrite the modal equation (6.1.1) into the following form:

$$[\rho^{-1}(x)(EJ(x)v'(x))'']' = \lambda v(x), \tag{6.4.13}$$

where $v(x) = u'(x)$. Set $\phi(x) = \tau'(x)$. The above equation can be further reformatted to

$$[\rho^{-1}(x)\phi'(x)]' = \lambda v(x), \tag{6.4.14}$$

$$[EJ(x)v'(x)]' = \phi(x). \tag{6.4.15}$$

The corresponding boundary conditions now become:

$$v(0) = v(l) = 0 \text{ and } \phi(0) = \phi(l) = 0. \tag{6.4.16}$$

Notice that the differential operators on the left-hand sides of Eqs. (6.4.14) and (6.4.15) happen to be of the Sturm-Liouville type with $q(x) = 0$. Under boundary conditions in Eq. (6.4.16), they are both positive and self-adjoint differential operators. Thus, the Green's function of the "beam", whose behavior is governed by Eq. (6.4.13) and boundary conditions in (6.4.16), can be expressed as

$$G(x, s) = \int_0^l G_1(x, t)G_2(t, s)\mathrm{d}t, \tag{6.4.17}$$

where $G_1(x, s)$ and $G_2(x, s)$ are Green's functions of Eqs. (6.4.14) and (6.4.15), respectively, under boundary conditions in Eq. (6.4.16). Additionally, $G_1(x, s)$ and $G_2(x, s)$ are both oscillatory kernels. We know from Property (1) of compound kernels in Sect. 2.8 that

$$G\begin{pmatrix} x_1 & x_2 & \ldots & x_p \\ s_1 & s_2 & \ldots & s_p \end{pmatrix} = \int_0^l \int_0^{t_p} \cdots \int_0^{t_2} G_1\begin{pmatrix} x_1 & x_2 & \ldots & x_p \\ t_1 & t_2 & \ldots & t_p \end{pmatrix} G_2\begin{pmatrix} t_1 & t_2 & \ldots & t_p \\ s_1 & s_2 & \ldots & s_p \end{pmatrix} \cdot \mathrm{d}t_1\mathrm{d}t_2 \ldots \mathrm{d}t_p. \tag{6.4.18}$$

With this result, it is easy to verify that the Green's function defined by Eq. (6.4.17) satisfies all three conditions in the definition of an oscillatory kernel. Therefore, $G(x, s)$ is also an oscillatory kernel, and we may obtain the following oscillatory properties of a beam with sliding supports at both ends:

(1) Its non-zero natural frequencies are all distinct. Since there is only one zero frequency in this case, we may arrange all frequencies of a sliding-sliding beam in an ascending order,

$$0 = f_1 < f_2 < \cdots < f_n < \cdots.$$

(2) The mode shape of the angle of rotation corresponding to f_i has $i - 2$ sign reversals in the interval $[0, l]$, where $i = 2, 3, \ldots$.

(3) Since none of the proofs for Lemmas 6.1–6.3 requires that the related system be well-constrained, we may also reach the conclusion that all mode shapes of the displacement, bending moment, shear force, etc. have fixed numbers of sign reversals, by applying Property (2) given above as well as an argument similar to that used in the derivation of Theorem 6.10. The results are detailed in Row 10 of Table 6.1.

4. An Additional Oscillatory Property of the Three Under-Constrained Beams with Well-Constrained Conjugates

We have obtained some oscillatory properties of beams with rigid-body motion in the discussion above. In the remaining two parts of Sect. 6.4.2, we are going to prove that Property (2)(c) in the corollary of Theorem 6.9 should also be valid for the under-constrained beams (See Ref. [9]). Let us look at beams in free-pinned, free-free, and free-sliding configurations here.

Obviously, all these beams have rigid-body motion. In addition, the "displacement mode shapes" $\tau_i^*(x)\,(i = 1, 2, \ldots)$ of their respective conjugate systems are all eigen-functions of some integral equations with oscillatory kernels, where the superscript * indicates the function to be related to non-zero natural frequencies of the original system. Thus, all of these three conjugate beams have Property (2)(c) in the corollary of Theorem 6.9. Specifically, for a set of real numbers not vanishing simultaneously $c_i\,(i = p, p+1, \ldots, q)$, the function

$$\tau(x) = c_p\tau_p^*(x) + c_{p+1}\tau_{p+1}^*(x) + \cdots + c_q\tau_q^*(x), \quad 1 \le p \le q, \tag{6.4.19}$$

has no less than $p-1$ nodes and no more than $q-1$ zeroes. Or put it in another way[1]:

$$p - 1 \le S_\tau^- \le S_\tau^+ \le q - 1. \tag{6.4.20}$$

Since $\tau(x) = EJ(x)u''(x)$, Eq. (6.4.20) is equivalent to

$$p - 1 \le S_{u''}^- \le S_{u''}^+ \le q - 1. \tag{6.4.21}$$

Since the function $u(x)$ in Eq. (6.4.21) depends on specific boundary conditions of the beam, it is necessary to separate the discussion based on the types of boundary conditions.

First, for a beam in the free-pinned configuration, the "displacement mode shapes" of its conjugate system are $\tau_i^*(x)\,(i = 1, 2, \ldots)$. The i-th natural frequencies f_i^* of the conjugate system, which is related to the i-th mode shape $\tau_i^*(x)$, corresponds to the $(i+1)$-th frequency f_{i+1} of the original system, where $i = 1, 2, \ldots$. Thus, Eq. (6.4.19) is equivalent to

[1]A continuous function in the interval $(0, l)$ can have only two types of zeroes: nodes and null anti-nodes. Based on the concept of the number of sign reversals of a function, the count of null anti-nodes has no contribution to the minimal number of sign reversals, but a null anti-node is treated as two distinct zeroes in the calculation of the maximal number of sign reversals. As a result, the count of nodes of a function in the interval $(0, l)$ equals the minimal number of sign reversals, while the count of its zeroes, including those at the end points of the interval, is identical to the maximal number of sign reversals.

$$u''(x) = c_p u''_{p+1}(x) + c_{p+1} u''_{p+2}(x) + \cdots + c_q u''_{q+1}(x), \quad 1 \le p \le q.$$

Start from the rightmost inequality in Eq. (6.4.21), apply Lemma 6.1 to it and then to the resulting inequality, and notice that $u(0) = 0$. We then get

$$S^+_{u'} \le S^+_{u''} + 1 = q - 1 + 1 = q, \quad S^+_u \le S^+_{u'} + 1 - 1 = q.$$

Furthermore, from the reformatted modal equation of the free-pinned beam

$$\lambda \rho u = \tau''(x),$$

we know that functions $u(x)$ and $\tau''(x)$ share the same number of sign reversals. Starting from the leftmost inequality of Eq. (6.4.20), applying Lemma 6.3 twice, and noticing that $\tau(0) = 0 = \tau(l)$ and $\tau'(0) = \phi(0) = 0$, we obtain

$$S^-_\phi \ge S^-_\tau - 1 + 2 = p - 1 + 1 = p, \quad S^-_u = S^-_{\tau''} \ge S^-_\phi - 1 + 1 = p.$$

Putting it together, we get

$$p \le S^-_u \le S^+_u \le q. \tag{6.4.22}$$

Since $\tau^*_i(x) = EJ(x)u''_{i+1}(x)$ for a free-pinned beam, the function $u(x)$ corresponding to Eq. (6.4.19) ought to be of the form:

$$u(x) = c_p u_{p+1}(x) + c_{p+1} u_{p+2}(x) + \cdots + c_q u_{q+1}(x), \quad 1 \le p \le q. \tag{6.4.23}$$

Consequently, Eq. (6.4.22) is what needs to be proved.

Next, for a beam unconstrained at both ends, the "displacement mode shapes" of its conjugate system are $\tau^*_i(x)\,(i = 1, 2, \ldots)$. The i-th natural frequencies f^*_i of the conjugate system, which is related to the i-th mode shape $\tau^*_i(x)$, corresponds to the $(i+2)$-th frequency f_{i+2} of the original system, where $i = 1, 2, \ldots$. We may continue with an argument almost identical to the one for the free-pinned beam. The only difference between the two is that here we may no longer have $u(0) = 0$ when applying Lemma 6.1. In the end, we get

$$S^+_{u'} \le q, \quad S^+_u \le q + 1.$$

Again, by applying Lemma 6.3 twice as well as noticing $\tau(0) = 0 = \tau(l)$ and $\phi(0) = \phi(l) = 0$, we obtain

$$S^-_\phi \ge p, \quad S^-_u \ge p + 1.$$

Summarizing the results, we conclude

$$p + 1 \le S_u^- \le S_u^+ \le q + 1. \tag{6.4.24}$$

Since $\tau_i^*(x) = EJ(x)u_{i+2}''(x)$ for a beam unconstrained at both ends, the function $u(x)$ corresponding to Eq. (6.4.19) is of the form:

$$u(x) = c_p u_{p+2}(x) + c_{p+1} u_{p+3}(x) + \cdots + c_q u_{q+2}(x), \quad 1 \le p \le q. \tag{6.4.25}$$

Therefore, Eq. (6.4.24) is what needs to be proved.

Finally, we can have a similar discussion on the beam in the free-sliding configuration by following the same process used above. The results are identical to those of the beam subject to free-pinned constraints.

5. ***An Additional Oscillatory Property of the Beam in Sliding-Sliding Configuration***

Now let us turn to the beam subject to the sliding constraint at both ends. We may refer to the system represented by Eq. (6.4.13) as the transformed system of the original beam. The "displacement mode shapes" $v_i^{\text{tr}}(x)\,(i = 1, 2, \ldots)$ of the transformed system are eigenfunctions of some integral equation with an oscillatory kernel, where the superscript tr indicates that the function is related to the transformed system. Hence, Property (2)(c) in the corollary of Theorem 6.9 remains valid for the transformed system. Specifically, for an arbitrary set of real numbers not vanishing simultaneously, $c_i\,(i = p, p + 1, \ldots, q)$, the function

$$v(x) = c_p v_p^{\text{tr}}(x) + c_{p+1} v_{p+1}^{\text{tr}}(x) + \cdots + c_q v_q^{\text{tr}}(x), \quad 1 \le p \le q \tag{6.4.26}$$

has no less than $p - 1$ nodes and no more than $q - 1$ zeroes, namely,

$$p - 1 \le S_v^- \le S_v^+ \le q - 1.$$

Since $v(x) = u'(x)$, the above inequalities are equivalent to

$$p - 1 \le S_{u'}^- \le S_{u'}^+ \le q - 1. \tag{6.4.27}$$

Starting from the right end of Eq. (6.4.27) and applying Lemma 6.1, we get $S_u^+ \le q$, where

$$u(x) = c_p u_{p+1}(x) + c_{p+1} u_{p+2}(x) + \cdots + c_q u_{q+1}(x), \quad 1 \le p \le q. \tag{6.4.28}$$

Moreover, the modal equation of a beam with the sliding support at both ends can be rewritten as

$$[EJ(x)(\rho^{-1}(x)\phi'(x))'']' = \lambda\phi(x), \tag{6.4.29}$$

where $\phi(x) = [EJ(x)u''(x)]'$. This is another form of the transformed system of the beam in the sliding-sliding configuration. Since the new transformed system satisfies the same type of the modal equation and boundary conditions as the one characterized by Eq. (6.4.13), we should have inequalities of the similar form, i.e.,

$$p - 1 \le S_\phi^- \le S_\phi^+ \le q - 1, \qquad (6.4.30)$$

for the "displacement mode shapes" $\phi_i^{\mathrm{tr}}(x)\,(i = 1, 2, \ldots)$ of the new transformed system. Next, it is known from the modified version of the modal equation associated with the sliding-sliding beam, i.e.,

$$\lambda \rho u = \phi'(x),$$

that the two functions $u(x)$ and $\phi'(x)$ share the same number of sign reversals. By starting from the left side of Eq. (6.4.30), noticing that $\phi(0) = 0 = \phi(l)$, and applying Lemma 6.3, we get

$$S_u^- = S_{\phi'}^- \ge S_\phi^- - 1 + 2 = p.$$

Putting things together, we have

$$p \le S_u^- \le S_u^+ \le q, \qquad (6.4.31)$$

where the function $u(x)$ in Eq. (6.4.31) is still given by Eq. (6.4.28), Thus, Eq. (6.4.31) is what we need to prove.

At the conclusion of this subsection, we would like to make two comments:

(1) We have showed how mode shapes are affected by boundary supports in three corollaries at the end of Sect. 6.4.1. These three corollaries are also applicable to beams with rigid-body motion.
(2) It has been proved above that Property (2)(c) in the corollary of Theorem 6.9 remains valid for displacement mode shapes of all ten types of beams listed in Table 6.1. This conclusion can be easily generalized to mode shapes of the angle of rotation, bending moment, and shear force associated with the ten types of beams shown in Table 6.1, by taking advantage of the relationship among different mode shape functions. Detailed discussion and proof have to be left out due to space limitation.

6.4.3 Interlacement of Nodes Associated with Various Mode Shapes

As was pointed out in Chap. 5, in order to prove nodes of functions $\varphi_i(x)$ and $\varphi_{i+1}(x)$ from a Markov function sequence to be interlacing in Theorem 2.22, we have only utilized Properties (2) and (3) of the Markov function sequence, which are just the oscillatory properties 2(b) and 2(c) in the corollary of Theorem 6.9 in the context of a well-constrained beam system. Notice that it was explained in Sect. 6.4.2 that Oscillatory Properties 2(b) and 2(c) remain valid for mode shapes of the displacement, angle of rotation, strain, and shear force when the associated beam is subject to any type of boundary constraints. Additional conclusions can be drawn based on this result, regarding interlacement of nodes of various mode shapes. Specifically, for an arbitrarily constrained beam, nodes of the following pairs of mode shapes interlace:

(1) two successive mode shapes of the displacement, i.e., $u_i(x)$ and $u_{i+1}(x)$;
(2) two successive mode shapes of the angles of rotation, i.e., $u_i'(x)$ and $u_{i+1}'(x)$;
(3) two successive mode shapes of the moment, i.e., $\tau_i(x)$ and $\tau_{i+1}(x)$; and
(4) two successive mode shapes of the shear force, i.e., $\phi_i(x)$ and $\phi_{i+1}(x)$;

where $i = 2, 3, \ldots$.

Again, consider the arbitrarily constrained beam and focus on its different types of mode shapes associated with the same natural frequency. Similar to the approach used in Chap. 5, we may prove nodes of the following pairs of mode shapes interlace, by applying Rolle's Theorem and taking advantage of the rules of sign reversals:

(1) mode shapes of the displacement and angle of rotation, i.e., $u_i(x)$ and $u_i'(x)$;
(2) mode shapes of the angle of rotation and bending moment, i.e., $u_i'(x)$ and $\tau_i(x) = EJu_i''(x)$; and
(3) mode shapes of the bending moment and shear force, i.e., $\tau_i(x)$ and $\phi_i(x) = \tau_i'(x)$;

where $i = 2, 3, \ldots$ and mode shapes in each pair above are associated with the same frequency.

In fact, let us look at the first result regarding mode shapes of the displacement and angle of rotation. Assume that nodes of its i-th displacement mode shape $u_i(x)$ are $\{\xi_r\}_1^{i-1}$, where $i = 1, 2, \ldots$. These points divide the interval $[0, l]$ into i subintervals. According to Corollary 2 in Sect. 6.4.1, there is one and only one extremum of $u_i(x)$ at x_r in each internal subinterval $(\xi_r,\ \xi_{r+1})$, where $r = 1, 2, \ldots, i-2$. As for the first subinterval $(0, \xi_1)$, we have $u_i(0) = 0$ when the beam is clamped or pinned at the left end, while $u_i(0) \neq 0$ when it is sliding or free. Thus, $u_i'(x)$ has one node inside $(0, \xi_1)$ in the former case, while it has none in the latter case. Similar observation can be made for the subinterval (ξ_{i-1}, l). Thus, we may conclude: As long as mode shapes of the displacement $u_i(x)$ and the angle of rotation $u_i'(x)$ correspond to the same frequency, their nodes must interlace.

By the same token, we may derive the other two results.

6.5 Number of Independent Modes of a Beam

6.5.1 Necessary and Sufficient Conditions for the Mode Shape of a Beam

In this subsection, we are going to derive the necessary and sufficient conditions for a function $u(x)$ to be a mode shape of the displacement associated with one of the ten beams listed in Table 6.1. These conditions are summarized below:

(1) The function $u(x)$ satisfies the respective boundary conditions;
(2) it has at least the fourth derivative; and
(3) node counts of $u_i(x)$ and $u_i''(x)$ are as those given in Table 6.1 for the specific boundary conditions.

We have shown that the necessary part in Sects. 6.4.1 and 6.4.2. Now let us prove, by actually constructing a beam, that they are also sufficient conditions (See Ref. [6]). Specifically, suppose that the function $u(x)$ has at least the fourth derivative and also satisfies boundary conditions of one of the beams listed in Table 6.1. If node counts of u and u'' also meet the respective requirements presented in Table 6.1, then there exists a real beam (may not be unique though) such that $u(x)$ serves as its i-th displacement mode shape, where $i = S_u + 1$.

Denote $\{\xi_m\}_1^{i-1}$ and $\{x_m\}_{N_1}^{N_2}$ as the ordered nodes of $u(x)$ and $u''(x)$, respectively. Here, $N_1 = 0$, 1, and 2, when the left end of the beam is clamped, pinned or sliding, and free, respectively; while $N_2 = i$, $i - 1$, and $i - 2$, when its right end is clamped, pinned or sliding, and free, respectively. By integrating the modal equation (6.1.1) twice on the interval $[x_k,\ x_{k+1}]$, we get

$$EJ(x)u''(x) = EJ(x_k)u'''(x_k)(x - x_k) + \lambda \int_{x_k}^{x} \mathrm{d}z \int_{x_k}^{z} \rho(s)u(s)\mathrm{d}s, \qquad (6.5.1)$$

where $k = N_1, N_1 + 1, \ldots, N_2 - 1$. Set

$$x_k' = \min(x_k, \xi_k) \ \text{ and } \ x_k'' = \max(x_k, \xi_k).$$

Then, we should have

$$\begin{aligned} &u''(x)u'''(x_k) > 0, \quad x_k < x < x_{k+1}, \\ &u''(x)u(x) < 0, \quad x_k'' < x < x_{k+1}', \\ &u''(x)u(x) > 0, \quad x_k' < x < x_k'' \ \text{ or } \ x_{k+1}' < x < x_{k+1}''. \end{aligned}$$

In order to assign a positive function to $EJ(x)$, let us first define $\rho(x)$ by considering the following four possible cases as follows:

Case 1 When $\xi_k \le x_k < x_{k+1} \le \xi_{k+1}$, set

$$\rho(x) = d_k, \quad x_k < x \le x_{k+1}.$$

Case 2 When $x_k < \xi_k < x_{k+1} \le \xi_{k+1}$, set

$$\rho(x) = \begin{cases} \varepsilon_{k1} d_k, & x_x < x < \xi_k, \\ d_k, & \xi_k \le x \le x_{k+1}. \end{cases}$$

Case 3 When $\xi_k \le x_k < \xi_{k+1} < x_{k+1}$, set

$$\rho(x) = \begin{cases} d_k, & x_x < x \le \xi_{k+1}, \\ \varepsilon_{k2} d_k, & \xi_{k+1} < x \le x_{k+1}. \end{cases}$$

Case 4 When $x_k < \xi_k < \xi_{k+1} < x_{k+1}$, set

$$\rho(x) = \begin{cases} \varepsilon_{k1} d_k, & x_x < x \le \xi_k, \\ d_k, & \xi_k < x \le \xi_{k+1}, \\ \varepsilon_{k2} d_k, & \xi_{k+1} < x \le x_{k+1}. \end{cases}$$

Here, d_k, ε_{k1}, and ε_{k2} are all positive constants to be determined.

For $\rho(x)$ defined above, we can sketch the function

$$F(x) = \lambda \int_{x_k}^{x} \mathrm{d}z \int_{x_k}^{z} \rho(s) u(s) \mathrm{d}s, \tag{6.5.2}$$

as shown in Fig. 6.4a–d. It is assumed in the graphs that $u'''(x_k) < 0$. If this is not true, i.e., $u'''(x_k) > 0$, then the shape characteristics of the function $F(x)$ should remain unchanged but the graphs would be upside down. Because both $F(x)$ and $F'(x)$ are directly proportional to d_k, by making ε_{k1} and ε_{k2} sufficiently small and properly adjusting the value of d_k, we may satisfy the following equations:

$$\begin{aligned} &EJ(x_k)u'''(x_k)(x_{k+1} - x_k) + F(x_{k+1}) = 0, \\ &EJ(x) = [EJ(x_k)u'''(x_k)(x - x_k) + F(x)]/u''(x) > 0, \quad x_k < x < x_{k+1}, \end{aligned} \tag{6.5.3}$$

$$EJ(x_{k+1}) = [EJ(x_k)u'''(x_k) + F'(x_{k+1})]/u'''(x_{k+1}) > 0. \tag{6.5.4}$$

The above discussion is applicable not only to subintervals $[x_k, x_{k+1}]$ $(k = N_1, N_1 + 1, \ldots, N_2 - 1)$ in the interior but to the last segment $[x_{N_2}, l]$ as well, where $u''(x)$ does not change sign. When the beam is unconstrained or pinned at the left end ($x = 0$), the above argument is also suitable for the segment $[0, x_{N_1}]$ and in this case, any positive number can be assigned to $EJ(0)$. However, it requires different treatment when the left end of the beam is either sliding or clamped, due to

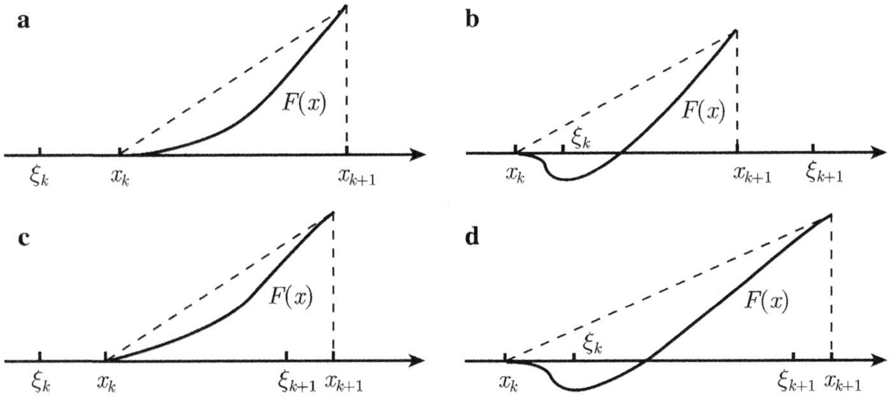

Fig. 6.4 Illustration of possible shapes of the function $F(x)$, defined in Eq. (6.5.2)

peculiarities of these two types of boundary conditions. This will be covered in the next two paragraphs.

When the left end of the beam ($x = 0$) is sliding, the integral form of Eq. (6.1.1) is

$$EJ(x)u''(x) = EJ(0)u''(0) + \lambda \int_0^x \mathrm{d}z \int_0^z \rho(s)u(s)\mathrm{d}s. \tag{6.5.5}$$

Since $u''(0)u(x) < 0$ $(0 \le x \le x_1)$, we may define $\rho(x)$ as in either Case 1 or Case 3 above. Then, the function

$$f(x) = \lambda \int_0^x \mathrm{d}z \int_0^z \rho(s)u(s)\mathrm{d}s \tag{6.5.6}$$

is of the shape depicted in either Fig. 6.4a or Fig. 6.4c. Furthermore, $A = EJ(0)u''(0)$ is a constant. Therefore, as long as ε_{02} is made sufficiently small and d_0 is properly chosen, we will have

$$A + f(x_1) = 0;$$

$$EJ(x) = [A + f(x)]/u''(x) > 0, \quad 0 < x < x_1; \tag{6.5.7}$$

$$EJ(x_1) = f'(x_1)/u'''(x_1) > 0. \tag{6.5.8}$$

When the beam is clamped at the left end, the integral form of Eq. (6.1.1) becomes

$$EJ(x)u''(x) = A + Bx + f(x).$$

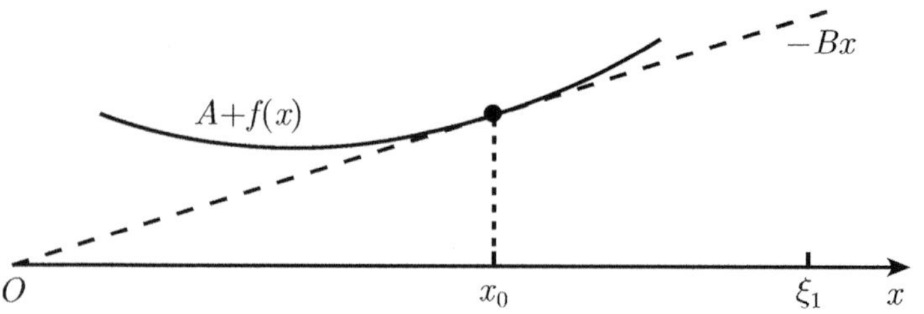

Fig. 6.5 Graphs of functions $A+f(x)$ and $-Bx$, shown in solid and dash lines, respectively

Here, we have $u''(0)u(x) > 0$ $(0 < x < x_0)$ and $u''(0)u'''(x_0) < 0$. Set $\rho(x) = d_{-1}$ $(0 \leq x \leq x_0)$, let $EJ(0)$ be any positive number, and choose $EJ'(0)$ such that the sign of $B = [EJ(x)u''(x)]'_{x=0}$ is opposite to that of A. Then, we may plot the graphs of $A+f(x)$ and $-Bx$, as depicted in Fig. 6.5. Thus, it is possible to satisfy the following conditions by adjusting the value of d_{-1}:

$$\begin{aligned} A + Bx_0 + f(x_0) &= 0; \\ EJ(x) = [A + Bx + f(x)]/u''(x) &> 0, \quad 0 < x < x_0; \\ EJ(x_0) = [B + f'(x_0)]/u'''(x_0) &> 0. \end{aligned}$$

As of now, we have defined a positive function $EJ(x)$ for the entire beam, and completed the proof of the sufficient conditions for the displacement mode shape of a beam.

6.5.2 *Inverse Problem of a Beam with Two Given Modes and the Number of Independent Modes*

In Chaps. 3, 5, and 4, we discussed problems of determining the spring-mass system, the continuous system of a bar, and the finite difference model of a beam, respectively, using two sets of modal data. Both theoretical results and numerical algorithms were presented there. However, the problem of determining the continuous system of a beam using the modal data is much more complicated than those studied in early chapters. As a result, although it captured much attention, the inverse problem of a beam with two given modes has not been fully resolved for a long time. In 2014, Zijun Zheng provided some excellent results on the subject in his doctoral thesis (See Ref. [10]). He showed that the problem could be converted into one involving the solution of a singular differential equation. In addition, he created an algorithm for solving the inverse problem. Below is a brief introduction of his work.

Given two sets of modes $(\lambda, u(x))$ and $(\mu, v(x))$, where $\lambda < \mu$ and both $u(x)$ and $v(x)$ are functions that are sufficiently smooth, we want to determine parameters $EJ(x)$ and $\rho A(x)$ of a beam.

From the two known modes as well as the modal equation (6.1.1), we get the ordinary differential equation satisfied by $EJ(x)$:

$$\gamma EJ + \beta(EJ)' + \alpha(EJ)'' = 0, \quad 0 < x < l, \tag{6.5.9}$$

where

$$\alpha = \mu vu'' - \lambda uv'', \quad \beta = 2(\mu vu''' - \lambda uv'''), \quad \gamma = \mu vu^{(4)} - \lambda uv^{(4)}.$$

Based on the theory of the ordinary differential equations, if $\alpha \neq 0$ $(0 < x < l)$, the above equation has two linearly independent solutions. Consequently, $EJ(x)$ is not unique and, as a matter of fact, may not even remain positive all the times. Thus, generally speaking, the modal data specified should not allow this case to occur. However, there are exceptions to this rule:

Exception 1 $\alpha \neq 0, \beta \equiv 0,$ *and* $\gamma \equiv 0$ in Eq. (6.5.9), where $0 < x < l$. In this case, the solution to Eq. (6.5.9) is of the form $EJ(x) = Ax + B$.

Exception 2 $\alpha \neq 0, \beta \neq 0,$ *and* $\gamma \equiv 0$ in Eq. (6.5.9), where $0 < x < l$. In this case, one solution to Eq. (6.5.9) is for $EJ(x)$ to be a constant B.

In sequent discussion, we primarily focus on the case where the parameter function $\alpha(x)$ has a number of isolated zeroes in the interval $(0, l)$.

If x_s satisfies the equation $\alpha(x) = 0$, then it is called a singular point of the equation. If limits of β/α and γ/α exist at the singular point, x_s is called a removable singular point. Otherwise, it is referred to as a non-removable singular point. Set

$$\frac{\beta}{\alpha} = \frac{f(x)}{x - x_s}.$$

If

$$f(x_s) < -1, \; f(x_s) > -1, \quad \text{and } f(x_s) = -1,$$

then the non-removable singular point x_s is termed as the first, second, and third type of singular point, respectively.

Without going into the complicated process of solving for $EJ(x)$, here is a brief summary of the final conclusion:

(1) If the two sets of modes specified allow the first and second types of singular points to interlace, then there exists a unique pair of functions $EJ(x)$ and $\rho(x)$ that are positive and have continuous second derivatives.
(2) $EJ(x)$ has no unique solution, if and only if, the two sets of modal data specified lead to continuous appearance of the first type of singular points, rather than interlacement with the second type of singular points.

These results lead to the following important theorem:

Theorem 6.11 *Among the infinite set of modes of a beam, one pair of them (two modes) may be used to unequely identify the beam, and all others can be later derived. However, certain pair of modes may not be sufficient to uniquely identify this beam.*

6.6 Additional Properties of Natural Frequencies of a Beam

6.6.1 Dependency of Natural Frequencies on Section and Boundary Parameters of a Beam

Let us begin by introducing the following bilinear forms:

$$J(u, v) = \int_0^l EJ(x)u''(x)v''(x)\mathrm{d}x + h_1 u(0)v(0) + h_2 u(l)v(l) + \beta_1 u'(0)v'(0) + \beta_2 u'(l)v'(l), \tag{6.6.1}$$

$$(\rho u, v) = \int_0^l \rho(x)u(x)v(x)\mathrm{d}x. \tag{6.6.2}$$

An alternative definition of the eigenvalue of a beam, based on the variational principle, is

$$\lambda_i = \max_{v_1, v_2, \ldots, v_{i-1}} \min_{(u, v_j)=0(j=1,2,\ldots,i-1)} \frac{J(u, u)}{(\rho u, u)}. \tag{6.6.3}$$

Apparently, $J(u,\ u)$ increases as the value of $EJ(x)$ goes up everywhere in the interval I and when constants h_r and β_r grow bigger, while $(\rho u,\ u)$ also becomes larger if the value of $\rho(x)$ rises everywhere in I. Thus, the first set of conclusions can be drawn below, based on Eq. (6.6.3):

(1) $\lambda_i\ (i = 1, 2, \ldots)$ increase as $EJ(x)$ rises everywhere in I, but decreases when $\rho(x)$ goes up at every point in I.
(2) $\lambda_i\ (i = 1, 2, \ldots)$ become bigger as h_r and β_r grow larger.

Next, suppose h_r and β_r are of finite value. We are going to prove that the following inequalities are satisfied when $h_r < h'_r$ and $\beta_r < \beta'_r$:

$$\lambda_i(h_1, h_2, \beta_1, \beta_2) < \lambda_i(h'_1, h_2, \beta_1, \beta_2) < \lambda_{i+1}(h_1, h_2, \beta_1, \beta_2), \tag{6.6.4}$$

$$\lambda_i(h_1, h_2, \beta_1, \beta_2) < \lambda_i(h_1, h'_2, \beta_1, \beta_2) < \lambda_{i+1}(h_1, h_2, \beta_1, \beta_2), \tag{6.6.5}$$

$$\lambda_i(h_1, h_2, \beta_1, \beta_2) < \lambda_i(h_1, h_2, \beta'_1, \beta_2) < \lambda_{i+1}(h_1, h_2, \beta_1, \beta_2), \tag{6.6.6}$$

$$\lambda_i(h_1, h_2, \beta_1, \beta_2) < \lambda_i(h_1, h_2, \beta_1, \beta'_2) < \lambda_{i+1}(h_1, h_2, \beta_1, \beta_2). \tag{6.6.7}$$

In other words, natural frequencies of a beam subject to various boundary constraints interlace. In order to establish Eqs. (6.6.4)–(6.6.7), let us denote the eigenpairs of

Eqs. (6.1.1), (6.1.3), and (6.1.4) as $\{\lambda_i, u_i(x)\}_1^\infty$. The squared norm of the mode shape function $u_i(x)$ is

$$\int_0^l \rho(x)u_i^2(x)\mathrm{d}x = \rho_i,$$

where $i = 1, 2, \ldots$. Then, the solution $u(x)$ to Eq. (6.1.1) under the following new set of boundary conditions

$$\left[EJ(x)u''(x)\right]'_{x=0} + h'_1 u(0) = 0 = \left[EJ(x)u''(x)\right]'_{x=l} - h_2 u(l), \tag{6.6.8}$$

$$EJ(0)u''(0) - \beta_1 u'(0) = 0 = EJ(l)u''(l) + \beta_2 u'(l), \tag{6.6.9}$$

may be expressed as

$$u(x) = \sum_{i=1}^{\infty} c_i u_i(x). \tag{6.6.10}$$

Since

$$J(u_i, u_j) = \lambda_i \rho_i \delta_{ij}, \quad i, j = 1, 2, \ldots, \tag{6.6.11}$$

we have, on the one hand,

$$\begin{aligned} J^*(u, u) &= \int_0^\infty EJ(x)[u''(x)]^2\mathrm{d}x + h'_1 u^2(0) + h_2 u^2(l) + \beta_1(u'(0))^2 + \beta_2(u'(l))^2 \\ &= \sum_{i=0}^{\infty} \lambda_i c_i^2 \rho_i + (h'_1 - h_1)u^2(0). \end{aligned}$$

On the other hand,

$$J^*(u, u) = \lambda(\rho u, u) = \sum_{i=1}^{\infty} \lambda c_i^2 \rho_i.$$

By combining the two equations above, we obtain

$$c_i = \frac{(h'_1 - h_1)u(0)u_i(0)}{\rho_i(\lambda - \lambda_i)}.$$

Substituting the calculated c_i into Eq. (6.6.10) and setting $x = 0$, we get

$$1 = (h_1' - h_1)\sum_{i=1}^{\infty}\frac{u_i^2(0)}{\rho_i(\lambda - \lambda_i)}. \tag{6.6.12}$$

This is the characteristic equation of Eq. (6.1.1) under boundary conditions (6.6.8) and (6.6.9). Based on Eq. (6.6.12), as well as Proposition 4.6, we may get Inequality (6.6.4).

By the same token, it can be shown that Eqs. (6.6.5)–(6.6.7) are also satisfied when $h_2 < h_2'$ or $\beta_r < \beta_r'$ $(r = 1, 2)$.

Different from the bar, a beam has four boundary parameters. By using the previous method, we may derive two additional results:

$$\lambda_i(h_2, \beta_1) < \lambda_i(h_2', \beta_1') < \lambda_{i+1}(h_2, \beta_1),$$
$$\lambda_i(h_2, \beta_2) < \lambda_i(h_2', \beta_2') < \lambda_{i+1}(h_2, \beta_2), \ldots$$

Notice that in the above inequalities, unchanged parameters have been removed from the argument lists of eigenvalues to make these inequalities more compact. It also needs to be pointed out that Eqs. (6.6.4)–(6.6.7) as well as the two inequalities above are applicable only to cases where h_r and β_r are either zero or of finite value. If at least one of h_r and β_r becomes infinity, some related inequalities may no longer be valid.

6.6.2 *Interlacement of Natural Frequencies of Beams Clamped at One End but Subject to Various Constraints at the Other End*

Consider the continuous system of a beam that is always clamped at the left end but subject to the free, sliding, pinned, anti-resonant, or clamped constraint at the right end. Denote natural angular frequencies of the same beam in clamped-free, clamped-sliding, clamped-pinned, clamped–anti-resonant, and clamped-clamped configurations as $\{\omega_i\}_1^\infty$, $\{\sigma_i\}_1^\infty$, $\{\mu_i\}_1^\infty$, $\{\nu_i\}_1^\infty$, and $\{\eta_i\}_1^\infty$, respectively. We may find the following relationship among these frequencies (see Refs. [2, 3]):

$$\omega_i < \sigma_i < \nu_i < \mu_i < (\eta_i, \omega_{i+1}) < \sigma_{i+1}, \quad i = 1, 2, \ldots. \tag{6.6.13}$$

Obviously, these results are similar to those obtained for the discrete system of the beam.

6.6.3 Interlacement of Natural Frequencies of a Beam Pinned at One End but Subject to Various Constraints at the Other End

Consider a beam subject to pinned-pinned constraints and under the action of a concentrated harmonic force couple $\tau \sin \omega t$ at the right end, where $\omega = \sqrt{\lambda}$ is the angular frequency. The equation of vibration and boundary conditions are

$$[EJ(x)u''(x)]'' = \lambda\rho(x)u(x), \quad 0 < x < l, \tag{6.6.14}$$

$$u(0) = 0,\ EJ(0)u''(0) = 0,\ u(l) = 0,\ EJ(l)u''(l) = \tau. \tag{6.6.15}$$

In order to solve for the response of the forced vibration, let us consider the problem of finding the stationary point of the functional:

$$J(u) = \frac{1}{2}\left[\int_0^l EJ(x)[u''(x)]^2 \mathrm{d}x - \lambda \int_0^l \rho(x)u^2(x)\mathrm{d}x\right] - \tau u'(l), \tag{6.6.16}$$

where $u(x)$ only needs to satisfy the displacement boundary conditions, i.e., the first and third equations in (6.6.15). Following the usual approach, get the first variation of Eq. (6.6.16) and then integrate it by parts twice. The result is shown below:

$$\delta J = \int_0^l [(EJu'')'' - \lambda\rho u]\delta u\,\mathrm{d}x + [EJ(l)u''(l) - \tau]\delta u'(l) - EJu''(0)\delta u'(0). \tag{6.6.17}$$

Thus, the displacement function $u(x)$ that makes the functional $J(u)$ stationary should satisfy Eqs. (6.6.14) and (6.6.15). In other words, it is the response of the beam pinned at both ends and under the load of the concentrated harmonic force couple $\tau \sin \omega t$ at $x = l$. Since the eigenfunctions $\{u_i(x)\}_1^\infty$ of a beam pinned at both ends constitute a set of orthonormal and complete basis functions, it is possible to express any displacement function $u(x)$ as

$$u(x) = \sum_{i=1}^{\infty} c_i u_i(x), \quad 0 \le x \le l. \tag{6.6.18}$$

For the beam pinned at both ends, we have

$$\int_0^l EJ(x)u_i''(x)u_j''(x)\mathrm{d}x = \omega_i^2 \rho_i \delta_{ij}, \quad i \text{ and } j = 1, 2, \ldots, \tag{6.6.19}$$

where $\omega_i = 2\pi f_i = \sqrt{\lambda_i}$ and $u_i(x)$ $(i = 1, 2, \ldots)$ are, respectively, natural angular frequencies and the corresponding mode shapes of the beam in the pinned-pinned configuration. Without loss of generality, we may choose $u_i(x)$ such that

$$\rho_i = \int_0^l \rho(x)u_i^2(x)\mathrm{d}x = 1, \quad i = 1, 2, \ldots.$$

Substituting Eq. (6.6.18) into Eq. (6.6.16), the functional $J(u)$ becomes

$$J(u) = \frac{1}{2}\sum_{i=1}^{\infty}(\lambda_i - \lambda)c_i^2 - \sum_{i=1}^{\infty} c_i \tau u_i'(l). \tag{6.6.20}$$

By setting $\delta J = 0$, we get $(\lambda_i - \lambda)c_i = \tau\, u_i'(l)$. After plugging the calculated c_i into Eq. (6.6.18), we obtain

$$u(x) = \sum_{i=1}^{\infty} \frac{\tau u_i'(l)}{\lambda_i - \lambda} u_i(x), \quad 0 \le x \le l.$$

Accordingly, we have

$$u'(x) = \sum_{i=1}^{\infty} \frac{\tau\, u_i'(l)}{\lambda_i - \lambda} u_i'(x), \quad 0 \le x \le l. \tag{6.6.21}$$

From these, we may derive the interlacement of natural frequencies associated a beam always pinned at the left end but subject to the pinned, clamped, or anti-resonant constraint at the right end, as follows:

(1) For a beam in the pinned-clamped configuration, the boundary condition at the right end is $u'(l) = 0$. Thus, we may establish the following frequency equation for this beam, based on Eq. (6.6.21):

$$\sum_{i=1}^{\infty} \frac{\theta_i^2(l)}{\lambda_i - \lambda} = 0. \tag{6.6.22}$$

It is known that $u_i'(l) = \theta_i(l) \neq 0$ $(i = 1, 2, \ldots)$ when this beam is pinned at both ends. Denote roots of Eq. (6.6.22) as $\{\mu_i^2\}_1^{\infty}$. Then,

$$\omega_i < \mu_i < \omega_{i+1}, \quad i = 1, 2, \ldots, \tag{6.6.23}$$

namely, natural angular frequencies of a beam pinned at both ends and its pinned-clamped counterpart interlace.

(2) For a beam in the pinned–anti-resonant configuration, its support conditions at the right end $x = l$ include: $u(l) = 0$ and the total reaction force vanishes.

An alternative way to realize the anti-resonant condition is to apply an external force couple τ at the right end of a simply supported beam, where τ satisfies the equation $\tau'(l) + \tau/l = 0$. This alternative way leads to the following frequency equation for the beam in the pinned–anti-resonant configuration:

$$\sum_{i=1}^{\infty} \frac{\theta_i(l)q_i(l)}{\lambda_i - \lambda} = -\frac{1}{l}, \tag{6.6.24}$$

where $q_i(l) = [EJ(x)u_i''(x)]'\big|_{x=l}$. Additionally, it can be shown that $\theta_i(l)q_i(l) < 0$ $(i = 1, 2, \ldots)$. From these results, we may derive the interlacing relationship between natural angular frequencies of a pinned–anti-resonant beam and the corresponding pinned-pinned, which is shown below:

$$\xi_i < \omega_i < \xi_{i+1}, \quad i = 1, 2, \ldots, \tag{6.6.25}$$

where $\{\xi_i\}_1^\infty$ represent natural angular frequencies of the beam in the pinned–anti-resonant configuration.

6.6.4 *Reconstruction of the Continuous System of a Beam from Three Sets of Spectra*

Based on interlacement of frequencies established in Sect. 6.6.2, Gladwell discussed in Ref. [3] the inverse problem in vibration, where a continuous system of the beam is reconstructed from three sets of its spectra obtained after imposing various boundary constraints on the beam. Specifically, he considered a beam clamped at the left end but is under three different types of support at the right end. If any of the three sets of frequencies $\{\omega_i, \sigma_i, \mu_i\}_1^\infty$, $\{\omega_i, \sigma_i, \nu_i\}_1^\infty$, or $\{\omega_i, \mu_i, \nu_i\}_1^\infty$ is known, and these frequencies satisfy certain required conditions, then it is possible to reconstruct parameters $EJ(x)$ and $\rho(x)$ of this beam in the clamped-free configuration.

Since the frequency can be measured through acoustic methods, the physical significance of this type of inverse problem is: Physical parameters of the beam may be determined by "hearing" them.

6.7 Modal Qualitative Properties of the Continuous System of an Overhang Beam

Consider an overhang beam with length l, linear density $\rho(x)$, and bending stiffness $r(x) = EJ(x)$. Its modal equation for undamped and free vibration in the transverse direction is of the following form:

$$[r(x)u''(x)]'' = \omega^2 \rho\, u(x), \quad 0 < x < l, \tag{6.7.1}$$

where ω is the natural angular frequency and $u(x)$ the displacement mode shape. Support conditions of the two-span overhang beam shown in Fig. 6.6a include

$$u(0) = 0, \quad r(0)u''(0) = 0, \quad r(l)u''(l) = 0, \quad [r(x)u''(x)]'|_{x=l} = 0, \tag{6.7.2}$$

and

$$u(c) = 0 \text{ while both } u'(c) \text{ and } u''(c) \text{ are continuous,} \tag{6.7.3}$$

where $0 < c < l$. In comparison, support conditions of the three-span overhang beam sketched in Fig. 6.6b are

$$r(0)u''(0) = 0,\ [r(x)u''(x)]'|_{x=0} = 0, \quad r(l)u''(l) = 0, \quad [r(x)u''(x)]'|_{x=l} = 0, \tag{6.7.4}$$

and

$$u(d_i) = 0 \text{ while both } u'(d_i) \text{ and } u''(d_i) \text{ are continuous,} \tag{6.7.5}$$

where $i = 1, 2$ and $0 < d_1 < d_2 < l$.

In order to establish the qualitative properties of natural frequencies and mode shapes of an overhang beam, let us first derive its Green's function. By referencing Sect. 6.2, we may get the governing equation and support conditions for the Green's function of a two-span overhang beam as follows:

$$\begin{cases} [r(x)G''(x,s)]'' = \delta(x-s), \\ G(0,s) = G(c,s) = 0,\ G''(0,s) = 0, \\ G'(c-0,s) = G'(c+0,s),\ G''(c-0,s) = G''(c+0,s), \\ G''(l,s) = 0,\ [r(x)G''(x,s)]'_{x=l} = 0, \end{cases} \tag{6.7.6}$$

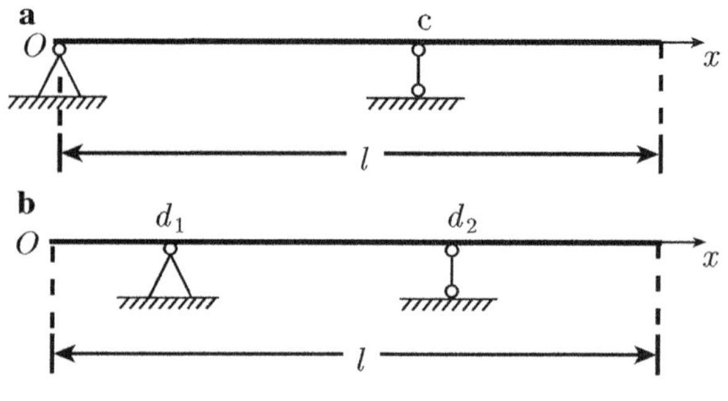

Fig. 6.6 Illustration of multi-span overhang beams: **a** with two spans, and **b** with three spans

where $0 < x,\ s < l$ and $0 < c < l$.

By directly integrating Eq. (6.7.6), we obtain the following expression for the Green's function of the two-span overhang beam:

$$G(x,s)=\begin{cases}\frac{s-c}{c}\int_0^x\frac{t(x-t)}{r(t)}\mathrm{d}t+\frac{xs}{c^2}\int_0^c\frac{(c-t)^2}{r(t)}\mathrm{d}t+\frac{x}{c}\int_0^s\frac{(t-c)(s-t)}{r(t)}\mathrm{d}t, & 0\le x<s,\\ \frac{s}{c}\int_0^x\frac{(t-c)(x-t)}{r(t)}\mathrm{d}t+\frac{xs}{c^2}\int_0^c\frac{(c-t)^2}{r(t)}\mathrm{d}t+\frac{x-c}{c}\int_0^s\frac{t(s-t)}{r(t)}\mathrm{d}t, & s<x\le c,\\ \frac{x-c}{c}\left[\frac{s-c}{c}\int_0^s\frac{t^2}{r(t)}\mathrm{d}t+\frac{s}{c}\int_s^c\frac{t(t-c)}{r(t)}\mathrm{d}t\right], & c<x\le l,\end{cases}\tag{6.7.7}$$

if $0 < s < c$; while

$$G(x,s)=\begin{cases}\frac{s-c}{c}\left[\frac{x-c}{c}\int_0^x\frac{t^2}{r(t)}\mathrm{d}t+\frac{x}{c}\int_x^c\frac{t(t-c)}{r(t)}\mathrm{d}t\right], & 0\le x\le c,\\ \int_0^x\frac{(x-t)(s-t)}{r(t)}\mathrm{d}t-\frac{xs}{c^2}\int_0^c\frac{c^2-t^2}{r(t)}\mathrm{d}t+\frac{x+s}{c}\int_0^c\frac{t(c-t)}{r(t)}\mathrm{d}t, & c<x<s,\\ \int_0^s\frac{(x-t)(s-t)}{r(t)}\mathrm{d}t-\frac{xs}{c^2}\int_0^c\frac{c^2-t^2}{r(t)}\mathrm{d}t+\frac{x+s}{c}\int_0^c\frac{t(c-t)}{r(t)}\mathrm{d}t, & s<x\le l,\end{cases}\tag{6.7.8}$$

if $c < s \le l$.

Similarly, we can derive the Green's function of a three-span overhang beam.

6.7.1 *Oscillation of the Green's Function of an Overhang Beam*

Just like Sect. 6.3, in order to show the Green's function of an overhang beam to be oscillatory, we only need to verify the oscillatory properties in static deformation, A and B, which are presented in Sect. 2.10, are satisfied and the strain energy of the overhang beam is positive definite.

1. Oscillation of the Green's Function Associated with a Two-Span Overhang Beam

The Green's function of a two-span beam defined by Eqs. (6.7.7) and (6.7.8) is apparently not oscillatory, because it does not satisfy Condition (2.7.4) in the definition of an oscillatory kernel. In order to study the qualitative properties of the overhang beam, let us introduce the mathematically transformed system, whose kernel is called the mathematically transformed kernel and defined by

$$\tilde{G}(x,s)=\varepsilon(x)\varepsilon(s)G(x,s),\tag{6.7.9}$$

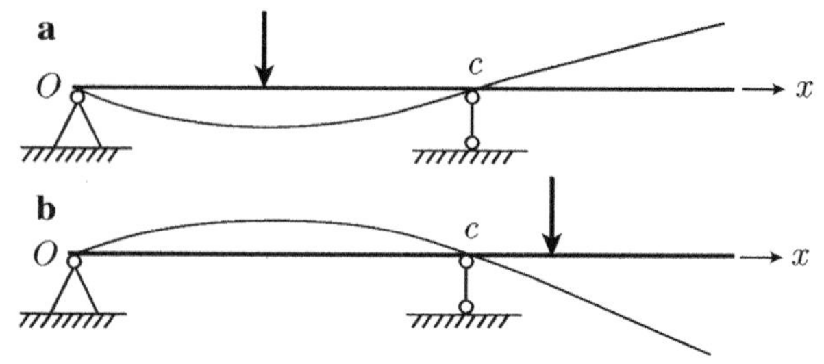

Fig. 6.7 Deflection curves of a two-span overhang beam under the action of a single external load: **a** The load is applied to the left span, and **b** the load is applied to the right span

where the function $\varepsilon(x) = (-1)^{m-1}$ and the superscript m represents the span number of the part of the overhang beam at the location x. Thus, for the two-span beam under consideration, $\varepsilon(x) = 1$ when $0 < x < c$, and $\varepsilon(x) = -1$ when $c < x \leq l$.

Now let us check whether the mathematically transformed kernel $\tilde{G}(x, s)$ is oscillatory.

First, based on the theory of Mechanics of Materials, if there is only one concentrated force acting on the beam at a point, then the deflection curve will be of shapes shown in Fig. 6.7. Thus, we would obviously have $\tilde{G}(x, s) > 0$, no matter which span s is located in, namely, Condition (2.7.4) in the definition of an oscillatory kernel is always satisfied.

Next, since the overhang beam is a statically determinate system, its strain energy

$$V = \sum G(s_i, s_j)F_iF_j/2$$

must be greater than zero, as long as the n concentrated loads acting on the beam are not all zero simultaneously. Thus, the right-hand side of the equation above is a positively definite bilinear form. Furthermore, per the criterion for a positively definite bilinear form, all of its leading principal minors should be positive, i.e.,

$$G\begin{pmatrix} s_1 & s_2 & \dots & s_n \\ s_1 & s_2 & \dots & s_n \end{pmatrix} = \tilde{G}\begin{pmatrix} s_1 & s_2 & \dots & s_n \\ s_1 & s_2 & \dots & s_n \end{pmatrix} > 0.$$

Thus, the mathematically transformed kernel $\tilde{G}(x, s)$ satisfies Condition (2.7.6) in the definition of an oscillatory kernel.

Finally, we are going to put a significant amount of effort into finding the number of sign reversals of the deflection $u(x)$ when the overhang beam is under the action of n concentrated forces. We are going to proceed by considering three cases.

Case 1 All of the n concentrated forces $\{F_i\}_1^n$ are positioned at the left span of the beam, i.e., $0 < s_i < c$ $(i = 1, 2, \dots, n)$, and the displacement of the beam can be expressed as

$$u(x) = \sum_{i=1}^{n} F_i G(x, s_i).$$

It is known from Mechanics of Materials that $u(x)$ satisfy the following equation of the shear force:

$$\tau'(x) = \left[EJ(x)u''(x)\right]' = \begin{cases} C_0, \ 0 \le x < s_1, \\ C_i, \ s_i < x < s_{i+1} \ \text{ and } \ i = 1, 2, \ldots, n-1, \\ C_n, \ s_n < x < c, \\ 0, \ \ c < x \le l, \end{cases}$$

where

$$C_i = C_0 + \sum_{j=1}^{i} F_j, \quad i = 1, 2, \ldots, n.$$

The equation above indicates that $\tau'(x)$ may change sign only at places where external forces are applied and furthermore, at each of these locations, it can have at most one sign reversal. Thus, the total count of sign reversals of $\tau'(x)$ must not exceed n, and all these sign reversals have to occur within the interval $(0, c)$. Because of the conditions $\tau(0) = 0 = \tau(c)$, we get from Lemma 6.1 that the number of sign reversals of $\tau(x)$ is no bigger than $n-1$ in the interval $(0, c)$; while the numbers of sign reversals of $u'(x)$ and $u(x)$ do not exceed n and $n+1$, respectively, within the same interval. But since $u(0) = 0$ and $u(c) = 0$, each of these conditions causes the number of sign reversals of $u(x)$ in $(0, c)$ to drop by 1. Thus, the actual count of sign reversals of $u(x)$ in $(0, c)$ should be no more than $n-1$. In addition, under the current circumstance, the deflection curve of the overhang portion of the beam on the right ($c < x \le l$) is a straight line. Because the slope of the deflection curve at the support c has to be continuous, the function $u(x)$ must be of opposite sign in $(c, l]$ and (η_{n-1}, c), where η_{n-1} is the last zero of $u(x)$ in $(0, c)$. This implies that the total number of sign reversals of $u(x)$ within the interval $(0, l]$ does not exceed n. Hence, the number of sign reversals of the function

$$\tilde{u}(x) = \varepsilon(x)u(x), \tag{6.7.10}$$

which corresponds to the mathematically transformed kernel $\tilde{G}(x, s)$, is no greater than $n-1$. In other words, the oscillatory property in static deformation, B, is satisfied.

Note: A little clarification is needed here. In the above discussion, we only stated that the number of sign reversals of $u(x)$ does not exceed $n-1$ in the interval $(0, c)$. But the actual count can be smaller than that. See Fig. 6.8. If the deflection curve has a positive local minimum or negative local maximum, this should create two additional sign reversals of $u'(x)$. Consequently, the number of sign reversals of the corresponding $u(x)$ in the interval $(0, c)$ must decrease by 2. Otherwise, according to Rolle's theorem, it will lead to the contradiction that the count of sign reversals of $u'(x)$ in $(0, c)$ goes beyond n. The point d in Fig. 6.8 is called a null anti-node. Similar to the special types of extrema just mentioned, existence of a null anti-node

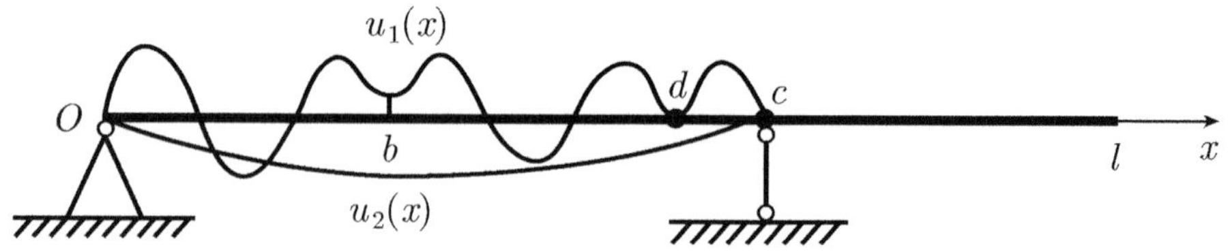

Fig. 6.8 Superposition of $u_1(x)$ and $u_2(x)$

causes a reduction in the number of sign reversals of $u(x)$ in the interval $(0, c)$ by 2. Apparently, the situations discussed here can occur only when $n \geq 3$.

Case 2 All of the n concentrated forces $\{F_i\}_1^n$ are applied to the right span of the beam, i.e., $c < s_i \leq l$ $(i = 1, 2, \ldots, n)$. In this case, the displacement of the beam can still be expressed as

$$u(x) = \sum_{i=1}^{n} F_i G(x, s_i).$$

But the equation of the shear force now assumes the form:

$$\tau'(x) = \left[EJ(x)u''(x)\right]' = \begin{cases} C_0, & 0 \leq x < c, \\ C_0 + C_c, & c < x < s_1, \\ C_i, & s_{i-1} < x < s_i \quad \text{and} \quad i = 2, 3, \ldots, n, \\ 0, & s_n < x \leq l, \end{cases}$$

where

$$C_i = C_0 + C_c + \sum_{j=1}^{i-1} F_j, \quad i = 1, 2, \ldots, n,$$

and C_0 and C_c are reaction forces at supports $x = 0$ and $x = c$, respectively. As with the previous case, the above equation also shows that the number of sign reversals of $\tau'(x)$ does not exceed $n-1$ in the interval $(c, l]$. Since $\tau(l) = 0$, we know from Lemma 6.1 that the number of sign reversals of $\tau(x)$ is no more than $n - 1$ in the interval (c, l). Furthermore, from Mechanics of Materials, $\tau(c)u'(c) > 0$. Thus, counts of sign reversals of $u'(x)$ and $u(x)$ inside the interval (c, l) are no greater than $n - 1$ and n, respectively. Finally, because $u(c) = 0$, the number of sign reversals of $u(x)$ in the interval (c, l) will decrease by 1. Consequently, the final count of sign reversals of $u(x)$ within the interval (c, l) should not exceed $n - 1$. Under the current situation, the deformed shape of the left span of the beam $(0 < x < c)$ must lie on the same side of its original undeformed position. Due to the continuity of the slope of the deflection curve at the support c, the function $u(x)$ must be of opposite sign in the two subintervals $(0, c)$ and (c, η_1), where η_1 is the first zero of $u(x)$ in (c, l). Thus, the total count of sign reversals of $u(x)$ should not go beyond n in the interval $(0, l]$. This leads to the conclusion that the number of sign reversals of the function $\tilde{u}(x) = \varepsilon(x)u(x)$, which corresponds to the mathematically transformed kernel $\tilde{G}(x, s)$, does not exceed $n - 1$ in the interval $(0, l]$. Hence, the oscillatory property in static deformation, B, still holds true in this case.

By the way, the note to the previous case remains valid here.

Case 3 The n concentrated forces are distributed between two spans of the beam. Suppose n_1 and n_2 of them are applied to the left $(0 < x < c)$ and right $(c < x \leq l)$ spans, respectively, where $n_1 + n_2 = n$. Then, in the context of the linear theory, the n_1 and n_2 forces should create two independent displacements $u_1(x)$ and $u_2(x)$, respectively. According to the superposition principle, the deflection of the beam under the action of the n concentrated forces is given by

$$u(x) = u_1(x) + u_2(x).$$

Based on previous discussions, the first n_1 forces produce no more than $n_1 - 1$ and precisely 0 sign reversals of $u_1(x)$ at the left and right spans, respectively; while the latter n_2 forces generate exactly 0 and no greater than $n_2 - 1$ sign reversals of $u_2(x)$ at the left and right spans, respectively.

We would like to point out that counts of sign reversals of the superpositioned deflection curve $u(x)$ at the left and right spans of the beam do not exceed $n_1 - 1$ and $n_2 - 1$, respectively. Take the left span as an example. Only when there exists a positive local minimum, negative local maximum, or null anti-node of $u_1(x)$ at the left span $(0 < x < c)$, can it be possible that the number of sign reversals of the superpositioned deflection curve changes on this part of the beam. To illustrate impact of these factors, we consider the situation depicted in Fig. 6.8, where $u_2(x) < 0$ while $u_1(x)$ has a positive local minimum at $x = b$ and a null anti-node at $x = d$ on the left span. On the one hand, existence of a positive local minimum may cause the number of sign reversals of the superpositioned deflection curve to increase by 2, while the null anti-node should definitely lead to 2 additional sign reversals of the superpositioned curve. But on the other hand, as has been pointed out in the earlier note, this type of extremum and the null anti-node will each help reduce the count of sign reversals of $u_1(x)$ by 2 at the left span. Thus, the two opposing effects cancel each other out, and the total net count of sign reversals at the left span stays the same before and after the superposition, which is always equal to or less than $n_1 - 1$. Similar argument can be applied to the right span.

Thus, the superpositioned deflection curve of the beam $u(x)$ has no more than $n_1 - 1$ and $n_2 - 1$ sign reversals at the left and right spans, respectively. Per the definition of $\tilde{u}(x)$, it is identical to $u(x)$ at the left span, while they are equal in value but opposite in sign at the right span. Consequently, the numbers of sign reversals of $\tilde{u}(x)$ are no greater than $n_1 - 1$ and $n_2 - 1$ at the left and right spans, respectively. Due to continuity, the function $\tilde{u}(x)$ may be of opposite signs on the two sides of the mid-support. We thus reach the conclusion that the number of sign reversals of $\tilde{u}(x)$

does not exceed $n-1$ in the interval $(0, l]$ when the external forces act on both spans of the beam. This again shows that the oscillatory property in static deformation, B, is valid.

Based on the above results, we may conclude that the mathematically transformed kernel $\tilde{G}(x, s)$ of the two-span beam is indeed oscillatory.

2. ***Oscillation of the Green's Function Associated with a Three-Span Overhang Beam***

With regard to the three-span overhang beam, its Green's function is also not an oscillatory kernel. Thus, it is again needed to introduce the mathematically transformed kernel

$$\tilde{G}(x, s) = \varepsilon(x)\varepsilon(s)G(x, s).$$

In order to prove this mathematically transformed kernel to be oscillatory, we may mimic the proof for the two-span beam but have to deal with some additional cases. Due to space limitation, we have to skip detailed discussion here.

6.7.2 Oscillatory Properties of Natural Vibration of an Overhang Beam

In the previous subsection, we established the fact that $\tilde{G}(x, s)$, the mathematically transformed kernel of the Green's function $G(x, s)$ of either a two-span overhang beam with one overhang or a three-span overhang beam with two overhangs, is oscillatory. Now we will further study the qualitative properties of natural frequencies and mode shapes of these overhang beams.

By utilizing the Green's function $G(x, s)$, the modal equation (6.7.1) governing the natural vibration of an overhang beam as well as related support conditions, either Eqs. (6.7.2) and (6.7.3) or Eqs. (6.7.4) and (6.7.5), can be converted into the following eigenvalue problem of the integral equation:

$$u(x) = \omega^2 \int_0^l G(x, s)\rho(s)u(s)\mathrm{d}s. \tag{6.7.11}$$

Multiplying $\varepsilon(x)$ on both sides of the above equation, we get

$$\tilde{u}(x) = \omega^2 \int_0^l \tilde{G}(x, s)\rho(s)\tilde{u}(s)\mathrm{d}s. \tag{6.7.12}$$

Since $\tilde{G}(x, s)$ is known to be an oscillatory kernel, the product of it and a positive function $\sqrt{\rho(x)\rho(s)}$ should remain oscillatory, i.e.,

$$K(x, s) = \tilde{G}(x, s)\sqrt{\rho(x)\rho(s)}$$

is still an oscillatory kernel. Thus, the mathematically transformed system automatically inherits all properties pertaining to the eigenvalue problem of an integral equation with an oscillatory kernel. Specifically, we have:

(1) Eigenvalues $\lambda_i = \omega_i^2$ $(i = 1, 2, ...)$ of the mathematically transformed system are all positive and distinct. Furthermore, they can be sorted from small to big as

$$(0 <)\lambda_1 < \lambda_2 < \lambda_3 < \cdots.$$

(2) The eigenfunction $\tilde{u}_i(x)$ corresponding to λ_i of the mathematically transformed system has exactly $i - 1$ nodes in the interval I, where $i = 1, 2, 3, \ldots$.
(3) Nodes of two successive eigenfunctions $\tilde{u}_i(x)$ and $\tilde{u}_{i+1}(x)$ interlace, where $i = 2, 3, \ldots$.

According to Eqs. (6.7.10) – (6.7.12), all properties listed above for the mathematically transformed system can be passed on to the respective overhang beam as follows:

Property 1 *Since Eq. (6.7.12) comes from Eq. (6.7.11) as a result of several simple algebraic operations, they should share an identical set of eigenvalues. Thus, natural frequencies of two- and three-span overhang beams* $f_i = \omega_i / 2\pi$ $(i = 1, 2, 3, \ldots)$ *are all distinct. They may be arranged in an increasing order as*

$$(0 <)f_1 < f_2 < f_3 < \cdots.$$

Property 2 *The first mode shape of an overhang beam happens to have exactly* p *nodes, where* $p = 1$ *for the two-span overhang beam and* 2 *for the three-span overhang beam. These nodes all overlap with the supports.*

In fact, per Eq. (6.7.10), the displacement mode shape $u_i(x)$ of an overhang beam and its mathematically transformed counterpart $\tilde{u}_i(x)$ are identical in odd-numbered spans and of the same value but opposite signs at even-numbered spans. Consequently, the displacement mode shape should have either a node or a null anti-node at each mid-support. In the case of the first mode shape, the zero at the mid-support obviously has to be a node.

Property 3 *The i-th* $(i \geq 2)$ *mode shape* $u_i(x)$ *of the overhang beam has precisely* $i-1+p-2t$ *nodes, where* t *represents the number of times nodes of* $\tilde{u}_i(x)$ *overlapping with the supports and furthermore,* $t \leq 1$ *and* $t \leq 2$ *for two- and three-span overhang beams, respectively.*

In fact, the function $\tilde{u}_i(x)$ has $i - 1$ nodes. If t of them overlap with the supports, then the node count of the i-th $(i \geq 2)$ mode shape $u_i(x)$ of the overhang beam decreases by t, and yet there will be $p - t$ additional nodes at the rest of the $p - t$

supports. Thus, the final node count for the i-th ($i \geq 2$) mode shape $u_i(x)$ of this overhang beam becomes

$$(i-1)-t+(p-t)=i-1+p-2t.$$

Notice that the node count of the displacement mode shape $u_i(x)$ decreases when a node of the function $\tilde{u}_i(x)$ overlaps with one of the supports. However, as pointed out earlier, this zero at the mid-support must be a null anti-node of $u_i(x)$. If a null anti-node is counted as two distinct zeros when calculating the total number of zeros of the mode shape $u_i(x)$, then we still get $i+p-1$ zeroes, which include $i-1+p-2t$ nodes and t null anti-nodes.

Property 4 *When none of the nodes of functions $\tilde{u}_i(x)$ and $\tilde{u}_{i+1}(x)$ overlaps with any mid-support, nodes of the two successive displacement mode shapes $u_i(x)$ and $u_{i+1}(x)$ of an overhang beam, excluding those located right at the mid-supports, shall still interlace, where $i=2,3,\ldots$. However, when one of the nodes of the function $\tilde{u}_i(x)$ or $\tilde{u}_{i+1}(x)$ overlaps with a mid-support, nodes of displacement mode shapes $u_i(x)$ and $u_{i+1}(x)$ may no longer interlace, where $i=2,3,\ldots$.*

Similar to its discrete counterpart, this type of beams has four other somehow important properties, i.e., Properties 5–8. Because it is more complicated to prove them, they will be introduced later in Sects. 6.7.4 and 6.7.5.

6.7.3 The Conjugate of an Overhang Beam

In order to further study modal qualitative properties of the angle of rotation, bending moment, and shear force, we introduce the conjugate of an overhang beam by following the example of Sect. 6.4.2. To this end, denote the bending moment of an overhang beam as

$$\bar{u}(x)=\tau(x)=r(x)u''(x). \tag{6.7.13}$$

Then, Eq. (6.7.1) may be rewritten as

$$[r^*(x)\bar{u}''(x)]''=\lambda\rho^*\bar{u}(x). \tag{6.7.14}$$

This may be viewed as the modal equation of a beam defined on the interval $[0, l]$ and having parameters

$$r^*(x)=\rho^{-1}(x),\quad \rho^*(x)=r^{-1}(x) \tag{6.7.15}$$

We refer to this beam as the conjugate of the original beam. Apparently, the "displacement" mode shape associated with a conjugate beam is none other than the mode shape of the bending moment associated with the original beam.

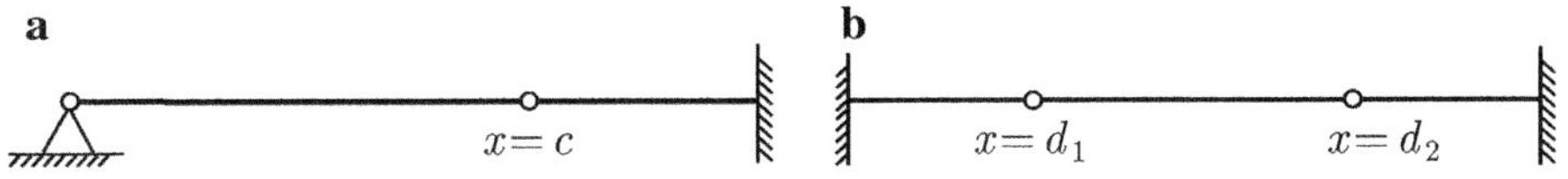

Fig. 6.9 Conjugate systems of overhang beams: **a** with two spans, and **b** with three spans

From the transform (6.7.13) as well as the following equations

$$\bar{u}''(x) = \lambda\rho(x)u(x), \quad \bar{u}'(x) = [r(x)u''(x)]',$$

we know that the conjugate of the two-span overhang beam remains to be pinned on the left, but its right end becomes clamped while the mid-support is converted to a hinge $\left(\bar{u}''(c) = 0\right)$. Thus, the conjugate system of a two-span overhang beam is a two-span continuous beam illustrated in Fig. 6.9a. A similar discussion reveals that the conjugate system of a three-span overhang beam is a three-span continuous beam clamped at both ends and with two hinges at $x = d_1$ and $x = d_2$. A sketch of the conjugate system is given in Fig. 6.9b.

The above derivation also shows that the overhang beam and its conjugate system share a common set of spectra.

6.7.4 Oscillatory Properties of the Conjugate System of an Overhang Beam

1. Oscillation of the Green's function $\bar{G}(x, s)$ Associated with the Conjugate System of a Two-Span Overhang Beam

It can be shown that the Green's function $\bar{G}(x, s)$ of the conjugate system of an overhang beam is an oscillatory kernel. We will again demonstrate this with a two-span overhang beam.

First, when the conjugate beam is under the action of a single concentrated load, its deflection curve should be of one of the shapes shown in Fig. 6.10, based on our knowledge of Structural Mechanics. Thus, it is apparent that $\bar{G}(x, s) > 0$, i.e., Condition (2.7.4) in the definition of an oscillatory kernel is satisfied.

Second, since the two-span continuous beam shown in Fig. 6.9a is statically determinate, its strain energy

$$V = \sum \bar{G}(s_i, s_j)F_iF_j/2$$

must be positive as long as the n concentrated forces $F_1, F_2, \ldots,$ and F_n do not vanish simultaneously. Thus, the right-hand side of the above equation is a positive definite bilinear form, and $\bar{G}(x, s)$ satisfies Condition (2.7.6) in the definition of an oscillatory kernel.

Third, when studying the number of sign reversals of the deflection $\bar{u}(x)$ associated with a two-span continuous beam under the joint action of n concentrated forces, we need to discuss three separate cases, much like what we have done with the two-span overhang beam. The final outcome is that the total number of sign reversals of the displacement function $\bar{u}(x)$ in the interval $(0, l)$ does not exceed $n - 1$. Therefore, the oscillatory property in static deformation, B, also holds true for the two-span continuous beam.

Based on the aforementioned three results, we conclude that the Green's function $\bar{G}(x, s)$ associated with the conjugate system of the overhang beam is an oscillatory kernel.

2. ***Oscillation of the Green's Function $\bar{G}(x, s)$ Associated with the Conjugate System of a Three-Span Overhang Beam***

With an argument similar to the one applied to the two-span overhang beam, we can prove that the Green's function of the conjugate system associated with a three-span overhang beam is also an oscillatory kernel.

3. ***Qualitative Properties of Mode Shapes Associated with the Bending Moment of an Overhang Beam***

Since Green's functions associated with conjugate systems of both two- and three-span overhang beams are oscillatory kernels, the product of any of them and a positive function continues to be oscillatory. Denote the mode shape of the bending moment corresponding to the frequency f_i as $\tau_i(x)$, where $i = 1, 2, \ldots$. Then, according to Theorem 2.23 on properties of the integral equation with an oscillatory kernel, mode shapes of the bending moment of an overhang beam has the following qualitative properties, in addition to Properties 1–4 introduced in Sect. 6.7.2:

Property 5 *There are exactly $i - 1$ nodes or sign reversals of $\tau_i(x) = \bar{u}_i(x)$ within the interval $(0, l)$, where $i = 1, 2, \ldots$. The mode shape has no other zero in this interval.*

Property 6 *Nodes of $\tau_i(x)$ and $\tau_{i+1}(x)$ interlace, where $i = 2, 3, \ldots$.*

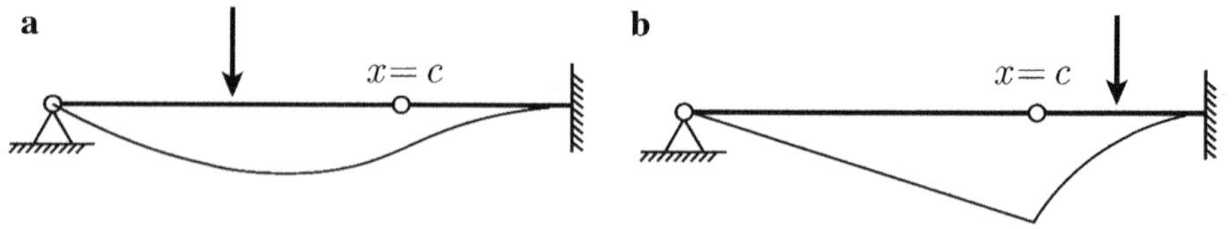

Fig. 6.10 Deflection curves of a two-span continuous beam subject to a single external load: **a** the load is applied to the left span, and **b** the load applied to the right span

6.7.5 Other Qualitative Properties of the Overhang Beam

In this subsection, we continue the discussion on qualitative properties of the overhang beam, started in Sects. 6.7.2 and 6.7.4.

1. Some Properties of the Extrema of a Displacement Mode Shape

We would like to state an important fact: If $u(x_0)$ is an internal extremum of a mode shape, then x_0 has to be an isolated point, rather than a line segment. Otherwise, we would have $u'(x) = u''(x) = 0$ in a subinterval, which contradicts Property 5 introduced in Sect. 6.7.4.

Thus, we know from the basic knowledge of Calculus: At an internal point x_0 where the mode shape function $u_i(x)$ reaches the local maximum, there must be $\tau_i(x_0) < 0$; while at an internal point x_1 where $u_i(x)$ becomes a local minimum, there must be $\tau_i(x_1) > 0$.

2. Node Counts of Mode Shapes Associated with the Angle of Rotation

For an overhang beam, we can establish the following property of mode shapes associated with the angle of rotation:

Property 7 *The i-th mode shape $u_i'(x)$ of the angle of rotation associated with an overhang beam has exactly i nodes within the interval* $(0, l)$, *where* $i = 1, 2, \ldots$.

In fact, based on the definition of nodes as well as the result that the i-th mode shape $u_i(x)$ of an overhang beam has precisely $i - 1 + p - 2t$ $(p = 1, 2)$ nodes, it is possible to divide the mode shape $u_i(x)$ into $i + p - 2t$ segments, within each of which the function retains its sign. There should be at least one extremum in every common-sign segment, excluding those that contain a free end (there are p free ends altogether). Thus, the mode shape $u_i(x)$ has at least $i - 2t$ internal extrema. As before, let us denote the i-th eigenfunction of the mathematically transformed system of the overhang beam as $\tilde{u}_i(x)$, which corresponds to $u_i(x)$. If none of the nodes of $\tilde{u}_i(x)$ overlaps with any mid-support, then $t = 0$. However, if one of the nodes of $\tilde{u}_i(x)$ overlaps with a support, then, as stated before, there is a null anti-node of the mode shape $u_i(x)$ at the mid-support and it is naturally the location of an extremum of $u_i(x)$. Thus, $u_i(x)$ reaches extremum at three points in the interval between the two nodes of $\tilde{u}_i(x)$ that are immediately adjacent to the mid-support from both sides. As a result, the mode shape $u_i(x)$ should have, in any case, at least i internal extrema. From a different perspective, the mode shape $\tau_i(x)$ of the bending moment of an overhang beam happens to have $i - 1$ nodes, i.e., it has exactly $i - 1$ sign reversals, where $i = 1, 2, \ldots$. Based on this as well as Lemma 6.1, there exist, at most, i internal extrema of the mode shape $u_i(x)$. By combining the results above, we may conclude that the mode shape $u_i(x)$ has and only has i internal extrema. Notice that the mode shape of the angle of rotation changes its sign once at two sides of each of the internal location where $u_i(x)$ achieves an extremum. Thus, the point of an internal extremum is none other than a node of the mode shape of the angle of rotation. This completes the proof of Property 7.

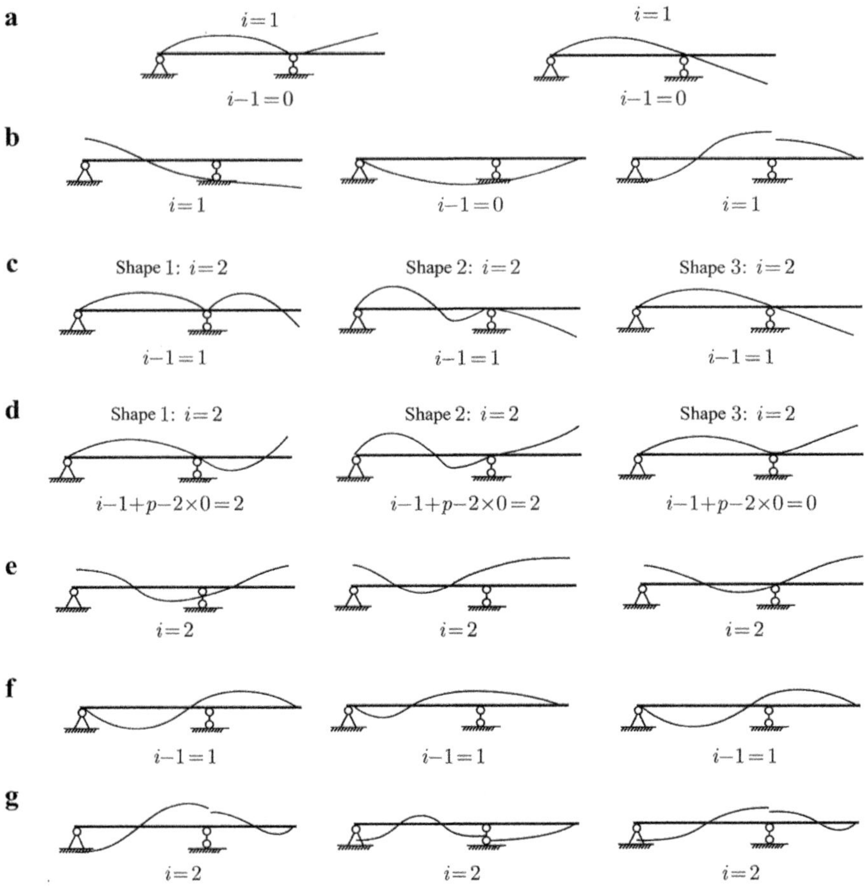

Fig. 6.11 First and second mode shapes of a two-span overhang beam: **a** the first eigenfunction of the mathematically transformed system (*left*) and the first mode shape of the corresponding two-span overhang beam (*right*), **b** first mode shapes of the angle of rotation (*left*), bending moment (*middle*), and shear force (*right*) associated with the two-span overhang beam, **c** three possible shapes for the second eigenfunction of the mathematically transformed system, **d** three possible shapes for the second mode shape of the displacement of the two-span overhang beam, **e** three possible shapes for the second mode shape of the angle of rotation associated with the two-span overhang beam, **f** three possible shapes for the second mode shape of the bending moment of the two-span overhang beam, and **g** three possible shapes for the second mode shape of the shear force of the two-span overhang beam

Now consider the common-sign segments of the mode shape $u_i(x)$, in each of which the function $u_i(x)$ retains its sign. The discussion above also reveals: Excluding the two segments, one on each side of the middle support, there is only one local maximum in a positive segment of the mode shape $u_i(x)$, while there is only one local minimum in a negative segment.

3. *Node Counts of Mode Shapes of the Shear Force*

Starting from the node count of $\tau_i(x)$, the number of zeroes of $u_i(x)$, as well as the modified form of Eq. (6.7.1) $\phi_i'(x) = \omega_i^2 \rho(x) u_i(x)$, we get the corresponding result for the mode shape of the shear force, with an argument similar to the one used earlier.

Property 8 *The i-th mode shape of the shear force* $\phi_i(x) = [r(x)u_i''(x)]'$ *has precisely i nodes, where* $i = 1, 2, \ldots$.

Based on results established thus far, we may sketch the first and second mode shapes of the continuous system of a two-span overhang beam, as shown in Fig. 6.11.

Finally, we would like to make three concluding remarks: (1) Discussion in this section has, in fact, covered the subject of qualitative properties of vibration associated with two- and three-span continuous beams. (2) The approach used here can be generalized to beams with more than three spans, as along as their conjugate systems are well-constrained. (3) By comparing results here with those in Sect. 4.9 of Chap. 4, we see that discrete and continuous systems of an overhang beam share very similar qualitative properties of vibration.

The content of this section is taken primarily from Ref. [8].

References

1. Gantmacher FP, Krein MG (1961) Oscillation matrices and kernels and small vibrations of mechanical systems. US Atomic Energy Commission, Washington
2. Gladwell GML (1985) Qualitative properties of vibrating systems. Proc Royal Soc London A(401):299–315
3. Gladwell GML (2004) Inverse Problems in Vibration. 2nd edn. Springer, Dordrecht (1986, 1st edn, Martinus Nijhoff Publishers, Dordrecht)
4. Wang QS, Wang DJ (1996) The flexibility matrices and its limits of difference discrete system for a rod and beam. Mech Prac 18(5):43–47 (in Chinese)
5. Wang QS, Wang DJ (1997) Supplemental definition of beam's positive systems and oscillatory properties of Green's functions. J AQTC (Nat Sci Ed) 3(1):14–16 (in Chinese)
6. Wang QS, Wang DJ (1997) Qualitative properties of frequency spectrum and modes of arbitrary supported beams in vibration. Acta Mech Sin 29(5):540–547 (in Chinese)
7. Wang QS, Wang DJ (1998) The flexibility coefficients and Green functions of statically determinate and indeterminate beams. J AQTC (Nat Sci Ed) 4(2):25–32 (in Chinese)
8. Wang QS, Wang DJ, He M et al (2012) Some qualitative properties of the vibration modes of the continuous system of a beam with one or two overhangs. J Eng Mech 138(8):945–952
9. Wang QS, Wang DJ (2014) The supplementary proof of some oscillation property for continuous systems of rod and beam having rigid modes. J AQTC (Nat Sci Ed) 20(1):1–5 (in Chinese)
10. Zheng ZJ (2014) The qualitative vibrational property and modal inverse problems of rods and Euler beams [D]. Department of Mechanics and Engineering Science, College of Engineering, Peking University (in Chinese)

Chapter 7
Qualitative Properties of Vibration and Static Deformation of Repetitive Structures

The focus of this chapter is qualitative properties of vibrational modes, static deformation, vibration control, etc. of repetitive structures. Types of repetitive structures covered here include structures with mirror symmetry (abbreviated as symmetric structures in subsequent discussion), rotationally periodic structures (also referred to as cyclic periodic or cyclic symmetric structures in the literature), linearly periodic structures (called linear periodic structures by some authors), chain structures (also known as linking or linked structures), and axisymmetric structures.

Repetitive structures, consisting of a group of identical substructures arranged in certain fashion, are commonly seen in both the natural environment and engineering application. The repetitiveness manifests itself through various attributes of the substructures, such as geometric parameters, physical parameters, boundary conditions, and interconnection among substructures. Solutions of static deformation, vibrational modes, and forced vibration depend, under general conditions, on physical and geometric properties of the entire structure. But for a repetitive structure, its repetitiveness leads to some special and important qualitative properties that make the solution rely solely on characteristics of just one substructure as well as the relationship with its neighboring counterparts. By taking advantage of this property or behavior, we should be able to significantly decrease the effort to obtain solutions of static deformation, vibrational modes, and forced vibration associated with a repetitive structure in numerical analyses and experimental studies.

As of now, there has been a lot of published research on vibration of repetitive structures in the literature, among which are Refs. [1–10, 12, 13, 16–19]. C. W. Cai, etc. have conducted a systematic and comprehensive study on the subject. They summarized their work in a monograph (see Ref. [5]), which is of great theoretical significance and practical value. Nevertheless, the majority of the existing work has focused on numerical analysis and consequently, involved primarily discrete systems of repetitive structures. So far as studying vibrational problems of repetitive structures from the perspective of continuous systems is concerned, there has been relatively few publications. In 2002, P. Chen (see Ref. [6]) applied the discrete Fourier transformation to equilibrium equations of the continuous system of a rotationally

D. Wang et al., *Qualitative Theory in Structural Mechanics*,
https://doi.org/10.1007/978-981-13-1376-9_7

periodic structure, and reduced the static problem of the entire structure to those of the corresponding substructures. The approach is clean and concise, and may be generalized to handle vibration problems of continuous systems associated with a broader group of repetitive structures. D. J. Wang, C. Y. Zhou and J. Rong proposed an alternative method in Ref. [15] that can be used to decouple vibration problems of continuous systems associated with multiple types of repetitive structures into those at the substructure level. Since the structures we study are often continuous in the physical world, the approach based on continuous systems is better suited for directly, accurately, and thoroughly revealing qualitative properties of vibrational modes and static deformation.

As indicated earlier, the main objective of this chapter is to discuss qualitative properties of vibration modes and static deformations of various repetitive structures. This is accomplished by studying the respective continuous systems. The approach has the benefit of better revealing the physical nature of the problem. In addition, we are going to discuss applications of these qualitative properties in numerical calculation and experimental measurement of vibrational modes and static deformation. Finally, we will briefly introduce related results for discrete systems of the repetitive structures, with the intention to facilitate application of qualitiative properties in numerical analysis.

By taking advantage of qualitative properties of vibrational modes and static deformation, not only is it possible to simplify numerical computation and develop highly efficient experimental plans, but also this helps the effort to analyze and screen numerical and testing results in many structural problems, such as categorization of mode shapes, detection of missing modes, etc.

7.1 Modal Qualitative Properties of Symmetric Structures

7.1.1 The Continuous System and Its Governing Equations

A symmetric structure refers to one whose geometric parameters, physical parameters, and boundary conditions all have mirror symmetry about one plane. This plane is called the plane of symmetry. If the system under consideration is planar, the plane of symmetry degenerates to a line; while if the system is linear, the plane of symmetry reduces to a points. Fig. 7.1 shows an example of a symmetric structure, which consists of a uniform three-dimensional elastic solid fixed at the bottom, a thin plate L_1 resting securely on the solid, two additional thin plates L_2 and L_3 attached perpendicularly to L_1, a bar of circular cross-section connecting the two plates L_2 and L_3 at two symmetric locations labeled s_3, a rigid link between the two plates at two symmetric locations labeled s_4, as well as two identical springs joining plates L_2 and L_3 at two sets of symmetric locations s_1 and s_2 and forming an X shape. Here, the object at s_3 is modeled as a bar along the x-direction, while it is treated as a beam in the y- and z-directions. Furthermore, the existence of a rigid link causes the

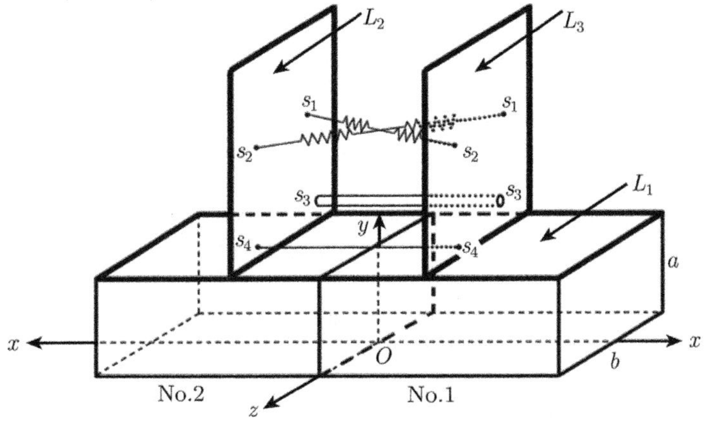

Fig. 7.1 An example of the symmetric structure

two points labeled s_4 to share the same displacement. The structure as a whole has one plane of symmetry, which divides it into two substructures of identical shape, physical parameters, and boundary conditions. These two substructures are labeled No.1 and No.2 in Fig. 7.1.

The modal equation of the entire structure can be given in the form of the following eigenvalue problem of differential equations

$$\begin{cases} \boldsymbol{L}\boldsymbol{u} = \omega^2 \boldsymbol{M}\boldsymbol{u} & \text{in } \tilde{\Omega}, \\ \boldsymbol{B}\boldsymbol{u} = \boldsymbol{0} & \text{on } \partial\tilde{\Omega}, \end{cases} \tag{7.1.1}$$

where ω is the natural angular frequency of the structure, $\boldsymbol{u}$ the mode shape function, $\boldsymbol{L}$ the elasticity differential operator, $\boldsymbol{M}$ the inertia differential operator, $\boldsymbol{B}$ the boundary condition operator, $\tilde{\Omega}$ the spatial domain occupied by the whole system, and $\partial\tilde{\Omega}$ the boundary of $\tilde{\Omega}$. If the system is composed of multiple components of different types, such as beams, plates, and three-dimensional elastic solids, then Eq. (7.1.1) will include equations for all components, boundary conditions, inter-component continuity requirements, as well as other possible constraints within a component and among components. These details are encapsulated in the simple and unified form given in Eq. (7.1.1).

By taking advantage of the symmetry of the structure, the eigenvalue problem (7.1.1) of the entire system may be expressed in terms of those of its substructures. The process is shown in this and next a few paragraphs. We begin by setting up coordinate systems at both substructures, making sure that they are of the same type and also symmetric with respect to the original plane of symmetry. Without loss of generality, let us use the Cartesian coordinate system in our discussion. In the case of Fig. 7.1, choose the Oyz plane of both substructures, i.e., the plane at $x = 0$, to be located right at the plane of symmetry of the complete structure. The coordinate systems in Substructures No.1 and No.2 need to be right- and left-handed, respectively, which ensures the two coordinate systems to be symmetric with respect to the plane of symmetry. Denote the generalized displacements in Substructures No.1 and No.2 as $\boldsymbol{u}_1$ and $\boldsymbol{u}_2$, respectively. The mode shape of the whole system $\boldsymbol{u} = (\boldsymbol{u}_1, \boldsymbol{u}_2)^{\mathrm{T}}$ satisfies the following modal equations and boundary conditions

$$\begin{cases} \boldsymbol{L}\boldsymbol{u}_k - \omega^2 \boldsymbol{M}\boldsymbol{u}_k = \boldsymbol{0} & \text{in } \Omega \text{ and } k = 1, 2, \\ \boldsymbol{B}\boldsymbol{u}_k = \boldsymbol{0} & \text{on } \partial\Omega \text{ and } k = 1, 2, \end{cases} \tag{7.1.2}$$

where Ω represents the spatial domain of one substructure while $\partial\Omega$ is its boundary excluding the portion shared by the two substructures.

It is worth noting that at the common boundary b_0 between Substructures No. 1 and No. 2, i.e., the intersection between Ω and the plane Oyz, the generalized displacements and internal forces should both satisfy continuity conditions, which may be represented by the following general form of differential equations, respectively,

$$\boldsymbol{J}_1\boldsymbol{u}_1 = -\boldsymbol{J}_1\boldsymbol{u}_2 \quad \text{on } b_0, \tag{7.1.3a}$$

$$\boldsymbol{J}_2\boldsymbol{u}_1 = \boldsymbol{J}_2\boldsymbol{u}_2 \quad \text{on } b_0, \tag{7.1.3b}$$

where $\boldsymbol{J}_1$ and $\boldsymbol{J}_2$ are both differential operators.

In addition, there may be elastic and rigid constraints between Substructures No. 1 and No. 2, which may be written as

$$\boldsymbol{J}_{r_j}\boldsymbol{u}_1\big|_{s_j} = \overline{\boldsymbol{J}}_{r_j}\boldsymbol{u}_2\big|_{\overline{s}_j}, \quad j = 1, 2, \ldots, l, \tag{7.1.4a}$$

$$\overline{\boldsymbol{J}}_{r_j}\boldsymbol{u}_1\big|_{s_j} = \boldsymbol{J}_{r_j}\boldsymbol{u}_2\big|_{\overline{s}_j}, \quad j = 1, 2, \ldots, l, \tag{7.1.4b}$$

where $\boldsymbol{J}_{r_j}$ and $\overline{\boldsymbol{J}}_{r_j}$ are differential operators and the subscript r indicates that the operator is related to a constraint.

Specifically for the case shown in Fig. 7.1, Eq. (7.1.2) has implicitly incorporated all modal equations associated with the three-dimensional elastic solid, elastic thin plates, beam, and elastic bar as well as the boundary constraints. The bottom of the structure is a three-dimensional elastic solid. Next, we look at the specific forms of Eqs. (7.1.3a) and (7.1.3b) for the case of Fig. 7.1. Let u_k, v_k, and w_k denote displacement components along x, y, and z directions, respectively, within the k-th substructure of the solid, where $k = 1, 2$. Continuity conditions of the displacement and stress at b_{01} ($x = 0, 0 \le y \le a, -b/2 \le z \le b/2$), which is the interface between lower parts of the two substructures, may be expressed as

$$\begin{bmatrix} 1 & 0 & 0 \\ \partial/\partial y & \partial/\partial x & 0 \\ \partial/\partial z & 0 & \partial/\partial x \end{bmatrix} \begin{bmatrix} u_1 \\ v_1 \\ w_1 \end{bmatrix} = - \begin{bmatrix} 1 & 0 & 0 \\ \partial/\partial y & \partial/\partial x & 0 \\ \partial/\partial z & 0 & \partial/\partial x \end{bmatrix} \begin{bmatrix} u_2 \\ v_2 \\ w_2 \end{bmatrix} \quad \text{on } b_{01}, \tag{7.1.5a}$$

$$\begin{bmatrix} 0 & 1 & 0 \\ 0 & 0 & 1 \\ \frac{\partial}{\partial x} & 0 & 0 \end{bmatrix}\begin{bmatrix} u_1 \\ v_1 \\ w_1 \end{bmatrix} = \begin{bmatrix} 0 & 1 & 0 \\ 0 & 0 & 1 \\ \frac{\partial}{\partial x} & 0 & 0 \end{bmatrix}\begin{bmatrix} u_2 \\ v_2 \\ w_2 \end{bmatrix} \quad \text{on } b_{01}. \tag{7.1.5b}$$

Let u_{bk}, v_{bk}, and w_{bk} be displacement components of the elastic bar $(y = y_3, z = z_3)$ connecting points labeled s_3 on plates L_2 and L_3, where the subscript k indicates the portion of the bar belongs to Substructure No. k and $k = 1, 2$. At the joint b_{02} $(x = 0, y = y_3, z = z_3)$ of the two substructures, i.e., the intersection between the bar and the Oyz plane, the displacement, resultant force, and resultant moment are continuous. Thus, Eqs. (7.1.3a) and (7.1.3b) may be written, respectively, as

$$\begin{bmatrix} 1 & 0 & 0 \\ 0 & \partial/\partial x & 0 \\ 0 & 0 & \partial/\partial x \\ 0 & \partial^3/\partial x^3 & 0 \\ 0 & 0 & \partial^3/\partial x^3 \end{bmatrix}\begin{bmatrix} u_{b1} \\ v_{b1} \\ w_{b1} \end{bmatrix} = -\begin{bmatrix} 1 & 0 & 0 \\ 0 & \partial/\partial x & 0 \\ 0 & 0 & \partial/\partial x \\ 0 & \partial^3/\partial x^3 & 0 \\ 0 & 0 & \partial^3/\partial x^3 \end{bmatrix}\begin{bmatrix} u_{b2} \\ v_{b2} \\ w_{b2} \end{bmatrix}, \quad \text{on } b_{02}, \tag{7.1.5c}$$

$$\begin{bmatrix} \partial/\partial x & 0 & 0 \\ 0 & 1 & 0 \\ 0 & 0 & 1 \\ 0 & \partial^2/\partial x^2 & 0 \\ 0 & 0 & \partial^2/\partial x^2 \end{bmatrix}\begin{bmatrix} u_{b1} \\ v_{b1} \\ w_{b1} \end{bmatrix} = \begin{bmatrix} \partial/\partial x & 0 & 0 \\ 0 & 1 & 0 \\ 0 & 0 & 1 \\ 0 & \partial^2/\partial x^2 & 0 \\ 0 & 0 & \partial^2/\partial x^2 \end{bmatrix}\begin{bmatrix} u_{b2} \\ v_{b2} \\ w_{b2} \end{bmatrix}, \quad \text{on } b_{02}. \tag{7.1.5d}$$

Equations (7.1.5a) and (7.1.5c) represent continuity conditions for antisymmetric deformation of the two substructures with respect to the plane $x = 0$, while Eqs. (7.1.5b) and (7.1.5d) denote the related continuity conditions for symmetric deformation with respect to the same plane. Notice that here we do not provide details of the continuity condition for the horizontal plate L_1 at its intersection with the symmetry plane. Finally, there are two elastic springs and one rigid bar connecting plates L_2 and L_3 on the structure depicted in Fig. 7.1. The specific forms of Eqs. (7.1.4a) and (7.1.4b) are, respectively:

$$\begin{cases} Q_1(s_1) + k_1 \sin^2\theta\, u_1(s_1) = -k_1 \sin^2\theta\, u_2(s_2), \\ Q_1(s_2) + k_1 \sin^2\theta\, u_1(s_2) = -k_1 \sin^2\theta\, u_2(s_1), \\ u_1(s_4) = -u_2(s_4), \end{cases} \tag{7.1.6a}$$

$$\begin{cases} Q_2(s_1) + k_1 \sin^2\theta\, u_2(s_1) = -k_1 \sin^2\theta\, u_1(s_2), \\ Q_2(s_2) + k_1 \sin^2\theta\, u_2(s_2) = -k_1 \sin^2\theta\, u_1(s_1), \\ u_2(s_4) = -u_1(s_4), \end{cases} \tag{7.1.6b}$$

where $Q_1(s_i)$ and $Q_2(s_i)$ $(i = 1,\ 2)$ represent spring forces applied to Plates L_2 and L_3 in Substructures No.1 and No.2 at points labeled s_1 and s_2, respectively, k_1 is the spring constant, and θ stands for the angle between the plate and the spring.

In summary, modal equations of the system are given by Eqs. (7.1.2)–(7.1.4) in terms of its substructures. The specific forms of the continuity conditions (7.1.3) and constraints between substructures (7.1.4) are shown in Eqs. (7.1.5) and (7.1.6), respectively. Due to the continuity conditions and inter-component constraints, displacements of the substructures, $\boldsymbol{u}_1$ and $\boldsymbol{u}_2$, become coupled.

7.1.2 Reduction of the Eigenvalue Problem and Discussion on Its Modal Qualitative Properties

Since Substructures No.1 and No.2 are symmetric with respect to the plane of symmetry, it makes sense to assume that the mode shape $\boldsymbol{u}_1$ of Substructure No.1 is α times of the mode shape $\boldsymbol{u}_2$ of Substructure No.2, and vice versa, where α is a constant. Thus,

$$\boldsymbol{u}_2 = \alpha \boldsymbol{u}_1 = \alpha(\alpha \boldsymbol{u}_2) = \alpha^2 \boldsymbol{u}_2.$$

The above equation yields $\alpha = \pm 1$. In other words, there can be only two possibilities for mode shapes:

$$(\boldsymbol{u}_1, \boldsymbol{u}_2)^{\mathrm{T}} = (\boldsymbol{q}_1, \boldsymbol{q}_1)^{\mathrm{T}},$$
$$(\boldsymbol{u}_1, \boldsymbol{u}_2)^{\mathrm{T}} = (\boldsymbol{q}_2, -\boldsymbol{q}_2)^{\mathrm{T}}.$$

Inspired by this observation, we may introduce the following transformation on the generalized displacement of the structure:

$$\boldsymbol{u} = \begin{bmatrix} \boldsymbol{u}_1 \\ \boldsymbol{u}_2 \end{bmatrix} = \boldsymbol{P}\boldsymbol{q} = \frac{1}{\sqrt{2}}\begin{bmatrix} \boldsymbol{I} & \boldsymbol{I} \\ \boldsymbol{I} & -\boldsymbol{I} \end{bmatrix}\begin{bmatrix} \boldsymbol{q}_1 \\ \boldsymbol{q}_2 \end{bmatrix} = \frac{1}{\sqrt{2}}\begin{bmatrix} \boldsymbol{I} \\ \boldsymbol{I} \end{bmatrix}\boldsymbol{q}_1 + \frac{1}{\sqrt{2}}\begin{bmatrix} \boldsymbol{I} \\ -\boldsymbol{I} \end{bmatrix}\boldsymbol{q}_2, \tag{7.1.7}$$

where $\boldsymbol{I}$ is the identity matrix whose order is identical to the dimension of the displacement variable $\boldsymbol{u}_1$. The first and second terms on the right-hand side of the last equality in Eq. (7.1.7) represent, respectively, the symmetric and antisymmetric mode shapes. Obviously, the transformation matrix $\boldsymbol{P}$ is orthogonal, i.e.,

$$\boldsymbol{P}^{\mathrm{T}}\boldsymbol{P} = \boldsymbol{I}, \tag{7.1.8}$$

where the order of the identity matrix $\boldsymbol{I}$ here is twice the dimension of $\boldsymbol{u}_1$. Thus, Eqs. (7.1.2)–(7.1.4) may be rewritten as

$$\begin{bmatrix} \boldsymbol{L} & \boldsymbol{0} \\ \boldsymbol{0} & \boldsymbol{L} \end{bmatrix}\begin{bmatrix} \boldsymbol{u}_1 \\ \boldsymbol{u}_2 \end{bmatrix} - \omega^2 \begin{bmatrix} \boldsymbol{M} & \boldsymbol{0} \\ \boldsymbol{0} & \boldsymbol{M} \end{bmatrix}\begin{bmatrix} \boldsymbol{u}_1 \\ \boldsymbol{u}_2 \end{bmatrix} = \boldsymbol{0}, \qquad \boldsymbol{u}_i \text{ in } \Omega, \tag{7.1.9a}$$

$$\begin{bmatrix} \boldsymbol{B} & \boldsymbol{0} \\ \boldsymbol{0} & \boldsymbol{B} \end{bmatrix}\begin{bmatrix} \boldsymbol{u}_1 \\ \boldsymbol{u}_2 \end{bmatrix} = \boldsymbol{0}, \qquad \boldsymbol{u}_i \text{ on } \partial\Omega, \tag{7.1.9b}$$

$$\begin{bmatrix} \boldsymbol{J}_1 & \boldsymbol{J}_1 \\ \boldsymbol{J}_2 & -\boldsymbol{J}_2 \end{bmatrix}\begin{bmatrix} \boldsymbol{u}_1 \\ \boldsymbol{u}_2 \end{bmatrix} = \boldsymbol{0}, \qquad \boldsymbol{u}_i \text{ on } b_0, \tag{7.1.9c}$$

$$\left.\begin{bmatrix} \boldsymbol{J}_{r_j} & \boldsymbol{0} \\ \boldsymbol{0} & \boldsymbol{J}_{r_j} \end{bmatrix}\begin{bmatrix} \boldsymbol{u}_1 \\ \boldsymbol{u}_2 \end{bmatrix}\right|_{s_j} = \left.\begin{bmatrix} \overline{\boldsymbol{J}}_{r_j} & \boldsymbol{0} \\ \boldsymbol{0} & \overline{\boldsymbol{J}}_{r_j} \end{bmatrix}\begin{bmatrix} \boldsymbol{0} & \boldsymbol{I} \\ \boldsymbol{I} & \boldsymbol{0} \end{bmatrix}\begin{bmatrix} \boldsymbol{u}_1 \\ \boldsymbol{u}_2 \end{bmatrix}\right|_{\overline{s}_j}, \qquad j = 1, 2, \ldots, l, \tag{7.1.9d}$$

where $\boldsymbol{L}, \boldsymbol{M}, \boldsymbol{B}, \boldsymbol{J}_1, \boldsymbol{J}_2, \boldsymbol{J}_{r_j}$ and $\overline{\boldsymbol{J}}_{r_j}$ are all matrices of differential operators while the order of the identity matrix $\boldsymbol{I}$ is the same as the dimension of $\boldsymbol{u}_1$. Substituting Eq. (7.1.7) into Eqs. (7.1.9a)–(7.1.9d), left multiplying modified versions of Eqs. (7.1.9a), (7.1.9b), and (7.1.9d) by $\boldsymbol{P}^T$, and also taking into account Eq. (7.1.8), we then obtain

$$\begin{cases} \boldsymbol{L}\boldsymbol{q}_i - \omega^2 \boldsymbol{M}\boldsymbol{q}_i = \boldsymbol{0} & \text{in } \Omega, \\ \boldsymbol{B}\boldsymbol{q}_i = \boldsymbol{0} & \text{on } \partial\Omega, \\ \boldsymbol{J}_i\boldsymbol{q}_i = \boldsymbol{0} & \text{on } b_0, \\ \boldsymbol{J}_{r_j}\boldsymbol{q}_i\big|_{s_j} = (-1)^{i+1}\overline{\boldsymbol{J}}_{r_j}\boldsymbol{q}_i\big|_{\overline{s}_j}, & \end{cases} \tag{7.1.10}$$

where $i = 1, 2$ and $j = 1, 2, \ldots, l$. These are the two sets of decoupled systems of differential equations as well as related boundary conditions for $\boldsymbol{q}_1$ and $\boldsymbol{q}_2$. Here, the two sets of equations are identical, but the connection conditions at the interface, namely the symmetry plane, differ and reveal the characteristics of symmetry and antisymmetry.

We are now ready to draw some conclusions: The natural vibration problem (7.1.9) of a symmetric structure may be reduced to two separate natural vibration problems (7.1.10) of one substructure. Here is the solution process: (1) When $i = 1$, solve for $\boldsymbol{q}_1$ from Eq. (7.1.10) and set $\boldsymbol{q}_2 = 0$. Substituting the solutions into Eq. (7.1.7), we obtain symmetric mode shapes of the whole structure. (2) When $i = 2$, solve for $\boldsymbol{q}_2$ from Eq. (7.1.10) and set $\boldsymbol{q}_1 = 0$. Inserting the results into Eq. (7.1.7), we get the antisymmetric mode shapes of the entire system. In other words, mode shapes of a symmetric structure have the following properties:

The mode shapes of a symmetric structure can be divided into two groups, one symmetric while the other antisymmetric. If a mode shape is neither symmetric nor antisymmetric, it must be related to a repeated frequency, and has to be a linear combination of the symmetric and antisymmetric mode shapes associated with this repeated frequency.

7.1.3 Applications

By taking advantage of modal qualitative properties, we can simplify numerical calculation and experimental measurement of vibrational modes. This is elaborated in the next two paragraphs.

First, in order to calculate natural frequencies and mode shapes of a symmetric structure, it only needs to solve two eigenvalue problems for one half of the structure, where boundary conditions include the original constraints, if any, on the substructure as well as the newly added continuity constraints at the interface on the symmetry plane. There are two types of constraints at the interface, one causing the entire system to deform in a symmetric manner while the other leading to antisymmetric deformation. The substructure and structure share the same set of natural frequencies, while mode shapes of the whole structure can be obtained by extending mode shapes of the substructure to the other side of the symmetry plane either symmetrically or anti-symmetrically. The advantage of this method in numerical analysis is a 50% reduction in the number of degrees of freedom, compared to the approach of using the entire structure. Since the amount of calculation required to solve a matrix eigenvalue problem is about the number of degrees of freedom to the third power, the existence of one symmetry plane can reduce the amount of calculation to 1/4 of the alternative approach, while the availability of three symmetry planes may further decrease the amount of computation to 1/64 of the method ignoring any symmetry.

Second, to determine the mode shape corresponding to a natural frequency through experimental means, it only needs to acquire the mode shape data on one half of the structure, plus those on the symmetry plane or at a point whose movement is nonzero on the other half of the structure. If it is decided that the deformation is symmetric (antisymmetric), then the mode shape for the entire system can be obtained by symmetrically (anti-symmetrically) extending the measured mode shape on one half of the structure to the other half. If, however, the mode shape is neither symmetric nor antisymmetric, then this must be a mode shape associated with a repeated frequency. In this scenario, we need to modify the experimental conditions to expose the two mode shapes, one symmetric while the other antisymmetric, that correspond to the repeated frequency.

Example A diamond-shaped symmetric beam.
The natural vibration problem of an unconstrained diamond-shaped symmetric beam shown in Fig. 7.2a may be reduced to two subproblems: One involves a substructure of the beam in the sliding-free configuration depicted in Fig. 7.2b, which is used to find symmetric mode shapes; while the other deals with a substructure of the beam in the pinned-free configuration illustrated in Fig. 7.2c, which is for obtaining anti-symmetric mode shapes. Analytical solutions can be derived for both subproblems.

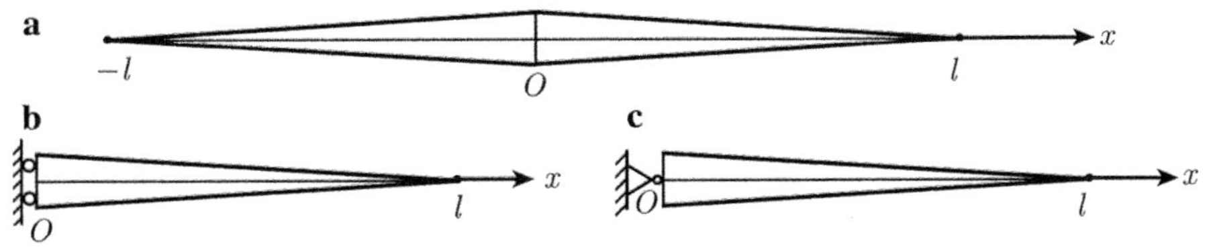

Fig. 7.2 Simplification of a diamond-shaped beam: **a** the original diamond-shaped symmetric beam, **b** the substructure producing symmetric mode shapes, and **c** the substructure producing antisymmetric mode shapes

7.1.4 Discrete System and Its Modal Qualitative Properties

To obtain vibrational modes through numerical analysis, we need to discretize the structure. For a discrete system resulting from some valid simplification, its frequencies and mode shapes can be obtained by solving the generalized matrix eigenvalue problem of the form:

$$\boldsymbol{K}\boldsymbol{u} = \omega^2 \boldsymbol{M}\boldsymbol{u}, \tag{7.1.11}$$

where ω stands for the natural angular frequency, $\boldsymbol{u}$ is the vector representing the displacement mode shape, and $\boldsymbol{K}$ and $\boldsymbol{M}$ denote, respectively, stiffness and mass matrices of the system. Here, both $\boldsymbol{K}$ and $\boldsymbol{M}$ are real symmetric matrices. Furthermore, $\boldsymbol{M}$ is positive definite, while $\boldsymbol{K}$ is either positive definite or semi-positive definite.

If the structural symmetry has been taken into account during discretization, the discrete system of the structure may be put in a form that can be easily simplified. This will be elaborated in the paragraphs below.

The generalized displacements may include both displacements and angles. For discretization nodes on the plane of symmetry, i.e., interface between the two substructures on the left to right, there are two types of generalized displacements, one in-plane while the other out-of-plane. Suppose the in-plane and out-of-plane generalized displacements are denoted by vectors $\boldsymbol{u}_2$ of b_1 dimensions and $\boldsymbol{u}_3$ of b_2 dimensions, respectively, where b_1 designates the number of in-plane displacements and angles while b_2 represents the count of out-of-plane displacements and angles. It needs to be pointed out that in physics $\boldsymbol{u}_2$ is the symmetric generalized displacement while $\boldsymbol{u}_3$ corresponds to the antisymmetric generalized displacement. However, from the perspective of mathematical representation, the angle components in $\boldsymbol{u}_2$ is antisymmetric, while their counterparts in $\boldsymbol{u}_3$ are symmetric.

To help readers better understand the notation, let us look at some examples: (1) For the discrete model of a symmetric bar, i.e., a symmetric mass-spring system, the axial displacement at a point on the symmetry plane is a scalar u_3. (2) For the discrete model of a symmetric string, i.e., a symmetric system comprised of masses and massless strings, the transverse displacement at a point on the symmetry plane is a scalar u_2. (3) For the discrete model of a symmetric bar in torsion, i.e., a symmetric system consisting of rotational springs and lumped moments of inertia, the angle of rotation at a point on the plane of symmetry is a scalar u_2. (4) For the finite element model of a symmetric beam, the generalized displacement at a node includes the nodal deflection and bending angle. Suppose there is a node on the plane of symmetry. The deflection and bending angle at this node are scalars u_2 and u_3, respectively. (5) For the finite element model of a three-dimensional elastic solid, the generalized displacement at a node consists of nodal displacement components in three directions. Suppose there are l nodes on the plane of symmetry. Then, the in-plane generalized displacement vector $\boldsymbol{u}_2$ is of $2l$ dimensions, while the generalized displacement vector $\boldsymbol{u}_3$ perpendicular to the symmetry plane is of l dimensions.

Now let us return to the discussion started earlier on building discrete models. Suppose there are p generalized displacements inside each of the two adjacent substructures. Denote the corresponding generalized displacement vectors within the left and right substructures as $\boldsymbol{u}_1$ and $\boldsymbol{u}_4$, respectively. Apparently, both of these vectors are of p dimensions. Thus, we have separated the displacement vector of the entire structure into four pieces: $\boldsymbol{u}_1, \boldsymbol{u}_2, \boldsymbol{u}_3$, and $\boldsymbol{u}_4$. Next, number components of these generalized displacements $\boldsymbol{u}_1, \boldsymbol{u}_2, \boldsymbol{u}_3$, and $\boldsymbol{u}_4$ in the sequence of 1 to p, $p+1$ to $p+b_1$, $p+b_1+1$ to $p+b_1+b_2$, and $p+b_1+b_2$ to $2p+b_1+b_2$, respectively. It is critical to ensure that the i-th and $(2p+b_1+b_2+1-i)$-th generalized displacements are symmetric with respect to the symmetry plane in terms of both location and direction, where $i=1,2,\ldots,p$. After these steps, Matrices $\boldsymbol{K}$ and $\boldsymbol{M}$ of the whole structure assume the following special forms:

$$\boldsymbol{K}=\begin{bmatrix}\boldsymbol{K}_{11} & \boldsymbol{K}_{12} & \boldsymbol{K}_{13} & \boldsymbol{K}_{14}\\ \boldsymbol{K}_{12}^{\mathrm{T}} & \boldsymbol{K}_{22} & \boldsymbol{0} & \boldsymbol{K}_{12}^{\mathrm{T}}\boldsymbol{S}\\ \boldsymbol{K}_{13}^{\mathrm{T}} & \boldsymbol{0} & \boldsymbol{K}_{33} & -\boldsymbol{K}_{13}^{\mathrm{T}}\boldsymbol{S}\\ \boldsymbol{K}_{14}^{\mathrm{T}} & \boldsymbol{S}\boldsymbol{K}_{12} & -\boldsymbol{S}\boldsymbol{K}_{13} & \boldsymbol{K}_{11}^{\mathrm{T}'}\end{bmatrix} \quad \text{and} \quad \boldsymbol{M}=\begin{bmatrix}\boldsymbol{M}_{11} & \boldsymbol{M}_{12} & \boldsymbol{M}_{13} & \boldsymbol{M}_{14}\\ \boldsymbol{M}_{12}^{\mathrm{T}} & \boldsymbol{M}_{22} & \boldsymbol{0} & \boldsymbol{M}_{12}^{\mathrm{T}}\boldsymbol{S}\\ \boldsymbol{M}_{13}^{\mathrm{T}} & \boldsymbol{0} & \boldsymbol{M}_{33} & -\boldsymbol{M}_{13}^{\mathrm{T}}\boldsymbol{S}\\ \boldsymbol{M}_{14}^{\mathrm{T}} & \boldsymbol{S}\boldsymbol{M}_{12} & -\boldsymbol{S}\boldsymbol{M}_{13} & \boldsymbol{M}_{11}^{\mathrm{T}'}\end{bmatrix}, \tag{7.1.12}$$

where the superscript T′ indicates transpose about the second diagonal of the matrix.

If $\boldsymbol{A}=(a_{ij})_{p\times q}$, then

$$\boldsymbol{A}^{\mathrm{T}'}=(a_{p-j+1,q-i+1})_{q\times p}.$$

Define the following matrix of order p for switching rows and columns of other matrices

$$S_p=\begin{bmatrix}0 & & 1\\ & \cdot^{\cdot^{\cdot}} & \\ 1 & & 0\end{bmatrix}. \tag{7.1.13}$$

Then,

$$\boldsymbol{A}^{\mathrm{T}'} = \boldsymbol{S}_q \boldsymbol{A}^{\mathrm{T}} \boldsymbol{S}_p.$$

For ease of writing, we may denote S_p as S, whenever there is no confusion, in the rest of this subsection.

$\boldsymbol{K}_{ij}$ in Eq. (7.1.12) is the stiffness matrix associated with generalized displacements $\boldsymbol{u}_i$ and $\boldsymbol{u}_j$, while $\boldsymbol{M}_{ij}$ in the same equation represents the mass matrix related to $\boldsymbol{u}_i$ and $\boldsymbol{u}_j$. Furthermore, $\{\boldsymbol{K}_{ij}\}$ have the following properties:

$$\boldsymbol{K}_{11} = \boldsymbol{K}_{11}^{\mathrm{T}}, \boldsymbol{K}_{22}^{\mathrm{T}} = \boldsymbol{K}_{22}, \boldsymbol{K}_{33}^{\mathrm{T}} = \boldsymbol{K}_{33}, \boldsymbol{K}_{14}^{\tilde{\mathrm{T}}} = \boldsymbol{K}_{14}; \tag{7.1.14}$$

while elements of $\{\boldsymbol{M}_{ij}\}$ share similar properties. By the way, the general form of the matrix $\boldsymbol{M}$ is presented in Eq. (7.1.12), but in most finite element implementations the mass matrix is reduced to the much simpler diagonal form.

It has been shown that mode shapes of the continuous system of a symmetric structure can be separated into two groups, one symmetric and the other antisymmetric. It is reasonable to expect the corresponding discrete system to also possess this property. In other words, there are following two types of mode shapes:

$$\begin{cases} (\boldsymbol{u}_1, \boldsymbol{u}_2, \boldsymbol{u}_3, \boldsymbol{u}_4)^{\mathrm{T}} = (\boldsymbol{q}_1, \boldsymbol{q}_2, \boldsymbol{0}, \boldsymbol{S}\boldsymbol{q}_1)^{\mathrm{T}}, \\ (\boldsymbol{u}_1, \boldsymbol{u}_2, \boldsymbol{u}_3, \boldsymbol{u}_4)^{\mathrm{T}} = (\boldsymbol{q}_4, \boldsymbol{0}, \boldsymbol{q}_3, -\boldsymbol{S}\boldsymbol{q}_4)^{\mathrm{T}}. \end{cases} \tag{7.1.15}$$

Based on this hypothesis, we introduce the transformation

$$\begin{bmatrix} \boldsymbol{u}_1 \\ \boldsymbol{u}_2 \\ \boldsymbol{u}_3 \\ \boldsymbol{u}_4 \end{bmatrix} = \frac{1}{\sqrt{2}} \begin{bmatrix} \boldsymbol{I}_p & \boldsymbol{0} & \boldsymbol{0} & \boldsymbol{I}_p \\ \boldsymbol{0} & \sqrt{2}\boldsymbol{I}_{b_1} & \boldsymbol{0} & \boldsymbol{0} \\ \boldsymbol{0} & \boldsymbol{0} & \sqrt{2}\boldsymbol{I}_{b_2} & \boldsymbol{0} \\ \boldsymbol{S} & \boldsymbol{0} & \boldsymbol{0} & -\boldsymbol{S} \end{bmatrix} \begin{bmatrix} \boldsymbol{q}_1 \\ \boldsymbol{q}_2 \\ \boldsymbol{q}_3 \\ \boldsymbol{q}_4 \end{bmatrix} = \boldsymbol{P}\boldsymbol{Q}, \tag{7.1.16}$$

where $\boldsymbol{I}$ represents the identity matrix and the subscripts indicate orders of the respective identity matrices. It is easy to verify that $\boldsymbol{P}$ is an orthogonal matrix, i.e., $\boldsymbol{P}^{\mathrm{T}}\boldsymbol{P} = \boldsymbol{I}$.

Substitute the transformation (7.1.16) into Eq. (7.1.11), and then left multiply the resulting equation by $\boldsymbol{P}^{\mathrm{T}}$. After further taking into account of Eqs. (7.1.12) and (7.1.14), we get

$$\begin{bmatrix} \boldsymbol{K}_{11}+\boldsymbol{K}_{14}\boldsymbol{S} & \sqrt{2}\boldsymbol{K}_{12} & \boldsymbol{0} & \boldsymbol{0} \\ \sqrt{2}\boldsymbol{K}_{12}^{\mathrm{T}} & \boldsymbol{K}_{22} & \boldsymbol{0} & \boldsymbol{0} \\ \boldsymbol{0} & \boldsymbol{0} & \boldsymbol{K}_{33} & \sqrt{2}\boldsymbol{K}_{13}^{\mathrm{T}} \\ \boldsymbol{0} & \boldsymbol{0} & \sqrt{2}\boldsymbol{K}_{13} & \boldsymbol{K}_{11}-\boldsymbol{K}_{14}\boldsymbol{S} \end{bmatrix} \begin{bmatrix} \boldsymbol{q}_1 \\ \boldsymbol{q}_2 \\ \boldsymbol{q}_3 \\ \boldsymbol{q}_4 \end{bmatrix}$$
$$= \omega^2 \begin{bmatrix} \boldsymbol{M}_{11}+\boldsymbol{M}_{14}\boldsymbol{S} & \sqrt{2}\boldsymbol{M}_{12} & \boldsymbol{0} & \boldsymbol{0} \\ \sqrt{2}\boldsymbol{M}_{12}^{\mathrm{T}} & \boldsymbol{M}_{22} & \boldsymbol{0} & \boldsymbol{0} \\ \boldsymbol{0} & \boldsymbol{0} & \boldsymbol{M}_{33} & \sqrt{2}\boldsymbol{M}_{13}^{\mathrm{T}} \\ \boldsymbol{0} & \boldsymbol{0} & \sqrt{2}\boldsymbol{M}_{13} & \boldsymbol{M}_{11}-\boldsymbol{M}_{14}\boldsymbol{S} \end{bmatrix} \begin{bmatrix} \boldsymbol{q}_1 \\ \boldsymbol{q}_2 \\ \boldsymbol{q}_3 \\ \boldsymbol{q}_4 \end{bmatrix}. \quad (7.1.17)$$

Thus, the original generalized matrix eigenvalue problem (7.1.11) can be decoupled into two smaller eigenvalue problems given by Eqs. (7.1.18) and (7.1.19) below:

$$\begin{bmatrix} \boldsymbol{K}_{11}+\boldsymbol{K}_{14}\boldsymbol{S} & \sqrt{2}\boldsymbol{K}_{12} \\ \sqrt{2}\boldsymbol{K}_{12}^{\mathrm{T}} & \boldsymbol{K}_{22} \end{bmatrix} \begin{bmatrix} \boldsymbol{q}_1 \\ \boldsymbol{q}_2 \end{bmatrix} = \omega_{\mathrm{s}}^2 \begin{bmatrix} \boldsymbol{M}_{11}+\boldsymbol{M}_{14}\boldsymbol{S} & \sqrt{2}\boldsymbol{M}_{12} \\ \sqrt{2}\boldsymbol{M}_{12}^{\mathrm{T}} & \boldsymbol{M}_{22} \end{bmatrix} \begin{bmatrix} \boldsymbol{q}_1 \\ \boldsymbol{q}_2 \end{bmatrix}, \quad (7.1.18)$$

where the subscript s of ω_{s} indicates symmetry. After solving the equation above for both $\boldsymbol{q}_1$ and $\boldsymbol{q}_2$ and also taking into account $\boldsymbol{q}_3 = \boldsymbol{q}_4 = \boldsymbol{0}$, we may get the symmetric mode shape for the entire structure from the first equation in (7.1.15).

$$\begin{bmatrix} \boldsymbol{K}_{33} & \sqrt{2}\boldsymbol{K}_{13}^{\mathrm{T}} \\ \sqrt{2}\boldsymbol{K}_{13} & \boldsymbol{K}_{11}-\boldsymbol{K}_{14}\boldsymbol{S} \end{bmatrix} \begin{bmatrix} \boldsymbol{q}_3 \\ \boldsymbol{q}_4 \end{bmatrix} = \omega_{\mathrm{a}}^2 \begin{bmatrix} \boldsymbol{M}_{33} & \sqrt{2}\boldsymbol{M}_{13}^{\mathrm{T}} \\ \sqrt{2}\boldsymbol{M}_{13} & \boldsymbol{M}_{11}-\boldsymbol{M}_{14}\boldsymbol{S} \end{bmatrix} \begin{bmatrix} \boldsymbol{q}_3 \\ \boldsymbol{q}_4 \end{bmatrix}, \quad (7.1.19)$$

where the subscript a of ω_{a} indicates anti-symmetry. Substituting the solutions $\boldsymbol{q}_3$ and $\boldsymbol{q}_4$ to Eq. (7.1.19) as well as the condition $\boldsymbol{q}_1 = \boldsymbol{q}_2 = \boldsymbol{0}$ into the second equation in (7.1.15), we obtain the antisymmetric mode shape of the whole structure.

Notice that the matrix $\boldsymbol{K}$ is real, symmetric, and either positive definite or semi-positive definite while the matrix $\boldsymbol{M}$ is real, symmetric, and positive definite. Consequently, the generalized matrix eigenvalue problem (7.1.11) should have nonnegative and real eigenvalues as well as real mode shapes. It is not difficult to verify that stiffness and mass matrices of the two sets of decoupled eigenvalue problems (7.1.18) and (7.1.19) are also real, symmetric, and positive definite (the stiffness matrix may be semi-positive definite). Thus, they have nonnegative and real eigenvalues as well as real mode shapes, too.

Under most circumstances, it is not required to compute generalized displacements right on the symmetry plane of the structure. As a result, if discretization nodes are placed inside the two substructures and none of them is located on the symmetry plane, the above eigenvalue problem can be greatly simplified. In this case, the mode shape $\boldsymbol{u} = (\boldsymbol{u}_1, \boldsymbol{u}_2)^{\mathrm{T}}$ becomes a vector of $2p$ dimensions, and the corresponding stiffness and mass matrices assume the following forms, respectively,

$$\boldsymbol{K} = \begin{bmatrix} \boldsymbol{K}_{11} & \boldsymbol{K}_{12} \\ \boldsymbol{K}_{12}^{\mathrm{T}} & \boldsymbol{K}_{11}^{\mathrm{T}'} \end{bmatrix} \quad \text{and} \quad \boldsymbol{M} = \begin{bmatrix} \boldsymbol{M}_{11} & \boldsymbol{M}_{12} \\ \boldsymbol{M}_{12}^{\mathrm{T}} & \boldsymbol{M}_{11}^{\mathrm{T}'} \end{bmatrix}.$$

Thus, the generalized matrix eigenvalue problem (7.1.11), which is now of $2p$-dimension, can be decomposed to two p-dimensional ones as follow:

$$[\boldsymbol{K}_{11} + \boldsymbol{K}_{12}\boldsymbol{S}]\boldsymbol{q}_1 = \omega_{\mathrm{s}}^2[\boldsymbol{M}_{11} + \boldsymbol{M}_{12}\boldsymbol{S}]\boldsymbol{q}_1,$$

and

$$[\boldsymbol{K}_{11} - \boldsymbol{K}_{12}\boldsymbol{S}]\boldsymbol{q}_2 = \omega_{\mathrm{a}}^2[\boldsymbol{M}_{11} - \boldsymbol{M}_{12}\boldsymbol{S}]\boldsymbol{q}_2.$$

The above equations yield, respectively, one group of symmetric mode shapes

$$(\boldsymbol{u}_1, \boldsymbol{u}_2)^{\mathrm{T}} = (\boldsymbol{q}_1, \boldsymbol{S}\boldsymbol{q}_1)^{\mathrm{T}}$$

and another group of antisymmetric mode shapes

$$(\boldsymbol{u}_1, \boldsymbol{u}_2)^{\mathrm{T}} = (\boldsymbol{q}_2, -\boldsymbol{S}\boldsymbol{q}_2)^{\mathrm{T}}.$$

Based on previous discussions, we may reach the following two important conclusions:

(1) Mode shapes of discrete and continuous systems of a symmetric structure have consistent qualitative properties.
(2) By taking advantage of qualitative properties of symmetric and antisymmetric mode shapes, we may decouple the original generalized matrix eigenvalue problem of $2p$ (or $2p + b_1 + b_2$ in the presence of generalized displacements on the symmetry plane) dimensions into two smaller problems of p (or one $p + b_1$ while the other $p + b_2$) dimensions. This helps greatly reduce the amount of computation needed and work load required.

7.2 Modal Qualitative Properties of Rotationally Periodic Structures

7.2.1 The Continuous System and Its Governing Equations

Consider a structure with a center axis. Suppose its geometric parameters, physical parameters, and boundary conditions can be expressed as functions of the angle of

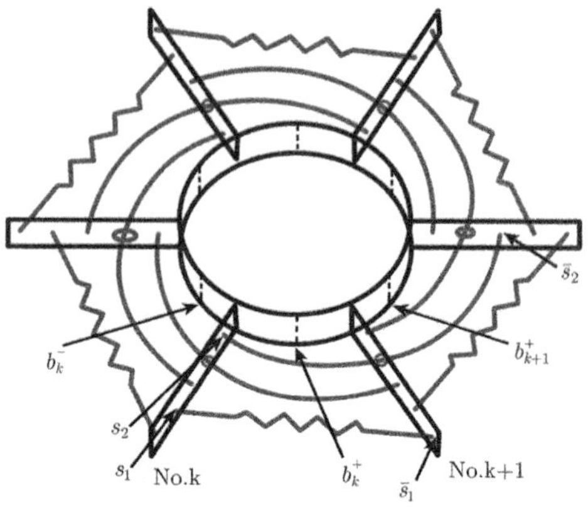

Fig. 7.3 An example of a rotationally periodic structure

rotation about the axis. If all of these functions have the same period $\psi = 2\pi/n$, where n is a positive integer greater than 1, then this structure is called a rotationally periodic structure with n sectors. The rotationally periodic structure is also referred to as structures with cyclic symmetry or period. Figure 7.3 is an illustration of a rotationally periodic structure with 6 sectors, which consists of a circular ring and 6 pieces of thin rectangular plates. There are elastic and rigid constraints between the rectangular plates as well.

For convenience of theoretical analysis, let us adopt the cylindrical coordinate system $Or\theta\, z$ and take the z-axis to be the center axis of the structure. The modal equations of the whole structure may be expressed as the following eigenvalue problem of differential equations

$$\begin{cases} \boldsymbol{L}\boldsymbol{u} = \omega^2 \boldsymbol{M}\boldsymbol{u} & \text{in } \tilde{\Omega}, \\ \boldsymbol{B}\boldsymbol{u} = 0 & \text{on } \partial\tilde{\Omega}, \end{cases} \tag{7.2.1}$$

where ω is the natural angular frequency of the structure, $\boldsymbol{u}$ the mode shape function, $\boldsymbol{L}$ and $\boldsymbol{M}$ the matrices of elastic and inertia differential operator, respectively, $\boldsymbol{B}$ the matrix of the boundary condition operator, $\tilde{\Omega}$ the domain occupied by the structure, and $\partial\tilde{\Omega}$ the boundary of $\tilde{\Omega}$. If the structure contains different types of structural components, such as beams, plates, and three-dimensional elastic solids, then Eq. (7.2.1) includes all equations for the components, boundary conditions, the continuity requirements at their interfaces, as well as other possible inter-component constraints. All of these details are encapsulated in a simple and unified system of equations, i.e., Eq. (7.2.1). The structure should have positive eigenvalues and real eigenfunctions, if it is a valid structure, namely, its mechanical parameters and exterior shape are both smooth and related components are properly connected. See Chap. 8 for details.

Due to cyclic symmetry of the structure, it is possible to reformulate the eigenvalue problem (7.2.1) of the whole structure in terms of those of its substructures.

Set up coordinate systems of the same type in all substructures, and make sure that locations of their origins and orientations of their coordinate axes all preserve the cyclic symmetry of the original structure. In addition, suppose that the k-th substructure is represented by Ω_k, its shared borders with the $(k-1)$-th and $(k+1)$-th substructures are designated, respectively, by b_k^- and b_k^+, and the rest of its boundary is denoted by $\partial\Omega_k$. In the absence of confusion, we may also use Ω, b^-, b^+, and $\partial\Omega$ in lieu of Ω_k, b_k^-, b_k^+, and $\partial\Omega_k$, respectively, to simplify the representation. Here, $k = 1, 2, \ldots, n$, $k+1$ is set to 1 when $k = n$; and $k-1$ is assigned to n when $k = 1$.

Let $\boldsymbol{u}_k$ be the generalized displacement of the k-th substructure. Then, the modal equation and boundary conditions of the whole structure may be rewritten as

$$\begin{cases} \boldsymbol{L}\boldsymbol{u}_k = \omega^2 \boldsymbol{M}\boldsymbol{u}_k & \text{in } \Omega, \\ \boldsymbol{B}\boldsymbol{u}_k = \boldsymbol{0} & \text{on } \partial\Omega, \end{cases} \tag{7.2.2}$$

where $k = 1, 2, \ldots, n$. The continuity conditions for generalized displacements and internal forces at the interface between two adjacent substructures may be expressed as

$$\boldsymbol{J}_0\boldsymbol{u}_k|_{b^+} = \boldsymbol{J}_0\boldsymbol{u}_{k+1}|_{b^-}, \quad k = 1, 2, \ldots, n, \tag{7.2.3}$$

where $\boldsymbol{u}_{n+1} \equiv \boldsymbol{u}_1$ while $\boldsymbol{J}_0$ is a differential operator matrix of the same dimension as $\boldsymbol{u}_k$.

If there exist rigid and elastic constraints between the substructure and others that are either adjacent or nonadjacent, these conditions may be represented by

$$\boldsymbol{J}_{pj}\boldsymbol{u}_k|_{s_{pj}} = \bar{\boldsymbol{J}}_{pj}\boldsymbol{u}_{k+p}|_{\bar{s}_{pj}}, \tag{7.2.4}$$

where $\boldsymbol{J}_{pj}$ and $\bar{\boldsymbol{J}}_{pj}$ are differential operator matrices of the same dimension as $\boldsymbol{u}_k$, $k = 1, 2, \ldots, n$, $p = 1, 2, \ldots, n$, and $j = 0, 1, \ldots, l_p$. Equation (7.2.4) shows constraints between two point sets s_{pj} and $\bar{s}_{pj}$ in the k-th and $(k+p)$-th substructures, respectively, where s_{pj} and $\bar{s}_{pj}$ can be either two points or two one-, two-, or three-dimensional regions. When $k + p = n + r$, set $k + p = r$. If $p = n$, Eq. (7.2.4) represents a self-constraint on the substructure. If, for some p, there is no constraint between the k-th and $(k+p)$-th substructures, i.e., $j = l_p = 0$, then equations in (7.2.4) with this particular subscript p will disappear. Based on the above discussion, there should be a total of $\sum_{p=1}^{n} l_p$ constraints in Eq. (7.2.4). By the way, notations and assumptions used here are also applicable to Eq. (7.2.10) below.

As an example, there is a spring connecting the point s_1 on the k-th substructure and the point $\bar{s}_1$ on the $(k+1)$-th substructure in Fig. 7.3. Thus, the first equation in (7.2.4) represents the spring force at the point s_1 induced by relative movement between points s_1 and $\bar{s}_1$. Furthermore, the point s_2 on the k-th substructure is rigidly connected to the point $\bar{s}_2$ on the $(k+2)$-th substructure. The second equation in (7.2.4) stipulates that displacements at points s_2 and $\bar{s}_2$ must be identical. Other equations in (7.2.4) should vanish. In the case of Fig. 7.3, $p = 1, 2$ while $l_1 = l_2 = 1$.

Because of the continuity condition (7.2.3) as well as the constraint between substructures (7.2.4), the general displacements of substructures $\{\boldsymbol{u}_k\}_1^n$ are all coupled. To obtain natural frequencies and mode shapes directly from Eqs. (7.2.1) to (7.2.4), it is required to solve the coupled systems of equations involving $\boldsymbol{u}_1, \boldsymbol{u}_2, \ldots, \boldsymbol{u}_n$.

7.2.2 Reduction of the Eigenvalue Problem and Discussion on Its Modal Qualitative Properties

As pointed out earlier, the geometric parameters, physical parameters, and boundary conditions of a rotationally periodic structure are all periodic functions of the angle of rotation about the center axis, and they have a common period of $\psi = 2\pi/n$. Thus, it is reasonable to assume that mode shape components of the substructures have the following relationship:

$$\boldsymbol{u}_n = \alpha \boldsymbol{u}_{n-1} = \alpha(\alpha \boldsymbol{u}_{n-2}) = \cdots = \alpha^{n-1}\boldsymbol{u}_1 = \alpha^n \boldsymbol{u}_n,$$

where α is a complex number. The above equations yield

$$\alpha^n = 1.$$

By solving the above equation, we get its n complex roots as follows:

$$\alpha_r = \mathrm{e}^{\mathrm{i}r\psi} = \cos r\psi + \mathrm{i}\sin r\psi, \quad r = 1, 2, \ldots, n,$$

where $\mathrm{i} = \sqrt{-1}$ and $\psi = 2\pi/n$. Notice that $\alpha_n = 1$ while $\alpha_{n/2} = -1$. The r-th group of mode shapes may be expressed as

$$\begin{aligned}\boldsymbol{u}^{(r)} &= (\boldsymbol{u}_1^{(r)}, \boldsymbol{u}_2^{(r)}, \ldots, \boldsymbol{u}_n^{(r)})^{\mathrm{T}} \\ &= \frac{1}{\sqrt{n}}(\boldsymbol{I}, \mathrm{e}^{\mathrm{i}r\psi}\boldsymbol{I}, \mathrm{e}^{\mathrm{i}2r\psi}\boldsymbol{I}, \ldots, \mathrm{e}^{\mathrm{i}(n-1)r\psi}\boldsymbol{I})^{\mathrm{T}}\boldsymbol{q}_r, \quad r = 1, 2, \ldots, n,\end{aligned}$$

where the order of the identity matrix $\boldsymbol{I}$ is the same as the dimension of $\boldsymbol{u}_k$.

Inspired by this, we may introduce the following transformation on the original generalized displacement in order to decouple the modal equation:

$$\boldsymbol{u} = (\boldsymbol{u}_1, \boldsymbol{u}_2, \ldots, \boldsymbol{u}_n)^{\mathrm{T}} = \begin{bmatrix} \boldsymbol{R}_1 & \boldsymbol{R}_2 & \ldots & \boldsymbol{R}_n \end{bmatrix} \begin{bmatrix} \boldsymbol{q}_1 \\ \boldsymbol{q}_2 \\ \vdots \\ \boldsymbol{q}_n \end{bmatrix} = \boldsymbol{R}\boldsymbol{q}, \tag{7.2.5a}$$

$$\boldsymbol{R}_r = \frac{1}{\sqrt{n}}\left[\boldsymbol{I}, \mathrm{e}^{\mathrm{i}r\psi}\boldsymbol{I}, \mathrm{e}^{\mathrm{i}2r\psi}\boldsymbol{I}, \ldots, \mathrm{e}^{\mathrm{i}r(n-1)\psi}\boldsymbol{I}\right]^{\mathrm{T}}, \quad r = 1, 2, \ldots, n, \tag{7.2.5b}$$

where $\boldsymbol{I}$ is an identity matrix whose order equals the dimension of $\boldsymbol{u}_k$. It is easy to demonstrate that $\boldsymbol{R}$ is a unitary matrix, i.e.,

$$\overline{\boldsymbol{R}}^{\mathrm{T}}\boldsymbol{R} = \boldsymbol{I}, \tag{7.2.6}$$

where the order of the identity matrix $\boldsymbol{I}$ here is n times of the dimension of $\boldsymbol{u}_k$, and $\bar{\boldsymbol{R}}$ is the conjugate matrix of $\boldsymbol{R}$.

The modal equation of the whole structure as well as its boundary conditions, continuity requirements, and inter-component constraints, i.e., Equations (7.2.2)–(7.2.4), may be rewritten as

$$\boldsymbol{L}'\boldsymbol{u} - \omega^2\boldsymbol{M}'\boldsymbol{u} = 0 \quad \text{for } \boldsymbol{u}_k \text{ inside } \Omega, \tag{7.2.7}$$

$$\boldsymbol{B}'\boldsymbol{u} = 0 \qquad \text{for } \boldsymbol{u}_k \text{ on } \partial\Omega, \tag{7.2.8}$$

$$\boldsymbol{J}_0'\boldsymbol{u}|_{b^+} = \boldsymbol{J}_0'\boldsymbol{Y}\boldsymbol{u}|_{b^-}, \tag{7.2.9}$$

$$\boldsymbol{J}'_{pj}\boldsymbol{u}|_{s_{pj}} = \overline{\boldsymbol{J}'}_{pj}\boldsymbol{Y}^p\boldsymbol{u}|_{\bar{s}_{pj}}, \quad p = 1, 2, \ldots, n, \quad j = 0, 1, \ldots, l_p, \tag{7.2.10}$$

where $\boldsymbol{L}'$, $\boldsymbol{M}'$, $\boldsymbol{B}'$, $\boldsymbol{J}'_0$, $\boldsymbol{J}'_{pj}$, and $\overline{\boldsymbol{J}}'_{pj}$ are, respectively, block diagonal matrices formed by $\boldsymbol{L}$, $\boldsymbol{M}$, $\boldsymbol{B}$, $\boldsymbol{J}_0$, $\boldsymbol{J}_{pj}$, and $\overline{\boldsymbol{J}}_{pj}$, respectively, while

$$\boldsymbol{Y} = \begin{bmatrix} \mathbf{0} & \overset{2}{\boldsymbol{I}} & & & \\ & \ddots & \ddots & & \\ & & \ddots & \ddots & \\ & & & \mathbf{0} & \boldsymbol{I} \\ \underset{1}{\boldsymbol{I}} & & & & \mathbf{0} \end{bmatrix}, \quad \boldsymbol{Y}^p = \begin{bmatrix} & & \overset{p+1}{\boldsymbol{I}} & & \\ & & & \ddots & \\ & & & & \boldsymbol{I} \\ \boldsymbol{I} & & & & \\ & \ddots & & & \\ & & \underset{p}{\boldsymbol{I}} & & \end{bmatrix}. \tag{7.2.11}$$

Here, the order of the identity matrix $\boldsymbol{I}$ is the same as the dimension of $\boldsymbol{u}_k$, and all block submatrices in $\boldsymbol{Y}$ (namely $\boldsymbol{Y}^1$) and $\boldsymbol{Y}^p$ that are not shown in Eq. (7.2.11) contain zero element only. $\boldsymbol{Y}^p$ is a row-permutation matrix for a block of rows. It is easy to derive the following equation:

$$\overline{\boldsymbol{R}}^{\mathrm{T}}\boldsymbol{Y}^p\boldsymbol{R} = \mathrm{diag}(\mathrm{e}^{\mathrm{i}p\psi}\boldsymbol{I}, \mathrm{e}^{\mathrm{i}2p\psi}\boldsymbol{I}, \ldots, \mathrm{e}^{\mathrm{i}np\psi}\boldsymbol{I}). \tag{7.2.12}$$

By inserting the coordinate transformation (7.2.5a) into Eqs. (7.2.7)–(7.2.10), left multiplying the resulting equations by $\overline{\boldsymbol{R}}^{\mathrm{T}}$, and also taking into account Eqs. (7.2.6) and (7.2.12), we get

$$\begin{cases} \boldsymbol{L}\boldsymbol{q}_r - \omega^2 \boldsymbol{M}\boldsymbol{q}_r = \boldsymbol{0} & \text{in } \Omega, \\ \boldsymbol{B}\boldsymbol{q}_r = \boldsymbol{0} & \text{on } \partial\Omega, \\ \boldsymbol{J}_0 \boldsymbol{q}_r|_{b^+} = \boldsymbol{J}_0\, \mathrm{e}^{\mathrm{i}r\psi} \boldsymbol{q}_r|_{b^-},\ \boldsymbol{J}_{pj}\boldsymbol{q}_r|_{s_{pj}} = \overline{\boldsymbol{J}}_{pj}\, \mathrm{e}^{\mathrm{i}pr\psi} \boldsymbol{q}_r|_{\bar{s}_{pj}}, \end{cases} \tag{7.2.13}$$

where $r = 1, 2, \ldots, n$; $p = 1, 2, \ldots, n$; and $j = 0, 1, \ldots, l_p$. Notice that $\boldsymbol{q}_r (r = 1, 2, \ldots, n)$ are decoupled in the system of equations above, and their eigenvalues are all positive. Since $\boldsymbol{q}_r$ may be a complex eigenfunction under general conditions, we set $\boldsymbol{q}_r = \boldsymbol{q}_r^{\mathrm{r}} + \mathrm{i}\boldsymbol{q}_r^{\mathrm{i}}$, where the superscripts r and i indicate the real and imaginary parts, respectively, and the same notation applies to subsequent discussion. After plugging the expression for $\boldsymbol{q}_r$ into Eq. (7.2.13), we get

$$\boldsymbol{L}\boldsymbol{q}_r^{\mathrm{r}} - \omega^2 \boldsymbol{M}\boldsymbol{q}_r^{\mathrm{r}} = \boldsymbol{0},\quad \boldsymbol{L}\boldsymbol{q}_r^{\mathrm{i}} - \omega^2 \boldsymbol{M}\boldsymbol{q}_r^{\mathrm{i}} = \boldsymbol{0},\quad \text{in } \Omega, \tag{7.2.14}$$

$$\boldsymbol{B}\boldsymbol{q}_r^{\mathrm{r}} = \boldsymbol{0},\quad \boldsymbol{B}\boldsymbol{q}_r^{\mathrm{i}} = \boldsymbol{0},\quad \text{on } \partial\Omega, \tag{7.2.15}$$

$$\begin{cases} \boldsymbol{J}_0 \boldsymbol{q}_r^{\mathrm{r}}|_{b^+} = \boldsymbol{J}_0 (\cos r\psi\, \boldsymbol{q}_r^{\mathrm{r}} - \sin r\psi\, \boldsymbol{q}_r^{\mathrm{i}})|_{b^-}, \\ \boldsymbol{J}_0 \boldsymbol{q}_r^{\mathrm{i}}|_{b^+} = \boldsymbol{J}_0 (\sin r\psi\, \boldsymbol{q}_r^{\mathrm{r}} + \cos r\psi\, \boldsymbol{q}_r^{\mathrm{i}})|_{b^-}, \end{cases} \tag{7.2.16}$$

$$\begin{cases} \boldsymbol{J}_{pj} \boldsymbol{q}_r^{\mathrm{r}}|_{s_{pj}} = \bar{\boldsymbol{J}}_{pj} (\cos rp\psi\, \boldsymbol{q}_r^{\mathrm{r}} - \sin rp\psi\, \boldsymbol{q}_r^{\mathrm{i}})|_{\bar{s}_{pj}},\ r = 1, 2, \ldots, n, \\ \boldsymbol{J}_{pj} \boldsymbol{q}_r^{\mathrm{i}}|_{s_{pj}} = \bar{\boldsymbol{J}}_{pj} (\sin rp\psi\, \boldsymbol{q}_r^{\mathrm{r}} + \cos rp\psi\, \boldsymbol{q}_r^{\mathrm{i}})|_{\bar{s}_{pj}},\ p = 1, 2, \ldots, n, \end{cases} \quad j = 0, 1, \ldots, l_p. \tag{7.2.17}$$

In the system of equations above, $\boldsymbol{q}_r^{\mathrm{r}}$ and $\boldsymbol{q}_r^{\mathrm{i}}$ are coupled. It may be verified from Eq. (7.2.13) that the r-th and $(n - r)$-th real eigenvalues are identical while their corresponding complex eigenfunctions $\boldsymbol{q}_r = \boldsymbol{q}_r^{\mathrm{r}} + \mathrm{i}\, \boldsymbol{q}_r^{\mathrm{i}}$ and $\boldsymbol{q}_{n-r} = \boldsymbol{q}_r^{\mathrm{r}} - \mathrm{i}\, \boldsymbol{q}_r^{\mathrm{i}}$ are a pair of conjugates. By adding and subtracting the two complex functions $\boldsymbol{q}_r$ and $\boldsymbol{q}_{n-r}$, we get two real functions $\boldsymbol{q}_r^{\mathrm{r}}$ and $\boldsymbol{q}_r^{\mathrm{i}}$, respectively. Thus, it only needs to find solutions corresponding to $r = 1, 2, \ldots, (n - 1)/2$ when n is an odd number, or to get those solutions associated with $r = 1, 2, \ldots, n/2$ when n is an even number. In the special cases of $r = n$ and $n/2$ (only when n is an even number), the solution $\boldsymbol{q}_r$ to Eq. (7.2.13) ought to be a real function.

Now we are ready to draw the following conclusions:

First, for a rotationally periodic structure, its natural vibration problem (7.2.1)–(7.2.4) may be reduced to n natural vibration problems (7.2.14)–(7.2.17) of one substructure. After $\boldsymbol{q}_r^{\mathrm{r}}$ and $\boldsymbol{q}_r^{\mathrm{i}}$ are calculated, the mode shape of the entire structure $\boldsymbol{u}^{(r)} = \boldsymbol{v}^{(r)} + \mathrm{i}\boldsymbol{w}^{(r)}$ may be obtained from the equation below:

$$\begin{bmatrix} \boldsymbol{v}^{(r)} \\ \boldsymbol{w}^{(r)} \end{bmatrix} = \begin{bmatrix} \boldsymbol{v}_1^{(r)} \\ \vdots \\ \boldsymbol{v}_n^{(r)} \\ \boldsymbol{w}_1^{(r)} \\ \vdots \\ \boldsymbol{w}_n^{(r)} \end{bmatrix} = \begin{bmatrix} \boldsymbol{I} & \boldsymbol{0} \\ \cos r\psi \boldsymbol{I} & -\sin r\psi \boldsymbol{I} \\ \vdots & \vdots \\ \cos r(n-1)\psi \boldsymbol{I} & -\sin r(n-1)\psi \boldsymbol{I} \\ \boldsymbol{0} & \boldsymbol{I} \\ \sin r\psi \boldsymbol{I} & \cos r\psi \boldsymbol{I} \\ \vdots & \vdots \\ \sin r(n-1)\psi \boldsymbol{I} & \cos r(n-1)\psi \boldsymbol{I} \end{bmatrix} \begin{bmatrix} \boldsymbol{q}_r^{\mathrm{r}} \\ \boldsymbol{q}_r^{\mathrm{i}} \end{bmatrix}, \tag{7.2.18}$$

where the identify matrix $\boldsymbol{I}$ is of the order equal to the dimension as $\boldsymbol{u}_k$.

Second, the n groups of mode shapes of a rotationally periodic structure all have the property shown in Eq. (7.2.5), i.e., the following relationship holds for mode shape components of two adjacent substructures:

$$\boldsymbol{u}_{k+1}^{(r)} = \mathrm{e}^{\mathrm{i}r\psi} \boldsymbol{u}_k^{(r)}. \tag{7.2.19}$$

Results similar to Eq. (7.2.19) can also be established for other mechanical variables such as stresses and internal forces. Since continuity conditions at the two shared boundaries of a substructure, i.e., b^- and b^+, often involve the displacement and other mechanical variables, Eq. (7.2.19) and the like provide the basis for us to set up the continuity equations. We will proceed by dividing the mode shapes into three groups, based on the value of r:

Group 1 For the case $r = n$, all substructures share the same mode shape component, namely,

$$\boldsymbol{u}^{(n)} = (\boldsymbol{q}_n, \boldsymbol{q}_n, \ldots, \boldsymbol{q}_n)^{\mathrm{T}}. \tag{7.2.20}$$

This implies that components of the displacement mode shape are identical at the two shared boundaries of the k-th substructure, i.e., $\boldsymbol{u}_k^{(n)}\big|_{b^+} = \boldsymbol{u}_k^{(n)}\big|_{b^-}$. Furthermore, components of other mechanical variables involved in the continuity conditions should also be identical at the shared boundaries b^- and b^+. In this case, we have $\mathrm{e}^{\mathrm{i}r\psi} = 1$ in the modal equation (7.2.13).

Group 2 When n is an even number and $r = n/2$, components of the displacement mode shape of adjacent substructures are of equal magnitude but opposite signs, namely,

$$\boldsymbol{u}^{(n/2)} = (\boldsymbol{q}_{n/2}, -\boldsymbol{q}_{n/2}, \ldots, \boldsymbol{q}_{n/2}, -\boldsymbol{q}_{n/2})^{\mathrm{T}}. \tag{7.2.21}$$

This implies that mode shape components have the same magnitude but different signs at the two shared boundaries of the k-th substructure, i.e., $\left.\boldsymbol{u}_k^{(n/2)}\right|_{b^+} = -\left.\boldsymbol{u}_k^{(n/2)}\right|_{b^-}$. Furthermore, components of other mechanical variables involved in the continuity conditions should also be identical in magnitude but opposite in sign at the shared boundaries b^- and b^+. In this case, we get $\mathrm{e}^{\mathrm{i}r\psi} = -1$ in the modal equation (7.2.13).

Group 3 When r differs from either n or $n/2$ (only if n is an even number), eigenvalues of Eq. (7.2.13) are repeated, and for each repeated eigenvalue, there is a pair of complex eigenfunctions that are conjugate to each other. Since the sum or difference of two solutions to Eq. (7.2.13) remains a solution, real and imaginary parts of any of the two complex eigenfunctions should also be eigenfunctions. Thus, for each repeated frequency, we may get the following two sets of displacement mode shapes of the entire structure, which are real functions:

$$\boldsymbol{v}^{(r)} = \left(\boldsymbol{v}_1^{(r)}, \boldsymbol{v}_2^{(r)}, \ldots, \boldsymbol{v}_n^{(r)}\right)^{\mathrm{T}} \quad \text{and} \quad \boldsymbol{w}^{(r)} = \left(\boldsymbol{w}_1^{(r)}, \boldsymbol{w}_2^{(r)}, \ldots, \boldsymbol{w}_n^{(r)}\right)^{\mathrm{T}},$$

where mode shape components $\boldsymbol{v}_k^{(r)}$ and $\boldsymbol{w}_k^{(r)} (k = 1, 2, \ldots, n)$ are calculated from Eq. (7.2.18) for any given r, and $r = 1, 2, \ldots, (n-2)/2$ (n is even) or $(n-1)/2$ (n is odd). Components of the displacement mode shapes from two substructures next to each other are related by

$$\begin{bmatrix} \boldsymbol{v}_{k+1}^{(r)} \\ \boldsymbol{w}_{k+1}^{(r)} \end{bmatrix} = \begin{bmatrix} \cos r\psi & -\sin r\psi \\ \sin r\psi & \cos r\psi \end{bmatrix} \begin{bmatrix} \boldsymbol{v}_k^{(r)} \\ \boldsymbol{w}_k^{(r)} \end{bmatrix}. \tag{7.2.22}$$

Based on (7.2.22), we may easily identify conditions that must be satisfied by components of the displacement mode shapes at the two shared boundaries of the k-th substructure as follows

$$\begin{cases} \left.\boldsymbol{v}_k^{(r)}\right|_{b^+} = \cos r\psi \cdot \left.\boldsymbol{v}_k^{(r)}\right|_{b^-} - \sin r\psi \cdot \left.\boldsymbol{w}_k^{(r)}\right|_{b^-}, \\ \left.\boldsymbol{w}_k^{(r)}\right|_{b^+} = \sin r\psi \cdot \left.\boldsymbol{v}_k^{(r)}\right|_{b^-} + \cos r\psi \cdot \left.\boldsymbol{w}_k^{(r)}\right|_{b^-}. \end{cases}$$

Similar equations can be established for components of other mechanical variables involved in the continuity conditions at the shared boundaries b^- and b^+.

7.2.3 Applications

When computing or measuring modes of a rotationally periodic structure, we should take advantage of its qualitative properties.

First, natural frequencies and mode shapes of a rotationally periodic structure may be calculated in two steps as follows:

(1) Determine coupled vectors $\boldsymbol{q}_r^{\mathrm{r}}$ and $\boldsymbol{q}_r^{\mathrm{i}}$ by solving the real eigenvalue problem defined by Eqs. (7.2.14) to (7.2.17).
(2) The mode shape of the whole structure may be obtained by extending the solution associated with a substructure, $\boldsymbol{q}_r^{\mathrm{r}}$ and $\boldsymbol{q}_r^{\mathrm{i}}$, to the entire structure using Eq. (7.2.18). It is worth noting that when $r = n$ or $r = n/2$ (if n is an even number), both Eq. (7.2.13) and its solution are real.

By taking advantage of cyclic symmetry and the associated qualitative properties, we can avoid performing modal analysis on the entire structure. Since all analysis needs only to be done at the substructure level, this allows for a significant reduction in computation effort.

Depending on the value of the parameter r, we should follow different approach when performing analysis. Specifically, when $r = n$, it is clear from Eq. (7.2.20) that we only need to analyse one substructure with the requirement that displacements and some other mechanical variables are identical at its two shared borders; when $r = n/2$ and n is an even integer, it is known from Eq. (7.2.21) that we also need to evaluate just one substructure with the requirement that displacements and some other mechanical variables are "opposite" at its two shared borders; and when r is any different positive integer, we need to solve equations (7.2.14)–(7.2.17) of one substructure for two displacement mode shapes associated with a repeated frequency.

Second, many experimental methods are available to measure vibrational modes. Modal qualitative properties of rotationally periodic structures can be exploited in any of them. This helps achieve dramatic reduction in work load as well as significant saving in time and cost.

Example 1 A planar rigid frame consisting of four beams.
Consider the planar frame structure depicted in Fig. 7.4. It consists of four identical beams of uniform cross-section. These beams are rigidly connected and form the shape of a square. Suppose the length and bending stiffness of each beam are, respectively, l and EJ. The frame has pinned supports at four corners. In addition, there are two springs in series connecting the k -th and $(k + 2)$ -th beams at their mid-points. The elastic constant of each spring is K. Let $\boldsymbol{u}$ represent the planer displacement of the beam in the transverse direction. Given below are continuity and other support conditions between the k-th and other relevant beams:

$$
\begin{aligned}
&u_k(l) = u_{k+1}(0) = 0, \quad u_k'(l) = u_{k+1}'(0), \quad u_k''(l) = u_{k+1}''(0), \\
&EJ\left[u_k'''\left(\tfrac{l}{2}+0\right) - u_k'''\left(\tfrac{l}{2}-0\right)\right] + Ku_k\left(\tfrac{l}{2}\right) = -Ku_{k+2}\left(\tfrac{l}{2}\right).
\end{aligned}
$$

Here, $k = 1, 2, 3$, and 4; and whenever it occurs $k + i > 4$, $(k + i)$ should be reset to $(k + i - 4)$.

Based on the problem specification, mode shapes and frequencies of the whole structure ought to be divided into three groups and will be discussed in details in the next three paragraphs.

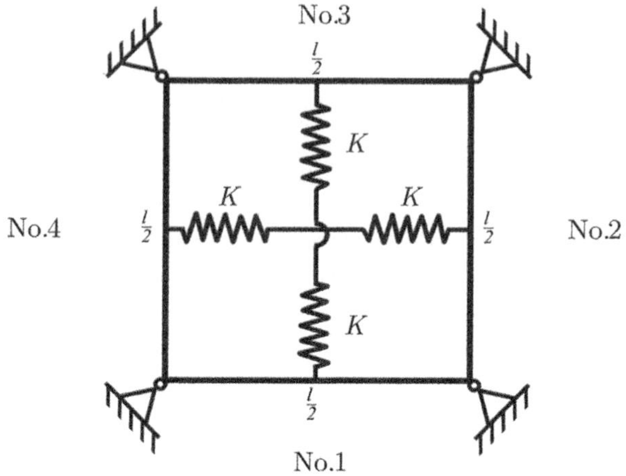

Fig. 7.4 The planar frame consisting of 4 beams. Only movement of the structure within the plane is considered

Fig. 7.5 Equivalent structures of the planar frame shown in Fig. 7.4

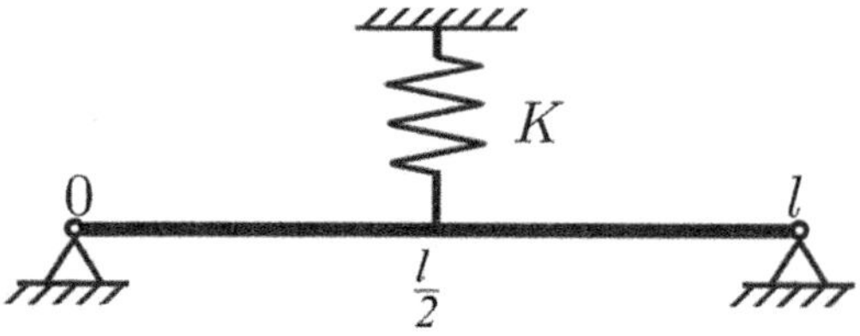

In the first group, mode shape components of adjacent substructures are related by Eq. (7.2.19) with $r = 4$. q_4 may be viewed as the mode shape of the beam shown in Fig. 7.5. Its boundary conditions and constraints are, respectively,

$$q_4(0) = q_4(l) = 0, \quad q_4'(0) = q_4'(l), \quad q_4''(0) = q_4''(l),$$
$$EJ\left[q_4'''\left(\tfrac{l}{2}+0\right) - q_4'''\left(\tfrac{l}{2}-0\right)\right] = -2Kq_4\left(\tfrac{l}{2}\right).$$

The corresponding mode shape of the entire structure is

$$\boldsymbol{u}^{(4)} = (u_1^{(4)}, u_2^{(4)}, u_3^{(4)}, u_4^{(4)})^{\mathrm{T}} = \begin{bmatrix} 1, & 1, & 1, & 1 \end{bmatrix}^{\mathrm{T}} q_4(x).$$

In the second group, mode shape components of adjacent substructures are related by Eq. (7.2.19) with $r = 2$. q_2 can be viewed as the mode shape of the beam illustrated in Fig. 7.5, whose boundary conditions and constraints are, respectively,

$$q_2(0) = q_2(l) = 0, \quad q_2'(0) = -q_2'(l), \quad q_2''(0) = -q_2''(l),$$
$$EJ\left[q_2'''\left(\tfrac{l}{2}+0\right) - q_2'''\left(\tfrac{l}{2}-0\right)\right] = -2Kq_2\left(\tfrac{l}{2}\right).$$

The corresponding mode shape of the whole structure is

$$\boldsymbol{u}^{(2)} = (u_1^{(2)}, u_2^{(2)}, u_3^{(2)}, u_4^{(2)})^{\mathrm{T}} = [1, -1, 1, -1]^{\mathrm{T}} q_2(x).$$

In the third and last group, mode shape components of adjacent substructures are related by Eq. (7.2.19) with $r = 1$. Here we are dealing with a complex eigenvalue problem with real eigenvalues. It may also be reformulated into two coupled real eigenvalue problems, where their mode shapes, q_1^{r} and q_1^{i}, represent real and imaginary parts of the aforementioned complex eigenvalue problem, respectively, and are subject to the following coupled boundary and constraint conditions:

$$q_1^{\mathrm{r}}(0) = q_1^{\mathrm{r}}(l) = 0, \quad q_1^{\mathrm{i}}(0) = q_1^{\mathrm{i}}(l) = 0, \quad (q_1^{\mathrm{r}})'|_l = -(q_1^{\mathrm{i}})'|_0, \quad (q_1^{\mathrm{i}})'|_l = (q_1^{\mathrm{r}})'|_0,$$
$$(q_1^{\mathrm{r}})''|_l = -(q_1^{\mathrm{i}})''|_0, \quad (q_1^{\mathrm{i}})''|_l = (q_1^{\mathrm{r}})''|_0.$$

Example 2 This case is obtained after removing all springs in Fig. 7.4, which is covered in Example 1 in this subsection. We can also obtain three groups of mode shapes, as will be discussed in the rest of this example.

The first group corresponds to $r = 4$ in Eq. (7.2.19). Its mode shape

$$\boldsymbol{u}^{(4)} = (u_1^{(4)}, u_2^{(4)}, u_3^{(4)}, u_4^{(4)})^{\mathrm{T}} = (q_4, q_4, q_4, q_4)^{\mathrm{T}}$$

is related to the natural angular frequency $\omega^{(4)}$. There are two sets of boundary conditions for the respective model consisting of a single beam, shown in the top two subfigures of Fig. 7.6:

$$q_4'(0) = q_4'(l) = 0, \quad q_4''(0) = q_4''(l),$$

and

$$q_4'(0) = q_4'(l), \quad q_4''(0) = q_4''(l) = 0.$$

The first four mode shapes of the entire structure are depicted in the bottom four subfigures of Fig. 7.6.

The second group corresponds to $r = 2$ in Eq. (7.2.19). Its mode shape $\boldsymbol{u}^{(2)} = (u_1^{(2)}, u_2^{(2)}, u_3^{(2)}, u_4^{(2)})^{\mathrm{T}} = (q_2, -q_2, q_2, -q_2)^{\mathrm{T}}$ is related to the natural angular frequency $\omega^{(2)}$. There are also two sets of boundary conditions for the respective model consisting of a single beam, shown in the top two subfigures of Fig. 7.7:

$$q_2'(0) = -q_2'(l), \quad q_2''(0) = -q_2''(l) = 0,$$

and

$$q_2'(0) = q_2'(l) = 0, \quad q_2''(0) = -q_2''(l).$$

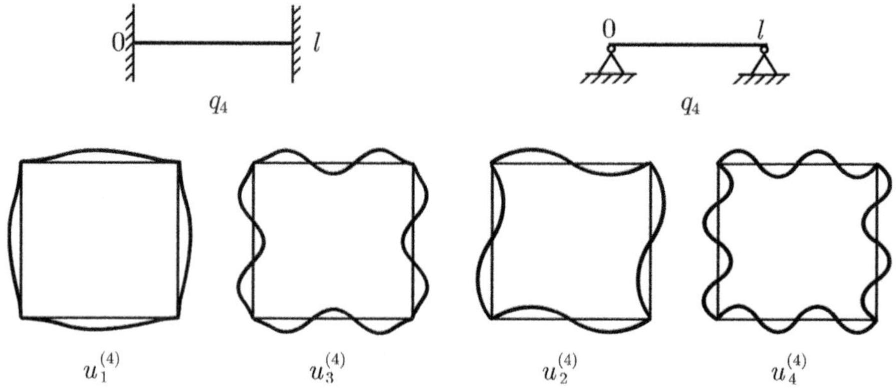

Fig. 7.6 The first group of mode shapes

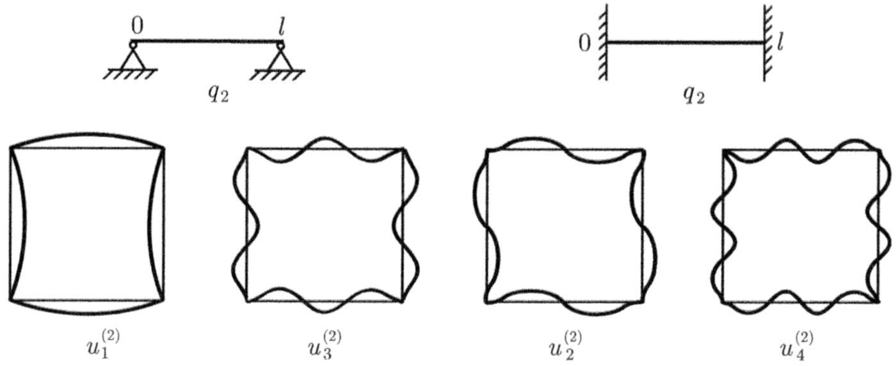

Fig. 7.7 The second group of mode shapes

The first four mode shapes for the whole frame are graphed in the bottom four subfigures of Fig. 7.7.

The third group should be further divided into two subgroups and are covered in the next two paragraphs.

The first subgroup corresponds to $r = 1$ in Eq. (7.2.19). Mode shapes

$$\boldsymbol{v}^{(1)} = \begin{bmatrix} v_1^{(1)} \\ v_2^{(1)} \\ v_3^{(1)} \\ v_4^{(1)} \end{bmatrix} = \begin{bmatrix} 1 & 0 \\ 0 & -1 \\ -1 & 0 \\ 0 & 1 \end{bmatrix} \begin{bmatrix} q_1^{\mathrm{r}} \\ q_1^{\mathrm{i}} \end{bmatrix} \quad \text{and} \quad \boldsymbol{w}^{(1)} = \begin{bmatrix} w_1^{(1)} \\ w_2^{(1)} \\ w_3^{(1)} \\ w_4^{(1)} \end{bmatrix} = \begin{bmatrix} 0 & 1 \\ 1 & 0 \\ 0 & -1 \\ -1 & 0 \end{bmatrix} \begin{bmatrix} q_1^{\mathrm{r}} \\ q_1^{\mathrm{i}} \end{bmatrix}$$

are related to the repeated frequency $\omega^{(1)}$, and we have $\omega^{(1)} = \omega^{(3)}$. Again, there are two sets of boundary conditions for the respective models, each of which are for two separate beams, as shown in the top four subfigures of Fig. 7.8.

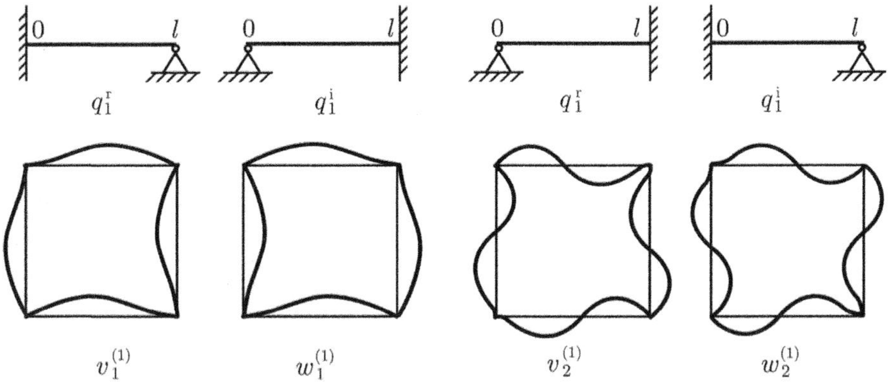

Fig. 7.8 The first subgroup of mode shapes in the third group

$$(q_1^r)'|_0 = (q_1^i)'|_l = 0, \quad (q_1^r)''|_l = -(q_1^i)''|_0 = 0,$$
$$(q_1^r)'|_l = -(q_1^i)'|_0, \quad (q_1^i)''|_l = (q_1^r)''|_0.$$

and

$$(q_1^r)'|_l = -(q_1^i)'|_0 = 0, \quad (q_1^r)''|_0 = (q_1^i)''|_l = 0,$$
$$(q_1^r)'|_0 = (q_1^i)'|_l, \quad (q_1^r)''|_l = -(q_1^i)''|_0.$$

The first two mode shapes for the entire frame are depicted in the bottom four subfigures of Fig. 7.8.

The second subgroup corresponds to $r = 3$ in Eq. (7.2.19). Mode shapes

$$\boldsymbol{v}^{(3)} = \begin{bmatrix} v_1^{(3)} \\ v_2^{(3)} \\ v_3^{(3)} \\ v_4^{(3)} \end{bmatrix} = \begin{bmatrix} 1 & 0 \\ 0 & 1 \\ -1 & 0 \\ 0 & -1 \end{bmatrix} \begin{bmatrix} q_3^r \\ q_3^i \end{bmatrix} \quad \text{and} \quad \boldsymbol{w}^{(3)} = \begin{bmatrix} w_1^{(3)} \\ w_2^{(3)} \\ w_3^{(3)} \\ w_4^{(3)} \end{bmatrix} = \begin{bmatrix} 0 & 1 \\ -1 & 0 \\ 0 & -1 \\ 1 & 0 \end{bmatrix} \begin{bmatrix} q_3^r \\ q_3^i \end{bmatrix}$$

are related to the repeated frequency $\omega^{(3)}$, and we have $\omega^{(3)} = \omega^{(1)}$. Boundary conditions for the corresponding single-beam models as well as the first two mode shapes of the entire structure are shown in Fig. 7.9.

It is clear from Figs. 7.8 and 7.9 that within a constant factor, $\boldsymbol{v}^{(3)}$ and $\boldsymbol{w}^{(3)}$ are identical to $\boldsymbol{v}^{(1)}$ and $-\boldsymbol{w}^{(1)}$, respectively.

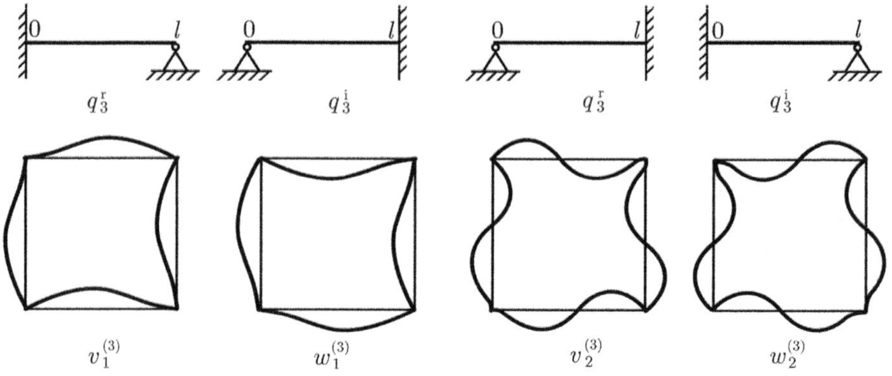

Fig. 7.9 The second subgroup of mode shapes in the third group

7.2.4 The Discrete System and Its Modal Qualitative Properties

When the continuous system is discretized, care should be taken to preserve the cyclic symmetry of the structure. This ensures the resulting discrete system is in a form that can be easily simplified. Consider a rotationally periodic structure comprised of n substructures. Suppose the number of discretization nodes as well as their layout and ordering are identical among all substructures. Also assume the number of degrees of freedom of generalized displacements at one substructure to be p. Then, the modal equation of the whole structure is

$$\boldsymbol{K}\boldsymbol{u} = \omega^2 \boldsymbol{M}\boldsymbol{u}, \tag{7.2.23}$$

where ω is the natural angular frequency, the vector

$$\boldsymbol{u} = (\boldsymbol{u}_1, \boldsymbol{u}_2, \ldots, \boldsymbol{u}_n)^{\mathrm{T}} \tag{7.2.24}$$

represents the mode shape of the entire system, $\boldsymbol{u}_k$ $(k = 1, 2, \ldots, n)$ denotes the mode shape component at the k-th substructure, and $\boldsymbol{K}$ and $\boldsymbol{M}$ are stiffness and mass matrices, respectively. Both $\boldsymbol{K}$ and $\boldsymbol{M}$ are circulant matrices that are real, symmetric, and either positive or semi-positive definite. Specifically,

$$\boldsymbol{K} = \begin{bmatrix} \boldsymbol{K}_{11} & \boldsymbol{K}_{12} & \cdots & \boldsymbol{K}_{1n} \\ \boldsymbol{K}_{1n} & \boldsymbol{K}_{11} & \cdots & \boldsymbol{K}_{1,n-1} \\ \vdots & \vdots & & \vdots \\ \boldsymbol{K}_{12} & \boldsymbol{K}_{13} & \cdots & \boldsymbol{K}_{11} \end{bmatrix}, \tag{7.2.25}$$

where the submatrices of order p satisfy

$$\boldsymbol{K}_{11}^{\mathrm{T}} = \boldsymbol{K}_{11}, \quad \boldsymbol{K}_{1j}^{\mathrm{T}} = \boldsymbol{K}_{1,(n+2-j)}, \quad j = 2, 3, \ldots, n. \tag{7.2.26}$$

The mass matrix $\boldsymbol{M}$ is of the same structure as $\boldsymbol{K}$ in Eq. (7.2.25), and its submatrices also have properties given in Eq. (7.2.26).

Since the complete structure can be generated by rotating one substructure at the angle $\psi = 2\pi/n$ for n times, we may hypothesize that the following relationship holds for mode shape components at the substructure level:

$$\boldsymbol{u}_n = \alpha \boldsymbol{u}_{n-1} = \alpha^2 \boldsymbol{u}_{n-2} = \cdots = \alpha^{n-1} \boldsymbol{u}_1 = \alpha^n \boldsymbol{u}_n. \tag{7.2.27}$$

Thus, $\alpha^n = 1$ and the equation has n complex roots:

$$\alpha_r = \mathrm{e}^{\mathrm{i}r\psi} = \mathrm{e}^{\mathrm{i}r2\pi/n}, \quad r = 1, 2, \ldots, n, \tag{7.2.28}$$

where $\mathrm{i} = \sqrt{-1}$. The r-th group of mode shape may be expressed as

$$\begin{aligned} \boldsymbol{u}^{(r)} &= (\boldsymbol{u}_1^{(r)}, \boldsymbol{u}_2^{(r)}, \ldots, \boldsymbol{u}_n^{(r)})^{\mathrm{T}} \\ &= \frac{1}{\sqrt{n}}(\boldsymbol{I}, \mathrm{e}^{\mathrm{i}r\psi}\boldsymbol{I}, \mathrm{e}^{\mathrm{i}2r\psi}\boldsymbol{I}, \ldots, \mathrm{e}^{\mathrm{i}(n-1)r\psi}\boldsymbol{I})^{\mathrm{T}}\boldsymbol{q}_r, \quad r = 1, 2, \ldots, n, \end{aligned} \tag{7.2.29}$$

where $\boldsymbol{I}$ is an identity matrix of order p while $\boldsymbol{q}_r$ is a p-dimensional vector. Equation (7.2.29) shows that the r-th group of mode shape $\boldsymbol{u}^{(r)}$ for the whole structure, which is a vector of the size $(n \times p)$, depends only on the mode shape component $\boldsymbol{q}_r$ at one substructure, which is a p dimensional vector.

Apply the following transformation to the mode shape $\boldsymbol{u}$:

$$\boldsymbol{u} = \frac{1}{\sqrt{n}} \begin{bmatrix} \boldsymbol{I} & \cdots & \boldsymbol{I} & \cdots & \boldsymbol{I} \\ \mathrm{e}^{\mathrm{i}\psi}\boldsymbol{I} & \cdots & \mathrm{e}^{\mathrm{i}r\psi}\boldsymbol{I} & \cdots & \mathrm{e}^{\mathrm{i}n\psi}\boldsymbol{I} \\ \vdots & & \vdots & & \vdots \\ \mathrm{e}^{\mathrm{i}(n-1)\psi}\boldsymbol{I} & \cdots & \mathrm{e}^{\mathrm{i}(n-1)r\psi}\boldsymbol{I} & \cdots & \mathrm{e}^{\mathrm{i}(n-1)n\psi}\boldsymbol{I} \end{bmatrix} \begin{bmatrix} \boldsymbol{q}_1 \\ \boldsymbol{q}_2 \\ \vdots \\ \boldsymbol{q}_n \end{bmatrix} = \boldsymbol{R}\boldsymbol{q}, \tag{7.2.30}$$

where the complex matrix $\boldsymbol{R}$ is unitary, satisfying

$$\bar{\boldsymbol{R}}^{\mathrm{T}} \boldsymbol{R} = \boldsymbol{I}. \tag{7.2.31}$$

Plug Eq. (7.2.30) into the modal equation (7.2.23), left multiply the resulting equation by $\bar{\boldsymbol{R}}^{\mathrm{T}}$, and take into account Eq. (7.2.26) as well as the equality

$$\sum_{j=1}^{n} \mathrm{e}^{\mathrm{i}j\psi} = 0. \tag{7.2.32}$$

We may finally obtain

$$\boldsymbol{K}_r\boldsymbol{q}_r = \omega^2\boldsymbol{M}_r\boldsymbol{q}_r, \quad r = 1, 2, \ldots, n, \tag{7.2.33}$$

where

$$\boldsymbol{K}_r = \sum_{j=1}^{n}\boldsymbol{K}_{1j}\mathrm{e}^{\mathrm{i}(j-1)r\psi}, \quad \boldsymbol{M}_r = \sum_{j=1}^{n}\boldsymbol{M}_{1j}\mathrm{e}^{\mathrm{i}(j-1)r\psi}, \quad r = 1, 2, \ldots, n. \tag{7.2.34}$$

In the next a few paragraphs, we are going to further analyze the eigenvalue problem defined by Eqs. (7.2.33) and (7.2.34).

First, when $r = n$ or $r = n/2$ (if n is an even number), both $\boldsymbol{K}_r$ and $\boldsymbol{M}_r$ are real symmetric matrices. Therefore, the eigenvalue ω^2 is a real number, while the related eigenvector $\boldsymbol{q}_r$ is a real vector. It is known from Eq. (7.2.29) that the corresponding eigenvectors $\boldsymbol{u}^{(n)}$ and $\boldsymbol{u}^{(n/2)}$ of the entire structure must be real vectors, too.

Second, since $\mathrm{e}^{\mathrm{i}(n-j)r\psi} = \mathrm{e}^{\mathrm{i}nr\psi}\mathrm{e}^{-\mathrm{i}jr\psi} = \mathrm{e}^{-\mathrm{i}jr\psi} = \overline{\mathrm{e}^{\mathrm{i}jr\psi}}$ $(j = 1, 2, \ldots, n)$, we get

$$\boldsymbol{K}_{n-r} = \overline{\boldsymbol{K}}_r, \quad \boldsymbol{M}_{n-r} = \overline{\boldsymbol{M}}_r.$$

Hence, for the $(n-r)$-th and r-th groups of modes, their eigenvalues are identical while their eigenvectors are conjugates of each other, i.e., $\boldsymbol{q}_{n-r} = \overline{\boldsymbol{q}}_r$.

Third, when $r \neq n$ and $r \neq n/2$ (if n is an even number), we get a complex eigenvector $\boldsymbol{q}_r = \boldsymbol{q}_r^{\mathrm{r}} + \mathrm{i}\boldsymbol{q}_r^{\mathrm{i}}$. The complex eigenvectors $\boldsymbol{u}^{(r)} = \boldsymbol{v}^{(r)} + \mathrm{i}\boldsymbol{w}^{(r)}$ of the eigenvalue problem (7.2.23) associated with the entire structure may be obtained from Eq. (7.2.29). Specifically,

$$\boldsymbol{v}^{(r)} = \frac{1}{\sqrt{n}}\begin{bmatrix} \boldsymbol{q}_r^{\mathrm{r}} \\ \cos r\psi\boldsymbol{q}_r^{\mathrm{r}} - \sin r\psi\boldsymbol{q}_r^{\mathrm{i}} \\ \vdots \\ \cos(n-1)r\psi\boldsymbol{q}_r^{\mathrm{r}} - \sin(n-1)r\psi\boldsymbol{q}_r^{\mathrm{i}} \end{bmatrix},$$

$$\boldsymbol{w}^{(r)} = \frac{1}{\sqrt{n}}\begin{bmatrix} \boldsymbol{q}_r^{\mathrm{i}} \\ \sin r\psi\boldsymbol{q}_r^{\mathrm{r}} + \cos r\psi\boldsymbol{q}_r^{\mathrm{i}} \\ \vdots \\ \sin(n-1)r\psi\boldsymbol{q}_r^{\mathrm{r}} + \cos(n-1)r\psi\boldsymbol{q}_r^{\mathrm{i}} \end{bmatrix}.$$

Since the stiffness and mass matrices, $\boldsymbol{K}$ and $\boldsymbol{M}$, respectively, of the whole structure are both real and symmetric, all eigenvalues must be real. In addition, $\overline{\boldsymbol{u}^{(r)}} = \boldsymbol{v}^{(r)} - \mathrm{i}\boldsymbol{w}^{(r)}$ and $\boldsymbol{u}^{(r)}$ are eigenvectors associated with the same eigenvalue. By adding or subtracting the two eigenvectors, we may find $\boldsymbol{v}^{(r)}$ and $\boldsymbol{w}^{(r)}$ to be real eigenvectors associated with the repeated eigenvalue of Eq. (7.2.23).

Finally, when the vector $\boldsymbol{u}$ is set to $\boldsymbol{u}^{(r)}$ specified in Eq. (7.2.29), bilinear forms associated with eigenvalue problems (7.2.23) and (7.2.33) are related by

$$\overline{\boldsymbol{u}^{(r)}}^{\mathrm{T}} \boldsymbol{K} \boldsymbol{u}^{(r)} = \overline{\boldsymbol{q}_r}^{\mathrm{T}} \boldsymbol{K}_r \boldsymbol{q}_r, \quad \overline{\boldsymbol{u}^{(r)}}^{\mathrm{T}} \boldsymbol{M} \boldsymbol{u}^{(r)} = \overline{\boldsymbol{q}_r}^{\mathrm{T}} \boldsymbol{M}_r \boldsymbol{q}_r.$$

Thus, $\boldsymbol{K}_r$, like its counterpart $\boldsymbol{K}$ of the whole structure, is either semi-positive definite or positive definite. Similarly, $\boldsymbol{M}_r$, like its counterpart $\boldsymbol{M}$ of the entire structure, is positive definite. As a result, the eigenvalue ω^2 associated with Eqs. (7.2.23) and (7.2.33) is non-negative, and thus we can find a physical solution $\omega \geq 0$. In order to obtain the r-th group of mode shape of the entire system, we need to solve for $\boldsymbol{q}_r$ from Eqs. (7.2.33) and (7.2.34), and then substitute the result into Eq. (7.2.29).

Based on the above discussion, we can obtain the following modal qualitative properties of the discrete system of a rotationally periodic structure:

(1) Mode shape components at any two adjacent substructures are identical, i.e., mode shape components at all substructures are the same. This corresponds to the case where $r = n$ in Eq. (7.2.29).
(2) When n is an even number, mode shape components of any two adjacent substructures are of the same magnitude but opposite signs. This corresponds to the case of $r = n/2$ in Eq. (7.2.29).
(3) Let

$$\boldsymbol{v}^{(r)} = (\boldsymbol{v}_1^{(r)}, \boldsymbol{v}_2^{(r)}, \ldots \boldsymbol{v}_n^{(r)})^{\mathrm{T}} \quad \text{and} \quad \boldsymbol{w}^{(r)} = (\boldsymbol{w}_1^{(r)}, \boldsymbol{w}_2^{(r)}, \ldots \boldsymbol{w}_n^{(r)})^{\mathrm{T}}$$

denote two sets of mode shapes corresponding to a repeated frequency. They and their counterparts of an adjacent substructure have the following relationship:

$$\begin{aligned} \boldsymbol{v}_{k+1}^{(r)} &= \cos r\psi\, \boldsymbol{v}_k^{(r)} - \sin r\psi\, \boldsymbol{w}_k^{(r)}, \\ \boldsymbol{w}_{k+1}^{(r)} &= \sin r\psi\, \boldsymbol{v}_k^{(r)} + \cos r\psi\, \boldsymbol{w}_k^{(r)}. \end{aligned}$$

The two sets of mode shapes, $\boldsymbol{v}^{(r)}$ and $\boldsymbol{w}^{(r)}$, correspond to the case of $r \neq n$ and $r \neq n/2$ (when n is an even number) in Eq. (7.2.29).

Some important conclusions can be drawn based on the qualitative properties listed above. First, modal qualitative properties of a discrete system are in agreement with those of its continuous counterpart. Second, by taking advantage of modal qualitative properties, we find a way to reduce the problem size when calculating natural frequencies and mode shapes of a discrete system. Specifically, for a rotationally periodic structure with n substructures, we may reduce the original generalized matrix eigenvalue problem, involving real symmetric, positive definite ($\boldsymbol{K}$ may be semi-positive definite though), and square matrices of order $(n \times p)$, to a series of smaller generalized matrix eigenvalue problems, consisting of either 1 (n is odd) or 2 (n is even) p-dimensional problems plus either $(n-1)/2$ (n is odd) or $(n-2)/2$ (n is even) problems of $2p$ dimensions. Here, just like the original problem, all these reduced problems have real symmetric and positive definite ($\boldsymbol{K}_r$ may be semi-positive definite) matrices. When n becomes large, this approach leads to a very significant reduction in problem size and computation effort.

7.3 Modal Qualitative Properties of Linearly Periodic Structures

Suppose geometric parameters, physical parameters, and boundary conditions of a structure are all functions of the location variable along a straight line. If these functions share the same period of length l, then this structure is called a linearly periodic structure.

In certain situations, a linearly periodic structure may be extended, and the extended version may be treated as if it is a rotationally periodic structure, which is referred to as the extended rotationally periodic structure of the original system. If some mode shapes of the extended rotationally periodic structure happen to satisfy boundary conditions of the original linearly periodic system, then they are also mode shapes of this linearly periodic structure. Furthermore, qualitative properties associated with these common mode shapes should be shared by the extended rotationally periodic structure and the original linearly periodic structures.

Example Consider a multi-span beam, where the spans are of equal length and identical shape and have the same bending stiffness. Suppose there are pinned supports at the ends of every span. As an example, we will conduct analysis on a two-span beam clamped at the left end and pinned at the right end. The original two-span structure is illustrated in Fig. 7.10a, while the extended eight-span beam is displayed in Fig. 7.10b.

The extended rotationally periodic structure is shown in Fig. 7.11, which consists of 8 substructures. In this case, those mode shapes of the extended rotationally periodic structure that are symmetric about the diagonal line AC and anti-symmetric about the other diagonal line BD satisfy clamped and pinned constraints at points A and B, respectively. Therefore, components of these types of mode shapes at the first and second substructures should be mode shapes of the original linearly periodic structure.

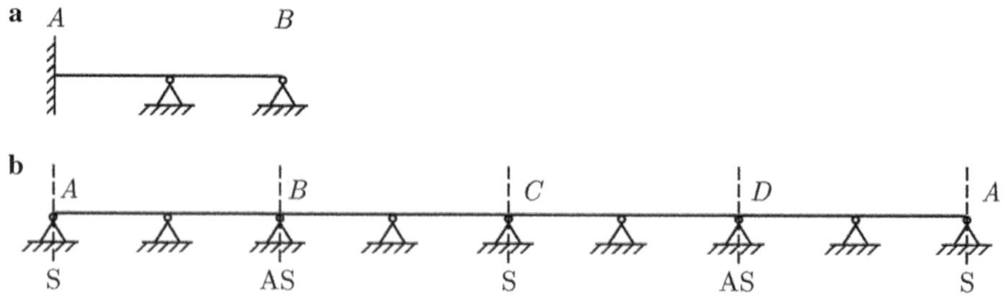

Fig. 7.10 **a** The original two-span linearly periodic structure, **b** the extended eight-span linearly periodic structure

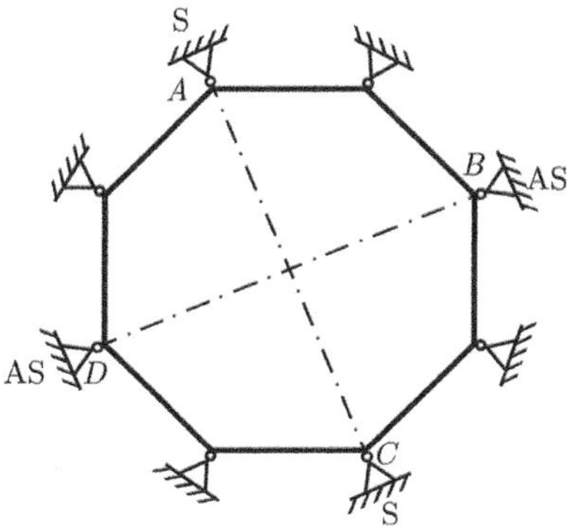

Fig. 7.11 Extended rotationally periodic structure

7.4 Modal Qualitative Properties of Chain Structures

7.4.1 *The Model and Governing Equations*

A chain structure refers to the system consisting of multiple identical substructures that are sequentially connected together in a specific way: There is no direct interface or connection boundary between substructures. However, there exist rigid and/or elastic linkages between adjacent substructures, and these linkages are identical for all neighboring substructures. Finally, the whole system is fixed at is two external ends. This type of structures resembles a chain. The typical and simplest chain structure is depicted in Fig. 7.12, where n identical particles and $n + 1$ identical springs are connected in series, the two external ends are fixed, and the system vibrates in the axial direction. The mass of each particle is m, while the elastic constant of each spring is k.

Figure 7.13 shows another example of the chain structures: There is a group of identical beams in the pinned-pinned configuration. Adjacent beams are connected with two springs of the same type at points s_1 and s_2, and these springs are oriented at an angle α with respect to the vertical beam. In addition, a horizontal spring joins beams next to each other at points s_3, while a massless rigid bar is pinned to two neighboring beams at points s_4.

The chain structure may be viewed as one type of linearly periodic structures. Thus, we may again resort to the idea of an extended rotationally periodic structure

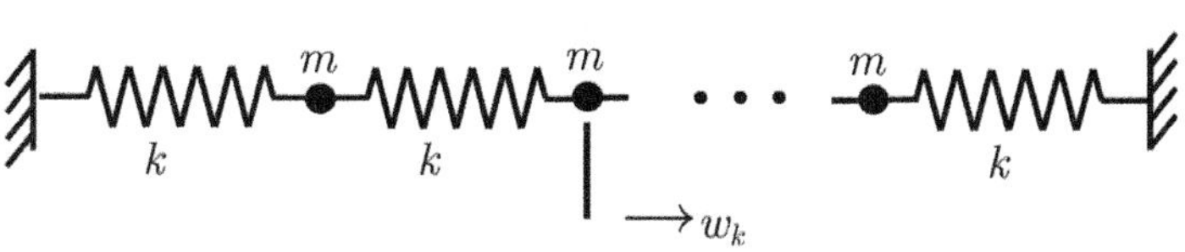

Fig. 7.12 A series system comprised of springs and masses

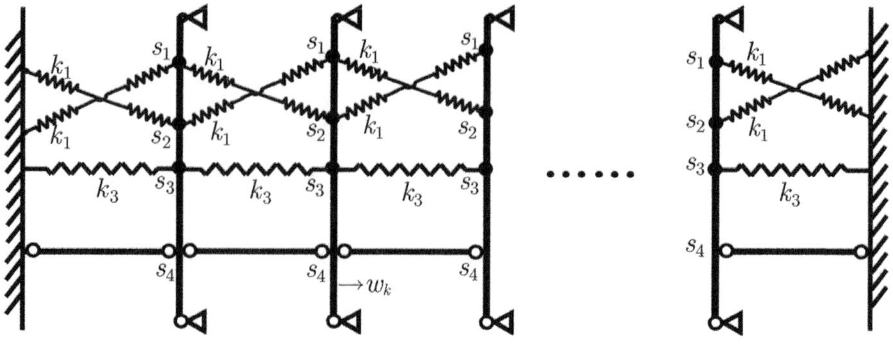

Fig. 7.13 Another example of the chain structure

to perform modal analysis. However, due to the special form of the structure, an even simpler and more effective method can be developed to compute its natural frequencies and mode shapes. This will be elaborated in the next subsection, but the governing equations of a chain structure are going to be introduced in the rest of this subsection.

Divide the whole structure into substructures, and create a local coordinate system at each substructure. Make sure that all these substructures are identical, and so are their coordinate systems. Let us denote the domain and boundary of a substructure as Ω and $\partial\Omega$, respectively.

The natural frequency and mode shape of a chain structure consisting of n substructures satisfy the following equations:

$$\boldsymbol{L}\boldsymbol{u}_k - \omega^2\boldsymbol{M}\boldsymbol{u}_k = \boldsymbol{0} \quad \text{in } \Omega \text{ and } k = 1, 2, \ldots, n, \tag{7.4.1}$$

$$\boldsymbol{B}\boldsymbol{u}_k = \boldsymbol{0} \quad \text{on } \partial\Omega \text{ and } k = 1, 2, \ldots, n, \tag{7.4.2}$$

$$\boldsymbol{J}_j\boldsymbol{u}_k\big|_{s_j} = \overline{\boldsymbol{J}}_j\boldsymbol{u}_{k+1}\big|_{\overline{s_j}} + \overline{\boldsymbol{J}}_j\boldsymbol{u}_{k-1}\big|_{\overline{s_j}}, \quad k = 1, 2, \ldots, n, \quad j = 1, 2, \ldots, l, \tag{7.4.3}$$

where $\boldsymbol{u}_k$ denotes the mode shape component associated with the k-th substructure, $\boldsymbol{u}_0 \equiv \boldsymbol{u}_{n+1} = \boldsymbol{0}$, and Eq. (7.4.3) is for the elastic and rigid connections. Notice that since there is no direct interface or connection boundary between substructures of a chain structure, the continuity conditions are missing in the system of equations (7.4.1)–(7.4.3).

For the specific system shown in Fig. 7.13, $l = 4$ because there are 4 connections between substructures, and Eq. (7.4.3) represents the four constraints on the k-th substructure:

$$Q_k(s_1) + 2k_1\sin^2\alpha \cdot u_k(s_1) = k_1\sin^2\alpha\big[u_{k+1}(s_2) + u_{k-1}(s_2)\big], \tag{7.4.4}$$

$$Q_k(s_2) + 2k_1\sin^2\alpha \cdot u_k(s_2) = k_1\sin^2\alpha\big[u_{k+1}(s_1) + u_{k-1}(s_1)\big], \tag{7.4.5}$$

$$Q_k(s_3) + 2k_3u_k(s_3) = k_3u_{k+1}(s_3) + k_1u_{k-1}(s_3), \tag{7.4.6}$$

$$Q_k(s_4) + 2k_4u_k(s_4) = k_4u_{k+1}(s_4) + k_4u_{k-1}(s_4), \tag{7.4.7}$$

where $Q(s_r)$ denotes the spring force at the point s_r and k_i $(i = 1, 3, 4)$ are the spring constants. Here, as $k_4 \to \infty$, Eq. (7.4.7) yields

$$u_k(s_4) = 0, \quad k = 1, 2, \ldots, n,$$

which indicates a rigid connection.

7.4.2 Reduction of the Eigenvalue Problem and Discussion on Its Modal Qualitative Properties

Let us begin by studying modal properties of the simplest chain structure shown in Fig. 7.12, i.e., the series system consisting of springs and masses. The work should give us some inspiration in the bigger effort to establish modal properties of the general form of chain structures.

The eigenvalue problem associated with a spring-mass series system can be expressed as

$$\boldsymbol{K u} = \omega^2 \boldsymbol{M u}, \tag{7.4.8}$$

where the stiffness matrix is of the tridiagonal form, i.e.,

$$\boldsymbol{K} = \begin{bmatrix} 2k & -k & & & & \\ -k & 2k & -k & & & \\ & \ddots & \ddots & \ddots & & \\ & & \ddots & \ddots & \ddots & \\ & & & -k & 2k & -k \\ & & & & -k & 2k \end{bmatrix}$$

while the mass matrix $\boldsymbol{M}$ is the product of the mass m and identify matrix $\boldsymbol{I}$.

The discrete system may be viewed as the finite difference model of the continuous system of a bar of uniform cross-section, in longitudinal vibration, and subject to fixed-fixed boundary constraints. Suppose the bar of length l_0 has constant tension stiffness p and linear density ρ. Further assume that the bar is divided into $n + 1$ segments. Then, by setting the mass $m = \rho l_0/(n + 1)$ and the spring constant $k = p(n + 1)/l_0$, we get a spring-mass series system with n degrees of freedom. As n approaches the infinity, the discrete system will become, in the limit, the continuous system of a bar of uniform cross-section and subject fixed-fixed constraints. Thus, we can anticipate that as n goes to the infinity, the r-th mode shape vector $\boldsymbol{u}^{(r)}$ of the discrete system should evolve to become the r-th mode shape function of the continuous system of the bar i.e., $u_r = q_r \sin r\pi x/l_0$, where q_r is an arbitrary constant. It is then natural to assume the mode shape of the discrete system to be of the form:

$$\boldsymbol{u}^{(r)} = \left(u_1^{(r)}, u_2^{(r)}, \ldots, u_n^{(r)}\right)^{\mathrm{T}} = \left[\sin r\psi,\ \sin 2r\psi,\ \ldots,\ \sin nr\psi\right]^{\mathrm{T}} q_r, \qquad (7.4.9)$$

where $\psi = \pi/(n+1)$ and $r = 1, 2, \ldots, n$. Substitute Eq. (7.4.9) into Eq. (7.4.8), and also take into account the trigonometric identity:

$$\sin(s-1)r\psi + \sin(s+1)r\psi = 2\cos(r\psi)\sin(sr\psi), \quad s = 1, 2, \ldots, n.$$

We then find that $\boldsymbol{u}^{(r)}$ defined by Eq. (7.4.9) satisfies Eq. (7.4.8) indeed and the corresponding eigenvalue is

$$\omega_r^2 = 2(1 - \cos r\psi)k/m.$$

Here, ω_r and $\boldsymbol{u}^{(r)}$ are the r-th natural angular frequency and mode shape of the structure, respectively, where $r = 1, 2, \ldots, n$.

Because the mode shape $\boldsymbol{u}^{(r)}$ of the simplest chain structure, i.e., the series system of springs and masses, is of the form given in Eq. (7.4.9), where q_r is a constant, it is natural to expect the mode shape of the general type of chain structures to have the following form

$$\boldsymbol{u}_r = (\boldsymbol{u}_{r1}, \boldsymbol{u}_{r2}, \ldots, \boldsymbol{u}_{rn})^{\mathrm{T}} = [\sin r\psi \boldsymbol{I}, \sin 2r\psi \boldsymbol{I}, \ldots, \sin nr\psi \boldsymbol{I}]^{\mathrm{T}} \boldsymbol{q}_r,$$

where $\boldsymbol{q}_r$ is a vector function in Ω, $\boldsymbol{I}$ is an identity matrix of order identical to the dimension of $\boldsymbol{q}_r$, and $r = 1, 2, \ldots, n$.

Thus, introduce the following transformation on the generalized displacement of a chain structure

$$\boldsymbol{u} = \begin{bmatrix} \boldsymbol{u}_1 \\ \boldsymbol{u}_2 \\ \vdots \\ \boldsymbol{u}_n \end{bmatrix} = \sqrt{\frac{2}{n+1}} \begin{bmatrix} \sin\psi \boldsymbol{I} & \ldots & \sin r\psi \boldsymbol{I} & \ldots & \sin n\psi \boldsymbol{I} \\ \sin 2\psi \boldsymbol{I} & \ldots & \sin 2r\psi \boldsymbol{I} & \ldots & \sin 2n\psi \boldsymbol{I} \\ \vdots & & \vdots & & \vdots \\ \sin n\psi \boldsymbol{I} & \ldots & \sin rn\psi \boldsymbol{I} & \ldots & \sin nn\psi \boldsymbol{I} \end{bmatrix} \begin{bmatrix} q_1 \\ q_2 \\ \vdots \\ q_n \end{bmatrix} = \boldsymbol{C}\boldsymbol{q}, \qquad (7.4.10)$$

where $\boldsymbol{C}$ is an orthogonal matrix and satisfies

$$\boldsymbol{C}^{\mathrm{T}}\boldsymbol{C} = \boldsymbol{I} \qquad (7.4.11)$$

and

$$\boldsymbol{C}^{\mathrm{T}}\left(\boldsymbol{Y} + \boldsymbol{Y}^{n-1}\right)\boldsymbol{C} = \mathrm{diag}(2\cos\psi \boldsymbol{I}, 2\cos 2\psi \boldsymbol{I}, \ldots, 2\cos n\psi \boldsymbol{I}). \qquad (7.4.12)$$

Here, $\boldsymbol{Y}$ and $\boldsymbol{Y}^{n-1}$ are row-permutation matrices defined in Eq. (7.2.11). As a result, Eqs. (7.4.1)–(7.4.3) may be rewritten as

$$\boldsymbol{L}'\boldsymbol{u} - \omega^2 \boldsymbol{M}'\boldsymbol{u} = \boldsymbol{0}, \quad \boldsymbol{u}_k \text{ in } \Omega \text{ and } k = 1, 2, \cdots, n, \tag{7.4.13}$$

$$\boldsymbol{B}'\boldsymbol{u} = \boldsymbol{0}, \quad \boldsymbol{u}_k \text{ on } \partial\Omega \text{ and } k = 1, 2, \cdots, n, \tag{7.4.14}$$

$$\boldsymbol{J}'_j \boldsymbol{u}\big|_{s_j} = \overline{\boldsymbol{J}}'_j \left(\boldsymbol{Y}\boldsymbol{u}\big|_{\bar{s}_j} + \boldsymbol{Y}^{n-1}\boldsymbol{u}\big|_{\bar{s}_j} \right), \quad r = 1, 2, \ldots, n, \quad j = 1, 2, \ldots, l, \tag{7.4.15}$$

where $\boldsymbol{L}'$, $\boldsymbol{M}'$, $\boldsymbol{B}'$, $\boldsymbol{J}'_j$, and $\overline{\boldsymbol{J}}'_j$ are, respectively, block diagonal matrices formed by $\boldsymbol{L}, \boldsymbol{M}, \boldsymbol{B}, \boldsymbol{J}_j$, and $\overline{\boldsymbol{J}}_j$ while $j = 1, 2, \ldots, l$. Notice that components $\boldsymbol{u}_1, \boldsymbol{u}_2, \ldots, \boldsymbol{u}_n$ of the vector $\boldsymbol{u}$ are coupled in the above equations. By substituting Eq. (7.4.10) into Eqs. (7.4.13)–(7.4.15), left multiplying the resulting equations by $\boldsymbol{C}^{\mathrm{T}}$, and taking into account Eqs. (7.4.11) and (7.4.12), we obtain

$$\boldsymbol{L}\boldsymbol{q}_r - \omega^2 \boldsymbol{M}\boldsymbol{q}_r = \boldsymbol{0} \quad \text{in } \Omega, \tag{7.4.16}$$

$$\boldsymbol{B}\boldsymbol{q}_r = \boldsymbol{0} \quad \text{on } \partial\Omega, \tag{7.4.17}$$

$$\boldsymbol{J}_j \boldsymbol{q}_r\big|_{s_j} = \overline{\boldsymbol{J}}_j 2\cos r\psi \boldsymbol{q}_r\big|_{\bar{s}_j}, \quad r = 1, 2, \ldots, n, \quad j = 1, 2, \ldots, l. \tag{7.4.18}$$

Now we are ready to draw the following conclusions:

For a chain structure, the natural vibration problem (7.4.1)–(7.4.3) of the entire system may be broken down into n decoupled natural vibration problems (7.4.16)–(7.4.18) involving only one substructure. After the component solutions are obtained, they may be assembled to get the mode shape of the whole structure using the equation below:

$$\boldsymbol{u}_r = (\boldsymbol{u}_{r1}, \boldsymbol{u}_{r2}, \ldots, \boldsymbol{u}_{rn})^{\mathrm{T}} = [\sin r\psi \boldsymbol{I}, \sin 2r\psi \boldsymbol{I}, \ldots, \sin nr\psi \boldsymbol{I}]^{\mathrm{T}} \boldsymbol{q}_r, \quad r = 1, 2, \ldots, n. \tag{7.4.19}$$

7.4.3 Applications

To calculate modes of chain structures or obtain them through experimental means, we may take advantage of the qualitative properties of this type of structures. Specifically, we should do the following:

(1) To find frequencies and mode shapes of a chain structure numerically, we only need to solve n smaller eigenvalue problems of a single substructure, each of which is of the form shown in Eqs. (7.4.16)–(7.4.18). After this step, mode shapes of the entire structure may be calculated from Eq. (7.4.19). This approach greatly reduces the computational effort required.

(2) To obtain frequencies and mode shapes through experiments, we only need to complete two sets of measurements involving two substructures. First, find the mode shape component $\boldsymbol{u}_1$ at the first substructure. Second, locate the point s on the first substructure such that $\boldsymbol{u}_1(s) \neq 0$, and then measure the component value of $\boldsymbol{u}_2$ at the linearly periodic point of s on the second substructure. Thus, the value of r may be determined from the equation $u_2(s)/u_1(s) = \sin 2r\psi / \sin r\psi$.

Since r must be from the set $\{1, 2, \ldots, n\}$, we may need to round the calculated value to the nearest integer. Next, set $\boldsymbol{q}_r = \boldsymbol{u}_1$, where the two may differ only by a constant; and the corresponding mode shape for the whole structure is computed from Eq. (7.4.19). Repeating this process n times, we will obtain n groups of mode shapes of the whole system. Of course, the value of r can be more accurately estimated if we measure mode shape components at multiple points that are linearly periodic to the point s on several substructures.

Example Consider the structure illustrated in Fig. 7.13. Remove springs at points s_1 and s_2, but retain the spring constraint and rigid link at points s_3 and s_4, respectively. The constraint equation (7.4.18) now assumes the following form:

$$Q_r(s_3) = -2k_3(1 - \cos r\psi)q_r(s_3), \quad q_r(s_4) = 0.$$

Under the above conditions, the substructure becomes one depicted in Fig. 7.14. Numerical analysis and experimental testing are performed primarily on this substructure only.

7.4.4 Discrete System and Its Modal Qualitative Properties

Consider a chain structure consisting of n substructures. Suppose that the generalized displacement component at the k-th substructure is denoted by the p-dimensional vector $\boldsymbol{u}_k$, where $k = 1, 2, \ldots, n$. Then, the generalized displacement $\boldsymbol{u} = (\boldsymbol{u}_1, \boldsymbol{u}_2, \ldots, \boldsymbol{u}_n)^{\mathrm{T}}$ of the whole structure is a vector of $(n \times p)$ dimensions and satisfies the following modal equation:

$$\boldsymbol{K}\boldsymbol{u} = \omega^2 \boldsymbol{M}\boldsymbol{u}. \tag{7.4.20}$$

Further assume that the location and ordering of generalized displacements are identical among substructures. Since each substructure directly interacts with at most two adjacent substructures, one ahead of it while the other behind it, and the nature of the interaction is the same for all substructures, the stiffness and mass matrices are of the following block tridiagonal forms:

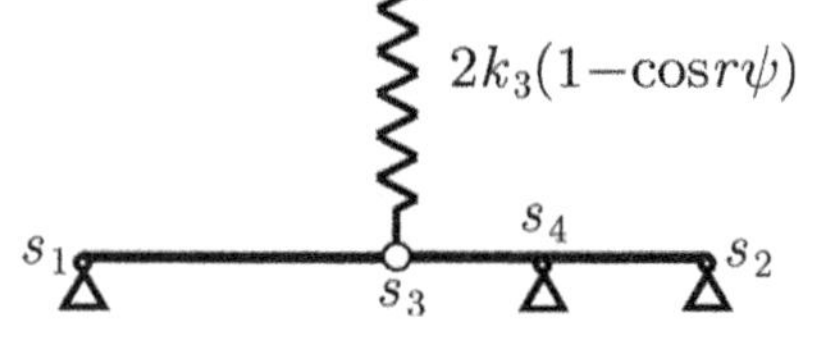

Fig. 7.14 Substructure of a chain structure

$$
\boldsymbol{K} = \begin{bmatrix} \boldsymbol{K}_{11} & \boldsymbol{K}_{12} & & & \\ \boldsymbol{K}_{12} & \boldsymbol{K}_{11} & \boldsymbol{K}_{12} & & \\ & \ddots & \ddots & \ddots & \\ & & \boldsymbol{K}_{12} & \boldsymbol{K}_{11} & \boldsymbol{K}_{12} \\ & & & \boldsymbol{K}_{12} & \boldsymbol{K}_{11} \end{bmatrix},
$$

$$
\boldsymbol{M} = \begin{bmatrix} \boldsymbol{M}_{11} & \boldsymbol{M}_{12} & & & \\ \boldsymbol{M}_{12} & \boldsymbol{M}_{11} & \boldsymbol{M}_{12} & & \\ & \ddots & \ddots & \ddots & \\ & & \boldsymbol{M}_{12} & \boldsymbol{M}_{11} & \boldsymbol{M}_{12} \\ & & & \boldsymbol{M}_{12} & \boldsymbol{M}_{11} \end{bmatrix}, \tag{7.4.21}
$$

where $\boldsymbol{K}_{11}$, $\boldsymbol{K}_{12}$, $\boldsymbol{M}_{11}$, and $\boldsymbol{M}_{12}$ are all symmetric matrices of order p, $\boldsymbol{K}$ and $\boldsymbol{M}$ are both symmetric matrices of order $(n \times p)$, and the blocks of zero submatrices are not shown in Eq. (7.4.21). To solve the generalized matrix eigenvalue problem (7.4.20), whose coefficient matrices are of the special forms given above, we may follow the same approach that works well for the continuous system of a chain structure, i.e., decoupling the original problem into n generalized matrix eigenvalue problems of order p.

Introduce a parameter $\psi = \pi/(n+1)$. Suppose the mode shape of the discrete system is of the form

$$
\boldsymbol{u} = (\sin r\psi, \sin 2r\psi, \ldots, \sin nr\psi)^{\mathrm{T}} \boldsymbol{q}_r, \quad r = 1, 2, \ldots, n, \tag{7.4.22}
$$

where $\boldsymbol{q}_r$ is a p-dimensional vector. Substitute Eq. (7.4.22) into Eq. (7.4.20), take advantage of the special structure of $\boldsymbol{K}$ and $\boldsymbol{M}$ given in Eq. (7.4.21), and also apply the trigonometric identities below:

$$
\sin(s-1)r\psi + \sin(s+1)r\psi = 2\cos(r\psi)\sin(sr\psi), \quad s = 1, 2, \ldots, n.
$$

After these steps, we may decouple the eigenvalue problem (7.4.20) into

$$
\boldsymbol{K}_r \boldsymbol{q}_r = \omega^2 \boldsymbol{M}_r \boldsymbol{q}_r, \quad r = 1, 2, \ldots, n, \tag{7.4.23}
$$

where

$$
\boldsymbol{K}_r = \boldsymbol{K}_{11} + 2\boldsymbol{K}_{12}\cos r\psi, \quad \boldsymbol{M}_r = \boldsymbol{M}_{11} + 2\boldsymbol{M}_{12}\cos r\psi, \quad r = 1, 2, \ldots, n. \tag{7.4.24}
$$

Solve Eqs. (7.4.23) and (7.4.24) for $\boldsymbol{q}_r$, and then calculate mode shapes of the whole structure using Eq. (7.4.22).

It is easy to verify

$$u^{\mathrm{T}} K u = (\sin^2 r\psi + \sin^2 2r\psi + \cdots + \sin^2 nr\psi) q_r{}^{\mathrm{T}} K_r q_r.$$

Mass matrices M and M_r have this kind of relationship, too. Thus, if K and M, the stiffness and mass matrices, respectively, of the entire structure, are real, symmetric, and positive definite (in reality, K may be semi-positive definite) matrices, then K_r and M_r, the stiffness and mass matrices, respectively, of a substructure, should have the same properties, too.

Some important conclusions may be reached based on the discussion above. First, modal qualitative properties of discrete and continuous systems modeling a chain structure are consistent. Second, by taking advantage of the modal qualitative properties, we may derive modal equations of the discrete system that involve significantly fewer number of degrees of freedom. Specifically, we can reduce the original generalized matrix eigenvalue problem of $(n \times p)$ dimensions to n decoupled problems of p dimensions. This helps dramatically decrease the work required to solve for natural frequencies and mode shapes.

7.5 Modal Qualitative Properties of Axisymmetric Structures

7.5.1 The Model and Governing Equations

If the geometric parameters, physical properties, and boundary conditions of a structure are all axisymmetric about a line, then the system is referred to as an axisymmetric structure. The straight line is called the axis of symmetry. If we make it the z-axis of a cylindrical coordinate system $Or\theta z$, then the geometric parameters, physical properties, and boundary conditions of this structure are all independent of the angle θ. Consider the continuous system of this structure and let u, v, and w denote the displacements along the r (radial), θ (circumferential), and z (axial) directions, respectively. The modal equations and boundary conditions of the continuous system modeling a three-dimensional axisymmetric structure are of the following form:

$$\begin{cases} L_{r,\theta,z}(r,z)(u(r,\theta,z), v(r,\theta,z), w(r,\theta,z))^{\mathrm{T}} \\ \qquad -\omega^2 M_{r,\theta,z}(r,z)(u(r,\theta,z), v(r,\theta,z), w(r,\theta,z))^{\mathrm{T}} = \mathbf{0} \quad \text{in } \Omega, \\ B_{r,\theta,z}(r,z)(u(r,\theta,z), v(r,\theta,z), w(r,\theta,z))^{\mathrm{T}} = \mathbf{0} \qquad \text{on } \partial\Omega, \end{cases} \tag{7.5.1}$$

where Ω is the three-dimensional domain occupied by the structure, while $L_{r,\theta,z}$, $M_{r,\theta,z}$, and $B_{r,\theta,z}$ are operators of elasticity, inertia, and boundary conditions, respectively. Due to axisymmetry, coefficients of these operators are all independent of θ.

For two-dimensional problems such as circular planar membranes, circular plates, and axisymmetric shells, we only deal with two coordinates, either (r, θ) or (θ, z);

while for one-dimensional problems such rings, we use only one coordinate θ. Furthermore, the displacement vector sometimes contains u and v components only, e.g., in the case of a planar membrane; or may include just the w component, e.g., in the case of a plate.

7.5.2 Modal Qualitative Properties

Let us consider the most complex case, where the structure is a three-dimensional axisymmetric elastic solid and all three displacement components u, v, and w are involved in the analysis. Due to axisymmetry, the displacement has a period of 2π. Thus, its components can be expanded into a Fourier series:

$$\begin{aligned} u(r,\theta,z) &= \sum_{n=0}^{\infty}\left[U_n(r,z)\cos n\theta + U_n'(r,z)\sin n\theta\right], \\ v(r,\theta,z) &= \sum_{n=0}^{\infty}\left[V_n(r,z)\cos n\theta + V_n'(r,z)\sin n\theta\right], \\ w(r,\theta,z) &= \sum_{n=0}^{\infty}\left[W_n(r,z)\cos n\theta + W_n'(r,z)\sin n\theta\right]. \end{aligned} \tag{7.5.2}$$

Since $\boldsymbol{L}_{r,\theta,z}$, $\boldsymbol{M}_{r,\theta,z}$, and $\boldsymbol{B}_{r,\theta,z}$ are all linear operators, their coefficients are independent of θ. Because of this and also due to the fact that functions $\cos n\theta$ and $\sin n\theta$ are orthogonal to each other, we may decouple different harmonic waves by substituting Eq. (7.5.2) into Eq. (7.5.1). In the end, we get

$$\begin{cases} \boldsymbol{L}_{r,\theta,z}(r,z)\big(U_n\cos n\theta + U_n'\sin n\theta,\ V_n\cos n\theta + V_n'\sin n\theta,\ W_n\cos n\theta + W_n'\sin n\theta\big)^{\mathrm{T}} \\ \qquad -\omega^2\boldsymbol{M}_{r,\theta,z}(r,z)(U_n\cos n\theta + U_n'\sin n\theta,\ V_n\cos n\theta \\ \qquad + V_n'\sin n\theta,\ W_n\cos n\theta + W_n'\sin n\theta)^{\mathrm{T}} = 0 & \text{in } \Omega, \\ \boldsymbol{B}_{r,\theta,z}(r,z)(U_n\cos n\theta + U_n'\sin n\theta,\ V_n\cos n\theta \\ \qquad + V_n'\sin n\theta,\ W_n\cos n\theta + W_n'\sin n\theta) = 0 & \text{on } \partial\Omega. \end{cases} \tag{7.5.3}$$

Thus, the mode shape of this axisymmetric structure is of the form

$$\boldsymbol{U}_n = \begin{bmatrix} U(r,\theta,z) \\ V(r,\theta,z) \\ W(r,\theta,z) \end{bmatrix} = \begin{bmatrix} U_n(r,z) \\ V_n(r,z) \\ W_n(r,z) \end{bmatrix}\cos n\theta + \begin{bmatrix} U_n'(r,z) \\ V_n'(r,z) \\ W_n'(r,z) \end{bmatrix}\sin n\theta, \quad n = 0, 1, 2, \ldots. \tag{7.5.4}$$

Next, let us define a new cylindrical coordinate system $Or'\theta'z'$ by setting θ' to be $-\theta$ while keeping r' and z' identical to r and z, respectively. In the new coordinate system, modal equations and boundary conditions become

$$\begin{cases} \boldsymbol{L}_{r',\theta',z'}(r',z')\begin{pmatrix} u'(r',\theta',z') \\ v'(r',\theta',z') \\ w'(r',\theta',z') \end{pmatrix} - \omega^2 \boldsymbol{M}_{r',\theta',z'}(r',z')\begin{pmatrix} u'(r',\theta',z') \\ v'(r',\theta',z') \\ w'(r',\theta',z') \end{pmatrix} = \boldsymbol{0} & \text{in } \Omega, \\ \boldsymbol{B}_{r',\theta',z'}(r',z')\left(u'(r',\theta',z'),\ v'(r',\theta',z'),\ w'(r',\theta',z')\right)^{\mathrm{T}} = \boldsymbol{0} & \text{on } \partial\Omega, \end{cases} \tag{7.5.5}$$

where u', v', and w' are displacements along r', θ', and z' directions respectively. Since an axisymmetric structure should also have mirror symmetry, the following relationship holds between displacement components in the original and new coordinate systems:

$$\begin{aligned} u'(r',\theta',z') &= u(r,-\theta,z), \\ v'(r',\theta',z') &= -v(r,-\theta,z), \\ w'(r',\theta',z') &= w(r,-\theta,z). \end{aligned} \tag{7.5.6}$$

Per Eqs. (7.5.5) and (7.5.6), if $\boldsymbol{U}_n$ in Eq. (7.5.4) is a mode shape of the structure, then the vector

$$\boldsymbol{U}_n^* = \begin{bmatrix} u(r,-\theta,z) \\ -v(r,-\theta,z) \\ w(r,-\theta,z \end{bmatrix} = \begin{bmatrix} U_n(r,z) \\ -V_n(r,z) \\ W_n(r,z) \end{bmatrix}\cos n\theta - \begin{bmatrix} U_n'(r,z) \\ -V_n'(r,z) \\ W_n'(r,z) \end{bmatrix}\sin n\theta \tag{7.5.7}$$

should also be a mode shape associated with the same frequency. The sum and difference of U_n and U_n^* are, respectively,

$$(\boldsymbol{U})_{ns} = \frac{1}{2}\left[\boldsymbol{U}_n + \boldsymbol{U}_n^*\right] = \begin{bmatrix} U_n(r,z)\cos n\theta \\ V_n'(r,z)\sin n\theta \\ W_n(r,z)\cos n\theta \end{bmatrix}, \quad n = 0, 1, 2, \ldots, \tag{7.5.8}$$

and

$$(\boldsymbol{U})_{na} = \frac{1}{2}\left[\boldsymbol{U}_n - \boldsymbol{U}_n^*\right] = \begin{bmatrix} U_n'(r,z)\sin n\theta \\ V_n(r,z)\cos n\theta \\ W_n'(r,z)\sin n\theta \end{bmatrix}, \quad n = 0, 1, 2, \ldots. \tag{7.5.9}$$

Equations (7.5.8) and (7.5.9) ought to give yet another two groups of mode shapes corresponding to the same frequency. In fact, Eq. (7.5.8) produces a group of symmetric mode shapes, as indicated by the subscript s; while Eq. (7.5.9) creates a group of antisymmetric mode shapes, as marked by the subscript a. Furthermore, it is possible to express any mode shape of the structure as a linear combination of the above two groups of mode shapes.

Now let us further examine the relationship between these symmetric and antisymmetric mode shapes specified in Eqs. (7.5.8) and (7.5.9), respectively. Introduce the transformation $\theta = \theta' - \pi/2n$. Then, the symmetric mode shape in (7.5.8) becomes

$$\overline{(\boldsymbol{U})}_{n\mathrm{a}} = \begin{bmatrix} U_n(r,z)\sin n\theta' \\ -V_n'(r,z)\cos n\theta' \\ W_n(r,z)\sin n\theta' \end{bmatrix}, \quad n = 0, 1, 2, \ldots. \tag{7.5.10}$$

It is clear that this is none other than the antisymmetric mode shape, given by Eq. (7.5.9), in the new cylindrical coordinate system $Or'\theta'z'$. However, the antisymmetric mode shapes in the new and old coordinate systems should be the same, due to axisymmetry. Therefore, the antisymmetric mode shape in Eq. (7.5.9) should be equivalent to

$$(\boldsymbol{U})_{n\mathrm{a}} = \begin{bmatrix} U_n(r,z)\sin n\theta \\ -V_n'(r,z)\cos n\theta \\ W_n(r,z)\sin n\theta \end{bmatrix}, \quad n = 0, 1, 2, \ldots. \tag{7.5.11}$$

Thus, we may now draw the following conclusions:

(1) Mode shapes of a three-dimensional axisymmetric elastic solid may be divided into two groups, one symmetric given by Eq. (7.5.8) while the other antisymmetric shown in Eq. (7.5.11). Both of these groups consist of the n-th harmonic waves ($n = 0, 1, 2, \ldots$). In addition, an antisymmetric mode shape can be obtained by rotating the respective symmetric mode shape by an angle of $\pi/2n$, and these two mode shapes correspond to the same frequency.

(2) Substituting Eq. (7.5.8) into Eq. (7.5.1), we get the following equations for U_n, V_n', and W_n with n as a parameter:

$$\begin{cases} \boldsymbol{L}_{r,z,n}(U_n(r,z), V_n'(r,z), W_n(r,z))^{\mathrm{T}} \\ \quad -\omega^2 \boldsymbol{M}_{r,z,n}(U_n(r,z), V_n'(r,z), W_n(r,z))^{\mathrm{T}} = \boldsymbol{0} & \text{in } \Omega, \\ \boldsymbol{B}_{r,z,n}(U_n(r,z), V_n'(r,z), W_n(r,z))^{\mathrm{T}} = \boldsymbol{0} & \text{on } \partial\Omega, \end{cases}$$

$$n = 0, 1, 2, \ldots. \tag{7.5.12}$$

Hence, the original three-dimensional eigenvalue problem (7.5.1) is replaced by an infinite number of two-dimensional eigenvalue problems, specified in Eq. (7.5.12).

As for two-dimensional axisymmetric problems such as circular planar membranes, circular plates, and revolving shells, as well as one-dimensional axisymmetric problems such as circular rings, their mode shapes are special cases of Eqs. (7.5.8) and (7.5.11). Thus, they should share qualitative properties presented above.

7.5.3 Applications

To obtain modes of an axisymmetric structure through numerical analysis or experimental means, we may take advantage of qualitative properties of the structure as follows:

(1) In engineering applications, often it needs only to find a finite number of mode shapes of an axisymmetric structure. When seeking numerical solution to the eigenvalue problem, we may use Eq. (7.5.12) rather than Eq. (7.5.1), which corresponds the original formulation of the problem. This approach helps decrease the dimension of the continuous problem by one, and thus significantly reduce the number of degrees of freedom of the related discrete system.
(2) When measuring a mode shape experimentally, we only need to obtain data for one component of the mode shape along a complete circle, and use it to estimate the order n of the harmonic waves. Then, acquire additional data of the mode shape on a plane passing through the center axis, along the meridian, or at a point for a three-, two-, or one-dimensional problem, respectively.

7.6 Qualitative Properties of Forced Vibration of Repetitive Structures

When dealing with forced vibration of a repetitive structure, we may again take advantage of the modal qualitative properties and introduce a transformation for the external forces, which is similar to the one applied to the generalized displacements. This helps break down the original forced vibration problem into a number of decoupled ones at the substructure level.

As an example, let us look at a simple case involving a symmetric structure comprised of two substructures. Suppose that $\boldsymbol{w}_1$ and $\boldsymbol{w}_2$ denote the generalized displacements of the first and second substructures, respectively; and let $\boldsymbol{F}_1$ and $\boldsymbol{F}_2$ represent the corresponding external forces applied to the two substructures. Here, the external forces are functions of the spatial coordinates as well as time t, but they may not have any symmetry. The governing equation for forced vibration of the whole system, boundary conditions, continuity requirements at the plane of symmetry, and other constraints between the two substructures may be expressed as

$$
\begin{cases}
\boldsymbol{L}\boldsymbol{w}_i + \boldsymbol{M}\ddot{\boldsymbol{w}}_i = \boldsymbol{F}_i & \text{in } \Omega \text{ and } i = 1, 2, \\
\boldsymbol{B}\boldsymbol{w}_i = \boldsymbol{0} & \text{on } \partial\Omega \text{ and } i = 1, 2, \\
\boldsymbol{J}_1\boldsymbol{w}_1|_{x=0} = -\boldsymbol{J}_1\boldsymbol{w}_2|_{x=0},\ \boldsymbol{J}_2\boldsymbol{w}_1|_{x=0} = \boldsymbol{J}_2\boldsymbol{w}_2|_{x=0}, & \\
\boldsymbol{J}_{rj}\boldsymbol{w}_1\big|_{s_j} = \overline{\boldsymbol{J}}_{rj}\boldsymbol{w}_2\big|_{\bar{s}_j},\quad \boldsymbol{J}_{rj}\boldsymbol{w}_2\big|_{s_j} = \overline{\boldsymbol{J}}_{rj}\boldsymbol{w}_1\big|_{\bar{s}_j}, & j = 1, 2, \ldots, l,
\end{cases}
\tag{7.6.1}
$$

where $\boldsymbol{L}$, $\boldsymbol{M}$, $\boldsymbol{B}$, $\boldsymbol{J}_1$, $\boldsymbol{J}_2$, and $\boldsymbol{J}_{rj}$ are all differential operator matrices.

Making use of modal qualitative properties of a symmetric structure, we may apply the same transformation to both generalized displacements and external forces in Eq. (7.6.1):

$$
\boldsymbol{w} = \begin{bmatrix} \boldsymbol{w}_1 \\ \boldsymbol{w}_2 \end{bmatrix} = \frac{1}{\sqrt{2}} \begin{bmatrix} \boldsymbol{I} & \boldsymbol{I} \\ \boldsymbol{I} & -\boldsymbol{I} \end{bmatrix} \begin{bmatrix} \boldsymbol{q}_1 \\ \boldsymbol{q}_2 \end{bmatrix}, \quad \boldsymbol{F} = \begin{bmatrix} \boldsymbol{F}_1 \\ \boldsymbol{F}_2 \end{bmatrix} = \frac{1}{\sqrt{2}} \begin{bmatrix} \boldsymbol{I} & \boldsymbol{I} \\ \boldsymbol{I} & -\boldsymbol{I} \end{bmatrix} \begin{bmatrix} \boldsymbol{f}_1 \\ \boldsymbol{f}_2 \end{bmatrix}. \tag{7.6.2}
$$

Plugging Eq. (7.6.2) into Eq. (7.6.1), we get the following results, which are similar to Eq. (7.1.10):

$$
\begin{cases}
\boldsymbol{L}\boldsymbol{q}_i + \boldsymbol{M}\ddot{\boldsymbol{q}}_i = \boldsymbol{f}_i & \text{in } \Omega, \\
\boldsymbol{B}\boldsymbol{q}_i = \boldsymbol{0} & \text{on } \partial\Omega, \\
\boldsymbol{J}_i\boldsymbol{q}_i|_{x=0} = \boldsymbol{0},\quad \boldsymbol{J}_{rj}\boldsymbol{q}_i\big|_{s_j} = (-1)^{i+1}\,\overline{\boldsymbol{J}}_{rj}\boldsymbol{q}_i\big|_{\bar{s}_j}, &
\end{cases}
\tag{7.6.3}
$$

where $i = 1, 2$ and $j = 0, 1, \ldots, l$. In addition, initial conditions should also be incorporated into Eqs. (7.6.1) and (7.6.3), and then be transformed with Eq. (7.6.2). After these steps, the forced vibration problem (7.6.1) of the entire structure can be reduced to two decoupled forced vibration problems involving one substructure only.

Other types of repetitive structures may be analyzed in a similar way. For instance, in the case of a rotationally periodic structure, we may reference Eq. (7.2.5a) and apply the transformation $\boldsymbol{F} = \boldsymbol{R}\boldsymbol{f}$ to convert the external force $\boldsymbol{F} = (\boldsymbol{F}_1, \boldsymbol{F}_2, \ldots, \boldsymbol{F}_n)^{\mathrm{T}}$ into an external load $\boldsymbol{f}$ for the decoupled system. As another example, consider a chain structure. Based on Eq. (7.4.10), we may introduce the transformation $\boldsymbol{F} = \boldsymbol{C}\boldsymbol{f}$ and turn the external load $\boldsymbol{F}$ into a force $\boldsymbol{f}$ acting on the decoupled system. After such treatment, we can reduce the forced vibration problem of the entire system to a series of decoupled problems at the substructure level.

7.7 Dimension Reduction in Vibration and Shape Control of Repetitive Structures

By using modal qualitative properties, it is possible to establish a procedure for dimension reduction in vibration control of repetitive structures. In this section, we primarily focus on the method of dimension reduction for discrete systems of repetitive structures. See Refs. [7, 8, 11]. Notice that discrete systems are often referred to as lumped parameter systems in the theory of vibration control.

Before proceeding further, we would like to point out that the method presented here only addresses certain aspects of dimension reduction in vibration control by taking advantage of the repetitive nature of a structure. It does not deal with issues such as the actual design, controllability, observability, etc.

If the original repetitive characteristics of a structure are preserved in the condensation of generalized displacements, layout of sensors and actuators, as well as the relationship between inputs and control forces, then the vibration control problem of the whole structure may be simplified to several smaller problems at the substructure level. This allows for a significant reduction in problem size.

Below we will use the symmetric structure as an example to illustrate the method for dimension reduction in vibration control.

Suppose the b_1-dimensional vector $\boldsymbol{w}_2$ and b_2-dimensional vector $\boldsymbol{w}_3$ represent the generalized displacement inside and perpendicular to the symmetry plane, respectively. Furthermore, it is assumed that the generalized displacements at the left and right parts of the symmetric structure are denoted by $\boldsymbol{w}_1$ and $\boldsymbol{w}_4$, respectively, where both vectors are of p dimensions. Then, vibration of the entire system under some control force is governed by the following equation

$$\boldsymbol{M}\ddot{\boldsymbol{w}} + \boldsymbol{K}\boldsymbol{w} + \boldsymbol{B}\boldsymbol{u} = 0, \tag{7.7.1}$$

where $\boldsymbol{w} = (\boldsymbol{w}_1, \boldsymbol{w}_2, \boldsymbol{w}_3, \boldsymbol{w}_4)^{\mathrm{T}}$ is a generalized displacement vector of $2p + b_1 + b_2$ dimensions, $\boldsymbol{u}$ represents the system input, and $\boldsymbol{B}\boldsymbol{u}$ denotes the control force. If both the location and numbering of the generalized displacements at left and right sides of the structure are defined in such a way that they are symmetric about the same symmetry plane, then the stiffness matrix $\boldsymbol{K}$ of the system assumes the following special form:

$$\boldsymbol{K} = \begin{bmatrix} \boldsymbol{K}_{11} & \boldsymbol{K}_{12} & \boldsymbol{K}_{13} & \boldsymbol{K}_{14} \\ \boldsymbol{K}_{12}^{\mathrm{T}} & \boldsymbol{K}_{22} & \boldsymbol{0} & \boldsymbol{K}_{12}^{\mathrm{T}}\boldsymbol{S}_p \\ \boldsymbol{K}_{13}^{\mathrm{T}} & \boldsymbol{0} & \boldsymbol{K}_{33} & -\boldsymbol{K}_{13}^{\mathrm{T}}\boldsymbol{S}_p \\ \boldsymbol{K}_{14}^{\mathrm{T}} & \boldsymbol{S}_p\boldsymbol{K}_{12} & -\boldsymbol{S}_p\boldsymbol{K}_{13} & \boldsymbol{K}_{11}^{\mathrm{T}'} \end{bmatrix}, \quad \boldsymbol{S}_p = \begin{bmatrix} 0 & & 1 \\ & 1 & \\ & \cdot^{\cdot^{\cdot}} & \\ 1 & & 0 \end{bmatrix}_{p\times p}. \tag{7.7.2}$$

The mass matrix $\boldsymbol{M}$ is of the same form as $\boldsymbol{K}$.

When actuators are placed inside the plane of symmetry as well as symmetrically on left and right parts of the structure, the input $\boldsymbol{u}$ has four components, i.e.,

$$\boldsymbol{u} = (\boldsymbol{u}_1, \boldsymbol{u}_2, \boldsymbol{u}_3, \boldsymbol{u}_4)^{\mathrm{T}},$$

where both $\boldsymbol{u}_1$ and $\boldsymbol{u}_4$ are m-dimensional vectors, $\boldsymbol{u}_2$ is of s_1 dimensions, $\boldsymbol{u}_3$ has s_2 dimensions, $m \le p$, and $s_i \le b_i$ $(i = 1, 2)$. Further assume that the control force on the structure generated by the input $\boldsymbol{u}$ is also symmetric about the symmetry plane. Specifically, components of the control force $\boldsymbol{B}_{11}\boldsymbol{u}_1$, $\boldsymbol{B}_{12}\boldsymbol{u}_2$, $\boldsymbol{B}_{13}\boldsymbol{u}_3$, and $\boldsymbol{B}_{14}\boldsymbol{u}_4$ created by input components $\boldsymbol{u}_1$, $\boldsymbol{u}_2$, $\boldsymbol{u}_3$, and $\boldsymbol{u}_4$ in the $\boldsymbol{w}_1$ direction are identical to those produced by $\boldsymbol{S}_m\boldsymbol{u}_4$, $\boldsymbol{u}_2$, $\boldsymbol{u}_3$, and $\boldsymbol{S}_m\boldsymbol{u}_1$ in the $\boldsymbol{S}_p\boldsymbol{w}_4$ direction, respectively; control forces induced by $\boldsymbol{u}_1$ and $\boldsymbol{S}_m\boldsymbol{u}_4$ in $\boldsymbol{w}_2$ and $\boldsymbol{w}_3$ directions, respectively, are the same; and $\boldsymbol{u}_2$ produces no control force in the $\boldsymbol{w}_3$ direction, while $\boldsymbol{u}_3$ has no influence in the $\boldsymbol{w}_2$ direction. Thus, the control matrix should be of the following form:

$$\boldsymbol{B} = \begin{bmatrix} \boldsymbol{B}_{11} & \boldsymbol{B}_{12} & \boldsymbol{B}_{13} & \boldsymbol{B}_{14} \\ \boldsymbol{B}_{21} & \boldsymbol{B}_{22} & \boldsymbol{0} & \boldsymbol{B}_{21}\boldsymbol{S}_m \\ \boldsymbol{B}_{31} & \boldsymbol{0} & \boldsymbol{B}_{33} & -\boldsymbol{B}_{31}\boldsymbol{S}_m \\ \boldsymbol{S}_p\boldsymbol{B}_{14}\boldsymbol{S}_m & \boldsymbol{S}_p\boldsymbol{B}_{12} & -\boldsymbol{S}_p\boldsymbol{B}_{13} & \boldsymbol{S}_p\boldsymbol{B}_{11}\boldsymbol{S}_m \end{bmatrix}, \tag{7.7.3}$$

where $\boldsymbol{S}_p$ is a square matrix of order p defined in Eq. (7.7.2) while $\boldsymbol{S}_m$ has the same form as $\boldsymbol{S}_p$ but is of order m.

In practice, it is difficult to measure a large number of generalized displacements, and even more so to implement control on high-dimensional systems. Thus, it often needs to condense the generalized displacement $\boldsymbol{w}$ of high dimension that is associated with the original structure to a variable $\boldsymbol{y}$ of low dimension that can be measured with installed sensors, in order to reduce the dimension in structural control. Suppose there are respectively, r_1 and r_2 sensors at the symmetry plane for measuring in-plane displacement $\boldsymbol{y}_2$ and out-of-plane displacement $\boldsymbol{y}_3$. In addition, there are l sensors on each of the left and right parts of the structure, and their locations and types are selected in such a way that the original structural symmetry is preserved. Here, $m < l \leq p$ while $s_i < r_i \leq b_i$ $(i = 1, 2)$. Let $\boldsymbol{y}_1$ and $\boldsymbol{y}_4$ represent, respectively, the generalized displacements measured by sensors on the left and right parts of the structure. Thus, the condensed generalized displacement of the whole system is

$$\boldsymbol{y} = (\boldsymbol{y}_1, \boldsymbol{y}_2, \boldsymbol{y}_3, \boldsymbol{y}_4)^{\mathrm{T}}.$$

To ensure that the condensed system preserves the original symmetry, we have the following relationship:

$$\boldsymbol{w} = \begin{bmatrix} \boldsymbol{w}_1 \\ \boldsymbol{w}_2 \\ \boldsymbol{w}_3 \\ \boldsymbol{w}_4 \end{bmatrix} = \boldsymbol{C}\boldsymbol{y} = \begin{bmatrix} \boldsymbol{C}_{11} & \boldsymbol{C}_{12} & \boldsymbol{C}_{13} & \boldsymbol{C}_{14} \\ \boldsymbol{C}_{21} & \boldsymbol{C}_{22} & \boldsymbol{0} & \boldsymbol{C}_{21}\boldsymbol{S}_l \\ \boldsymbol{C}_{31} & \boldsymbol{0} & \boldsymbol{C}_{33} & -\boldsymbol{C}_{31}\boldsymbol{S}_l \\ \boldsymbol{S}_p\boldsymbol{C}_{14}\boldsymbol{S}_l & \boldsymbol{S}_p\boldsymbol{C}_{12} & -\boldsymbol{S}_p\boldsymbol{C}_{13} & \boldsymbol{S}_p\boldsymbol{C}_{11}\boldsymbol{S}_l \end{bmatrix} \begin{bmatrix} \boldsymbol{y}_1 \\ \boldsymbol{y}_2 \\ \boldsymbol{y}_3 \\ \boldsymbol{y}_4 \end{bmatrix}. \tag{7.7.4}$$

Substituting Eq. (7.7.4) into (7.7.1) and left multiplying the resulting equation by $\boldsymbol{C}^{\mathrm{T}}$, we get a control system of $2l + r_1 + r_2$ dimensions as follows:

$$\tilde{\boldsymbol{M}}\ddot{\boldsymbol{y}} + \tilde{\boldsymbol{K}}\boldsymbol{y} + \tilde{\boldsymbol{B}}\boldsymbol{u} = \boldsymbol{0}, \tag{7.7.5}$$

where

$$\tilde{K} = \begin{bmatrix} \tilde{K}_{11} & \tilde{K}_{12} & \tilde{K}_{13} & \tilde{K}_{14} \\ \tilde{K}_{12}^{\mathrm{T}} & \tilde{K}_{22} & \mathbf{0} & \tilde{K}_{12}^{\mathrm{T}} S_l \\ \tilde{K}_{13}^{\mathrm{T}} & \mathbf{0} & \tilde{K}_{33} & -\tilde{K}_{13}^{\mathrm{T}} S_l \\ \tilde{K}_{14}^{\mathrm{T}} & S_l \tilde{K}_{12} & -S_l \tilde{K}_{13} & \tilde{K}_{11}^{\mathrm{T}'} \end{bmatrix}, \quad \tilde{B} = \begin{bmatrix} \tilde{B}_{11} & \tilde{B}_{12} & \tilde{B}_{13} & \tilde{B}_{14} \\ \tilde{B}_{21} & \tilde{B}_{22} & \mathbf{0} & \tilde{B}_{21} S_m \\ \tilde{B}_{31} & \mathbf{0} & \tilde{B}_{33} & -\tilde{B}_{31} S_m \\ S_l \tilde{B}_{14} S_m & S_l \tilde{B}_{12} & -S_l \tilde{B}_{13} & S_l \tilde{B}_{11} S_m \end{bmatrix}, \tag{7.7.6}$$

$$\tilde{K}_{11}^{\mathrm{T}} = \tilde{K}_{11}, \quad \tilde{K}_{14}^{\tilde{\mathrm{T}}} = \tilde{K}_{14},$$

and $\tilde{M}$ is of the same form as $\tilde{K}$. This indicates that $\tilde{M}$, $\tilde{K}$, and $\tilde{B}$ of the condensed system have indeed kept the characteristics of the original symmetric structure, which are shown in Eqs. (7.7.2) and (7.7.3).

In order to simplify Eq. (7.7.5), introduce the following transformations

$$y = \begin{bmatrix} y_1 \\ y_2 \\ y_3 \\ y_4 \end{bmatrix} = \frac{1}{\sqrt{2}} \begin{bmatrix} I_l & \mathbf{0} & \mathbf{0} & I_l \\ \mathbf{0} & \sqrt{2} I_{r_1} & \mathbf{0} & \mathbf{0} \\ \mathbf{0} & \mathbf{0} & \sqrt{2} I_{r_2} & \mathbf{0} \\ S_l & \mathbf{0} & \mathbf{0} & -S_l \end{bmatrix} \begin{bmatrix} q_1 \\ q_2 \\ q_3 \\ q_4 \end{bmatrix} = H_1 q, \tag{7.7.7}$$

$$u = \begin{bmatrix} u_1 \\ u_2 \\ u_3 \\ u_4 \end{bmatrix} = \frac{1}{\sqrt{2}} \begin{bmatrix} I_m & \mathbf{0} & \mathbf{0} & I_m \\ \mathbf{0} & \sqrt{2} I_{s_1} & \mathbf{0} & \mathbf{0} \\ \mathbf{0} & \mathbf{0} & \sqrt{2} I_{s_2} & \mathbf{0} \\ S_m & \mathbf{0} & \mathbf{0} & -S_m \end{bmatrix} \begin{bmatrix} v_1 \\ v_2 \\ v_3 \\ v_4 \end{bmatrix} = H_2 v, \tag{7.7.8}$$

where I_l represents the identity matrix of order l. Notice that $H_i^{\mathrm{T}} H_i = I \; (i = 1, 2)$. Inserting Eqs. (7.7.7) and (7.7.8) into Eq. (7.7.5) and left multiplying the resulting equation by H_i^{T}, we get

$$\begin{bmatrix} \tilde{M}_{11} + \tilde{M}_{14} S_l & \sqrt{2} \tilde{M}_{12} \\ \sqrt{2} \tilde{M}_{12}^{\mathrm{T}} & \tilde{M}_{22} \end{bmatrix} \begin{bmatrix} \ddot{q}_1 \\ \ddot{q}_2 \end{bmatrix} + \begin{bmatrix} \tilde{K}_{11} + \tilde{K}_{14} S_l & \sqrt{2} \tilde{K}_{12} \\ \sqrt{2} \tilde{K}_{12}^{\mathrm{T}} & \tilde{K}_{22} \end{bmatrix} \begin{bmatrix} q_1 \\ q_2 \end{bmatrix}$$
$$+ \begin{bmatrix} \tilde{B}_{11} + \tilde{B}_{14} S_l & \sqrt{2} \tilde{B}_{12} \\ \sqrt{2} \tilde{B}_{21} & \tilde{B}_{22} \end{bmatrix} \begin{bmatrix} v_1 \\ v_2 \end{bmatrix} = \mathbf{0} \tag{7.7.9}$$

and

$$\begin{bmatrix} \tilde{M}_{33} & \sqrt{2}\tilde{M}_{13}^{\mathrm{T}} \\ \sqrt{2}\tilde{M}_{13} & \tilde{M}_{11} - \tilde{M}_{14}S_l \end{bmatrix} \begin{bmatrix} \ddot{q}_3 \\ \ddot{q}_4 \end{bmatrix} + \begin{bmatrix} \tilde{K}_{33} & \sqrt{2}\tilde{K}_{13}^{\mathrm{T}} \\ \sqrt{2}\tilde{K}_{13} & \tilde{K}_{11} - \tilde{K}_{14}S_l \end{bmatrix} \begin{bmatrix} q_3 \\ q_4 \end{bmatrix}$$
$$+ \begin{bmatrix} \tilde{B}_{33} & \sqrt{2}\tilde{B}_{31} \\ \sqrt{2}\tilde{B}_{13} & \tilde{B}_{11} - \tilde{B}_{14}S_l \end{bmatrix} \begin{bmatrix} v_3 \\ v_4 \end{bmatrix} = \mathbf{0}. \tag{7.7.10}$$

Thus, the original control system (7.7.5) of $2l+r_1+r_2$ dimensions has been reduced to two separate systems, one having $l + r_1$ dimensions while the other being of $l + r_2$ dimensions.

If there is no generalized displacement on the symmetry plane, the reduction process outlined above can be further simplified. As a matter of fact the vectors w_2, w_3, u_2, u_3, y_2, and y_3 all vanish in this case. Thus, Eq. (7.7.6) is reduced to

$$\tilde{M} = \begin{bmatrix} \tilde{M}_{11} & \tilde{M}_{14} \\ \tilde{M}_{14}^{\mathrm{T}} & \tilde{M}_{11}^{\mathrm{T}'} \end{bmatrix}, \quad \tilde{K} = \begin{bmatrix} \tilde{K}_{11} & \tilde{K}_{14} \\ \tilde{K}_{14}^{\mathrm{T}} & \tilde{K}_{11}^{\mathrm{T}'} \end{bmatrix}, \quad \tilde{B} = \begin{bmatrix} \tilde{B}_{11} & \tilde{B}_{14} \\ S_l\tilde{B}_{14}S_m & S_l\tilde{B}_{11}S_m \end{bmatrix}. \tag{7.7.11}$$

It can be seen from above that matrices $\tilde{M}$ and $\tilde{K}$ are symmetric with respect to both their respective main- and anti-diagonals. Moreover, Eqs. (7.7.7) and (7.7.8) now assume the following forms:

$$y = \begin{bmatrix} y_1 \\ y_4 \end{bmatrix} = \frac{1}{\sqrt{2}} \begin{bmatrix} I_l & I_l \\ S_l & -S_l \end{bmatrix} \begin{bmatrix} q_1 \\ q_4 \end{bmatrix} = U_1 q, \tag{7.7.12}$$

$$u = \begin{bmatrix} u_1 \\ u_4 \end{bmatrix} = \frac{1}{\sqrt{2}} \begin{bmatrix} I_m & I_m \\ S_m & -S_m \end{bmatrix} \begin{bmatrix} v_1 \\ v_4 \end{bmatrix} = U_2 v. \tag{7.7.13}$$

In the end, the original control system is decoupled into two smaller systems, each of which has l degrees of freedom:

$$(\tilde{M}_{11} + \tilde{M}_{14}S_l)\ddot{q}_1 + (\tilde{K}_{11} + \tilde{K}_{14}S_l)q_1 + (\tilde{B}_{11} + \tilde{B}_{14}S_m)v_1 = \mathbf{0}, \tag{7.7.14}$$
$$(\tilde{M}_{11} - \tilde{M}_{14}S_l)\ddot{q}_4 + (\tilde{K}_{11} - \tilde{K}_{14}S_l)q_4 + (\tilde{B}_{11} - \tilde{B}_{14}S_m)v_4 = \mathbf{0}. \tag{7.7.15}$$

Equations (7.7.12)–(7.7.15) provide the theoretical basis for implementing control with measured data. The procedure is outlined below:

(1) Based on Eq. (7.7.12), calculate q_1 and q_4 from the observed data $y = (y_1, y_4)^{\mathrm{T}}$ as follows:

$$q = \begin{bmatrix} q_1 \\ q_4 \end{bmatrix} = U_1^{\mathrm{T}} y = \frac{1}{\sqrt{2}} \begin{Bmatrix} y_1 + Sy_4 \\ y_1 - Sy_4 \end{Bmatrix}. \tag{7.7.16}$$

(2) Substitute $\boldsymbol{q}_1$ and $\boldsymbol{q}_4$ into Eqs. (7.7.14) and (7.7.15), respectively, and solve for the required system inputs $\boldsymbol{v}_1$ and $\boldsymbol{v}_4$.
(3) Combine $\boldsymbol{v}_1$ and $\boldsymbol{v}_4$ per Eq. (7.7.13) to determine the input $\boldsymbol{u}$ needed for the entire system. In this way, vibration control of the original 2 l-dimensional system is achieved through control of two smaller systems, each of which is of l dimensions.

Let us now consider a simple case. Suppose there are just one sensor and one actuator on each of the substructures, and a symmetric structure consists of substructures of this type only. By applying the dimension reduction method introduced in this section to the symmetric structure, the original control system involving dual-input and dual-output is broken down to two single-input and single-output control systems.

At the conclusion of this section, we would like to point out that the problem of static shape control of repetitive structures can be handled in a way completely analogous to the above problem of vibration control. The only difference between the two is that the inertia term is dropped in the case of static shape control; and because of this, the problem is actually made much simpler. See Ref. [11].

7.8 Qualitative Properties in Static Deformation of Repetitive Structures

In engineering application involving repetitive structures, the demand may be more widespread for solutions of static deformation than for vibration analysis. However, the former is much easier to deal with than the latter from the perspective of Structural Mechanics and Mathematics. In this section, we discuss qualitative properties in static deformation of various repetitive structures. These properties can be derived and used in a way similar to the corresponding qualitative properties in vibration. It needs to be pointed out that all notations used in equations of this section have the same physical or mathematical meaning as their counterparts in Sects. 7.1 to 7.5 of this chapter.

7.8.1 Qualitative Properties in Static Deformation of Symmetric Structures

Suppose a symmetric structure is under the action of the static force $\boldsymbol{F} = \{\boldsymbol{F}_1, \boldsymbol{F}_2\}^{\mathrm{T}}$, where $\boldsymbol{F}_1$ and $\boldsymbol{F}_2$ are the static forces applied to the No. 1 and No. 2 substructures, respectively. The governing equation for static deformation is

$$\begin{cases} \boldsymbol{L}\boldsymbol{u} = \boldsymbol{F} & \text{in } \tilde{\Omega}, \\ \boldsymbol{B}\boldsymbol{u} = \boldsymbol{0} & \text{on } \partial\tilde{\Omega}. \end{cases} \tag{7.8.1}$$

In terms of substructures, the equation may be rewritten as

$$\begin{cases} \boldsymbol{L}\boldsymbol{u}_k = \boldsymbol{F}_k & \text{in } \tilde{\Omega}, \\ \boldsymbol{B}\boldsymbol{u}_k = \boldsymbol{0} & \text{on } \partial\tilde{\Omega}. \end{cases} \tag{7.8.2}$$

Here, $k = 1, 2$; continuity conditions at the interface between substructures are still given by Eqs. (7.1.3a) and (7.1.3b); and there may also be other constraints between substructures, as specified in Eqs. (7.1.4a) and (7.1.4b).

Apply the following transformation to the original generalized displacement of the structure

$$\boldsymbol{u} = \begin{bmatrix} \boldsymbol{u}_1 \\ \boldsymbol{u}_2 \end{bmatrix} = \boldsymbol{P}\boldsymbol{q} = \frac{1}{\sqrt{2}}\begin{bmatrix} \boldsymbol{I} & \boldsymbol{I} \\ \boldsymbol{I} & -\boldsymbol{I} \end{bmatrix}\begin{bmatrix} \boldsymbol{q}_1 \\ \boldsymbol{q}_2 \end{bmatrix} = \frac{1}{\sqrt{2}}\begin{bmatrix} \boldsymbol{I} \\ \boldsymbol{I} \end{bmatrix}\boldsymbol{q}_1 + \frac{1}{\sqrt{2}}\begin{bmatrix} \boldsymbol{I} \\ -\boldsymbol{I} \end{bmatrix}\boldsymbol{q}_2. \tag{7.8.3}$$

Perform the same transformation on the external force $\boldsymbol{F}$

$$\begin{gathered} \boldsymbol{F} = \begin{bmatrix} \boldsymbol{F}_1 \\ \boldsymbol{F}_2 \end{bmatrix} = \boldsymbol{P}\boldsymbol{f} = \frac{1}{\sqrt{2}}\begin{bmatrix} \boldsymbol{I} \\ \boldsymbol{I} \end{bmatrix}\boldsymbol{f}_1 + \frac{1}{\sqrt{2}}\begin{bmatrix} \boldsymbol{I} \\ -\boldsymbol{I} \end{bmatrix}\boldsymbol{f}_2, \\ \boldsymbol{f}_1 = \frac{1}{\sqrt{2}}(\boldsymbol{F}_1 + \boldsymbol{F}_2), \quad \boldsymbol{f}_2 = \frac{1}{\sqrt{2}}(\boldsymbol{F}_1 - \boldsymbol{F}_2). \end{gathered} \tag{7.8.4}$$

The last equation in Eq. (7.8.4) shows that components the external force can be transformed to two groups, one symmetric while the other antisymmetric. Making use of the transformation and following the discussion in Sect. 7.1.2, we get in the end the following equations for the static deformation of the symmetric structure:

$$\begin{cases} \boldsymbol{L}\boldsymbol{q}_i = \boldsymbol{f}_i & \text{in } \Omega, \\ \boldsymbol{B}\boldsymbol{q}_i = \boldsymbol{0} & \text{on } \partial\Omega, \\ \boldsymbol{J}_i\boldsymbol{q}_i = \boldsymbol{0} & \text{on } b_0, \\ \boldsymbol{J}_{r_j}\boldsymbol{q}_i\big|_{s_j} = (-1)^{i+1}\overline{\boldsymbol{J}}_{r_j}\boldsymbol{q}_i\big|_{\overline{s}_j}, \end{cases} \tag{7.8.5}$$

where $i = 1, 2$ and $j = 1, 2, \ldots, l$.

The above two sets of differential equations and boundary conditions are decoupled for $\boldsymbol{q}_1$ and $\boldsymbol{q}_2$. Thus, we may conclude: For a symmetric structure, both the external force and static deformation may be separated into symmetric and antisymmetric groups with respect to the plane of symmetry. Consequently, in order to get the static deformation of the entire structure, we only need to solve two smaller problems involving one half of the structure. In the case of the external force being symmetric, the static deformation will also be symmetric. By the same token, if the external force is antisymmetric, so should be the static deformation. In both cases, we only need to calculate or measure the static deformation of one half of the structure.

7.8.2 Qualitative Properties in Static Deformation of Rotationally Periodic Structures

Suppose a rotationally periodic structure with n sectors is under the action of the static external force $\boldsymbol{F} = (\boldsymbol{F}_1, \boldsymbol{F}_2, \ldots, \boldsymbol{F}_k, \ldots, \boldsymbol{F}_n)$, where $\boldsymbol{F}_k$ is the force applied to the k-th substructure and $k = 1, 2, \ldots, n$. The governing equation for the static deformation is

$$\begin{cases} \boldsymbol{L}\boldsymbol{u} = \boldsymbol{F} & \text{in } \tilde{\Omega}, \\ \boldsymbol{B}\boldsymbol{u} = \boldsymbol{0} & \text{on } \partial\tilde{\Omega}. \end{cases} \tag{7.8.6}$$

In terms of substructures, the equation may be rewritten as

$$\begin{cases} \boldsymbol{L}\boldsymbol{u}_k = \boldsymbol{F}_k & \text{in } \tilde{\Omega}, \\ \boldsymbol{B}\boldsymbol{u}_k = \boldsymbol{0} & \text{on } \partial\tilde{\Omega}, \end{cases} \tag{7.8.7}$$

where $k = 1, 2, \ldots, n$. Eqs. (7.2.3) and (7.2.4) cover, respectively, the continuity conditions at the shared boundaries of a substructure and other constraints between substructures.

Introduce the following transformations for the generalized displacement $\boldsymbol{u}_k$ and generalized force $\boldsymbol{F}_k$:

$$\boldsymbol{u} = (\boldsymbol{u}_1, \boldsymbol{u}_2, \cdots, \boldsymbol{u}_n)^{\mathrm{T}} = \begin{bmatrix} \boldsymbol{R}_1 & \boldsymbol{R}_2 & \cdots & \boldsymbol{R}_n \end{bmatrix} \begin{bmatrix} q_1 \\ q_2 \\ \vdots \\ q_n \end{bmatrix} = \boldsymbol{R}\boldsymbol{q}, \tag{7.8.8a}$$

$$\boldsymbol{R}_r = \frac{1}{\sqrt{n}}(\boldsymbol{I}, \mathrm{e}^{\mathrm{i}r\psi}\boldsymbol{I}, \mathrm{e}^{\mathrm{i}2r\psi}\boldsymbol{I}, \cdots, \mathrm{e}^{\mathrm{i}(n-1)r\psi}\boldsymbol{I})^{\mathrm{T}}, \quad r = 1, 2, \ldots, n, \tag{7.8.8b}$$

$$\boldsymbol{F} = (\boldsymbol{F}_1, \boldsymbol{F}_2, \cdots, \boldsymbol{F}_n)^{\mathrm{T}} = \begin{bmatrix} \boldsymbol{R}_1 & \boldsymbol{R}_2 & \cdots & \boldsymbol{R}_n \end{bmatrix} \begin{bmatrix} \boldsymbol{f}_1 \\ \boldsymbol{f}_2 \\ \vdots \\ \boldsymbol{f}_n \end{bmatrix} = \boldsymbol{R}\boldsymbol{f}. \tag{7.8.9}$$

With the above transformations, the problem of solving for static deformation of a rotationally periodic structure with n substructures can be decomposed into n smaller static deformation problems of one substructure:

$$\begin{cases} \boldsymbol{L}\boldsymbol{q}_r = \boldsymbol{f}_r & \text{in } \Omega, \\ \boldsymbol{B}\boldsymbol{q}_r = \boldsymbol{0} & \text{on } \partial\Omega, \\ \boldsymbol{J}_0\boldsymbol{q}_r|_{b^+} = \boldsymbol{J}_0 e^{ir\psi}\boldsymbol{q}_r|_{b^-}, \ \boldsymbol{J}_{pj}\boldsymbol{q}_r\big|_{s_{pj}} = \overline{\boldsymbol{J}}_{pj} e^{ipr\psi}\boldsymbol{q}_r\big|_{\bar{s}_{pj}}, \end{cases} \tag{7.8.10}$$

where $r = 1, 2, \ldots, n$, $p = 1, 2, \ldots, n$, and $j = 0, 1, \ldots, l_p$. The parameter $\boldsymbol{f}_r$ as well as the variable $\boldsymbol{q}_r$ in the above system of equations are, in general, complex vectors, i.e., $\boldsymbol{f}_r = \boldsymbol{f}_r^{\mathrm{r}} + i\boldsymbol{f}_r^{\mathrm{i}}$ and $\boldsymbol{q}_r = \boldsymbol{q}_r^{\mathrm{r}} + i\boldsymbol{q}_r^{\mathrm{i}}$. Substituting these expressions into Eq. (7.8.10), we get

$$\boldsymbol{L}\boldsymbol{q}_r^{\mathrm{r}} = \boldsymbol{f}_r^{\mathrm{r}}, \quad \boldsymbol{L}\boldsymbol{q}_r^{\mathrm{i}} = \boldsymbol{f}_r^{\mathrm{i}}, \quad \text{in } \Omega. \tag{7.8.11}$$

The related boundary conditions, continuity conditions, and other constraints between substructures are still defined by (7.2.15)–(7.2.17). By following an argument similar to the one used in Sect. 7.2.2, we conclude that static deformations of a rotationally periodic structure can also be divided into three groups, and different type of continuity conditions should be specified for each group. These will be elaborated in the next three paragraphs:

Group 1 For the case $r = n$, the same external load $\boldsymbol{f}_n$ is applied to every substructure, and the static displacement of each substructure is the same. The displacement of the entire structure is $\boldsymbol{u}^{(n)} = (\boldsymbol{q}_n, \boldsymbol{q}_n, \cdots, \boldsymbol{q}_n)^{\mathrm{T}}$. The condition must be satisfied at the two shared borders of the k-th substructure includes $\mathbf{u}_k^{(n)}\Big|_{b^+} = \mathbf{u}_k^{(n)}\Big|_{b^-}$, where the notations of b^+ and b^- were introduced in Sect. 7.2.1. In case of $\boldsymbol{f}_n = (\boldsymbol{F}_1 + \boldsymbol{F}_2 + \cdots + \boldsymbol{F}_n) = \boldsymbol{0}$, we have $\boldsymbol{q}_n = \boldsymbol{0}$ and consequently, $\boldsymbol{u}^{(n)} = \boldsymbol{0}$.

Group 2 For the case $r = n/2$, which is possible only when n is an even number, loads applied to adjacent substructures are of the same magnitude but opposite signs, and so are the corresponding displacements. Thus, the displacement of the entire structure is of the form $\boldsymbol{u}^{(n/2)} = (\boldsymbol{q}_{n/2}, -\boldsymbol{q}_{n/2}, \cdots, \boldsymbol{q}_{n/2}, -\boldsymbol{q}_{n/2})^{\mathrm{T}}$. Conditions must be satisfied at the two shared borders of the k-th substructure include $\boldsymbol{u}_k^{(n/2)}\Big|_{b^+} = -\boldsymbol{u}_k^{(n/2)}\Big|_{b^-}$. In case of $\boldsymbol{f}_{n/2} = (\boldsymbol{F}_1 - \boldsymbol{F}_2 + \cdots + \boldsymbol{F}_{n-1} - \boldsymbol{F}_n) = \boldsymbol{0}$, we get $\boldsymbol{u}^{(n/2)} = \boldsymbol{0}$.

Group 3 For the case $r \neq n$ and $r \neq n/2$ (n is even), functions $\boldsymbol{f}_r$, $\boldsymbol{q}_r$, and $\boldsymbol{u}^{(r)}$ are conjugates of $\boldsymbol{f}_{n-r}$, $\boldsymbol{q}_{n-r}$, and $\boldsymbol{u}^{(n-r)}$, respectively. Thus, it is possible to express $\boldsymbol{u}^{(r)}$ and $\boldsymbol{u}^{(n-r)}$ as $\boldsymbol{u}^{(r)} = \boldsymbol{v}^{(r)} + i\boldsymbol{w}^{(r)}$ and $\boldsymbol{u}^{(n-r)} = \boldsymbol{v}^{(r)} - i\boldsymbol{w}^{(r)}$, respectively. By calculating the sum of and difference between $\boldsymbol{u}^{(r)}$ and $\boldsymbol{u}^{(n-r)}$, we get two sets of real displacement functions $\boldsymbol{v}_k^{(r)}$ and $\boldsymbol{w}_k^{(r)}$, where $k = 1, 2, \cdots, n$. They have the following relationship:

$$\begin{bmatrix} \boldsymbol{v}_{k+1}^{(r)} \\ \boldsymbol{w}_{k+1}^{(r)} \end{bmatrix} = \begin{bmatrix} \cos r\psi & -\sin r\psi \\ \sin r\psi & \cos r\psi \end{bmatrix} \begin{bmatrix} \boldsymbol{v}_k^{(r)} \\ \boldsymbol{w}_k^{(r)} \end{bmatrix}, \tag{7.8.12a}$$

where $r = 1, 2, \ldots, n/2 - 1$ if n is an even number, while $r = 1, 2, \ldots, (n-1)/2$ if n is an odd number. Based on Eq. (7.8.12a), conditions must be satisfied at the two shared borders of the k-th substructure include

$$\begin{cases} \boldsymbol{v}_k^{(r)}|_{b^+} = \cos r\psi \cdot \boldsymbol{v}_k^{(r)}|_{b^-} - \sin r\psi \cdot \boldsymbol{w}_k^{(r)}|_{b^-}, \\ \boldsymbol{w}_k^{(r)}|_{b^+} = \sin r\psi \cdot \boldsymbol{v}_k^{(r)}|_{b^-} + \cos r\psi \cdot \boldsymbol{w}_k^{(r)}|_{b^-}. \end{cases} \tag{7.8.12b}$$

Thus, to determine the static displacement of the entire structure, we should solve a series of static deformation problems at the substructure level and then recover the solution for the structure by applying Eqs. (7.8.8a) and (7.8.8b). This should lead to significant reduction in the computational effort. Depending on the value of the parameter r, different strategies should be followed in the analysis. Specifically, when $r = n$, we need to solve for static displacements $\boldsymbol{u}_k^{(n)}$ of one substructure with the condition that its two shared boundaries deform in the identical way; when $r = n/2$ and n is an even integer, we need to solve for static displacements $\boldsymbol{u}_k^{(n/2)}$ with the condition that its two shared boundaries deform in the "opposite" way; and finally, when r assumes all other integer values, we need to solve for two sets of static displacements $\boldsymbol{v}_k^{(r)}$ and $\boldsymbol{w}_k^{(r)}$ associated with one substructure, and conditions given by Eq. (7.8.12b) must be satisfied at the two boundaries where the substructure interfaces with its two adjacent counterparts.

7.8.3 *Qualitative Properties in Static Deformation of Chain Structures*

Consider a chain structure consisting of n substructures. Suppose the force acting on its k-th substructure is $\boldsymbol{F}_k$, and the entire structure is under the action of the force $\boldsymbol{F} = (\boldsymbol{F}_1, \boldsymbol{F}_2, \ldots, \boldsymbol{F}_n)^{\mathrm{T}}$. The governing equation for the static deformation is

$$\boldsymbol{L}\boldsymbol{u}_k = \boldsymbol{F}_k \quad \text{in } \Omega \text{ and } k = 1, 2, \ldots, n. \tag{7.8.13}$$

The boundary and connection conditions are still given in Eqs. (7.4.2) and (7.4.3), respectively.

Apply the following transformation to the generalized displacement $\boldsymbol{u} = (\boldsymbol{u}_1, \boldsymbol{u}_2, \ldots, \boldsymbol{u}_n)^{\mathrm{T}}$ and the external force $\boldsymbol{F} = (\boldsymbol{F}_1, \boldsymbol{F}_2, \ldots, \boldsymbol{F}_n)^{\mathrm{T}}$:

$$\begin{aligned} \boldsymbol{u} &= (\boldsymbol{u}_1, \boldsymbol{u}_2, \ldots, \boldsymbol{u}_n)^{\mathrm{T}} = \boldsymbol{C}\boldsymbol{q}, \\ \boldsymbol{F} &= (\boldsymbol{F}_1, \boldsymbol{F}_2, \ldots, \boldsymbol{F}_n)^{\mathrm{T}} = \boldsymbol{C}(\boldsymbol{f}_1, \boldsymbol{f}_2, \ldots, \boldsymbol{f}_n)^{\mathrm{T}} = \boldsymbol{C}\boldsymbol{f}, \end{aligned} \tag{7.8.14}$$

where $\boldsymbol{C}$ in the above equation is defined in Eq. (7.4.10). Thus, we can get the decoupled equations below:

$$\boldsymbol{L}\boldsymbol{q}_k = \boldsymbol{f}_k, \quad \text{in } \Omega, \quad k = 1, 2, \ldots, n, \tag{7.8.15}$$

as well as boundary conditions and external constraints specified in Eqs. (7.4.17) and (7.4.18).

After the decoupled vector $\boldsymbol{q}_r$ is determined, the first equation in Eq. (7.8.14) is used to obtain the static deformation of the entire chain structure $\boldsymbol{u} = (\boldsymbol{u}_1, \boldsymbol{u}_2, \ldots, \boldsymbol{u}_n)^{\mathrm{T}}$.

Thus, we may reach the following conclusion: For a chain structure composed of n substructures, its external force may be transformed, by using the second equation in Eq. (7.8.14), into n forces, each of which is applied to just one substructure. After calculating the static deformation of each substructure, we may assemble them using Eq. (7.4.19) to obtain the static deformation of the whole structure.

7.8.4 *Qualitative Properties in Static Deformation of Axisymmetric Structures*

Suppose a three dimensional axisymmetric structure is under the action of external forces $F_r(r, \theta, z)$, $F_\theta(r, \theta, z)$, and $F_z(r, \theta, z)$ in the r, θ, and z directions, respectively. The governing equation for the static deformation of this structure is

$$\begin{cases} \boldsymbol{L}_{r,\theta,z}(r, z)(u(r, \theta, z), v(r, \theta, z), w(r, \theta, z))^{\mathrm{T}} \\ \qquad = (F_r(r, \theta, z), F_\theta(r, \theta, z), F_z(r, \theta, z))^{\mathrm{T}} & \text{in } \Omega, \\ \boldsymbol{B}_{r,\theta,z}(r, z)(u(r, \theta, z), v(r, \theta, z), w(r, \theta, z))^{\mathrm{T}} = \boldsymbol{0} & \text{on } \partial\Omega. \end{cases} \tag{7.8.16}$$

Express the external forces as Fourier series:

$$\begin{aligned} F_r(r, \theta, z) &= \sum_{n=0}^{\infty} \left[f_{r,n}(r, z) \cos n\theta + f'_{r,n}(r, z) \sin n\theta \right], \\ F_\theta(r, \theta, z) &= \sum_{n=0}^{\infty} \left[f_{\theta,n}(r, z) \cos n\theta + f'_{\theta,n}(r, z) \sin n\theta \right], \\ F_z(r, \theta, z) &= \sum_{n=0}^{\infty} \left[f_{z,n}(r, z) \cos n\theta + f'_{z,n}(r, z) \sin n\theta \right]. \end{aligned} \tag{7.8.17}$$

Depending on the actual forces applied, the series may contain finite or infinite number of terms. In addition, also expand the displacements $u(r, \theta, z)$, $v(r, \theta, z)$, and $w(r, \theta, z)$ into Fourier series, as in Eq. (7.5.2). Plugging Eqs. (7.5.2) and (7.8.17) into Eq. (7.8.16) and taking account of the orthogonality relation between functions $\cos n\theta$ and $\sin n\theta$, we may decouple different harmonic waves and obtain

$$\begin{cases} \boldsymbol{L}_{r,z,n}(U_n(r,z),\ V_n(r,z),\ W_n(r,z))^{\mathrm{T}} = (f_{r,n}(r,z),\ f_{\theta,n}(r,z),\ f_{z,n}(r,z))^{\mathrm{T}} & \text{in } \Omega, \\ \boldsymbol{B}_{r,z,n}(U_n(r,z),\ V_n(r,z),\ W_n(r,z))^{\mathrm{T}} = \boldsymbol{0} & \text{on } \partial\Omega, \end{cases} \tag{7.8.18}$$

where $n = 0, 1, 2, \ldots$. It can be seen that U_n', V_n', and W_n' as well as $f_{r,n}'$, $f_{\theta,n}'$, and $f_{z,n}'$ also satisfy Eq. (7.8.18). Thus, the static deformation problem of a three-dimensional axisymmetric structure, represented by Eqs. (7.8.16), may be reduced to two groups of static deformation problems in two dimensions, one being symmetric while the other antisymmetric.

At the conclusion of this chapter, we would like to mention that there can be multiple level of repetition for repetitive structures. Here are some examples: (1) A uniform elastic cube with uniform boundary conditions has three mutually orthogonal planes of mirror symmetry. Hence, the structure has three levels of symmetry. Furthermore, the 1/8 of the cube, obtained by trimming the cube with three symmetry planes, also has a symmetry plane of its own when it deforms. (2) A ring torus or donut-shaped tube formed by rotating a circle 360 degrees is a shell with two-level of axisymmetry. (3) A two-level chain structure can be created by placing a number of identical chain structures shown in Fig. 7.12 in parallel and on a plane, making sure they are aligned and at equal distance, and then connecting respective masses on different chains with identical springs. By the same token, a three-level chain structure may be constructed by stacking two-level chains in the three-dimensional space. Additionally, the repetitive structure may also be an assembly structure.

There is another issue deserving some attention. Per discussion in this chapter, both eigenvalue and static deformation problems of a repetitive structure may be decomposed into a number of corresponding problems at the substructure level. But the success of this process depends a number of qualitative properties, including the existence and uniqueness of the solution of static deformation as well as the existence of modal solutions. These issues will be addressed in Chap. 8.

Finally, key references of this chapter include Refs. [14, 15].

References

1. Bao G (1982) Application of group theory in vibration analysis of shell structures with space rotating symmetry. J Shandong Inst Ind 12(1):14–26 (in Chinese)
2. Cai C, Wu F (1983) On the vibration of rotational periodic structures. Acta Scientiarum Naturalium Univ Sunyatseni 22(3):1–9
3. Cai C, Chung Y, Chan H (1990) Uncoupling of dynamic equations for periodic structures. J Sound Vibr 139(2):253–263
4. Cai CW, Liu JK, Chan HC (2002) Exact analysis of bi-periodic structures. World Scientific Publication, Singapore
5. Chan H, Cai C, Chung Y (1998) Exact analysis of structures with periodicity using U-transformation. World Scientific Publication, Singapore
6. Chen P (2002) A note on computation of rotationally periodic structures. Chin J Comp Mech 19(1):112–113 (in Chinese)
7. Chen WM, Wang DJ, Zhou CY et al (2001) The vibration control of repetitive structures. In: Proceedings of the Asia-Pacific vibration conference, vol III. Jilin Science & Technology Press, Changchun, pp 929–932
8. Chen WM, Sun DC, Wang DJ et al (2006) Reduction approaches for vibration control of repetitive structures. Appl Math Mech 27(5):575–582
9. Evensen DA (1976) Vibration analysis of multi-symmetric structures. AIAA J 14(4):446–453
10. Hu HY, Cheng DL (1988) Investigation on modal characteristics of cyclosymmetric structures. Chin J Appl Mech 5(3):1–8 (in Chinese)
11. Jin DK, Sun DC, Chen WM et al (2005) Static shape control of repetitive structures integrated with piezoelectric actuators. Smart Mater Struct 3:1410–1420
12. Thomas DL (1979) Dynamics of rotational periodic structure. Int J Num Methods Eng 1979(14):81–102
13. Timoshenko S (1955) Vibration problem in engineering, 3rd edn
14. Wang DJ, Wang CC (2000) Natural vibrations of repetitive structures. Chin J Mech 16(2):85–95
15. Wang DJ, Zhou CY, Rong J (2003) Free and forced vibration of repetitive structures. Int J Solids Struct 40:5477–5494
16. Wang DJ, Chen J, Wang HJ (2003) The feature of the dual tones of Chinese music bells. Mech Eng 25(4):12–16 (in Chinese)
17. Wang WL, Zhu NS, Xu JH (1990) Double compatible mode synthesis of symmetric structures with C_n group. J Aerospace Power 5(4):352–356 (in Chinese)
18. Zhang J, Wang WL, Chen XJ (1984) Natural mode analysis of N blades disc coupled system—modal synthesis of symmetry structure with C_{NV} group. Acta Mech Solida Sin 5(4):469–481 (in Chinese)
19. Zhong W (1991) The eigen-value problem of the chain of identical substructures and the expansion method solution lasted on the eigen-vectors. Acta Mech Sin 23(1):72–81

Chapter 8
Theory on the Existence of Solutions in Structural Mechanics

This chapter is devoted to the more fundamental subjects, such as the existence of solutions of static deformation and vibrational modes in the linear theory of Structural Mechanics and the validity of linear theoretical models of structures. Because the existence theory in linear Elasticity forms the theoretical foundation of the study, it will be heavily involved in our discussion, too. Since 1980s, Dajun Wang and Haichang Hu have published a series of research papers. They developed a brand new approach by integrating theories of Solid Mechanics and Functional Analysis, and established the unified existence theorem for solutions of static deformation and vibrational modes, which is applicable to shells, structures of composite materials, as well as assembly structures involving complex shapes. As part of his effort to study the validity of models in structural theories, Dajun Wang later made some important enhancement to the existence theorem. This chapter is primarily based on their related work.

8.1 Introduction

Regarding the existence of solutions in Solid Mechanics, we may identify three important and representative problems based on classification and evolution of the linear elastic theory: the existence of solutions of differential equations, primarily elliptic differential equations; existence of solutions in Elasticity; and existence of solutions in structural theories.

With the advent of the Hilbert space theory, some mathematicians have achieved tremendous success in systematically solving the first problem, i.e., the existence of solutions of differential equations. One representative work is Mikhlin's monograph *Minimization Problems of Quadratic Functionals* [24], published in 1952. A number of subjects were expounded in the book, including the existence and uniqueness of solutions of differential equations, the convergence of approximation methods such as the Ritz method, etc. The monograph has provided a theoretical framework for

D. Wang et al., *Qualitative Theory in Structural Mechanics*,
https://doi.org/10.1007/978-981-13-1376-9_8

many subsequent studies in Physics, among which is the existence of solutions in Elasticity.

Later, a few researchers solved various existence problems in Elasticity that are related to solutions of static deformation, modal analysis, and dynamic response, under a wide range of boundary conditions. See Refs. [6, 9, 11, 23, 24, 26]. In his monograph [24], Mikhlin summarized the process to prove the existence of solutions of static deformation and vibrational modes in Elasticity for a three-dimensional elastic solid subject to fixed, free, rigid contact conditions at the boundary, or a hybrid of the three. Fichera, in his dedicated study [10] published in 1972, provided a more refined analysis on solutions of static deformation, vibrational modes, and dynamic response in Elasticity, using the theory of Sobolev spaces.

Regarding the proof for the existence of solutions in Elasticity, there is an alternative to the approach based on the theory of Hilbert spaces. Kupradze proved the existence of static deformation solutions in Elasticity in 1954, by taking advantage of the theory of multidimensional singular potentials and the singular integral equations. In 1963, Kupradze summarized his work in a book [19], which was translated into English in 1965. In 1968, his monograph, *Three Dimensional Problems in Mathematical Theories of Elasticity and Thermal Elasticity* [20], was published, and its English version became available in 1979.

Comparing the two methods mentioned above, the approach based on the Hilbert space theory is well suited for handling equations with variable coefficients and subject to the general form of boundary constraints, while Kupradze's approach provides a convenient path for studying the differentiability and structure of a solution.

As for the work on the existence of solutions in structural theories, it is a somewhat different story. In the case of one-dimensional structures, membranes, and plates, it is relatively easy to deal with the related existence problems. But for shells, due to the geometric complexity of mid-surfaces as well as the diversity of their theoretical models, it is difficult to study the existence of solutions by starting directly from the shell equations. Moreover, when assembly structures are involved, the situation becomes even more unmanageable. Because of these reasons, the progress has been much slower in the study of the existence of solutions in structural theories.

Over the years, a number of researchers have proved the positive definiteness of theoretical operators in various shell theories. Here is a chronical listing of some authors and shell models they studied: Shoikhet [27] on the Novozhilov shell model and Gordegiani [13] on the Vekua shell model in 1974; Benadou and Ciarlet [2] on shells with a clamped boundary in 1975; J. K. Wu [39] on shells with a wide range of mid-surface shapes and subject to either hinged or clamped boundary conditions in 1981; Benadou and Lalanne [3] on Koiter's shallow shell in 1985; Ciarlet and Miara [5] on the linearized Marguerre-von Kánmán shallow shell in 1992; Benadou, Ciarlet, and Miara [4] on Koiter and Naghdi shells with general mid-surface shapes in 1994; and so on and so forth. All the work essentially followed the same approach, i.e., deriving the positive definiteness of shell theory operators directly from the respective governing equations. Furthermore, there was only limited involvement of the compactness of energy embedding operators in the discussion and consequently, dynamics problems were largely left untouched.

Since 1980s, Dajun Wang and Haichang Hu have published a series of articles [32–35]. They developed a brand new approach by integrating theories of Solid Mechanics and Functional Analysis. This leads to the establishment of the unified existence theorems for solutions of static deformation and vibrational modes that are applicable to complex shells, structures of composite materials, and assembly structures involving complicated shapes.

The theory they developed has been well received by many experts in this area. Some of their comments are included in a note attached to the end of this section.

In addition, the authors of this book further enhanced the work pioneered by Wang and Hu on the existence of solutions in structural theories. In particular, they expounded the validity of models in structural theories as well as the validity of elastic assembly structures. These results are given in Sects. 8.5 and 8.6 of this chapter.

In subsequent sections, we will present details of the existence and uniqueness of solutions in Elasticity and structural theories, and also introduce other subjects of interest in fundamental theories. The discussion will involve some mathematical knowledge in Calculus of Variations, Functional Analysis, Hilbert Spaces, Sobolev Spaces, etc. However, since the target readers of this book are primarily engineers as well as scholars and graduate students in the field of Applied Mechanics, we do not attempt to systematically introduce all prerequisites. Furthermore, some discussions involving Functional Analysis are handled in a less rigorous manner. If readers want to have a deeper understanding of certain subjects, they are encouraged to check out Refs. [1, 10, 12, 16, 18, 22, 24, 25, 38].

NOTE:

The well-known mathematician R. Valid of France stated in his monograph *The Nonlinear Theory of Shells through Variational Principles—From Elementary Algebra to Differential Geometry* [31], published in 1995: "A very important question remained unanswered by mathematician colleagues up to the publication of the following so-called 'Chinese theorems'. The question was 'Which general conditions would, in particular theories of plates and shells, by the application of specific stress and/or displacement assumptions to the 3D theory of linear elasticity, ensure the existence and uniqueness of solutions already shown in the 3D theory?' In 1985 the answer came in an article by WANG Dajun and HU Haichang of Peking University under the title 'Positive Definiteness and Compactness of Two Kinds of Operators in Theories of Elastic Structures'".

Member of Academy of Science of South Africa, Dr. Bohua Sun, pointed out in the disquisition *On Existence Theorem in Theories of Elastic Structures—Advances in Mechanics* [30] in 2012: "The research on the existence of solutions in structural theories by Dajun Wang and Haichang Hu is a seminal work. ... It has solved quite extensively the fundamental problem in the theory of Solid Mechanics, and is an achievement of long-term significance."

8.2 Variational Solutions for Three Categories of Problems in Structural Theories

8.2.1 Types of Solutions in Structural Theories

With regard to an elastic solid or structure, there are primarily three categories of problems in Elasticity and structural theories: Find the solution of static deformation, given a set of static external loads; solve for its vibration modes; and obtain dynamic response of the system to a set of specified dynamic external forces, initial displacements, and initial velocities. They usually correspond respectively to these three categories of mathematical problems: the boundary value, eigenvalue, and initial-boundary value problems of differential equations.

Furthermore, there are two types of solutions associated with each of the aforementioned three categories of problems. The first type is the classical solution, while the other type is the generalized solution.

First, those satisfying differential equations as well as boundary and initial conditions are referred to as classical solutions. This will be elaborated in the next three paragraphs.

Solutions of Static Deformation Solve the following boundary value problem of differential equations

$$\begin{cases} Au(x) = f(x) & \text{in } \Omega, \\ Bu(x) = 0 & \text{on } \partial\Omega, \end{cases} \tag{8.2.1}$$

where x represents the coordinate or coordinate vector for describing the structure or elastic solid, Ω denotes the region occupied by the structure or elastic solid, $\partial\Omega$ indicates the boundary of Ω, $u(x)$ represents the displacement or displacement vector of the object, A is the differential operator or differential operator matrix in the structural theory or Elasticity, B stands for the differential operator or differential operator matrix of boundary conditions, and f designates the external force or external force vector.

Modal Solutions in Vibration Solve the following eigenvalue problem of differential equations:

$$\begin{cases} Au = \lambda\rho u & \text{in } \Omega, \\ Bu = 0 & \text{on } \partial\Omega. \end{cases} \tag{8.2.2}$$

Here, ρ is the inertia operator, which is normally the mass density but may include other things under more complex situations, e.g., the moment of inertia of a section in the case of a Rayleigh beam; λ denotes the eigenvalue; and $\sqrt{\lambda}/2\pi$ designates the natural frequency.

Solution of Dynamic Response Solve the initial-boundary value problem of differential equations:

$$\begin{cases} Au + \rho u_{tt} = f(x,t) & \text{in } \Omega \text{ and } t > 0, \\ Bu = 0 & \text{on } \partial\Omega \text{ and } t \geq 0, \\ u(x,0) = g_0(x),\ u_t(x,0) = g_1(x) & \text{in } \Omega. \end{cases} \tag{8.2.3}$$

Second, solutions to variational equations, in the context of generalized differentiation, are referred to as generalized solutions. To facilitate subsequent discussion, let us introduce the space $L_2(\Omega)$, which is a Hilbert space[1] and consists of all square integrable vector functions $\{\varphi(p)\}$ in a bounded domain Ω. The following inner product is defined on the space:

$$(\varphi, \psi) = \int_{\Omega} \varphi(p)\bar{\psi}(p)\mathrm{d}x \tag{8.2.4}$$

with the squared norm

$$\|\varphi\|^2 = \int_{\Omega} |\varphi(p)|^2 \mathrm{d}x. \tag{8.2.5}$$

For the three categories of problems mentioned above, their variational formulations are given in the next three paragraphs.

Solutions of Static Deformation It minimizes the following functional of the potential energy

$$F(u) = \frac{1}{2}(Au, u) - (u, f). \tag{8.2.6}$$

Modal Solutions in Vibration Consider the following Rayleigh quotient

$$R(u) = \frac{(Au, u)}{(\rho u, u)}. \tag{8.2.7}$$

By minimizing it under different constraints, we get the solutions:

$$\begin{cases} \lambda_1 = R(u_1) = \min R(u), \\ \lambda_i = R(u_i) = \min R(u), \quad (\rho u, u_j) = 0, \quad j = 1, 2, \ldots, i-1. \end{cases} \tag{8.2.8}$$

Solution of Dynamic Response Find a solution that makes the variation of the Hamilton action functional vanish, i.e.,

$$\delta \int_0^{t_1} (\Pi - K - W)\mathrm{d}t = 0, \tag{8.2.9}$$

[1] All Hilbert spaces in this book refer to complete inner-product spaces.

where Π, K, and W represent, respectively, the potential energy of the system, its kinematic energy, and the work done by external forces.

Classical and generalized solutions have dissimilar differentiability. Specifically, for a differential equation of the $2k$-th order, its classical solution belongs to the function set of comparison displacements[2]; while the generalized solution only needs to have a minimum of k-th generalized derivatives. Our discussion on existence of solutions will be within the context of generalized solutions.

For a specific problem, the differentiability of its generalized solution can be deduced from that of the following factors: coefficients of the governing equation, border shapes, boundary conditions, as well as external forces. This subject is part of the regularity of solutions.

The concept of weak solutions can be found in some publications of theoretical Mechanics, which refers to solutions with k-th classical derivatives obtained through variational methods. These solutions are from the sets of admissible functions. We would like to point out that they are different from the generalized solutions based on the concept of generalized derivatives. It will be clear later in this chapter that the existence of solutions to the aforementioned variational problems can only be resolved in the context of generalized solutions. Thus, here we divide solutions into two types from the mathematical perspective, namely, the classical and generalized solutions.

8.2.2 *Solving Static Deformation Problems with Variational Methods*

Suppose an operator A is defined on a dense set D_A in the L_2 space. If

$$(Au, u) > 0$$

for any nonzero function u in the domain D_A, then A is called a positive operator. Furthermore, if

$$(Au, u) \geq \gamma^2 \|u\|^2, \tag{8.2.10}$$

where γ is a positive constant, then A is referred to as a positive definite operator. A positive operator must be a symmetric operator, and can be extended to become a self-adjoint operator.

---

[2]Consider a differential equation of $2k$-th order and its boundary conditions in Structural Mechanics. The functions are from the set of comparison functions, if they have $2k$-th classical derivatives and satisfy all boundary conditions; while the functions belong to the set of admissible functions, if they have k-th classical derivatives and satisfy the displacement boundary conditions only.

Theorem 8.1 *Suppose A to be a positive operator. If the equation*

$$Au = f \tag{8.2.11}$$

has a solution, it must be unique.

Proof If $Au_1 = Au_2 = f$, then $A(u_1 - u_2) = 0$ and consequently,

$$(A(u_1 - u_2),\ u_1 - u_2) = 0.$$

This leads to $u_1 - u_2 = 0$. In the L_2 space, $u_1 = u_2$ implies that u_1 and u_2 are equal almost everywhere; and additionally, if both u_1 and u_2 are also continuous functions, then $u_1 \equiv u_2$. ■

Theorem 8.2 *Let A be a positive operator. If* Eq. (8.2.11) *has a solution, then the solution makes the functional*

$$F(u) = (Au, u) - (u, f) - (f, u) \tag{8.2.12}$$

reach the minimum.

Theorem 8.3 *Suppose A represents a positive operator. Then, the function that minimizes the functional*

$$F(u) = (Au, u) - (u, f) - (f, u)$$

is the solution to the equation

$$Au = f.$$

Notice that the equality above is established in the L_2 space, i.e., in the sense of the square-integrable condition. If it is further required that both Au and f be continuous, then the equality is converted to one in the regular sense, i.e., u is now turned into a classical solution.

Theorem 8.2 provides the theoretical foundation to find the minimum of Functional (8.2.12) by solving Eq. (8.2.11); while conversely, Theorem 8.3 establishes the theoretical basis to obtain the solution to Eq. (8.2.11) by solving the minimization problem of Functional (8.2.12). The latter is referred to as the fundamental variational problem, and its counterpart in Solid Mechanics is to solve the static deformation problem by using the principle of minimum potential energy.

It needs to be emphasized that, in reality, we do not start from Eq. (8.2.12) to find the extremum of the functional $F(u)$. This is because the expression (Au, u) in Eq. (8.2.12) contains derivatives of the same order as the original equation, which imposes a stringent requirement on the function set eligible for the analysis and thus discounts the advantage of this approach. In practice, we normally solve the extremum problem in a different way. Specifically, integrate the expression (Au, u)

by parts for k times, during which the boundary constraints will be repeatedly applied. This leads to the following expression:

$$(Au, u) = \int_{\Omega} \bar{u}Au\mathrm{d}x = \Lambda(u, u) + \int_{\partial\Omega} \sum_{j=1}^{k} R_j u \tilde{R}_j u \mathrm{d}x, \tag{8.2.13}$$

where $\Lambda(u, u)$ is an integral on Ω. If Au in the differential equation contains $2k$-th derivatives, then the term $\Lambda(u, u)$ only contains k-th derivatives of u. Consequently, it is easier to solve the functional extremum problem by using Eq. (8.2.13).

Example 1 Consider a beam fixed at one end and unconstrained at the other. Let w denote its deflection and $EJ(x)$ represent its flexural stiffness. The governing equation and boundary conditions are:

$$\begin{cases} (EJ(x)w''(x))'' = f(x), \\ w(0) = 0, \quad w'(x)\big|_{x=0} = 0, \\ w''(x)\big|_{x=l} = 0, \quad (EJw'')'\big|_{x=l} = 0. \end{cases} \tag{8.2.14}$$

Here, A in Eq. (8.2.12) assumes the form of the differential operator on the left-hand side of the first equation in (8.2.14). Functional (8.2.12) becomes

$$\frac{1}{2}F(w) = \frac{1}{2}\int_0^l w(EJw'')''\mathrm{d}x - \int_0^l wf\,\mathrm{d}x. \tag{8.2.15}$$

Integrating by part the first term on the right-hand side of the equation above, we get

$$\begin{aligned} \frac{1}{2}F(w) &= \frac{1}{2}\int_0^l EJ\left(w''\right)^2\mathrm{d}x + \frac{1}{2}\left(w \cdot (EJw'')'\right)\big|_{x=l} \\ &- \frac{1}{2}\left(w' \cdot EJw''\right)\big|_{x=l} - \int_0^l wf\,\mathrm{d}x = \Pi + W, \end{aligned} \tag{8.2.16}$$

where Π and W represent, respectively, the first three terms and the last term between the two equal signs of the above equation. The physical meaning of Eq. (8.2.16) is that the total potential energy of the beam is the sum of its strain energy Π and potential energy W from external forces. This demonstrates that finding the minimum value of $F(u)$ leads to none other than the principle of minimum potential energy. In other words, the classical solution makes the total potential energy of the system reach the minimum.

It is clear that if Eq. (8.2.15) is used to find the extremum then the solution is part of the function set of comparison displacements. However, if Eq. (8.2.16) becomes the starting point for solving the extremum problem then the solution belongs to the function set of admissible displacements.

Thus, the advantage of the variational approach can be fulfilled only by starting from Eq. (8.2.16). For this reason, it has been tacitly assumed that the variational method is always applied to problems of the form shown in Eq. (8.2.16), whose differentiability requirement has been significantly loosened. However, according to Theorem 8.3, the solution obtained from Eq. (8.2.15) is indeed a solution to the original differential equation; while the one derived from Eq. (8.2.16) may not be. In this situation, an additional constraint has to be introduced, which requires the solution in the latter approach to have derivatives of the same order as the original differential equation, in order for it to be also a solution of the original differential equation.

Nevertheless, it will be seen in Sect. 8.3, where the existence of solutions is explained, that a solution with derivatives in the classical sense may often not exist, even when we start from Eq. (8.2.16) to solve the problem.

Example 2 Consider a thin elastic plate of uniform thickness. Suppose it occupies a finite planar region Ω, and is clamped at the border. The static deformation equation and boundary conditions are:

$$\begin{cases} D\Delta^2 w = f(x) & \text{in } \Omega, \\ w = \dfrac{\partial w}{\partial n} = 0 & \text{on } \partial\Omega, \end{cases} \tag{8.2.17}$$

where $D = Eh^3/12(1-\nu^2)$ is the bending stiffness of the thin plate, E represents the elastic modulus, ν denotes the Poisson ratio, h designates the thickness of the plate, Δ^2 is the bilaplacian operator, and w stands for the displacement. The potential energy of the plate is

$$\begin{aligned} \frac{1}{2}F(w) &= \frac{D}{2}\int\limits_{\Omega}\left\{(\Delta w)^2 + 2(1-\nu)\left[\left(\frac{\partial^2 w}{\partial x\partial y}\right)^2 - \frac{\partial^2 w}{\partial x^2}\frac{\partial^2 w}{\partial y^2}\right]\right\}\mathrm{d}x\mathrm{d}y \\ &\quad - \int\limits_{\Omega} wf\,\mathrm{d}x\mathrm{d}y = \Pi + W, \end{aligned} \tag{8.2.18}$$

where Δ denotes the two-dimensional Laplace operator.

For a thin plate, the differential operator involves *4*-th derivatives of the displacement, while the strain energy Π only includes 2-nd derivatives of the displacement.

Finally, we would like to point out that Operator A being positive is a precondition for Theorems 8.1–8.3. The physical meaning of this condition is the strain energy of the mechanical system under consideration to be positive.

8.2.3 Solving Modal Problems with Variational Methods

Starting from Eq. (8.2.2), we introduce another functional, the Rayleigh quotient, as follows:

$$R(u) = \frac{(Au, u)}{(\rho u, u)}. \tag{8.2.19}$$

It is known from Eq. (8.2.2) that the i-th natural frequency f_i and the corresponding mode shape ϕ_i satisfy the equation

$$\lambda_i = \frac{(A\phi_i, \phi_i)}{(\rho\phi_i, \phi_i)}, \tag{8.2.20}$$

where the eigenvalue λ_i is related to the natural frequency f_i by $\lambda_i = (2\pi f_i)^2$.

Theorem 8.4 *Suppose A is a positive operator and d is the infimum of the functional*

$$R(\varphi) = \frac{(A\varphi, \varphi)}{(\varphi, \varphi)}. \tag{8.2.21}$$

If there exists a function $\varphi_0 \neq 0$ such that

$$\frac{(A\varphi_0, \varphi_0)}{(\varphi_0, \varphi_0)} = d, \tag{8.2.22}$$

then d is the minimum eigenvalue of A while φ_0 is the corresponding mode shape.

The next theorem shows how to find eigenpairs of higher-orders.

Theorem 8.5 *Assume A to be a positive operator. Let* $\lambda_1 \leq \lambda_2 \leq \cdots \leq \lambda_n$ *and* $\{\varphi_1, \varphi_2, \ldots, \varphi_n\}$ *represent, respectively, its first n eigenvalues and the corresponding mode shapes. Furthermore, suppose there exists a function* $\varphi \neq 0$ *such that the functional* (8.2.21) *reaches the minimum under some additional constraints* $(\varphi, \varphi_i) = 0\,(i = 1, 2, \ldots, n)$. *Then,* φ *and* $\lambda = (A\varphi, \varphi)/(\varphi, \varphi)$ *are, respectively, the* $(n+1)$-*th mode shape* φ_{n+1} *and the corresponding eigenvalue* λ_{n+1}.

The proof of Theorems 8.2–8.5 can be found in Refs. [23, 24].

It is not hard to extend Theorems 8.4 and 8.5 to cover the case of Eq. (8.2.19), which is the modal equation.

Theorems 8.4 and 8.5 provide the theoretical basis to obtain modal solutions using the variational method. Specifically, they show, by solving for extrema of the functional (8.2.19), we get modal solutions of Eq. (8.2.2). However, the conclusion is drawn when the numerator in the expression (8.2.19) is of the form $(A\varphi, \varphi)$. This implicitly requires that the function set of selected functions $\{\varphi\}$ be differentiable up to the order of the original differential equation, i.e., they must belong to the comparison function set. In reality though, when the variational calculation is actually performed, the expression (8.2.13) will be utilized and consequently, the function set of $\{\varphi\}$ needs only to be an admissible function set. Notice that we have applied the same trick when dealing with static deformation problem in the previous subsection.

Example Consider a straight beam identical to the one appeared in Example 1 of Sect. 8.2.2. Suppose the mass per unit length of the beam to be m. The modal equation is

$$(EJw''(x))'' = \omega^2 mw. \tag{8.2.23}$$

By applying the variational calculus to

$$\omega^2 = \mathrm{st}\frac{\Pi}{T} = \mathrm{st}\frac{\int_0^l EJ(w''(x))^2\mathrm{d}x + \{w \cdot (EJw'')' - w' \cdot EJw''\}_{x=l}}{\int_0^l mw^2\mathrm{d}x}, \tag{8.2.24}$$

we get the equation

$$\delta\Pi - \omega^2\delta T = 0. \tag{8.2.25}$$

The modal solution can be found by solving Eq. (8.2.25). Here, the notation "st" in Eq. (8.2.24) represents the operation of finding the stationary value of the related functional. Notice that, in general, the order of derivatives of the function w involved in Eq. (8.2.25) drops to only half of that needed in the characteristic equation (8.2.23).

8.2.4 *Solving Dynamic Response Problems with Variational Methods*

Hamilton's Principle Suppose the time boundary conditions are

$$u = u_0 \text{ when } t = 0, \quad \text{and} \quad u = u_1 \text{ when } t = t_1.$$

Then, for any possible movement, the variational equation

$$\delta\int_0^{t_1}(\Pi - K - W)\mathrm{d}t = 0$$

is equivalent to the equation of motion as well as the initial and boundary conditions given in Eq. (8.2.3), where Π and K denote strain and kinetic energies of the structure, respectively, while W represents the work done by external forces.

Solving for the dynamic response of a structure is, in essence, a time initial-value problem or an initial-value problem in the time domain; while Hamilton's principle deals with boundary value problems in the time domain. Thus, the latter is not directly applicable to dynamic response problems. However, it is very convenient to use Hamilton's principle to establish the equations of motion. As a matter of fact, for some complex structures, it is difficult to set up their governing equations by directly applying Newton's second law. In comparison, it is relatively easy to obtain expressions of potential and kinetic energies for these systems, there is a lot of flexibility in the selection of generalized displacements, and the equations of motion of these structures can then be readily established by using Hamilton's principle.

Gurtin proposed Curtin's Variational Principle in 1964, which is suitable for setting up equations of motion with the time initial-value conditions. See Ref. [14]. This may become a very promising approach in solving dynamic response problems in engineering applications.

In summary, two methods for solving three categories of problems in Structural Mechanics have been discussed in Sect. 8.2: One works directly with differential equations, while the other is based on variational calculus. The former finds a solution in the set of comparison functions, which involve derivatives up to the order of the differential equation; while the latter obtains a solution within the set of admissible functions, whose level of differentiability requirement is reduced by half. Here are our primary concerns in the rest of this chapter: Are there solutions to the three categories of problems in Structural Mechanics? If they do exist, what types of solutions do they belong to? These questions will be addressed in subsequent discussion.

8.3 Existence of the Extremum of a Functional

8.3.1 Solution to the Fundamental Problem of Variational Calculus

The focus of this subsection is to discuss the existence of a solution that minimizes the following functional:

$$F(u) = (Au, u) - (u, f) - (f, u). \tag{8.3.1}$$

The discussion includes four parts.

First, suppose the operator A is defined on a dense linear set M within the Hilbert space H, and assume further that it is a positive definite operator, i.e., there exists a constant $\gamma > 0$ such that

$$(Au, u) \geq \gamma^2 \|u\|^2, \quad u \in M. \tag{8.3.2}$$

Second, define a new inner product on the function set M as

$$[u, v] = (Au, v), \quad u \text{ and } v \in M. \tag{8.3.3}$$

M is then turned into a new inner product space, represented by H_A. The norm is denoted by $|u|$ and calculated as follows,

$$|u|^2 = (Au, u), \quad u \in M. \tag{8.3.4}$$

Since A is a positive definite operator, we may get, from Eq. (8.3.2), the following important relationship between the two norms in the space M:

$$|u| \geq \gamma \|u\|, \quad u \in M. \tag{8.3.5}$$

Although it is possible that H_A is not complete, its completion may be constructed following the normal process, and this results in a new Hilbert space, which is still denoted by H_A for simplicity's sake. It can be shown that H_A is embedded into the original Hilbert space H, this complete space H_A consists of elements in the space H, and Eq. (8.3.5) remains true in H_A.

Third, a good comprehension of the structure of the complete space H_A is essential for correctly understanding the existence of solutions and properly interpreting the meaning of generalized solutions. However, due to space limitation, this important subject can only be lightly covered in the next a few paragraphs. Readers are referred to relevant books, such as Refs. [18], [25], and [38], for details.

Suppose Ω is an open set in the N-dimensional Euclidian space R_N and $x \in R_N$, where $x = (x_1, x_2, \ldots, x_N)$. Let $\alpha = (\alpha_1, \alpha_2, \ldots, \alpha_N)$ be a multi-index, where α_i $(i = 1, 2, \ldots, N)$ are all nonnegative integers. Set

$$x^\alpha = x_1^{\alpha_1} x_2^{\alpha_2} \cdots x_N^{\alpha_N}, \quad |\alpha| = \alpha_1 + \alpha_2 + \cdots + \alpha_N, \quad \text{and}$$

$$\mathrm{D}^\alpha = \mathrm{D}_1^{\alpha_1} \mathrm{D}_2^{\alpha_2} \cdots \mathrm{D}_N^{\alpha_N} = \frac{\partial^{|\alpha|}}{\partial x_1^{\alpha_1} \partial x_2^{\alpha_2} \cdots \partial x_N^{\alpha_N}}.$$

Further assume that the domain Ω is bounded and either its boundary is sufficiently smooth or the domain is a convex polyhedron. If the function $u \in C^1(\bar{\Omega})$, then, for any function $v \in C_0^\infty(\Omega)$, we obtain the following result by repeatedly applying integration by parts:

$$\int_\Omega v \mathrm{D}^\alpha u \mathrm{d}x = (-1)^{|\alpha|} \int_\Omega u \mathrm{D}^\alpha v \mathrm{d}x, \quad v \in C_0^\infty(\Omega).$$

The concept of generalized derivatives may be introduced by using the above equation.

Definition Consider a function $u \in L_2(\Omega)$. If there exists a function $\varphi \in L_2(\Omega)$ such that the following condition is satisfied for any function $v \in C_0^\infty(\Omega)$

$$\int_\Omega v\varphi \mathrm{d}x = (-1)^{|\alpha|} \int_\Omega u \mathrm{D}^\alpha v \mathrm{d}x, \quad v \in C_0^\infty(\Omega),$$

then φ is referred to as the $|\alpha|$-th generalized derivative of u, which is denoted by $\mathrm{D}^\alpha u$.

From the discussion before the definition, it is clear that classical derivatives must also be generalized derivatives, but the converse may not be correct.

Define the inner product on the function set $u \in C^k(\bar{\Omega})$ as

$$(u, v)_k = \int_\Omega \sum_{|\alpha|=0}^{k} \mathrm{D}^\alpha u \cdot \mathrm{D}^\alpha v \mathrm{d}x.$$

The norm is calculated by

$$\|u\|_k = \left[\int_\Omega \sum_{|\alpha|=0}^{k} (\mathrm{D}^\alpha u)^2 \mathrm{d}x \right]^{1/2}.$$

The function set $C^k(\bar{\Omega})$ forms an inner product space with the previously defined inner product. But this space is not complete. We may make function sets $C^k(\bar{\Omega})$ and $C_0^k(\Omega)$ complete with respect to the norm $\|\cdot\|_k$, and the resulting spaces are denoted by $H_k(\Omega)$ and $\overset{0}{H}_k(\Omega)$, respectively. They are both Hilbert spaces, and are often called Sobolev spaces. It is important to realize that the integrands in the above inner product and norm contain generalized derivatives, rather than classical derivatives.

For an operator A involving $2k$-th derivatives, the complete spaces corresponding to C^{2k} and C_0^{2k} are, respectively, H_k and $\overset{0}{H}_k$. The latter two spaces consist of functions with k-th generalized derivatives, and the generalized solution resides in one of these two spaces. For instance, the order of highest derivatives appearing in operators of beams and plates is four. Thus, their generalized solutions belong to H_2 spaces. In comparison, the order of highest derivatives in operators of elasticity, bars, and thin membranes is two. Consequently, their generalized solutions are in H_1 spaces.

Fourth and finally, with the fundamental problem in Calculus of Variations, we are supposed to find a function in the function set M such that the functional (8.3.1) is minimized. But there may be no solution to the problem under general circumstances. Therefore, it is necessary to expand the scope of the search for a solution. This is elaborated in the next two paragraphs.

Set $(Au, u) = [u, u]$ for $u \in M$. Next, let f be a fixed function in the space H, and u be an arbitrary function in the space H_A. We have, from the Schwarz inequality

$$|(u,f)| \leq \|f\| \cdot \|u\| \leq \frac{\|f\|}{\gamma} \mathbf{|u|}.$$

Thus, (u,f) is a bounded functional in the space H_A. According to the Riesz representation theorem, there exists a unique function $u_0 \in H_A$ such that

$$(u,f) = [u, u_0], \quad u \in H_A. \tag{8.3.6}$$

Hence, the functional (8.3.1) now becomes

$$F(u) = (Au, u) - (u, f) - (f, u) = [u, u] - [u, u_0] - [u_0, u]. \tag{8.3.7}$$

The above equation holds true for $u \in M$. But since all terms on the right-hand side of the second equal sign are also meaningful in the space H_A, the functional $F(u)$ may be extended to the space H_A by using Eq. (8.3.7), and the minimum of $F(u)$ in the space H_A can then be obtained. Now let us further reformat Eq. (8.3.7) into

$$F(u) = [u - u_0, u - u_0] - [u_0, u_0] = \mathbf{l}u - u_0\mathbf{l}^2 - \mathbf{l}u_0\mathbf{l}^2.$$

When $u = u_0$, $F(u)$ reaches the minimum, which is

$$\min F(u) = -\mathbf{l}u_0\mathbf{l}^2.$$

This result leads to the following theorem:

Theorem 8.6 *Given a positive definite operator A and a function f in the Hilbert space H, there exists a unique function in the space H_A such that the functional*

$$F(u) = (Au, u) - (u, f) - (f, u)$$

is minimized.

The solution in Space H_A is none other than the so-called generalized solution. The space H_A is a Sobolev space, whose member functions have generalized derivatives. For instance, if A is a differential operator of the $2k$-th order, then the term (Au, u) in the functional $F(u)$ contains the product of two differential operators of the zero-th up to the k-th order, and member functions in Space H_A have the first to the k-th generalized derivatives.

Example 1 Consider a beam clamped at $x = 0$. We will show that its differential operator is positive definite.

First, let us derive an inequality. Suppose $w(x)$ is a function on the interval $(0, \ l)$, and satisfies boundary conditions $w(0) = w'(0) = 0$. We have

$$w(x) = \int_0^x \frac{\mathrm{d}w(\xi)}{\mathrm{d}\xi}\mathrm{d}\xi.$$

Set

$$\varphi = 1 \quad \text{and} \quad \psi = \frac{\mathrm{d}w(\xi)}{\mathrm{d}\xi}$$

in the Schwarz inequality

$$|(\varphi, \psi)| \leq \|\varphi\| \cdot \|\psi\|. \tag{8.3.8}$$

Thus,

$$|w(x)| = |(\varphi, \psi)| \leq \sqrt{x} \cdot \sqrt{\int_0^x \left| \frac{\mathrm{d}w}{\mathrm{d}\xi} \right|^2 \mathrm{d}\xi}.$$

This further leads to

$$|w(x)|^2 \leq x \int_0^x \left| \frac{\mathrm{d}w}{\mathrm{d}\xi} \right|^2 \mathrm{d}\xi \leq l \int_0^l \left| \frac{\mathrm{d}w}{\mathrm{d}\xi} \right|^2 \mathrm{d}\xi. \tag{8.3.9}$$

Now we are ready to prove the operator of a beam to be positive definite. The operator of the beam is of the form

$$Aw = (EJw'')''. \tag{8.3.10}$$

Thus,

$$\begin{aligned}(Aw, w) &= \int_0^l (EJw'')''w\mathrm{d}x \\ &= \int_0^l EJ(w'')^2\mathrm{d}x + (EJw'')'w\Big|_0^l - (EJw')w'\Big|_0^l,\end{aligned}$$

where on the right-hand side of the second equal sign, the first term is the strain energy of the beam itself, while the second and third terms represent contribution from boundary supports to the strain energy. Based on the most general form of beam boundary conditions discussed in Chap. 6, the last two terms must be no less than zero. Consequently,

$$(Aw, w) \geq \int_0^l EJ(w'')^2\mathrm{d}x.$$

According to Eq. (8.3.9), we get

$$\begin{aligned}
\int_0^l w^2(x)\mathrm{d}x &\le l^2 \int_0^l \left(\frac{\mathrm{d}w}{\mathrm{d}\xi}\right)^2 \mathrm{d}\xi \le l^2 \left[l^2 \int_0^l \left(\frac{\mathrm{d}^2 w}{\mathrm{d}\xi^2}\right)^2 \mathrm{d}\xi \right] \\
&\le \frac{l^4}{EJ_0} \int_0^l EJ \left(\frac{\mathrm{d}^2 w}{\mathrm{d}\xi^2}\right)^2 \mathrm{d}\xi \\
&\le \frac{l^4}{EJ_0}(Aw, w),
\end{aligned} \tag{8.3.11}$$

where $EJ_0 = \min EJ(x)\,(0 < x < l)$. Eq. (8.3.11) implies

$$(Aw, w) \ge \gamma^2 \|w\|^2.$$

This shows that when $(EJw'')''$ is meaningful, i.e., when $EJ \in C^2$, the operator of a beam is positive definite.

Therefore, for static deformation of a beam, there is a solution to the fundamental problem in Calculus of Variation, which is the problem of minimizing the potential energy. In other words, the generalized solution to the static deformation problem of a beam exists.

Example 2 Consider a thin elastic plate of uniform thickness. Suppose it occupies a finite planar region, its boundary is sufficiently smooth, and the plate is clamped along the entire border. The governing differential equation and boundary conditions are given in Eq. (8.2.17), while the potential energy is provided in Eq. (8.2.18). After some derivation, we obtain

$$\begin{aligned}
\Pi &\ge \alpha \int_\Omega \left[w^2 + \left(\frac{\partial w}{\partial x}\right)^2 + \left(\frac{\partial w}{\partial y}\right)^2 + \left(\frac{\partial^2 w}{\partial x^2}\right)^2 + \left(\frac{\partial^2 w}{\partial x \partial y}\right)^2 + \left(\frac{\partial^2 w}{\partial y^2}\right)^2 \right] \mathrm{d}x\mathrm{d}y \\
&\ge \beta \int_\Omega w^2 \mathrm{d}x\mathrm{d}y.
\end{aligned}$$

The result above shows that the operator of the plate $A = D\Delta^2$ is positive definite, i.e.,

$$(Aw, w) \ge \gamma^2 \|w\|^2.$$

8.3.2 Existence of Solutions to the Eigenvalue Problem

It has been explained in Sect. 8.2.3 that the problem of finding modal solutions can be transformed to the problem of solving for the minima of a functional, specifically, the Rayleigh quotient. This subsection discusses the existence of local minima of the Rayleigh quotient, i.e., the modal solutions, as well as their structure.

Theorem 8.7 *If A is a positive definite operator and the embedding operator from the space H_A to the space H is compact, then the functional, the Rayleigh quotient $R(u)$, reaches its infimum in Space H_A, namely,*

$$m = \inf R(u_0) = \inf \frac{(Au_0, u_0)}{(u_0, u_0)},$$

where m and u_0 are, respectively, the first eigenvalue and the respective eigenfunction of Eq. (8.2.2).

Theorem 8.7 is about the existence of the first eigenpair of the generalized solutions to the eigenvalue problem (8.2.2); while the next theorem deals with the existence of higher-order eigenpairs of the generalized solutions and also reveals the overall structure of eigenvalues and eigenfunctions.

Theorem 8.8 *Let A be a positive definite operator and assume the embedding operator from the space H_A to the space H to be compact. Then,*

(1) *the operator A has infinitely many eigenvalues, if Space H is of infinite dimension;*
(2) *the only limit point of all these eigenvalues is the infinity; and*
(3) *the sequence of corresponding eigenfunctions form a complete and orthogonal set in both spaces H and H_A.*

In structural theories, the governing equation of an eigenvalue problem is usually of the form

$$Au = \lambda Bu, \tag{8.3.12}$$

which is the so-called generalized eigenvalue problem.

It is not difficult to generalize Theorem 8.8 to cover the generalized eigenvalue problem. We will present the following theorem without proof. The interested reader is referred to Ref. [24].

Theorem 8.9 *First, let both A and B be positive definite operators and $D_B \supset D_A$. Next, for any real number λ, the operator $A - \lambda B$ is self-adjoint. Finally, Operators A and B ensure that the embedding operator from H_A to H_B is compact. Then,*

(1) Eq. (8.3.12) *has countable but infinitely many eigenvalues, if the space H is of infinite dimension;*
(2) *the only limit point of all these eigenvalues is the infinity; and*

(3) *the sequence of corresponding eigenfunctions forms a complete set in spaces H, H_A, and H_B, while it constitutes an orthogonal set in both H_A and H_B.*

As of now, we have discussed the existence of solutions to both the boundary value problem of the differential equation (8.2.1) and the eigenvalue problem (8.2.2). Here is a brief summary of our findings:

(1) If the operator A is positive definite, then there exists a unique solution to the boundary value problem of the differential equation (8.2.1). In other words, a unique generalized solution exists.
(2) Suppose both operators A and B to be positive definite. In addition, the embedding operator from the space H_A to the space H_B is compact. Then, the generalized eigenvalue problem of the differential equations (8.3.12) has countable but infinitely many eigenvalues, the only limit point of these eigenvalues is the infinity, and the eigenfunction sequence is complete in all spaces H, H_A, and H_B. In other words, the problem has generalized solutions.
(3) For many problems in Solid Mechanics, Operator B corresponds to the mass density function ρ of an elastic solid or structure, which is a positive function with both lower and upper bounds and belongs to the L_2 space. Generally speaking, ρ has higher-order derivatives, and thus Operator B is positive definite. Furthermore, if the embedding operator from the space H_A to the space L_2 is compact, so should be the embedding operator from H_A to H_B.

8.3.3 The Method of Modal Expansion and Its Convergence

1. Modal Expansion Method for Solving Static Deformation Problems

There are many approximation methods for calculating static deformation of structures, among which is the highly effective Ritz method. But in situations where many vibration modes have already been obtained through either computational or experimental means, we can get solutions of static deformation much more quickly with the method of modal expansion.

Consider the following differential equation and boundary condition

$$\begin{cases} Au(x) = f(x) & \text{in } \Omega, \\ Bu = 0 & \text{on } \partial\Omega, \end{cases} \tag{8.3.13}$$

where f stands for the external force while A and B represent operators. The known eigenvalues λ_i and mode shapes $\varphi_i(x)$ $(i = 1, 2, \ldots)$ satisfy modal equations

$$A\varphi_i(x) = \lambda_i\varphi_i(x), \quad i = 1, 2, \ldots \tag{8.3.14}$$

as well as orthogonality conditions

$$(\varphi_i(x), \varphi_j(x)) = \delta_{ij}, \quad (A\varphi_i, \varphi_j) = \lambda_j \delta_{ij}, \quad i, j = 1, 2, \ldots. \tag{8.3.15}$$

Expanding the solution of Eq. (8.3.13), u_0, into a series of mode shapes

$$u_0 = \sum_{i=1} a_i \varphi_i(x),$$

substituting it into Eq. (8.3.13), and then multiplying the resulting equation by $\varphi_j(x)$ from the left, we obtain by taking advantage of the orthogonality condition (8.3.15):

$$a_j = \frac{1}{\lambda_j}(f(x), \varphi_j(x)) = \frac{f_j}{\lambda_j}, \quad j = 1, 2, \ldots. \tag{8.3.16}$$

Thus, the solution of static deformation u_0 is expressed as

$$u_0 = \sum_{i=1}^{\infty} \frac{f_i}{\lambda_i} \varphi_i(x). \tag{8.3.17}$$

Modal expansion method is, in fact, a special case of the Ritz method, where mode shapes serve as base functions. In Sect. 8.7, we will prove convergence of the Ritz method, which should guarantee convergence of the method of modal expansion.

2. ***Modal Expansion Method for Solving Dynamic Response Problems***

Modal expansion is also an effective method for obtaining dynamic responses of structures. Consider the following initial-boundary value problem of the differential equation:

$$\begin{cases} \rho(x)\ddot{u}(x,t) + Au(x,t) = f(x,t) & \text{in } \Omega \text{ and } t > 0, \\ Bu(x,t) = 0 & \text{on } \partial\Omega \text{ and } t > 0, \\ u(x,0) = g(x), \dot{u}(x,0) = h(x) & \text{on } \bar{\Omega}, \end{cases} \tag{8.3.18}$$

where superscripts "$\cdot$" and "$\cdot\cdot$" represent, respectively, the first and second derivatives with respect to the time variable t. Suppose the set of eigenvalues $\{\lambda_i\}$ and eigenfunctions $\{\varphi_i(x)\}$ have been determined, which satisfy the following conditions:

$$\begin{aligned} & A\varphi_i(x) = \lambda_i \rho(x)\varphi_i(x), \quad i = 1, 2, \ldots, \\ & (\rho(x)\varphi_i(x), \varphi_j(x)) = \delta_{ij}, \quad (A\varphi_i, \varphi_j) = \begin{cases} \lambda_i, \; i = j, \\ 0, \; i \neq j. \end{cases} \end{aligned} \tag{8.3.19}$$

Now, express the "external force per unit mass", initial displacement, initial velocity, and dynamic response all as series of $\{\varphi_i(x)\}$:

$$\begin{cases} \dfrac{f(x,t)}{\rho(x)} = \sum\limits_{i=1}^{\infty} f_i(t)\varphi_i(x), \\ g(x) = \sum\limits_{i=1}^{\infty} g_i\varphi_i(x), \\ h(x) = \sum\limits_{i=1}^{\infty} h_i\varphi_i(x), \end{cases} \tag{8.3.20}$$

$$u(x,t) = \sum_{i=1}^{\infty} a_i(t)\varphi_i(x). \tag{8.3.21}$$

Next, multiply all equations in (8.3.20) by $\rho(x)\varphi_j(x)$ and apply the orthogonality condition (8.3.19) of mode shapes. This leads to

$$\begin{cases} f_i(t) = (f(x,t), \varphi_i(x)), \\ g_i = (g(x), \rho(x)\varphi_i(x)), \\ h_i = (h(x), \rho(x)\varphi_i(x)). \end{cases} \tag{8.3.22}$$

Plugging Eqs. (8.3.20) and (8.3.21) into Eq. (8.3.18), we get

$$\begin{cases} \ddot{a}_i(t) + \lambda_i a_i(t) = f_i(t), \\ a_i(0) = g_i, \\ \dot{a}_i(0) = h_i. \end{cases} \tag{8.3.23}$$

The solution of Eq. (8.3.23) may be expressed in terms of the Duhamel integral as

$$a_i(t) = g_i \cos\omega_i t + \frac{h_i}{\omega_i}\sin\omega_i t + \frac{1}{\omega_i}\int\limits_0^t f_i(\tau)\sin\omega_i(t-\tau)\mathrm{d}\tau, \tag{8.3.24}$$

where $\omega_i = \sqrt{\lambda_i}$ is the angular frequency of the i-th single-degree-of-freedom system and $i = 1, 2, \ldots$. Plugging Eq. (8.3.24) into Eq. (8.3.21), we get the solution to Eq. (8.3.18) of the continuous system.

Convergence of the dynamic response solution obtained with the series expansion method has attracted much attention. However, since the discussion involves relatively complicated theory, it is difficult to go into the subject in depth within this book. Here, we only present some result for the simplest case, namely, the "C^∞-Theory", which is attributed to Fichera (see Ref. [10]). Specifically, let us consider the dynamic response problem represented by Eq. (8.3.18). If the domain Ω is C^∞ smooth and also bounded, A is a positive definite operator with C^∞ coefficients inside Ω, $f(x,t) \in C^\infty$ $(x \in \Omega,\ t > 0)$, $g(x) \in C^\infty(\bar{\Omega})$, $h(x) \in C^\infty(\bar{\Omega})$, and some additional conditions are also satisfied, then the series solution (8.3.21), with $a_i(t)$

defined by Eq. (8.3.24), converges to the solution of Eq. (8.3.18) in H_m, where $2m$ is the order of the differential equation.

In summary, an effective theoretical framework for solving differential equations has been introduced in Sects. 8.2 and 8.3. In subsequent sections 8.4 and 8.5, we will prove the existence of solutions to problems in Elasticity and structural theories. If it can be shown relevant operators to be positive definite and the embedding operators from the space H_A to the space H_B to be compact, then generalized solutions to problems of static deformation and vibrational modes should exist.

Major contents of Sects. 8.2 and 8.3 are taken from Ref. [24].

8.4 Existence of Solutions of Static Deformation and Vibrational Modes in Elasticity

8.4.1 Governing Equations and Boundary Conditions for Static Deformation and Vibrational Modes in Elasticity

The following equations are satisfied by the static deformation and vibrational modes, respectively,

$$\begin{cases} \boldsymbol{A}_{\mathrm{e}}\boldsymbol{u} = -\sum\limits_{i,k,l,m=1}^{3} \dfrac{\partial}{\partial x_i}(c_{iklm}\varepsilon_{lm}(\boldsymbol{u}))\boldsymbol{x}_k^{(0)} = \boldsymbol{f}(\boldsymbol{x}) & \text{in } \Omega, \\ \boldsymbol{B}_{\mathrm{e}}\boldsymbol{u} = \boldsymbol{0} & \text{on } \partial\Omega, \end{cases} \tag{8.4.1}$$

and

$$\begin{cases} \boldsymbol{A}_{\mathrm{e}}\boldsymbol{u} = \omega^2\rho\boldsymbol{u} & \text{in } \Omega, \\ \boldsymbol{B}_{\mathrm{e}}\boldsymbol{u} = \boldsymbol{0} & \text{on } \partial\Omega, \end{cases} \tag{8.4.2}$$

where $\boldsymbol{u}$ denotes the three-dimensional displacement vector of an elastic solid, ε_{lm} is the strain component, $\boldsymbol{x}_k^{(0)}$ represents the unit vector along the coordinate axis x_k $(k = 1, 2, 3)$, the elasticity operator $\boldsymbol{A}_{\mathrm{e}}$ is a second-order differential operator, and $\boldsymbol{B}_{\mathrm{e}}$ stands for the boundary differential operator.

For anisotropic and isotropic solids, the numbers of elastic coefficients are, respectively, 21 and 2. For an inhomogeneous material, its elastic coefficients are functions of the spatial coordinate $\boldsymbol{x}$; while for a homogeneous material, its elastic coefficients are all constants. As for the mass density ρ, it is a bounded positive function when the material is inhomogeneous, and a positive number when the material is homogeneous. Let Ω and $\partial\Omega$ denote, respectively, the bounded domain occupied by the elastic solid and the border of the domain. There are six types of common boundary conditions:

(1) fixed boundary, i.e., $\boldsymbol{u}|_{\partial\Omega} = 0$;
(2) free or unconstrained boundary, i.e., $\boldsymbol{t}(\boldsymbol{u})|_{\partial\Omega} = 0$, where $\boldsymbol{t}$ represents the stress vector at the boundary;
(3) rigid contact boundary, i.e., $\boldsymbol{u}_{(\nu)}|_{\partial\Omega} = 0$ and $\boldsymbol{t}(\boldsymbol{u})_{(s)}|_{\partial\Omega} = 0$, where ν and s denote the normal and tangential directions at the boundary, respectively;
(4) displacement at the boundary being unconstrained in the normal direction but fixed in the tangential directions, i.e., $\boldsymbol{u}_{(s)}|_{\partial\Omega} = 0$ and $\boldsymbol{t}(\boldsymbol{u})_{(\nu)}|_{\partial\Omega} = 0$;
(5) a hybrid of the above four types of conditions; and
(6) elastic boundary, i.e., $\boldsymbol{t}(\boldsymbol{u})|_{\partial\Omega} + \boldsymbol{K}\boldsymbol{u}|_{\partial\Omega} = 0$, where $\boldsymbol{K}$ is a 3×3 diagonal matrix with diagonal elements representing elastic coefficients of the elastic supports along x_1, x_2, and x_3 directions. These elastic coefficients are all positive functions.

When an elastic solid has a static displacement $\boldsymbol{u}$, its strain energy is $(\boldsymbol{A}_e\boldsymbol{u}, \boldsymbol{u})/2$. Furthermore, when the elastic solid is in a simple harmonic vibration $\boldsymbol{u}\sin(\omega t)$, its strain and kinetic energies are, respectively,

$$\frac{1}{2}(\boldsymbol{A}_e\boldsymbol{u}, \boldsymbol{u})\sin^2(\omega t) \quad \text{and} \quad \frac{1}{2}(\rho\boldsymbol{u}, \boldsymbol{u})\omega^2\cos^2(\omega t),$$

where ω represents the angular frequency and $\omega^2(\rho\boldsymbol{u}, \boldsymbol{u})/2$ is the magnitude of the kinetic energy. Thus, $(\boldsymbol{A}_e\boldsymbol{u}, \boldsymbol{u})$ is twice of the magnitude of the strain energy, while $(\rho\boldsymbol{u}, \boldsymbol{u})$ is two times the coefficient of the magnitude of the kinetic energy. We refer to the Hilbert space with the inner product $(\boldsymbol{A}_e\boldsymbol{u}, \boldsymbol{u})$ as the space with the strain energy norm and represent it as H_{A_e}; call the Hilbert space with the inner product $(\rho\boldsymbol{u}, \boldsymbol{u})$ the space with the kinetic energy norm and denote it as H_ρ; and assign the name of the energy embedding operator to the mapping from the space with the strain energy norm to the space with the kinetic energy norm.

8.4.2 *Existence of Solutions of Static Deformation and Vibrational Modes in Elasticity*

In order to study the existence of solutions of static deformation and vibrational modes in Elasticity, we only need to verify the positive definiteness of the elasticity operator $\boldsymbol{A}_e$ and the compactness of the energy embedding operator in Elasticity. This requires separate discussion for each type of the boundary conditions. A key step is to prove the Korn inequality, which involves a significant amount of effort. Due to space limitation, detailed derivation is omitted, and only relevant conclusions are presented in this book. If needed, please find the information in Chap. 4 of Ref. [24].

Since the existence and differentiability of a solution has much to do with the smoothness of the boundary, we would like to elaborate on the concept of boundary smoothness here: Choose any point along the boundary surface $\partial\Omega$ to be the origin of a local Cartesian coordinate system $z_1, z_2, \ldots, z_m$. Make the z_m axis align with the normal direction of the boundary $\partial\Omega$. In the neighborhood of this point, the equation of the surface $\partial\Omega$ can be expressed as $z_m = \phi(z_1, z_2, \ldots, z_{m-1})$. If the function ϕ has n-th continuous derivatives, then the boundary is of class C^n, denoted by $\Omega \in C^n$. If $n = 3$, we claim the boundary to be sufficiently smooth.

Theorem 8.10 (Theorem on the Positive Definiteness of the Elasticity Operator) *If elastic coefficients $\{c_{iklm}\}$ of an elastic solid have piecewise continuous first derivatives in a domain Ω that is bounded and with piecewise smooth border, and any one set of boundary conditions (1)–(5) given above is satisfied; then the elasticity operator $\mathbf{A}_\mathrm{e}$ is positive definite, i.e.,*

$$(\mathbf{A}_\mathrm{e}\boldsymbol{u}, \boldsymbol{u}) \geq \gamma^2 \|\boldsymbol{u}\|^2,$$

where γ is a positive number.

In Physics, the positive definiteness of the operator $\mathbf{A}_\mathrm{e}$ indicates the ratio between norms of strain energy and kinetic energy to be positive; while in Mathematics, it implies that the embedding operator from the space with the strain energy norm to the space with the kinetic energy norm is bounded.

Theorem 8.11 (Theorem on the Compactness of the Energy Embedding Operator in Elasticity) *If elastic coefficients $\{c_{iklm}\}$ of an elastic solid have piecewise continuous first derivatives in a domain Ω that is bounded and with piecewise smooth border, the mass density ρ is a bounded positive function and $\rho \in L_2$, and any one set of the aforementioned boundary conditions (1)–(5) is satisfied; then the energy embedding operator in Elasticity is compact.*

Here, the energy embedding operator being compact means that it maps a bounded set in the space with the strain energy norm to a compact set in the space with the kinetic energy norm.

Before proceeding further, we would like to make an important remark regarding Theorems 8.10 and 8.11: When an elastic solid is constrained by one of the boundary conditions among (2), (3), (4), and (5), rigid-body motion may occur. Under this circumstance, the external forces applied to the system must be such that the resultant force and moment vanish, and the displacement solution is unique only in the sense that one rigid-body translation and one rigid-body rotation have been excluded.

Based on the above two theorems as well as Theorems 8.6 and 8.8, we may obtain the following existence theorems for solutions in Elasticity.

Theorem 8.12 (Existence Theorem for the Solution of Static Deformation in Elasticity) *If elastic coefficients $\{c_{iklm}\}$ of an elastic solid have piecewise continuous first derivatives in the domain Ω that is bounded and with piecewise smooth border, the external force function belongs to $L_2(\Omega)$, and any one set of the aforementioned* *boundary conditions (1)–(5) is satisfied; then there exists a unique generalized solution to the static deformation problem in Elasticity.*

Theorem 8.13 (Existence Theorem for Modal Solutions in Elasticity) *If elastic coefficients $\{c_{iklm}\}$ of an elastic solid have piecewise continuous first derivatives in the domain Ω that is bounded and with piecewise smooth border, the mass density ρ is a bounded positive function and $\rho \in L_2$, and any one set of the aforementioned boundary conditions (1)–(5) is satisfied; then there exists generalized solutions to the modal problem in Elasticity. Specifically, there are countable but infinitely many natural frequencies, and the infinity is their only limit point; while the respective mode shapes form a complete and orthogonal set in both the space with the kinetic energy norm and the space with the strain energy norm.*

As pointed earlier, it requires separate treatment for each of the boundary conditions in order to prove Theorems 8.10 and 8.11. This process is relatively complex and tedious, especially in situations where rigid-body motion is involved.

Here, we only outline the major steps for proving Theorem 8.10, i.e., the theorem on the positive definiteness of the elasticity operator in the next two paragraphs.

First, prove the Korn inequality by using the stress–strain relationship and boundary conditions:

$$\Pi_{\mathrm{e}} \geq C \int_{\Omega} \sum_{i,k=1}^{3} \left(\frac{\partial u_i}{\partial x_k} \right)^2 \mathrm{d}\Omega,$$

where Π_{e} stands for the strain energy of the elastic solid. The Korn inequality was established by Korn in 1908. See Ref. [17].

Second, obtain the following result by starting from the Schwarz inequality and going through a series of derivations:

$$C \int_{\Omega} \sum_{i,k=1}^{3} \left(\frac{\partial u_i}{\partial x_k} \right)^2 \mathrm{d}\Omega \geq C_1 \|u\|^2.$$

Since $(\boldsymbol{A}_{\mathrm{e}}\boldsymbol{u}, \boldsymbol{u}) = 2\Pi_{\mathrm{e}}$, we may prove the positive definiteness of the elasticity operator:

$$(\boldsymbol{A}_{\mathrm{e}}\boldsymbol{u}, \boldsymbol{u}) \geq \gamma^2 \|u\|^2.$$

8.4.3 Examples of Generalized Solutions in Elasticity

Given below are three examples. The generalized solutions obtained in these examples have higher-order derivatives than needed, and may even be turned into classical solutions.

Example 1 Consider a homogeneous and isotropic elastic solid. Suppose its border is piecewise smooth and constrained in one of the following ways: (1) the displacement at the boundary is set to zero, (2) the boundary force is set to zero, (3) both the normal displacement and tangential force vanish, or (4) combine the previous three types of boundary conditions. Further assume that the body force is square integrable. Under these conditions, there exists a unique displacement that minimizes the potential energy, has square integrable generalized second derivatives, and satisfies the elasticity equation almost everywhere. See Ref. [24].

Example 2 Consider an inhomogeneous and anisotropic elastic object. Suppose its elastic coefficients $c_{iklm} \in C^{\infty}$, i.e., they have continuous derivatives of infinite order. Further assume that the two- or three-dimensional domain of the problem is bounded and C^{∞} smooth. Finally, the external body force $\boldsymbol{f} \in C^{\infty}(\bar{\Omega})$. Then, the static deformation equation in Elasticity has a unique solution under each of the following three types of boundary conditions: fixed, free or unconstrained, and a hybrid of the two. The solution belongs to $C^{\infty}(\bar{\Omega})$ and is thus a classical solution. See Ref. [9]. Notice that in the case of unconstrained boundary, the equilibrium state requires zero resultant force and moment, and the solution is unique after rigid-body motions are excluded.

Example 3 Consider the elastic object subject to the three types of boundary conditions discussed in Example 2 above. If the mass density $\rho \in L^2$, then there exist generalized solutions to the modal problem, and the mode shapes $u_i \in C^{\infty}$ $(i = 1, 2, \ldots)$. See Ref. [9].

8.5 Existence of Solutions of Static Deformation and Vibrational Modes in Structural Theories

There are significant differences among operators associated with various structures. One method for studying properties of these operators, which can be found in the existing literature, is to directly work with individual equations and each type of boundary conditions. However, the mathematics involved in this approach becomes quite complicated and tedious when we deal with certain structures, such as thin shells with complex shapes as well as assembly structures. Although the problem has caught the researchers' attention for a long time, no comprehensive solution has been found.

In the discussion below, we will follow the approach pioneered by Wang and Hu. See Refs. [32–35]. The method is based on connections of the displacement, strain energy, and kinetic energy between an elastic structural theory and the three-dimensional Elasticity. By taking advantage of properties related to the boundedness and compactness of operators in Hilbert Spaces, we are able to extend the positive definiteness of the elasticity operator and the compactness of the energy embedding operator to respective operators in the structural theory. As a result, we can establish a unified framework to discuss existence of solutions of the static deformation and vibrational modes associated with structural theories of shells subject to a wide range of boundary conditions, structures involving composite materials, various assembly structures, etc. Furthermore, the approach should remain valid for studying existence of solutions of static deformation and vibrational modes associated with new structural theories to be developed in the future.

Let us denote the generalized displacement in the structural theory as $\boldsymbol{w}(\boldsymbol{x})$, where $\boldsymbol{x}$ represents the independent spatial variable characterizing either the center line or surface of a structure, and the one- or two-dimensional domain of $\boldsymbol{x}$ is Ω. Here are some examples of this notation: For a slender beam, $\boldsymbol{w}(x)$ is the deflection $w(x)$ at the center line of the beam; for a thin plate, $\boldsymbol{w}(\boldsymbol{x})$ denotes the deflection $w(x_1, x_2)$ at the midplane of the plate; for a plate with shear deformation, $\boldsymbol{w}(\boldsymbol{x})$ stands for $(\psi_{x_1}(x_1, x_2), \psi_{x_2}(x_1, x_2), w(x_1, x_2))^{\mathrm{T}}$, where ψ_{x_1} and ψ_{x_2} represent, respectively, rotations of the midplane about x_1- and x_2-axes; and for a thin shell, $\boldsymbol{w}(x)$ assumes the form of $(u(x_1, x_2), v(x_1, x_2), w(x_1, x_2))^{\mathrm{T}}$, where u and v denote, respectively, displacements along directions of curvilinear coordinates x_1 and x_2 at the mid-surface while w is the displacement along the direction normal to the mid-surface.

The governing equations and boundary conditions for static deformation in a structural theory may be written as:

$$\begin{cases} \boldsymbol{A}_{\mathrm{s}}\boldsymbol{w}(\boldsymbol{x}) = \boldsymbol{f}(\boldsymbol{x}) & \text{in } \Omega, \\ \boldsymbol{B}_{\mathrm{s}}\boldsymbol{w}(\boldsymbol{x}) = \boldsymbol{0} & \text{on } \partial\Omega; \end{cases}$$

while the frequency and mode shape in the structural theory satisfy the following equation and boundary condition

$$\begin{cases} \boldsymbol{A}_{\mathrm{s}}\boldsymbol{w}(\boldsymbol{x}) - \omega^2 m\boldsymbol{w}(\boldsymbol{x}) = \boldsymbol{0} & \text{in } \Omega, \\ \boldsymbol{B}_{\mathrm{s}}\boldsymbol{w}(\boldsymbol{x}) = \boldsymbol{0} & \text{on } \partial\Omega, \end{cases}$$

where $\boldsymbol{A}_{\mathrm{s}}$ denotes the structural theory operator and $\boldsymbol{B}_{\mathrm{s}}$ represents the boundary differential operator.

The integral

$$\int_{\Omega} m\boldsymbol{w} \cdot \boldsymbol{w}\mathrm{d}\boldsymbol{x} = 2K_{\mathrm{s}}$$

is proportional to two times the coefficient of the magnitude of the kinetic energy T associated with the structure in natural vibration, where m is the mass density of the structure. Specifically, m may represent the mass per unit length of a beam, the mass per unit area of a plate, etc. However, K_{s} may contain additional terms in some structures. For instance, for a beam with its rotational inertia of the cross section being taken into account, twice of the density of the kinetic energy would be

$$mw^2 + I_s\left(\frac{\partial w}{\partial x}\right)^2,$$

where I_s stands for the cross-sectional moment of inertia. Additionally, the integral

$$\int_{\Omega} \boldsymbol{w}\boldsymbol{A}_s\boldsymbol{w}\mathrm{d}\boldsymbol{x} = 2\Pi_s$$

represents twice the magnitude of the strain energy associated with the structure.

8.5.1 Two Auxiliary Theorems

First, we would like to point out that all spaces within this chapter are Hilbert Spaces and all operators are linear. Denote the inner product of the space X_i as $(\cdot\ ,\ \cdot)_i$ and the norm as $\|\cdot\|_i$. Next, let us introduce three lemmas.

Lemma 8.1 *Consider two Hilbert spaces* X_1 *and* X_2*. If the operator* $T : \mathrm{X}_1 \to \mathrm{X}_2$ *is bounded, then the new operator obtained by restraining* T *to the linear subspace* $X_1^{\mathrm{r}} \subset X_1$*, namely,* $T^{\mathrm{r}} = T|_{X_1^{\mathrm{r}}} : X_1^{\mathrm{r}} \to X_2$*, is also bounded. In addition, the norm of the operator* T^{r} *is less or equal to that of the operator* T*, i.e.,*

$$\|T^{\mathrm{r}}\| \leq \|T\|.$$

Lemma 8.2 *Consider Hilbert spaces* X_1 *and* X_2*. If the operator* $T : X_1 \to X_2$ *is compact, then the operator, obtained by restricting* T *to the complete subspace* $X_1^{\mathrm{r}} \subset X_1$*, i.e.,* $T^{\mathrm{r}} = T|_{X_1^{\mathrm{r}}} : X_1^{\mathrm{r}} \to X_2$*, remains to be compact.*

Lemma 8.3 *Consider a finite number of operators* T_i*,* $i = 1, \ldots, n$*. If all of them are bounded, then their product* $T = T_n \cdot T_{n-1} \cdots T_1$ *remains to be bounded. In addition, if one of the given operators is compact, then* T *should also be a compact operator.*

Now consider three Hilbert spaces X_1, X_2, and X_3 as well as subspaces X_2^{r} and X_3^{r} of the latter two. Suppose spaces X_2 and X_3 contain identical elements while X_2^{r} and X_3^{r} also consist of identical elements. Furthermore, assume that there is a linear and positive operator $\boldsymbol{A}$ on X_1. Let us construct a new Hilbert space X_4, where the inner product between two arbitrary elements $\boldsymbol{x}$ and $\boldsymbol{y}$ in the domain of $\boldsymbol{A}$ is calculated as

$$(\boldsymbol{x}, \boldsymbol{y})_4 = (\boldsymbol{A}\boldsymbol{x}, \boldsymbol{y})_1.$$

Next, define operators

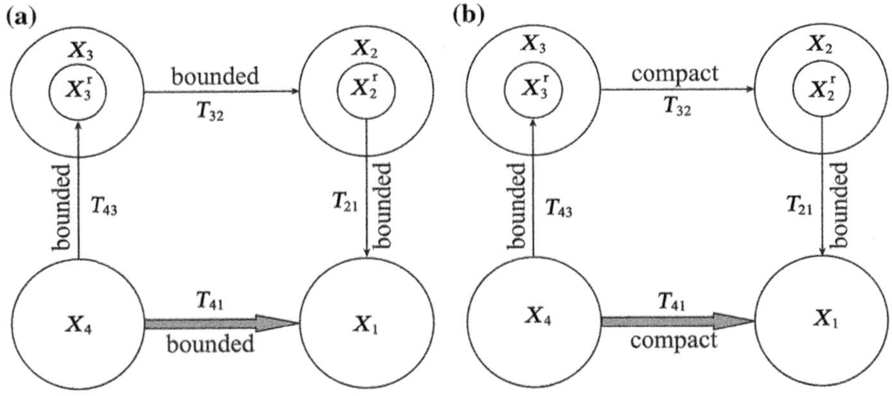

Fig. 8.1 **a** Illustrations of Auxiliary Theorem 8.1, **b** illustration of Auxiliary Theorem 8.2

$$T_{41} : X_4 \to X_1 \quad \text{and} \quad T_{32} : X_3 \to X_2$$

as mappings between identical elements. If there has been an operator with one-to-one correspondence

$$T_{21} : X_2^{\mathrm{r}} \to X_1$$

and there is also a restriction on Operator T_{32} put in place as

$$T_{32}^{\mathrm{r}} = T_{32}|_{X_3^{\mathrm{r}}} : X_3^{\mathrm{r}} \to X_2^{\mathrm{r}},$$

then the following one-to-one mapping exists

$$T_{43} = (T_{32}^{\mathrm{r}})^{-1} T_{21}^{-1} T_{41} : X_4 \to X_3^{\mathrm{r}}.$$

Auxiliary Theorem 8.1 *If operators T_{21}, T_{32}, and T_{43} are all bounded, then the operator*

$$T_{41} = T_{21} T_{32}^{\mathrm{r}} T_{43}$$

should also be bounded.

Proof It is known from Lemma 8.1 that the operator T_{32}^{r} is bounded. Furthermore, Lemma 8.3 shows that the operator T_{41} is also bounded. ■

This theorem is illustrated graphically in Fig. 8.1a.

Auxiliary Theorem 8.2 *If operators T_{21} and T_{43} are bounded while the operator T_{32} is compact, then the operator*

$$T_{41} = T_{21}T^{\mathrm{r}}_{32}T_{43}$$

is also compact.

Proof Lemma 8.2 leads to the result that the operator T^{r}_{32} is compact. Then, from Lemma 8.3, we further conclude that T_{41} is also compact. ■

This theorem is illustrated graphically in Fig. 8.1b.

8.5.2 Existence of Solutions of Static Deformation and Vibrational Modes

There are a variety of structural theories in Structural Mechanics, and they originated from different structural models. These theoretical models cover not only components such as bars, beams, plates, and shells, but also the more sophisticated assembly structures that have complex shapes and are made of advanced materials. Although they involve diverse physical characteristics and mathematical representations, all these models may be collectively viewed as derivatives of the elasticity model of a three-dimensional elastic solid by going through three types of simplifications: (1) simplification on deformation, realized primarily by incorporating displacement constraints; (2) simplification on stress, generally accomplished through relaxation of certain stress components; and (3) simplification on mass distribution, normally achieved via mass lumping.

Given a structure and its boundary conditions, we may identify the equivalent three-dimensional elastic solid and related boundary conditions. The latter is referred to as the elastic solid corresponding to the given structure. Furthermore, if simplification on deformation is applied to this elastic solid but there is no simplification on either its stress state or its mass distribution, then the resulting object is called the constrained elastic solid corresponding to the original structure.

Consider a structure that follows a specific structural theory. Suppose its center line or mid-surface occupies a region F and all of the comparison displacements $\{w\}$ constitute a function set $\boldsymbol{U}_{\mathrm{s}}$, which forms the domain of the operator $\boldsymbol{A}_{\mathrm{s}}$ of this structural theory. We may use the following methods to define three inner product spaces in $\boldsymbol{U}_{\mathrm{s}}$ and, after constructing their completions turn them into Hilbert spaces.

The first Hilbert space is represented by U_{ss}, which is a square integrable space L_2. Its inner product and squared norm are, respectively,

$$(\boldsymbol{w}_1, \boldsymbol{w}_2)_{\mathrm{s}} \quad \text{and} \quad \|\boldsymbol{w}\|^2_{\mathrm{s}} = (\boldsymbol{w}, \boldsymbol{w})_{\mathrm{s}} = \int\limits_F \boldsymbol{w} \cdot \boldsymbol{w} \mathrm{d}F. \tag{8.5.1}$$

U_{ss} is called the displacement space of the structural theory.

Denote the second Hilbert space as U_{ps}. The corresponding squared norm is defined as twice of the strain energy associated with the structure, namely,

$$\|\boldsymbol{w}\|_{\mathrm{ps}}^2 = 2\Pi_{\mathrm{s}}(\boldsymbol{w}) = (\boldsymbol{A}_{\mathrm{s}}\boldsymbol{w}, \boldsymbol{w})_{\mathrm{s}}, \tag{8.5.2}$$

where the quadratic form Π_{s} represents the strain energy of the structure. The value of Π_{s} is affected by the elastic parameters of the structure, which may be the bending stiffness EJ of a beam, bending stiffness D of a thin plate, etc. For all effective and valid structural theories, the operator $\boldsymbol{A}_{\mathrm{s}}$ is both linear and positive. The completion of U_{ps}, which is still represented by U_{ps} for the sake of simplicity, is referred to as the space with the strain energy norm of this structural theory.

Let the third Hilbert space be U_{ks}, and define the respective squared norm as twice of the coefficient of the kinetic energy associated with the structure, i.e.,

$$\|\boldsymbol{w}\|_{\mathrm{ks}}^2 = 2K_{\mathrm{s}}(\boldsymbol{w}) = (m\boldsymbol{w}, \boldsymbol{w})_{\mathrm{s}}, \tag{8.5.3}$$

where K_{s} represents the coefficient of the kinetic energy and m, which is a bounded and positive function and $m \in L_2$, stands for the mass density of the structure, such as the mass per unit length of a beam, the mass per unit area of a plate, etc. The completion of U_{ks}, which is denoted by U_{ks} also for the sake of simplicity, is called the space with the kinetic energy norm of the structure.

Consider mappings between identical elements in U_{ps} and U_{ss} as well as in U_{ps} and U_{ks}. They may be denoted respectively by

$$T_{\mathrm{ps,\,ss}} : U_{\mathrm{ps}} \to U_{\mathrm{ss}}, \tag{8.5.4a}$$

and

$$T_{\mathrm{ps,\,ks}} : U_{\mathrm{ps}} \to U_{\mathrm{ks}}. \tag{8.5.4b}$$

Here, $T_{\mathrm{ps,\,ks}}$ is referred to as the energy embedding operator in the structural theory.

The relationship between the aforementioned spaces and operators are shown in Fig. 8.2.

An important fact deserves some attention: From Eqs. (8.5.1), (8.5.2), and (8.5.4a), the condition for Operator $T_{\mathrm{ps,ss}}$ being bounded may be expressed as

$$(\boldsymbol{w}, \boldsymbol{w})_{\mathrm{s}} \le \alpha^2(\boldsymbol{A}_{\mathrm{s}}\boldsymbol{w}, \boldsymbol{w})_{\mathrm{s}},$$

where α is a constant. But this inequality gives none other than the definition of Operator $\boldsymbol{A}_{\mathrm{s}}$ being positive definite. Thus, in order to prove the operator $\boldsymbol{A}_{\mathrm{s}}$ to be positive definite, we only need to show the operator $T_{\mathrm{ps,ss}}$ is bounded.

There are two important mathematical problems in structural theories: First, is $T_{\mathrm{ps,ss}}$ a bounded operator? As explained in the previous paragraph, this is equivalent to the question whether the structural theory operator $\boldsymbol{A}_{\mathrm{s}}$ is positive definite. Second, is the energy embedding operator in the structural theory $T_{\mathrm{ps,ks}}$ compact? We will answer these questions below by taking advantage of the relationship between structural theories and Elasticity.

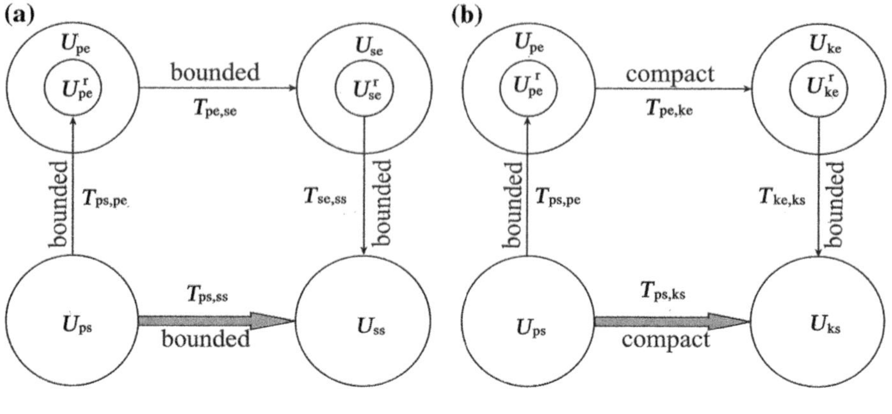

Fig. 8.2 **a** Illustration of the theorem on the positive definiteness of the structural theory operator, **b** illustration of the theorem on the compactness of the energy embedding operator in the structural theory

In fact, the structure that we are studying with the structural theory can be treated like an elastic solid, and put under a more elaborate analysis with Elasticity. In this approach, the displacement of the elastic solid may be represented by a three-dimensional vector field $\boldsymbol{u}(\Omega)$, where Ω denotes the region occupied by the elastic solid. All of the comparison displacements $\{\boldsymbol{u}\}$ constitute a function set $\boldsymbol{U}_\mathrm{e}$, which is the domain of the strain energy corresponding to the elasticity operator $\boldsymbol{A}_\mathrm{e}$. In parallel to earlier discussion on structural theories, we may establish three inner product spaces and find their completions, which are Hilbert spaces, as shown in the next three paragraphs.

The first Hilbert space is the displacement space U_se of Elasticity. It is an L_2 space, while its norm is defined by

$$\|\boldsymbol{u}\|^2_\mathrm{se} = (\boldsymbol{u}, \boldsymbol{u})_\mathrm{e} = \int\limits_\Omega \boldsymbol{u} \cdot \boldsymbol{u}\,\mathrm{d}\Omega. \tag{8.5.5}$$

The second space is U_pe, whose norm is of the form

$$\|\boldsymbol{u}\|^2_\mathrm{pe} = 2\Pi_\mathrm{e}(\boldsymbol{u}) = (\boldsymbol{A}_\mathrm{e}\boldsymbol{u}, \boldsymbol{u})_\mathrm{e}, \tag{8.5.6}$$

where the quadratic form Π_e is the strain energy of the elastic solid. The completion of the Hilbert space U_pe, which is still represented by U_pe for the sake of simplicity, is referred to as the space with the strain energy norm of Elasticity.

The third space is U_ke, and its norm is given by

$$\|\boldsymbol{u}\|^2_\mathrm{ke} = 2K_\mathrm{e}(\boldsymbol{u}) = (\rho\boldsymbol{u}, \boldsymbol{u})_\mathrm{e}, \tag{8.5.7}$$

where K_e and ρ are, respectively, the kinetic energy coefficient and mass density of the elastic solid. The completion of the Hilbert space U_{ke}, which is still represented by U_{ke} as before, is called the space with the kinetic energy norm of Elasticity.

For a real elastic solid, it often occurs that min $\rho \geq C > 0$ while max ρ is finite. Thus, we get

$$(\min \rho)(\boldsymbol{u}, \boldsymbol{u})_e \leq (\rho \boldsymbol{u}, \boldsymbol{u})_e \leq (\max \rho)(\boldsymbol{u}, \boldsymbol{u})_e.$$

As a result, U_{se} and U_{ke} are spaces with equivalent norms, and the two are used indistinguishably in many situations.

Consider mappings between identical elements in U_{pe} and U_{se} as well as in U_{pe} and U_{ke}. They are represented respectively by

$$T_{pe,se} : U_{pe} \rightarrow U_{se}, \tag{8.5.8a}$$

and

$$T_{pe,ke} : U_{pe} \rightarrow U_{ke}. \tag{8.5.8b}$$

Here, $T_{pe,ke}$ is called the energy embedding operator of Elasticity.

Now let us discuss the fundamental relationship between the structural theory and Elasticity. As we all know, each effective and valid structural theory may be viewed as an approximation of Elasticity under certain circumstances. Thus, generally speaking, a structural theory can be deduced from Elasticity after appropriate assumptions are introduced. In subsequent discussion, we will establish relationship between variables in the structural theory and their counterparts in Elasticity, where the variables includes the displacement and energy norms.

Relationship Between Displacements in the Structural Theory and Elasticity Introduce some constraints in U_{se} such that $\boldsymbol{u}$ becomes dependent on $\boldsymbol{w}$, i.e.,

$$\boldsymbol{u} = T_{ss,se} \boldsymbol{w}, \quad \forall \boldsymbol{w} \in U_{ss}. \tag{8.5.9}$$

As $\boldsymbol{w}$ exhausts the displacement set U_{ss} in the structural theory, $\boldsymbol{u}$ will go through all elements in the subset U_{se}^{r} of U_{se}. Thus, U_{ss} and U_{se}^{r} are, respectively, the domain and range of the operator $T_{ss,se}$. There is a one-to-one correspondence between their elements.

The boundary condition specified in the structural theory, namely,

$$\Gamma_s^j \boldsymbol{w}\big|_{\partial_j F} = 0 \tag{8.5.10}$$

can be mapped, through Eq. (8.5.9), to the corresponding boundary condition in Elasticity

$$\Gamma_e^j \boldsymbol{u}\big|_{\partial_j \Omega} = 0. \tag{8.5.11}$$

Relationship Between Strain Energies in the Structural Theory and Elasticity The strain energy in the structural theory may be obtained from that in Elasticity in two steps: First, by constraining the displacement to the subset $U^{\mathrm{r}}_{\mathrm{se}}$, we get the strain energy in Elasticity associated with this subset:

$$\Pi^{\mathrm{r}}_{\mathrm{e}} = \Pi^{\mathrm{r}}_{\mathrm{e}}(\boldsymbol{w}), \quad \forall \boldsymbol{w} \in U_{\mathrm{ss}}.$$

Next, this calculated $\Pi^{\mathrm{r}}_{\mathrm{e}}$ is, in general, considered excessive for real applications. As a result, some additional assumptions on stress distribution need to be introduced, and this leads to the strain energy $\Pi_{\mathrm{s}}(\boldsymbol{w})$ in the structural theory. Thus,

$$\Pi_{\mathrm{s}}(\boldsymbol{w}) \leq \Pi^{\mathrm{r}}_{\mathrm{e}}(\boldsymbol{w}), \quad \forall \boldsymbol{w} \in U_{\mathrm{ss}}.$$

Now we are going to further examine the relationship between $\Pi_{\mathrm{s}}(\boldsymbol{w})$ and $\Pi^{\mathrm{r}}_{\mathrm{e}}(\boldsymbol{w})$. For valid and commonly used structural theories, both $\Pi_{\mathrm{s}}(\boldsymbol{w})$ and $\Pi^{\mathrm{r}}_{\mathrm{e}}(\boldsymbol{w})$ may be viewed as being originated from the generalized strain energy $\Pi_3(\boldsymbol{\varepsilon}, \boldsymbol{\sigma}, \boldsymbol{u})$ with three variables in Elasticity. On the one hand, by finding the local minimum of Π_3 with respect to the stain $\boldsymbol{\varepsilon}$, we get the generalized strain energy $\Pi_2(\boldsymbol{\sigma}, \boldsymbol{u})$ with two variables in Elasticity; next, by computing the local maximum of Π_2 with respect to the stress $\boldsymbol{\sigma}$, we obtain the typical strain energy $\Pi_{\mathrm{e}}(\boldsymbol{u})$ in Elasticity; and finally, by enforcing the displacement constraint, i.e., Equation (8.5.9), we further deduce $\Pi^{\mathrm{r}}_{\mathrm{e}}(\boldsymbol{w})$ from $\Pi_{\mathrm{e}}(\boldsymbol{u})$ (see Ref. [15]). On the other hand, after approximating the strain $\boldsymbol{\varepsilon}$ based on the structural theory, we get $\Pi_{2\mathrm{s}}(\boldsymbol{\sigma}, \boldsymbol{u})$ by computing the local minimum of Π_3 with respect to the modified $\boldsymbol{\varepsilon}$; next, we obtain $\Pi_{\mathrm{s}}(\boldsymbol{u})$ by performing approximation on $\boldsymbol{\sigma}$ and then finding the local maximum of $\Pi_{2\mathrm{s}}$ with respect to the modified $\boldsymbol{\sigma}$; and finally, by applying the displacement approximation in Eq. (8.5.9) to $\Pi_{\mathrm{s}}(\boldsymbol{u})$, we end up with the strain energy $\Pi_{\mathrm{s}}(\boldsymbol{w})$ in the structural theory. Therefore, densities of $\Pi^{\mathrm{r}}_{\mathrm{e}}(\boldsymbol{w})$ and $\Pi_{\mathrm{s}}(\boldsymbol{w})$, i.e., integrands in the integrals of $\Pi^{\mathrm{r}}_{\mathrm{e}}(\boldsymbol{w})$ and $\Pi_{\mathrm{s}}(\boldsymbol{w})$, can both be expressed as positive definite quadratic forms of generalized strains of the structure, where the generalized strains are linear combinations of the structural displacement $\boldsymbol{w}$ and its derivatives. The only difference between the two is in their coefficients. In addition, since there is only a finite number of independent generalized strains, $\Pi_{\mathrm{s}}(\boldsymbol{w})$ and $\Pi^{\mathrm{r}}_{\mathrm{e}}(\boldsymbol{w})$ are equivalent when they are used to define norms. As a result, we get the following important inequality:

$$\Pi^{\mathrm{r}}_{\mathrm{e}} \geq \Pi_{\mathrm{s}} \geq a^2 \Pi^{\mathrm{r}}_{\mathrm{e}}, \tag{8.5.12}$$

where a is a nonzero constant.

Relationship Between Kinetic Energies in the Structural Theory and Elasticity The kinetic energy of the structural theory can be derived directly from its counterpart in Elasticity by constraining $\boldsymbol{u}$ to the subset $U^{\mathrm{r}}_{\mathrm{se}}$. Notice that this process may also involve dropping some positive terms. For instance, in the case of beams and plates, the kinetic energy associated with displacement along the normal direction of the center line or mid-surface is retained, but the kinetic energy associated with rotation is ignored. Under general circumstances, the following relationship exists between the two types of kinetic energy coefficients:

$$K_{\mathrm{e}}^{\mathrm{r}} \geq K_{\mathrm{s}}, \tag{8.5.13a}$$

i.e.,

$$(\rho \boldsymbol{u}, \boldsymbol{u})_{\mathrm{e}} \geq (m\boldsymbol{w}, \boldsymbol{w})_{\mathrm{s}}, \quad \forall \boldsymbol{w} \in U_{\mathrm{ss}}, \tag{8.5.13b}$$

where m denotes the mass density of the structure, such as the mass per unit length of a beam, the mass per unit area of a plate, etc. Considering there are many ways to redistribute the mass in various theoretical models of a structure, let us express the relationship between $K_{\mathrm{e}}^{\mathrm{r}}$ and K_{s} in a more general form:

$$K_{\mathrm{e}}^{\mathrm{r}} \geq K_{\mathrm{s}} \geq b^2 K_{\mathrm{e}}^{\mathrm{r}}, \tag{8.5.13c}$$

where b is a nonzero constant.

Now, we make use of the two auxiliary theorems in Sect. 8.5.1 by setting

$$X_1 = U_{\mathrm{ss}}, \quad X_2 = U_{\mathrm{se}}, \quad X_3 = U_{\mathrm{pe}}, \quad X_4 = U_{\mathrm{ps}},$$

$$\boldsymbol{A} = \boldsymbol{A}_{\mathrm{s}},$$

$$T_{21} = T_{\mathrm{se,ss}} = (T_{\mathrm{ss,se}})^{-1}, \quad T_{41} = T_{\mathrm{ps,ss}},$$

$$T_{43} = T_{\mathrm{ps,pe}} : U_{\mathrm{ps}} \to U_{\mathrm{pe}}^{\mathrm{r}}, \quad T_{32} = T_{\mathrm{pe,se}}.$$

Equation (8.5.13b) leads to

$$(\boldsymbol{u}, \boldsymbol{u})_{\mathrm{e}} \geq \frac{1}{\max \rho}(\rho \boldsymbol{u}, \boldsymbol{u})_{\mathrm{e}} \geq \frac{1}{\max \rho}(m\boldsymbol{w}, \boldsymbol{w})_{\mathrm{s}} \geq \frac{\min m}{\max \rho}(\boldsymbol{w}, \boldsymbol{w})_{\mathrm{s}}, \quad \forall \boldsymbol{w} \in U_{\mathrm{ss}},$$

which shows $T_{\mathrm{se,ss}}$ to be bounded.

Similarly, we know from Eq. (8.5.12) that the operator $T_{\mathrm{ps,pe}}$ is bounded.

Relationship among properties of the aforementioned operators is also shown in Fig. 8.2.

It should be pointed out that, for a valid structural theory, the relationship shown in Inequalities (8.5.12) and (8.5.13) exists. However, it is also possible that some structural theories do not have such a relationship.

Based on the above arguments, the following important theorem may be obtained from Auxiliary Theorems 8.1 and 8.2.

Theorem 8.14 (Theorem on the Positive Definiteness of the Structural Theory Operator) *For an elastic structure with given elastic parameters, geometric shape, and boundary conditions, if*

(1) *the corresponding elastic solid of the structure and its boundary conditions ensure the positive definiteness of the elasticity operator* $\mathbf{A}_e$, *and*
(2) *there is an equivalence of norms between the two spaces with the strain energy norm, one associated with the original structure while the other related to the corresponding constrained elastic solid,*

then the structural theory operator of this structure $\mathbf{A}_s$ *is positive definite.*

Theorem 8.15 (Theorem on the Compactness of the Energy Embedding Operator in the Structural Theory) *For an elastic structure with specified elastic parameters, mass density, geometric shape, and boundary conditions, if*

(1) *the corresponding elastic solid of the structure and its boundary conditions guarantee the energy embedding operator in Elasticity to be compact,*
(2) *there is an equivalence of norms between the two spaces with the strain energy norm, one associated with the original structure while the other related to the corresponding constrained elastic solid, and*
(3) *there also exists an equivalence of norms between the two spaces with the kinetic energy norm, one associated with the original structure while the other related to the corresponding constrained elastic solid,*

then the energy embedding operator in the structural theory, which is associated with the original structure, is compact.

Theorems 8.14 and 8.15 are graphically illustrated in Fig. 8.2.

From the above two theorems as well as Theorems 8.6 and 8.8, we get the following existence theorems for solutions to problems of static deformation and vibrational modes in the structural theory.

Theorem 8.16 (Existence Theorem for the Solution of Static Deformation in Structural Theory) *For an elastic structure with given elastic parameters, geometric shape, and boundary conditions, if*

(1) *the corresponding elastic solid of the structure and its boundary conditions ensure the positive definiteness of the elasticity operator* $\mathbf{A}_e$,
(2) *there is an equivalence of norms between the two spaces with the strain energy norm, one associated with the original structure while the other related to the corresponding constrained elastic solid, and*
(3) *the external force applied to the structure belongs to the* L_2 *space,*

then there exists a unique generalized solution to the static deformation problem of the structure.

On the one hand, this result may be referred to as **the preservation theorem that preserves the existence of the solution of static deformation from Elasticity to**

structural theories, if the structural theories are regarded as some simplification of the more sophisticated three-dimensional elasticity theory. On the other hand, it may be called **an extension theorem that extends the existence of the solution of static deformation from Elasticity to the structural theories**, if the more multifarious and practical structural theories are viewed as certain extension of Elasticity.

Theorem 8.17 (Existence Theorem for Modal Solutions in Structural Theory) *For an elastic structure with specified elastic parameters, mass density, geometric shape, and boundary conditions, if*

(1) *the corresponding elastic solid of the structure and its boundary conditions guarantee the positive definiteness of the elasticity operator* $\mathbf{A}_e$ *as well as the compactness of the energy embedding operator in Elasticity,*
(2) *there is an equivalence of norms between the two spaces with the strain energy norm, one associated with the original structure while the other related to the corresponding constrained elastic solid, and*
(3) *there is also an equivalence of norms between the two spaces with the kinetic energy norm, one associated with the original structure while the other related to the corresponding constrained elastic solid,*

then there exist generalized solutions to the modal problem of the structure. Specifically, there is a countable but infinite set of natural frequencies, with the infinity as the sole limit point; while the corresponding mode shapes formed a complete and orthogonal set in both the space with the kinetic energy norm and the space with the strain energy norm.

As with the previous theorem, this result is also known as, either **the preservation theorem that preserves or the extension theorem that extends, modal solutions from Elasticity to Structural Theories**, depending on a person's perspective.

Now let us take another look at conditions of the previous four theorems. Condition (2) shows relationship of the strain energy between the structure in the structural theory and its constrained solid in Elasticity, while Condition (3) in Theorems 8.15 and 8.17 indicates the relationship of the kinetic energy between the two systems. Both of these two conditions are about the energy relationship between models of the structural theory and Elasticity. As for Condition (1), which appears in all four theorems, it reveals requirements on physical parameters including stiffness and masses, geometric shape, and boundary conditions of specific structures.

Theorems 8.16 and 8.17 are none other than the two "Chinese theorems" that addressed an important problem in the theory of Solid Mechanics, as characterized by R. Valid of France in his monograph (see the note at the end of Sect. 8.1). They provide the theoretical foundation for studying the existence of solutions and the validity of models in structural theories.

8.6 Validity of Models in Structural Theories

8.6.1 Models in Structural Theories and Their Validity

A structural theory should have three aspects: the theoretical model and its governing equations, methods for problem solution, and various applications. Among these, the theoretical model is the most important aspect.

In the previous section, we discussed the existence of solutions of static deformation and vibrational modes. Our approach does not involve equations and boundary conditions of individual structures. As a matter of fact, we provide a unified proof for existence theorems of all types of structural theories. The approach and its outcome provide a basis that supports further research work, including characterization of models in structural theories, checking of their validity, etc.

(1) What is a model in a structural theory?

It may be regarded as the derivative of an elasticity model after the following simplification: (a) simplification on deformation, normally achieved by applying deformation constraints; (b) simplification on the state of stress, normally accomplished through relaxation of some stress components; or (c) simplification on mass distribution, normally fulfilled through mass lumping. Within the framework of Functional Analysis and in the context of the theory of Solid Mechanics, this process may be viewed as transforming the elasticity operator into the structural theory operator.

(2) What is a valid model in the structural theory?

According to Theorems 8.16 and 8.17, if energy properties of the model in the structural theory can be extended from those of the respective elasticity model, i.e., there is an equivalence between norms of strain energies and also between norms of kinetic energies that are associated with the structure and its corresponding constrained elastic solid, then existence of a solution in Elasticity leads to existence of a solution in the structural theory, i.e., the existence of a solution is extended from the model in Elasticity to its counterpart in the structural theory. Such model in the structural theory should be considered a valid one.

However, if a structural model does not satisfy Condition (2) in Theorem 8.16 as well as Conditions (2) and (3) in Theorem 8.17, a solution may still exist, because these conditions are only sufficient, rather than be both necessary and sufficient.

Existing models in structural theories fall roughly under the following three categories:

(a) For many commonly used models in structural theories, there is an equivalence between norms of strain energies associated with the structures and their counterparts associated with the corresponding constrained elastic solids. The same observation can be made on related norms of kinetic energies. As a result, these models in structural theories are valid. Some examples of the beam, plate, and shell models are shown in Sect. 8.6.2.

(b) There are also models in structural theories that obviously have no solution of static deformation and vibrational modes. Such models should be considered invalid. For a structure whose Green's function has singularity, if a concentrated rigid or elastic constraint is imposed right at the singular point and along the direction of the singularity, this would cause the strain energy to be unbounded; while if a concentrated mass is attached to the singular point, the kinetic energy would become unbounded. As a result, it is impossible to determine whether the structural theory operator to be positive definite and its energy embedding operator to be compact in these situations. Thus, there is no guarantee that solutions of static deformation and vibrational modes would exist. These problems will be discussed further in Sect. 8.6.3.

(c) Assembly structures deserve some attention in particular, because there is a huge interest in related theoretical studies and engineering applications. Some assembly structures are valid, while others are invalid. The problem will be tackled in Sect. 8.6.4.

8.6.2 Various Examples of Valid Models in Structural Theories

As explained earlier, for many commonly seen models in structural theories, there is an equivalence between norms of strain energies associated with the structure models and their corresponding constrained elastic solids. The same is true regarding the respective norms of kinetic energies. Thus, these theoretical models are valid.

Example 1 The theory of thin shells.
Consider a thin shell with its mid-surface represented by a curved surface F. Define an orthogonal curvilinear coordinate system (x_1, x_2) on F, and set z to be along the normal direction of the mid-surface. The mid-surface displacement $\boldsymbol{u}_0 = (u, v, w)^{\mathrm{T}}$ is a function of coordinates x_1 and x_2. Denote the theoretical operator of the shell theory as A_s.

Now let us treat the shell as a three-dimensional elastic solid taking up the space Ω and denote its boundary as $\partial\Omega$. Its displacement $\boldsymbol{u} = (u_1, u_2, u_3)^{\mathrm{T}}$ is a function of x_1, x_2, and z. See Fig. 8.3. Let us explore the relationship between the shell theory and three-dimensional Elasticity from three aspects:

Relationship Between Displacements in the Shell Theory and Elasticity Shell deformation is obtained by imposing the Kirchhoff's hypothesis on deformation of the three-dimensional elastic solid, where a line segment normal to the original undeformed mid-surface remains straight and normal to the deformed mid-surface and in addition, it undergoes negligible change in length. The relationship between $\boldsymbol{u}$ and the mid-surface displacement $\boldsymbol{u}_0$ is as follows:

$$u_1 = u + z\vartheta, \quad u_2 = v + z\psi, \quad \text{and} \quad u_3 = w, \tag{8.6.1}$$

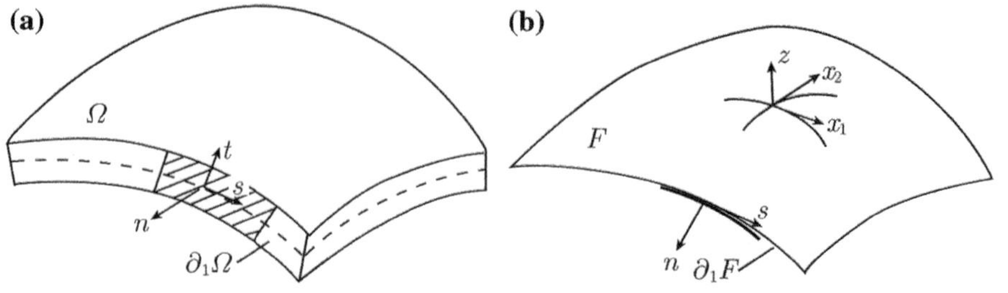

Fig. 8.3 **a** A shell, **b** the mid-surface of the shell

where the following two variables

$$\vartheta = -\frac{1}{A_1}\frac{\partial w}{\partial x_1} + \frac{u}{R_1} \quad \text{and} \quad \psi = -\frac{1}{A_2}\frac{\partial w}{\partial x_2} + \frac{v}{R_2} \tag{8.6.2}$$

designate rotational angles at the mid-surface, A_1 and A_2 are Lamé parameters of the curvilinear coordinate system at the mid-surface, and R_1 and R_2 denote the two principal radii of curvatures at the mid-surface.

Equation (8.6.1) is a specialization of Eq. (8.5.9) in the case of the shell theory. It defines the operator $T_{\text{ss, se}}$ for a one-to-one mapping. If a displacement $\boldsymbol{u}$ satisfies Eq. (8.6.1), then $\boldsymbol{u} \in U^{\text{r}}_{\text{se}}$.

Relationship Between Strain Energies in the Structural Theory and Elasticity Essentially, we need to verify Eq. (8.5.12). After displacement constraints are enforced, the strain components of the elastic solid become:

$$\varepsilon_{33} = \varepsilon_{13} = \varepsilon_{23} = 0, \tag{8.6.3a}$$

$$\varepsilon_{11} = \varepsilon_1 + z\kappa_1, \quad \varepsilon_{22} = \varepsilon_2 + z\kappa_2, \quad \varepsilon_{12} = \omega + 2z\tau, \tag{8.6.3b}$$

where the following three variables

$$\begin{aligned} \varepsilon_1 &= \frac{1}{A_1}\frac{\partial u}{\partial x_1} + \frac{v}{A_1 A_2}\frac{\partial A_1}{\partial x_2} + \frac{w}{R_1}, \\ \varepsilon_2 &= \frac{u}{A_1 A_2}\frac{\partial A_2}{\partial x_1} + \frac{1}{A_2}\frac{\partial v}{\partial x_2} + \frac{w}{R_2}, \\ \omega &= \frac{A_1}{A_2}\frac{\partial}{\partial x_2}\left(\frac{u}{A_1}\right) + \frac{A_2}{A_1}\frac{\partial}{\partial x_1}\left(\frac{v}{A_2}\right) \end{aligned} \tag{8.6.4a}$$

are normal and shear strains at the mid-surface, while the other three variables below

$$\kappa_1 = \frac{1}{A_1}\frac{\partial \vartheta}{\partial x_1} + \frac{\psi}{A_1 A_2}\frac{\partial A_1}{\partial x_2},$$
$$\kappa_2 = \frac{\vartheta}{A_1 A_2}\frac{\partial A_2}{\partial x_1} + \frac{1}{A_2}\frac{\partial \psi}{\partial x_2},$$
$$\tau = \frac{1}{A_1}\frac{\partial \psi}{\partial x_1} - \frac{\vartheta}{A_1 A_2}\frac{\partial A_1}{\partial x_2} + \frac{1}{R_1}\left(\frac{1}{A_2}\frac{\partial u}{\partial x_2} - \frac{1}{A_1 A_2}\frac{\partial A_2}{\partial x_1}v\right) \tag{8.6.4b}$$

represent rates of change in curvature and the torsion measured at the mid-surface.

The strain energy in Elasticity is

$$\Pi_e(\boldsymbol{u}) = \frac{1}{2}\int_{\Omega} \{\varepsilon\}^{\mathrm{T}}\{\sigma\}\mathrm{d}\Omega, \tag{8.6.5a}$$

or it may be rewritten as

$$\begin{aligned}\Pi_e(\boldsymbol{u}) &= \frac{1}{2}\int_{\Omega} \{\varepsilon\}^{\mathrm{T}}[D]\{\varepsilon\}\mathrm{d}\Omega \\ &= \frac{1}{2}\int_{\Omega}\left\{\frac{E\nu}{(1+\nu)(1-2\nu)}(\varepsilon_{11}+\varepsilon_{22}+\varepsilon_{33})^2\right. \\ &\quad \left.+\frac{E}{1+\nu}\left[\left(\varepsilon_{11}^2+\varepsilon_{22}^2+\varepsilon_{33}^2\right)+\frac{1}{2}\left(\varepsilon_{12}^2+\varepsilon_{23}^2+\varepsilon_{31}^2\right)\right]\right\}\mathrm{d}\Omega,\end{aligned} \tag{8.6.5b}$$

where $[D]$ represents the stiffness matrix of the stress–strain equation in Elasticity. It is known from Eq. (8.6.3) that the constrained elastic solid is in the state of plane strain, and the simplified representation of its strain energy may be obtained by plugging Eq. (8.6.3a) into Eq. (8.6.5b) as follows:

$$\Pi_e^{\mathrm{r}} = \frac{1}{2}\int_{\Omega} \frac{E}{2(1+\nu)(1-2\nu)}\left\{(\varepsilon_{11}+\varepsilon_{22})^2 + (1-2\nu)\left[(\varepsilon_{11}-\varepsilon_{22})^2+\varepsilon_{12}^2\right]\right\}\mathrm{d}\Omega. \tag{8.6.6}$$

In the shell theory, besides the aforementioned Kirchhoff's hypothesis, there is another assumption stipulating the stress along the normal direction at the mid-surface to be zero, i.e.,

$$\sigma_{33} = 0. \tag{8.6.7a}$$

The conditions $\varepsilon_{13} = \varepsilon_{23} = 0$ in Eq. (8.6.3a) yield $\sigma_{13} = \sigma_{23} = 0$. Thus, the structure in the shell theory is also in the state of plane stress. The stress–strain relationship is

$$\begin{bmatrix} \sigma_{11} \\ \sigma_{22} \\ \sigma_{12} \end{bmatrix} = \frac{E}{1-\nu^2}\begin{bmatrix} 1 & \nu & 0 \\ \nu & 1 & 0 \\ 0 & 0 & (1-\nu)/2 \end{bmatrix}\begin{bmatrix} \varepsilon_{11} \\ \varepsilon_{22} \\ \varepsilon_{12} \end{bmatrix}. \tag{8.6.7b}$$

Substituting Eqs. (8.6.7) and (8.6.3a) into Eq. (8.6.5a), we get the strain energy of the shell theory

$$\Pi_s = \frac{1}{2}\int_{\Omega} \frac{E}{2(1-\nu)}\left\{(\varepsilon_{11}+\varepsilon_{22})^2 + \frac{1-\nu}{1+\nu}\left[(\varepsilon_{11}-\varepsilon_{22})^2 + \varepsilon_{12}^2\right]\right\}\mathrm{d}\Omega. \tag{8.6.8}$$

Substituting Eq. (8.6.3b) into the above equation, we get, for a shell of uniform thickness h:

$$\begin{aligned}\Pi_s = &\frac{Eh}{2(1-\nu^2)}\int_F \left[(\varepsilon_1+\varepsilon_2)^2 - 2(1-\nu)\left(\varepsilon_1\varepsilon_2 - \omega^2/4\right)\right]A_1A_2\mathrm{d}x_1\mathrm{d}x_2 \\ &+ \frac{Eh^2}{24(1-\nu^2)}\int_F \left[(\kappa_1+\kappa_2)^2 - 2(1-\nu)\left(\kappa_1\kappa_2 - \tau^2\right)\right]A_1A_2\mathrm{d}x_1\mathrm{d}x_2.\end{aligned}$$

By inserting Eq. (8.6.4) into (8.6.8), we can express the strain energy Π_s of the shell in terms of the displacement $\boldsymbol{u}_0$ at the mid-surface.

Now we are ready to establish the relationship between Π_e^r and Π_s, which are calculated from Eqs. (8.6.6) and (8.6.8), respectively. Under the condition of an isothermal process, $-1 < \nu < 1/2$. Thus, $p = (1-\nu)/(1+\nu)(1-2\nu) \geq 1$. Multiplying the second term in the curly bracket of the integrand in Eq. (8.6.6) by the factor p, we find

$$\Pi_s \geq \frac{(1+\nu)(1-2\nu)}{1-\nu}\Pi_e^r.$$

Furthermore, if the first term in the curly bracket of the integrand in Eq. (8.6.6) is multiplied by the factor $1/p$, we get

$$\Pi_e^r \geq \Pi_s.$$

Thus, when showing the relationship between strain energies of a shell and its corresponding constrained elastic solid, Eq. (8.5.12) assumes the following form:

$$\Pi_e^r \geq \Pi_s \geq \frac{(1+\nu)(1-2\nu)}{1-\nu}\Pi_e^r. \tag{8.6.9a}$$

Relationship Between Kinetic Energies in the Structural Theory and Elasticity In fact, we need to derive Eq. (8.5.13c). Since

Table 8.1 Six typical boundary conditions of a shell and its corresponding elastic solid

	Boundary conditions of a shell	Related boundary conditions of the corresponding elastic solid
1	Clamped: $\boldsymbol{u}_0 = \mathbf{0},\ \theta_n = 0$	Fixed: $\boldsymbol{u} = \mathbf{0}$
2	Free	Free
3	$(\boldsymbol{u}_0)_n = \theta_n = 0$	$\boldsymbol{u}_n = 0$
4	$w_0 = (\boldsymbol{u}_0)_s = 0$	$\boldsymbol{u}_t = \boldsymbol{u}_s = 0$
5	Pinned: $w_0 = (\boldsymbol{u}_0)_s = (\boldsymbol{u}_0)_n = 0$	Stronger than the conditions $\boldsymbol{u}_t = \boldsymbol{u}_s = 0$
6	$(\boldsymbol{u}_0)_n = (\boldsymbol{u}_0)_s = \theta_n = 0$	Stronger than the condition $\boldsymbol{u}_n = 0$

$$\int_\Omega \rho \boldsymbol{u}^2 \mathrm{d}\Omega = \int_\Omega \rho\left[(u + z\vartheta)^2 + (v + z\psi)^2 + w^2\right]\mathrm{d}\Omega$$
$$= \int_\Omega \left[\rho(u^2 + v^2 + w^2) + \rho z^2(\vartheta^2 + \psi^2)\right]\mathrm{d}\Omega,$$

we obtain

$$K_\mathrm{e}^\mathrm{r} = \frac{1}{2}\int_\Omega \rho \boldsymbol{u}^2 \mathrm{d}\Omega = \frac{1}{2}\int_F \left[m\boldsymbol{u}_0^2 + I(\vartheta^2 + \psi^2)\right]\mathrm{d}F,$$

where I is the moment of inertia. In the shell theory, the kinetic energy associated with the moment of inertia is normally ignored. Thus, Eq. (8.5.13c), when showing the relationship between coefficients of the kinetic energy associated with the shell and its corresponding constrained elastic solid, takes the following form:

$$K_\mathrm{e}^\mathrm{r} \geq \frac{1}{2}\int_F m\boldsymbol{u}_0^2 \mathrm{d}F = K_\mathrm{s}. \tag{8.6.9b}$$

Six typical boundary conditions of a shell are listed in Table 8.1.

It has been proved that under boundary conditions listed on the right column of Table 8.1, the elasticity operator $\boldsymbol{A}_\mathrm{e}$ is positive definite and the respective energy embedding operator is compact. Thus, under the boundary conditions tabulated on the left column of Table 8.1, it is guaranteed that the operator $\boldsymbol{A}_\mathrm{s}$ of the shell theory is positive definite and the related energy embedding operator $T_{\mathrm{ps,ks}}$ is compact.

Example 2 The theory of thin plates.

A thin plate is a special case of a thin shell. Specifically, for a plate, $A_1 = A_2 = 1$ while R_1 and R_2 become infinity. In addition, there is only deflection at the midplane $w(x_1, x_2)$ and no in-plane movement, i.e., $u = v = 0$. The Kirchhoff's hypothesis represented by Eq. (8.6.1) now assumes the form:

$$u_1 = -z\frac{\partial w}{\partial x_1}, \quad u_2 = -z\frac{\partial w}{\partial x_2}, \quad u_3 = w.$$

For a plate of uniform thickness, strain energies corresponding to Eqs. (8.6.6) and (8.6.8) are, respectively,

$$\Pi_e^r = \frac{1}{2}\int_F \frac{Eh^3}{24(1+\nu)(1-2\nu)}\left\{(\kappa_1+\kappa_2)^2 + (1-2\nu)\left[(\kappa_1-\kappa_2)^2 + 4\tau^2\right]\right\}dx_1dx_2$$

and

$$\begin{aligned}\Pi_s &= \frac{1}{2}\int_F \frac{Eh^3}{24(1-\nu^2)}\left\{(\kappa_1+\kappa_2)^2 - 2(1-\nu)(\kappa_1\kappa_2-\tau^2)\right\}dx_1dx_2 \\ &= \frac{1}{2}\int_F \frac{Eh^3}{24(1-\nu^2)}\left\{\left(\frac{\partial^2 w}{\partial x_1^2}+\frac{\partial^2 w}{\partial x_2^2}\right)^2\right. \\ &\left. -2(1-\nu)\left[\frac{\partial^2 w}{\partial x_1^2}\cdot\frac{\partial^2 w}{\partial x_2^2} - \left(\frac{\partial^2 w}{\partial x_1\partial x_2}\right)^2\right]\right\}dx_1dx_2.\end{aligned}$$

Furthermore, Eqs. (8.5.12) and (8.5.13c) become, respectively,

$$\Pi_e^r \geq \Pi_s \geq \frac{(1+\nu)(1-2\nu)}{1-\nu}\Pi_e^r$$

and

$$\begin{aligned}K_e^r &= \frac{1}{2}\int_F \left\{mw^2 + I\left[\left(\frac{\partial w}{\partial x_1}\right)^2 + \left(\frac{\partial w}{\partial x_2}\right)^2\right]\right\}dx_1dx_2 \\ &\geq \frac{1}{2}\int_F mw^2 dx_1dx_2 = K_s.\end{aligned}$$

Example 3 The theory of straight beams.
Set the z-axis to be along the axis of the beam. Suppose the beam has no displacement in the z direction while its deflections in x and y directions are $u(z)$ and $v(z)$, respectively. The beam deformation follows the Euler–Bernoulli hypothesis, i.e., a plane section remains planar and normal to the neutral axis of the beam. Furthermore, it is assumed that there is no stress in the lateral directions, namely, $\sigma_x = \sigma_y = 0$. Thus, the beam is subject to unidirectional stress. The displacement constraints corresponding to Eq. (8.6.1) are

$$u_1 = u, \quad u_2 = v, \quad u_3 = -x\frac{du}{dz} - y\frac{dv}{dz}.$$

With the displacement constraints, the constrained elastic solid is also in the state of unidirectional strain. Equations (8.5.12) and (8.5.13c) become, respectively,

$$\Pi_{\mathrm{s}} = \frac{(1+\nu)(1-2\nu)}{1-\nu}\Pi_{\mathrm{e}}^{\mathrm{r}}$$

and

$$\begin{aligned} K_{\mathrm{e}}^{\mathrm{r}} &= \frac{1}{2}\int_0^l \left[m(u^2+v^2) + I_y\left(\frac{\mathrm{d}u}{\mathrm{d}x}\right)^2 + I_x\left(\frac{\mathrm{d}v}{\mathrm{d}y}\right)^2 \right]\mathrm{d}z \\ &\geq \frac{1}{2}\int_0^l m(u^2+v^2)\mathrm{d}z = K_{\mathrm{s}}. \end{aligned}$$

Similar results can be obtained for other one-dimensional structures such as bars and curved beams.

Example 4 The theory of anisotropic plates.
This type of models covers anisotropic plates with shear deformation as well as plates made of composite materials. An anisotropic plate can be viewed as an anisotropic elastic solid that is symmetric about its mid-plane. The stress–strain relationship is

$$\begin{bmatrix} \sigma_x \\ \sigma_y \\ \tau_{xy} \end{bmatrix} = \begin{bmatrix} d_{11} & d_{12} & d_{16} \\ d_{21} & d_{22} & d_{26} \\ d_{16} & d_{26} & d_{66} \end{bmatrix} \begin{bmatrix} e_x \\ e_y \\ \gamma_{xy} \end{bmatrix} = \boldsymbol{d\varepsilon},$$

$$\begin{bmatrix} \tau_{yz} \\ \tau_{zx} \end{bmatrix} = \begin{bmatrix} c_{44} & c_{45} \\ c_{45} & c_{55} \end{bmatrix} \begin{bmatrix} \gamma_{yz} \\ \gamma_{zx} \end{bmatrix} = \boldsymbol{c\gamma}.$$

Among the assumptions on the displacement, strain, and stress, we have:

$$\boldsymbol{u} = \begin{bmatrix} -z & 0 & 0 \\ 0 & -z & 0 \\ 0 & 0 & 1 \end{bmatrix} \begin{bmatrix} \psi_x \\ \psi_y \\ w \end{bmatrix} = T_{\mathrm{ss,se}}\boldsymbol{w}$$

and

$$\begin{bmatrix} \gamma_{yz} \\ \gamma_{zx} \end{bmatrix} = \begin{bmatrix} \gamma_y \\ \gamma_z \end{bmatrix} = \boldsymbol{\gamma},$$

while other conditions are identical to Eqs. (8.6.2) and (8.6.3).

Strain energies of the anisotropic plate and its constrained elastic solid are, respectively

$$\Pi_e^r = \frac{1}{2}\int_F \left(\frac{h^3}{12}\boldsymbol{\kappa}^T\boldsymbol{d}\boldsymbol{\kappa} + h\boldsymbol{\gamma}^T\boldsymbol{c}\boldsymbol{\gamma}\right)\mathrm{d}x_1\mathrm{d}x_2$$

and

$$\Pi_s = \frac{1}{2}\int_F (\boldsymbol{\kappa}^T\boldsymbol{D}\boldsymbol{\kappa} + \boldsymbol{\gamma}^T\boldsymbol{C}\boldsymbol{\gamma})\mathrm{d}x_1\mathrm{d}x_2,$$

where $\boldsymbol{D}$ and $\boldsymbol{C}$ are matrices of the bending and shear stiffness matrices of the anisotropic plate, respectively.

Similar to Example 1 of this subsection, we can prove that Eq. (8.5.12) holds true. See Refs. [33].

8.6.3 Validity of Models with Concentrated Mass and Support in Structural Theories

Because of practical needs and theoretical interests, vibration systems with concentrated parameters, such as concentrated masses, springs, dampers, and rigid supports, have been widely studied. One finding of theoretical significance is that certain structural theories do not allow concentrated parameters. Examples include the membrane theory, Mindlin theory for thick plates, and shell theory. This subject is our focus in this section. Also see Refs. [21, 36, 37].

1. Governing Equation and Its Solution

The governing equation and boundary condition for a vibrating system with concentrated masses, springs, and supports are of the forms, respectively:

$$\boldsymbol{A}\boldsymbol{w}(\boldsymbol{x},t) + \boldsymbol{\rho}\ddot{\boldsymbol{w}}(\boldsymbol{x},t) + \sum_{i=1}^{i_0}\delta(\boldsymbol{x}-\bar{\boldsymbol{x}}_i)\boldsymbol{M}_i\ddot{\boldsymbol{w}}(\bar{\boldsymbol{x}}_i,t)$$
$$+\sum_{j=i_0+1}^{j_0}\delta(\boldsymbol{x}-\bar{\boldsymbol{x}}_j)\boldsymbol{K}_j\boldsymbol{w}(\bar{\boldsymbol{x}}_j,t) - \sum_{k=j_0+1}^{k_0}\delta(\boldsymbol{x}-\bar{\boldsymbol{x}}_k)\boldsymbol{R}_k = \boldsymbol{p}(\boldsymbol{x},t) \quad \text{in } \Omega, \qquad (8.6.10)$$

$$\boldsymbol{B}\boldsymbol{w}(\boldsymbol{x},t) = \boldsymbol{0} \qquad \text{on } \partial\Omega. \qquad (8.6.11)$$

Here, Ω is the domain taken up by the system; $\boldsymbol{w}$ denotes the h-dimensional vector of the displacement function; $\bar{\boldsymbol{x}}_l$ $(l = i, j, k)$ represents the coordinate vector indicating the location of a concentrated parameter; $\boldsymbol{R}_k$ designates the h-dimensional vector of the reaction force at a support, where a zero r-th component of $\boldsymbol{R}_k$ indicates no support along the r-th direction; $\boldsymbol{A}$ and $\boldsymbol{B}$ are, respectively, matrix differential operators of the structure and boundary conditions; $\boldsymbol{\rho}$, $\boldsymbol{M}_i$, and $\boldsymbol{K}_j$ represent, respectively, the generalized mass density, the diagonal matrix of the concentrated generalized mass, and the diagonal matrix of the concentrated spring stiffness, where the r-th diagonal element K_{jr} of $\boldsymbol{K}_j$ to be zero, i.e., $K_{jr} = 0$, indicates no concentrated spring in the r-th direction and the same convention applies to elements of $\boldsymbol{M}_i$; and finally, $\delta(\cdot)$ stands for the δ function.

Now suppose the system is in natural vibration. Set $\boldsymbol{w}(\boldsymbol{x}, t) = \boldsymbol{\phi}(\boldsymbol{x}) \sin \omega t$ and $\boldsymbol{R}_k = \boldsymbol{p}_k \sin \omega t$. The natural angular frequency ω and mode shape $\boldsymbol{\phi}(\boldsymbol{x})$ satisfy the modal equation and boundary condition:

$$\begin{aligned} &\boldsymbol{A\phi}(\boldsymbol{x}) - \omega^2 \boldsymbol{\rho\phi}(\boldsymbol{x}) - \omega^2 \sum_{i=1}^{i_0} \delta(\boldsymbol{x} - \bar{\boldsymbol{x}}_i)\boldsymbol{M}_i \boldsymbol{\phi}(\bar{\boldsymbol{x}}_i) \\ &+ \sum_{j=i_0+1}^{j_0} \delta(\boldsymbol{x} - \bar{\boldsymbol{x}}_j)\boldsymbol{K}_j \boldsymbol{\phi}(\bar{\boldsymbol{x}}_j) - \sum_{k=j_0+1}^{k_0} \delta(\boldsymbol{x} - \bar{\boldsymbol{x}}_k)\boldsymbol{p}_k = \boldsymbol{0} \quad \text{in } \Omega, \end{aligned} \tag{8.6.12}$$

$$\boldsymbol{B\phi}(\boldsymbol{x}) = \boldsymbol{0} \qquad \text{on } \partial\Omega. \tag{8.6.13}$$

The above equation may be solved by using the harmonic dynamic Green's function. Consider a structure with no concentrated parameter. Apply a unit harmonic concentrated force with the angular frequency $\bar{\omega}$ and along the r-th direction to a location $\bar{\boldsymbol{x}}$ on the structure, where the r-th direction is the direction of the r-th the generalized coordinate. The force may be expressed as

$$\boldsymbol{p}(\bar{\boldsymbol{x}}, t) = \boldsymbol{I}_r \delta(\boldsymbol{x} - \bar{\boldsymbol{x}}) \sin \bar{\omega} t,$$

where $\boldsymbol{I}_r$ is an h-dimensional vector, with the r-th component and all other components being 1 and 0, respectively. Under the action of this concentrated load, the steady-state response of the structure with no concentrated parameter is the harmonic dynamic Green's function $\boldsymbol{G}_r(\boldsymbol{x}, \bar{\boldsymbol{x}}, \bar{\omega})$, which satisfies the following equations:

$$\boldsymbol{AG}_r(\boldsymbol{x}, \bar{\boldsymbol{x}}, \bar{\omega}) - \bar{\omega}^2 \boldsymbol{\rho G}_r(\boldsymbol{x}, \bar{\boldsymbol{x}}, \bar{\omega}) = \boldsymbol{I}_r \delta(\boldsymbol{x} - \bar{\boldsymbol{x}}) \quad \text{in } \Omega, \tag{8.6.14}$$

$$\boldsymbol{BG}_r(\boldsymbol{x}, \bar{\boldsymbol{x}}, \bar{\omega}) = \boldsymbol{0} \qquad \text{on } \partial\Omega, \tag{8.6.15}$$

where the angular frequency $\bar{\omega}$ of the concentrated force is a parameter and $\boldsymbol{G}_r(\boldsymbol{x}, \bar{\boldsymbol{x}}, \bar{\omega})$ is an h-dimensional vector.

Since both operators $\boldsymbol{A}$ and $\boldsymbol{\rho}$ are linear, we get the mode shape of the structure with concentrated parameters by comparing Eqs. (8.6.12) and (8.6.13) with Eqs. (8.6.14) and (8.6.15):

$$\begin{aligned} \boldsymbol{\phi}(\boldsymbol{x}) = \omega^2 \sum_{i=1}^{i_0} \sum_{r=1}^{h} M_{ir} \boldsymbol{\phi}_r(\bar{\boldsymbol{x}}_i) \boldsymbol{G}_r(\boldsymbol{x}, \bar{\boldsymbol{x}}_i, \omega) - \sum_{j=i_0+1}^{j_0} \sum_{r=1}^{h} K_{jr} \boldsymbol{\phi}_r(\bar{\boldsymbol{x}}_j) \boldsymbol{G}_r(\boldsymbol{x}, \bar{\boldsymbol{x}}_j, \omega) \\ + \sum_{j=j_0+1}^{k_0} \sum_{r=1}^{h} p_{kr} \boldsymbol{G}_r(\boldsymbol{x}, \bar{\boldsymbol{x}}_k, \omega). \end{aligned} \tag{8.6.16}$$

It is clear from the above equation that $\boldsymbol{\phi}(\boldsymbol{x})$ can be determined when all components of the harmonic dynamic Green's function

$$\boldsymbol{G}_r(\boldsymbol{x}, \bar{\boldsymbol{x}}_l, \bar{\omega}), \quad r = 1, 2, \ldots, h \text{ and } l = 1, 2, \ldots i_0, \ \ i_0 + 1, \ldots, j_0, \ \ j_0 + 1, \ldots, k_0 \tag{8.6.17}$$

are bounded. However, when there is at least one component of the function to be singular, then $\boldsymbol{\phi}(\boldsymbol{x})$ cannot be determined. This implies that it is not allowed to include a concentrated parameter at the location where $\boldsymbol{G}_r(\boldsymbol{x}, \bar{\boldsymbol{x}}_l, \bar{\omega})$ is singular. Otherwise, the structure obtained is invalid.

2. ***Relationship Between Singularities of Static and Harmonic Dynamic Green's Functions***

Source values of the harmonic dynamic Green's function and the static Green's function are denoted by $\boldsymbol{G}_r(\bar{\boldsymbol{x}}, \bar{\boldsymbol{x}}, \omega)$ and $\boldsymbol{G}_r(\bar{\boldsymbol{x}}, \bar{\boldsymbol{x}})$, respectively. Let us prove an important property: The two source values must either be both bounded or share singularities of the same order. See Refs. [21, 36, 37]. Following the convention, we refer to the static Green's function simply as the Green's function.

Equations (8.6.14) and (8.6.15) may be rewritten as

$$\begin{cases} \boldsymbol{A}\boldsymbol{G}_r(\boldsymbol{x}, \bar{\boldsymbol{x}}, \bar{\omega}) = \boldsymbol{I}_r\delta(\boldsymbol{x} - \bar{\boldsymbol{x}}) + \bar{\omega}^2\boldsymbol{\rho}\boldsymbol{G}_r(\boldsymbol{x}, \bar{\boldsymbol{x}}, \bar{\omega}) & \text{in } \Omega, \\ \boldsymbol{B}\boldsymbol{G}_r(\boldsymbol{x}, \bar{\boldsymbol{x}}, \bar{\omega}) = \boldsymbol{0} & \text{on } \partial\Omega. \end{cases} \tag{8.6.18}$$

The harmonic dynamic Green's function may be separated into two parts:

$$\boldsymbol{G}_r(\boldsymbol{x}, \bar{\boldsymbol{x}}, \bar{\omega}) = \boldsymbol{G}_{r1}(\boldsymbol{x}, \bar{\boldsymbol{x}}) + \boldsymbol{G}_{r2}(\boldsymbol{x}, \bar{\boldsymbol{x}}, \bar{\omega}), \tag{8.6.19}$$

where $\boldsymbol{G}_{r1}$ and $\boldsymbol{G}_{r2}$ satisfy equations, respectively,

$$\begin{cases} \boldsymbol{A}\boldsymbol{G}_{r1}(\boldsymbol{x}, \bar{\boldsymbol{x}}) = \boldsymbol{I}_r\delta(\boldsymbol{x} - \bar{\boldsymbol{x}}) & \text{in } \Omega, \\ \boldsymbol{B}\boldsymbol{G}_{r1}(\boldsymbol{x}, \bar{\boldsymbol{x}}) = \boldsymbol{0} & \text{on } \partial\Omega, \end{cases} \tag{8.6.20}$$

and

$$\begin{cases} \boldsymbol{A}\boldsymbol{G}_{r2}(\boldsymbol{x}, \bar{\boldsymbol{x}}, \bar{\omega}) - \bar{\omega}^2\boldsymbol{\rho}\boldsymbol{G}_{r2}(\boldsymbol{x}, \bar{\boldsymbol{x}}, \bar{\omega}) = \bar{\omega}^2\boldsymbol{\rho}\boldsymbol{G}_{r1}(\boldsymbol{x}, \bar{\boldsymbol{x}}) & \text{in } \Omega, \\ \boldsymbol{B}\boldsymbol{G}_{r2}(\boldsymbol{x}, \bar{\boldsymbol{x}}, \bar{\omega}) = \boldsymbol{0} & \text{on } \partial\Omega. \end{cases} \tag{8.6.21}$$

The function $\boldsymbol{G}_{r1}$ satisfying Eq. (8.6.20) is just the Green's function, and it is easy to see

$$\boldsymbol{G}_{r2}(\boldsymbol{x}, \bar{\boldsymbol{x}}, 0) = 0, \quad \boldsymbol{G}_{r1}(\boldsymbol{x}, \bar{\boldsymbol{x}}) = \boldsymbol{G}_r(\boldsymbol{x}, \bar{\boldsymbol{x}}, 0).$$

If $\boldsymbol{G}_{r1}(\boldsymbol{x}, \bar{\boldsymbol{x}})$ is bounded, then we find $\boldsymbol{G}_{r2}(\boldsymbol{x}, \bar{\boldsymbol{x}}, \bar{\omega})$ to be also bounded by integrating the first equation in (8.6.21). Consequently, $\boldsymbol{G}_r(\boldsymbol{x}, \bar{\boldsymbol{x}}, \bar{\omega})$ should be bounded. In contrast, if some components of $\boldsymbol{G}_{r1}(\boldsymbol{x}, \bar{\boldsymbol{x}})$ have singularity, then $\boldsymbol{G}_{r2}(\boldsymbol{x}, \bar{\boldsymbol{x}}, \bar{\omega})$ is either bounded or some of its components have singularities of order lower than those of $\boldsymbol{G}_{r1}(\boldsymbol{x}, \bar{\boldsymbol{x}})$. Hence, the harmonic dynamic Green's function $\boldsymbol{G}_r(\boldsymbol{x}, \bar{\boldsymbol{x}}, \bar{\omega})$ has the same order of singularity as the Green's function, which is $\boldsymbol{G}_{r1}(\boldsymbol{x}, \bar{\boldsymbol{x}})$ here.

We are now ready to reach some conclusions. For a structure with a bounded Green's function, it is valid to include concentrated parameters; while for a structure with some components of its Green's function being singular at certain points, it is invalid to include a concentrated parameter at any singular point along the direction associated with the component having singularity. For the latter case, the physical meaning is clear. Specifically, if a structure with concentrated masses, springs, and supports is set in motion, it will induce concentrated inertia forces, elastic forces, and reaction forces, respectively, at the singular points, and they, in turn, lead to unbounded displacements there. This is, of course, not allowed. A similar finding can be obtained for the case when a damper is applied to a singular point.

3. Singularities of Green's Functions Associated with Various Types of Structures

If a Green's function is singular along certain direction, then it has the same order of singularity as the fundamental solution. When a concentrated force is applied to the source, the first equation in (8.6.20) becomes

$$\boldsymbol{A}\boldsymbol{G}_r(\boldsymbol{x}, 0) = \boldsymbol{I}_r\delta(\boldsymbol{x}).$$

Suppose the dimension of the region occupied by the structure is m, which can be 1, 2, or 3. The equation

$$\int_C \cdots \int \boldsymbol{A}\boldsymbol{G}_r(x, 0)\mathrm{d}x_1\mathrm{d}x_2\cdots\mathrm{d}x_m = \boldsymbol{I}_r \tag{8.6.22}$$

indicates an equilibrium between the internal elastic force and the external concentrated force. The integration domain C is a curve when $m = 1$, a surface when $m = 2$, or a solid when $m = 3$. In any case, the region C contains the source. Since Eq. (8.6.22) involves integration of a δ function, the size of the domain C can be infinitely small.

Now let us look at one component of the displacement vector $\boldsymbol{w} = (u, v, w)^{\mathrm{T}}$, such as u. Suppose its highest derivative in the operator $\boldsymbol{A}$ is n. When $m = 1$, we get after integration

$$\left.\frac{\partial^{n-1}u}{\partial x^{n-1}}\right|_{-\varepsilon}^{\varepsilon} = c_1,$$

where c_1 is a constant. This reveals an equilibrium between internal forces at the two ends of the source and the external concentrated load. If $m = 2$, we can apply the formula converting surface integration to curve integration; while if $m = 3$, we may use the formula transforming a volume integral to a surface integral. Then, Eq. (8.6.22) yields

$$\int_{\partial C} \cdots \int \boldsymbol{R}\boldsymbol{G}_r(\boldsymbol{x}, 0)\mathrm{d}x_1\mathrm{d}x_2 \cdots \mathrm{d}x_m = \boldsymbol{I}_r, \tag{8.6.23}$$

where the integration is performed along the boundary of C, i.e., ∂C. Equation (8.6.23) shows an equilibrium between the internal force acting on ∂C and the concentrated load. In addition, it yields

$$\frac{\partial^{n-1}u}{\partial r^{n-1}}\varepsilon^{m-1} = c_1,$$

where

$$r = \left(\sum_i x_i^2\right)^{1/2}.$$

Thus, we get

$$\frac{\partial^{n-1}u}{\partial r^{n-1}} = O(r^{1-m}).$$

If $n \neq m$, then $u = O(r^{n-m})$.

Finally, we may draw the following conclusions: For a static Green's function,

(1) when $n > m$, it is bounded;
(2) when $n = m$, it has a singularity of the same order as the function $\ln r$; and
(3) when $n < m$, its singularity is of order $1/r^{m-n}$.

Table 8.2 tabulates the singularity of Green's functions found in commonly used structural theories as well as in Elasticity. It can be seen from Table 8.2 that Green's functions of all structural theories for one-dimensional structures are bounded. The Green's function of the thin plate theory is also bounded. But for plates taking into account transverse shear deformation, such as the Mindlin plate and various laminated plates, the Green's functions of these theories have singularity. As for the moment theory of shells, its Green's function has some very interesting properties: The component in the normal direction is bounded, while components along tangential directions are not.

It is worth noting that concentrated parameters are allowed along a line in all structural theories whose Green's functions involve a singularity of the order $\ln r$. However, the three-dimensional Elasticity is an exception. Because its Green's function has a singularity of the order $1/r$, concentrated parameters are not allowed along any line.

Table 8.2 Singularity of the Green's functions in structural theories and elasticity

Theory	Structure dimension m	Highest order n of the differential equation	Singularity of the Green's function
String	1	2	Bounded
Bar and shaft	1	2	Bounded
Euler–Bernoulli beam	1	4	Bounded
Rayleigh beam	1	4	Bounded
Timoshenko beam	1	2	Bounded
Deformation of a curved beam, normal direction	1	4	Bounded
Deformation of a curved beam, tangential direction	1	2	Bounded
Thin plate	2	4	Bounded
Lateral deformation of a membrane	2	2	$\ln r$
2D Elasticity	2	2	$\ln r$
Membrane theory of shells	2	2	$\ln r$
Mindlin Plate	2	2	$\ln r$
3D Elasticity	3	2	$1/r$
Moment theory of shells, normal direction	2	4	Bounded
Moment theory of shells, tangential direction	2	2	$\ln r$
Shells with transverse shear deformation	2	2	$\ln r$

8.6.4 Assembly Structures

1. About Assembly Structures

An assembly structure consists of various parts assembled together, including one-dimensional components such as bars, beams, and arches, two-dimensional components such as membranes, plates, and shells, three-dimensional elastic solids, as well as concentrated masses, isolated elastic or rigid supports, and other zero-dimensional objects. Assembly structures are commonplace in this world and have been playing an important role in engineering applications. However, it has been a huge challenge to develop effective mathematical models and algorithms for problems related to assembly structures, due to complexities involved such as different dimensions and physical properties of elastic components as well as dissimilar ways of connection. By the way, assembly structures may also be referred to as combined structures, elastic multi-structures, etc. in the literature.

In 1979, Feng proposed a system-level solution from the mathematics perspective. His research article *Elliptic Equations on Composite Manifold and Composite Elastic Structures* [7] provided a solid mathematical foundation for the study of assembly structures. In 1984, Feng and Shi presented an analysis algorithm for linear elastic assembly structures in their monograph, *The Mathematical Theory of Elastic Structure* [8]. Sun, in his Ph.D. thesis [28], included shells in his research on assembly structures, and also studied the nonlinear behavior of the elastic assembly structures. Built on this work, Sun and Yue [29] later formulated the nonlinear and linear–nonlinear hybrid problems of assembly structures involving shells, refined and perfected the analysis strategy for assembly structures, and further developed the mathematical theory of assembly structures.

2. *About the Existence of Solutions of Assembly Structures*

If we try to prove the existence of solutions of this type of structures by starting from their governing equations, it would be undoubtedly complicated. Feng [8] and Sun [28] have separately studied this important problem. The approach introduced in this chapter provides an alternative and effective way to deal with the issue of the existence of solutions in the context of assembly structures

It must be pointed out that not every structural combination can produce some assembly structure. In other words, there has to be some assembling criterion for the formation of assembly structures. Both Kang Feng and Bohua Sun pointed out that a three-dimensional elastic solid cannot be attached to a structural component at one point or line. For instance, the endpoint of a beam, the entire beam, or one side of a plate is not allowed to be connected to a three-dimensional elastic solid.

Here we propose a criterion for an assembly structure based on the compatibility of its components at the joints. Specifically, components of the assembly structure must share a finite and identical displacement at each and every location where they are connected. As can be seen in the analysis of Sect. 8.6.3, if the Green's function of one structural component is singular, then the displacement becomes infinite at the location and along the direction where the singularity occurs. Under this circumstance, it is impossible to form an assembly structure. In other words, the structural model is invalid.

Below are some examples of the invalid assembly structures:

(1) A three-dimensional elastic object is attached to another structure at a point or line. For instance, (a) one end of a beam is connected to the three-dimensional solid, (b) a plate joins the three-dimensional solid at one side of the plate; or (c) a circular curved beam hugs a three-dimensional cylinder of the same radius.
(2) A shell is connected to another structure at a point, e.g., the shell and a beam are joined together on the shell surface and at an endpoint of the beam. However, a bar is allowed to connect to a shell that follows the moment theory of shells along the normal direction of the shell.
(3) The endpoint of a beam or bar is affixed to a Mindlin plate.

Table 8.2 provides the basis for deciding whether a connection between structure components to be compatible.

The following results regarding the criterion of an assembly structure may be established based on the above analysis: If each component corresponds to a valid structural model and all connections of these components are compatible, then the structure formed should correspond to a valid structural model and thus can be referred to as an "assembly structure". In contrast, if any of the aforementioned two conditions is violated, then the combination of the structural components cannot produce an "assembly structure".

Furthermore, some important conclusions can be drawn with regard to the existence of solutions to problems of statific deformation and vibrational modes of the assembly structure: The strain energies and kinetic energies of an assembly structure are the sums of their respective counterparts associated with all structural components. Because of this, if Eqs. (8.5.12) and (8.5.13c) hold true for each component, then they should also be satisfied in the case of the assembly structure. Therefore, Theorems 8.16 and 8.17 are applicable to this assembly structure.

Finally, we would like to mention something of practical significance: Mid-surfaces of components of an assembly structure are not required to be smooth at the joints, because the boundary surface of the corresponding elastic solid is allowed to be just piecewise smooth. For the same reason, the mid-surface of a shell, as a standalone structure, only needs to be piecewise smooth, too.

8.6.5 Criteria for Existence of Solutions of Static Deformation and Vibrational Modes in the Context of Specific Structures

According to the theories presented in Sects. 8.4 and 8.5 as well as the earlier part of this section, two steps are involved in determining the existence of solutions of static deformation and vibrational modes for a specific structure. In the first step, decide whether this structure corresponds to a valid structural model using the principles introduced in Sect. 8.6.1. If the structure contains a concentrated mass or concentrated support, we must check whether the Green's function of the structure has singularities. Furthermore, if it involves a combination of structural components, then we have to find out whether this combination is valid, i.e., whether it leads to an assembly structure. If the specific structure does not correspond to a valid structural model, then, in general, it does not have solutions of either static deformation or vibrational modes. If it turns out that the structural model of this structure is valid, then we need to proceed to the second step of this process, where we further verify whether the elasticity operator of the corresponding elastic solid of the structure to be positive definite and the energy embedding operator to be compact. Only when positive answers are obtained in both checks of the second step, can we conclude that the structure has generalized solutions to problems of static deformation and vibrational modes.

As an example, let us consider the existence of a generalized solution to the static deformation problem of a beam clamped at one end and unconstrained at the other end. Suppose the beam is straight and has a rectangular cross section. Then, the static deformation equation and boundary conditions are:

$$\begin{cases} (EJ(x)w'')'' = f(x), \\ w(0) = w'(0) = 0, \quad w''(l) = (EJw''(x))'\big|_{x=l} = 0. \end{cases}$$

It is assumed that E is a constant and $J = b(x)h^3(x)/12$ does not have the second derivative. Thus, this equation does not have a classical solution.

Let us check whether there exists a generalized solution. First, the theoretical model of this beam is known to be valid. Second, we need to determine whether the elasticity operator of the corresponding elastic solid of the beam to be positive definite. We know the elastic solid is isotropic and homogeneous, and it is in the clamped-free configuration. If its border is piecewise smooth, the elasticity operator of this elastic solid ought to be positive definite, based on Theorem 8.10 (Theorem on the Positive Definiteness of the Elasticity Operator). This leads to the conclusion that the generalized solution exists for static deformation of this beam.

It needs to be pointed out particularly that the section parameter of the beam $b(x)h^3(x)/12 = J_0(x)$ may correspond to an infinite set of possible shapes of the beam cross-section along the neutral axis. Accordingly, the cross-section of the corresponding elastic solid may also assume infinitely many shapes. As long as one of them has a piecewise smooth boundary, the elasticity operator will be positive definite and thus, the operator of the beam will be positive definite.

This observation has some further implication. Suppose there is an elastic solid whose elasticity operator, according to the theory of Elasticity, fails to be positive definite. The conclusion from the previous paragraph implies: If this elastic solid is reduced to a structure, it may well be that the resulting structural theory operator is positive definite anyway.

The stated fact reveals the property that, after converting the elasticity theory to a structural theory through simplification and dimension reduction, we actually relax the requirements on the existence of a generalized solution. Specifically, the conditions for the existence of the generalized solution in a structural theory are weaker than those associated with the original elasticity theory.

As another example, consider a thin plate with a complex shape, variable thickness, and homogeneous material. It is subject to a hybrid of clamped and free boundary conditions. We are interested in modal solutions of the plate. First, per Sect. 8.6.2, this plate corresponds to a valid theoretical model. Next, treat this plate as an elastic solid. If its overall shape is piecewise smooth, then, according to Theorems 8.10 and 8.11, the elasticity operator of this elastic solid is positive definite and its energy embedding operator is compact. These results, in conjunction with Theorem 8.17, further lead to the conclusion that generalized solutions to the modal problem exist.

Now consider the situation where the thickness h of the plate makes it impossible for the shape of the exterior boundary to satisfy the smoothness requirement related to the generalized solution in Elasticity. Since the bending stiffness of the plate is determined by $D = Eh^3/[12(1-\nu^2)]$, we may be able to transfer the non-smoothness caused by the plate thickness h to the elastic modulus E while keeping the product Eh^3 unmodified. It is possible that such a change would cause the elasticity operator of the corresponding elastic solid to become positive definite and the energy embedding operator to be compact. In this scenario, the generalized solution of this plate could still exist.

8.6.6 Examples of Generalized Solutions in Structural Theories

In the following examples, we are going to verify the existence of generalized solutions by applying the criteria introduced in Sect. 8.6.5, and also show that the generalized solutions in some examples are actually of higher-order differentiability. The second part of the discussion is based on analysis using the regularity theory. Due to space limitation, we cannot not include details of the analysis.

Example 1 Consider the static deformation of a membrane subject to the displacement boundary constraint. See Ref. [18]. The governing equation and boundary condition may be expressed as

$$\begin{cases} -\Delta u = f(x, y) & \text{in } \Omega, \\ u = 0 & \text{on } \partial\Omega, \end{cases}$$

where Δ represents the two-dimensional Laplace operator. The generalized solution of the equations has the following properties:

(1) Let $\Omega \in C^{2+k}, f \in H_k(\Omega)$, and k be a positive integer or zero. Then,

$$u \in \overset{0}{H}_1(\Omega) \cap H_{2+k}(\Omega).$$

We also have the following estimate

$$\|u\|_{2+k} \leq M_k \|f\|_k,$$

which exposes the continuous dependence of u on the specified condition of the problem. Therefore, the above problem is well-posed in the sense of the generalized solution.

(2) Suppose that Ω is a convex polygon on a plane and $f \in L_2(\Omega)$. Then, $u \in H_2(\Omega)$.
(3) Assume that Ω is a polygon on a plane and at least one of its internal angles is greater than 180°. Then, the generalized solution u generally does not belong to $H_2(\Omega)$, even when $f \in L_2(\Omega)$.

Example 2 Consider the static deformation of a membrane subject to elastic constraints at the boundary. See Ref. [18]. The governing equation and boundary condition are:

$$\begin{cases} -\Delta u = f(x, y) & \text{in } \Omega, \\ \dfrac{\partial u}{\partial n} + \alpha u = \varphi & \text{on } \partial\Omega, \end{cases}$$

where $\alpha(x, y) \geq 0$ and $\alpha(x, y)$ is not always zero. The generalized solution of this problem has the following properties:

(1) If $\varphi \equiv 0$, $\alpha(x, y)$ is sufficiently smooth, $\Omega \in C^{(2+k)}$, and $f \in H_k(\Omega)$, then $u \in H_{2+k}(\Omega)$.
(2) Suppose that φ is not identical to zero, $\alpha(x, y)$ is sufficiently smooth, $\Omega \in C^{(2+k)}$, and $f \in H_k(\Omega)$. Moreover, there is some additional requirement on the smoothness of φ. Then, $u \in H_{2+k}(\Omega)$.

Example 3 Suppose $\Omega \in R^n$ is a bounded region. The eigenvalue problem

$$\begin{cases} -\Delta u = \lambda u & \text{in } \Omega, \\ u = 0 & \text{on } \partial\Omega \end{cases}$$

has generalized solutions, which include an infinite set of eigenvalues $\lambda_1, \lambda_2, \cdots \to \infty$ as well as mode shapes $\{\varphi_1, \varphi_2, \ldots\} \subset H_0^1(\Omega)$. The set of mode shapes is complete in both spaces $H_0^1(\Omega)$ and $L_2(\Omega)$. Furthermore, according to the regularity theory, the mode shapes have continuous derivatives of infinite order in Ω, i.e., $\varphi_i \in C^\infty(\Omega)$. If $\partial\Omega$ is smooth, then they should have continuous derivatives of infinite order in $\bar{\Omega}$, namely, $\varphi_i \in C^\infty(\bar{\Omega})$.

Notice that this conclusion is also applicable to the eigenvalue problem of the second-order elliptic equation with coefficients belonging to C^∞. See Ref. [12].

Example 4 Consider the static deformation of a plate with uniform thickness and clamped at the boundary. See Ref. [18]. The governing equation and boundary conditions are of the form:

$$\begin{cases} \Delta^2 u = f & \text{in } \Omega, \\ u = \dfrac{\partial u}{\partial n} = 0 & \text{on } \partial\Omega, \end{cases}$$

where Δ^2 denotes the two-dimensional biharmonic or bilaplacian operator. The generalized solution of this equation has the following properties:

(1) If $\Omega \in C^{(4+k)}$ and $f \in H_k(\Omega)$, then the generalize solution

$$u \in H_{4+k}(\Omega) \cap \overset{0}{H}_2(\Omega).$$

(2) If Ω represents a convex region on a plane and $f \in L_2(\Omega)$, then the generalized solution

$$u \in H_2(\Omega) \cap \overset{0}{H_2}(\Omega).$$

By comparing the above result with Property (3) in Example 1, we notice some significant difference between the membrane and plate regarding the differentiability of generalized solutions of static deformation, although the two systems share the identical domain Ω of the shape of a convex polygon as well as the same external load $f \in L_2(\Omega)$. Specifically, the generalized solution of the membrane has generalized derivatives of the same order as the corresponding differential equation, i.e., $u \in H_2(\Omega)$. In contrast, the generalized solution of the plate merely has second generalized derivatives, i.e., $u \in H_2(\Omega)$, but the respective differential equation is of the fourth order.

(3) If $\Omega \in C^\infty$ and $f \in C^\infty(\Omega)$, then the static deformation problem of the plate has one and only one solution $u \in C^\infty(\bar{\Omega})$ under clamped, free, or hybrid boundary conditions. See Ref. [9]. Notice that when the plate becomes unconstrained, there will be rigid-body movement. In this situation, the static deformation solution of the plate is unique after rigid-body displacement and rotation are both excluded.

8.7 Convergence of the Ritz Method for Problem Solving in Structural Theories

As shown in Sect. 8.2, an alternative way to find solutions to the boundary value problem of a differential equations is to convert it to the problem of finding extrema of a functional and then solve the newly formulated problem. This alternative approach has several advantages: First, as explained in Sect. 8.3, the issue on existence of a solution can be fully resolved only through the process of finding extrema of functionals. Second, some effective computational methods have been developed specifically for obtaining extrema of functionals, among which are the Ritz method and its derivative, the finite element method, are especially powerful. In this section, we explain procedures of applying the Ritz method to solve problems of static deformation and vibrational modes. In addition, we present, without proof, the convergence theorem for solutions of Ritz approximation. The related proof can be found in Refs. [22–24].

8.7.1 Ritz Method for Solving the Static Deformation Problem

In Sect. 8.2.2, we demonstrated that the static deformation problem can be transformed to the problem of finding the minimum of the functional (8.2.12), i.e.,

$$F(u) = (Au, u) - (u, f) - (f, u),$$

where the operator A is defined on a dense set in the Hilbert space H. In Sect. 8.3.1, we have proved that for a positive definite operator A, the minimization problem has a unique solution in the space H_A.

The space H under consideration is an L_2 space, the operator A is positive definite, and the space L_2 is a separable space. Thus, H_A should also be a separable space. Take a complete sequence $\{\psi_n\}$ in H_A and have them span an n-dimensional subspace. We attempt to find the minimum of $F(u)$ in this subspace using the Ritz method. Set

$$u_n = \sum_{k=1}^{n} a_k \psi_k, \tag{8.7.1}$$

where $\{a_k\}$ denote constants to be determined. Substituting Eq. (8.7.1) into the functional (8.2.12), we get

$$F(u_n) = \sum_{k,j=1}^{n} a_k \bar{a}_j (A\psi_k, \psi_j) - \sum_{k=1}^{n} a_k (\psi_k, f) - \sum_{k=1}^{n} \bar{a}_k (f, \psi_k). \tag{8.7.2}$$

Set $a_k = \alpha_k + i\beta_k$, and let

$$\frac{\partial F(u_n)}{\partial \alpha_k} = 0, \quad \frac{\partial F(u_n)}{\partial \beta_k} = 0, \quad k = 1, 2, \ldots, n.$$

Thus, the problem of finding the minimum of $F(u_n)$ is changed to the one of solving a system of linear equations:

$$\sum_{k=1}^{n} (A\psi_k, \psi_j)\, a_k = (f, \psi_j), \quad j = 1, 2, \ldots, n, \tag{8.7.3}$$

or

$$\sum_{k=1}^{n} [\psi_k, \psi_j]\, a_k = (f, \psi_j), \quad j = 1, 2, \ldots, n. \tag{8.7.4}$$

In fact, Eq. (8.7.4) is used primarily in the actual implementation of the Ritz method.

When the order of the differential equation $Au = f$ is $2k$, the basis functions $\{\psi_i\}$ need only to have k-th derivatives. Moreover, they are required to satisfy the displacement constraints at the border, but do not have to observe the force boundary conditions. In other words, the basis functions are chosen from the set of admissible functions. These are the advantages of the Ritz method. Of course, if Eq. (8.7.3) is used and the basis functions are taken from the domain of A, i.e., the set of comparison functions, then faster convergence may be achieved.

The matrix form of the system of algebraic equations (8.7.4) is

$$\boldsymbol{Ka} = \boldsymbol{p}, \tag{8.7.5}$$

where elements of the matrix $\boldsymbol{K} = (k_{ij})_{n\times n}$ is $k_{ij} = [\psi_i, \psi_j]$. Since the sequence of basis functions $\{\psi_i\}$ selected is linearly independent, the Gram determinant of the matrix $\boldsymbol{K}$ is nonzero. Consequently, Eq. (8.7.5) has a solution

$$\boldsymbol{a}_n = (a_1, a_2, \ldots, a_n)^{\mathrm{T}}, \tag{8.7.6}$$

and the Ritz approximation of the solution that minimizes $F(u)$ is obtained from Eq. (8.7.1).

The next theorem is about the convergence of the Ritz method, which answers the question: Under what condition will the above Ritz approximation u_n converge to the actual solution that minimizes the functional $F(u)$?

Theorem 8.18 *If the operator A is positive definite, then the sequence of approximate solutions, which are obtained from the Ritz method to minimize the functional*

$$F(u) = (Au, u) - (u, f) - (f, u),$$

converges to the accurate solution with respect to metrics on both H_A and H.

8.7.2 Ritz Method for Solving the Modal Problem

In Sect. 8.2.3, we converted the problem of finding modes of a structure to the one of computing local minima of the functional (8.2.19), i.e., the Rayleigh quotient.

The Ritz method offers an effective way to get approximate solutions of the above problem. For simplicity, let us minimize the Rayleigh quotient of the form

$$R(u) = \frac{(Au, u)}{(u, u)} = \frac{[u, u]}{(u, u)}.$$

It has been proved in Sect. 8.3.2: If the operator A is self-adjoint and positive definite while the embedding operator from the space H_A to the space H_B is compact, then there exist solutions that minimize the Rayleigh quotient $R(u)$. Specifically, when $u = \varphi_1$,

$$\lambda_1 = \inf \frac{[u, u]}{(u, u)} \tag{8.7.7}$$

and $R(u)$ reaches a local minimum. When $u = \varphi_{m+1}$,

$$\lambda_{m+1} = \inf \frac{[u, u]}{(u, u)}, \quad (u, \varphi_i) = 0, \quad i = 1, 2, \ldots, m \tag{8.7.8}$$

and $R(u)$ also reaches a local minimum.

Find a complete sequence ψ_i $(i = 1, 2, \ldots, n)$ in the space H_A, and set

$$u_n = \sum_{i=1}^{n} a_i \psi_i, \tag{8.7.9}$$

where $\{a_i\}$ are constants to be determined. Choose $\{a_i\}$ such that

$$(u_n, u_n) = \sum_{i,j=1}^{n} (\psi_i, \psi_j) a_i \bar{a}_j = 1 \tag{8.7.10}$$

and the expression

$$[u_n, u_n] = \sum_{i,j=1}^{n} [\psi_i, \psi_j] a_i \bar{a}_j \tag{8.7.11}$$

reaches minimum.

Now let us use the method of Lagrange multipliers to solve for $\{a_i\}$. Introduce the function

$$\Phi = [u_n, u_n] - \lambda (u_n, u_n), \tag{8.7.12}$$

where λ is a constant to be determined. Since a_i may be complex numbers, we separately take derivatives of Eq. (8.7.12) with respect to real and imaginary parts of a_i, set them to zero, and get after some simplification

$$\frac{\partial \Phi}{\partial \bar{a}_i} = 0, \quad i = 1, 2, \ldots, n.$$

Expanding the resulting equations, we get

$$\sum_{i=1}^{n} a_i \{[\psi_i, \psi_j] - \lambda (\psi_i, \psi_j)\} = 0, \quad j = 1, 2, \ldots, n. \tag{8.7.13}$$

The matrix form of Eq. (8.7.13) is

$$\boldsymbol{K}\boldsymbol{a} = \lambda \boldsymbol{M}\boldsymbol{a}, \tag{8.7.14}$$

where elements of matrices $\boldsymbol{K}$ and $\boldsymbol{M}$ are, respectively,

$$k_{ij} = [\psi_i, \psi_j] \quad \text{and} \quad m_{ij} = (\psi_i, \psi_j), \quad i,\ j = 1, 2, \ldots, n. \tag{8.7.15}$$

Since $\{\psi_i\}$ form a complete sequence in the space H_A and are linearly independent, the Gram determinant of the matrix $\boldsymbol{K}$ is nonzero. In addition, both $\boldsymbol{K}$ and $\boldsymbol{M}$ are real symmetric matrices. Thus, the solution of $\boldsymbol{a}$ must be a real vector. Hence, the matrix eigenvalue problem (8.7.14) is solvable, and the solutions of the eigenpairs are

$$\lambda_{ni}, \quad \boldsymbol{a}_{ni} = (a_{i1}, a_{i2}, \ldots, a_{in})^{\mathrm{T}}, \quad i = 1, 2, \ldots, n.$$

Plugging the results into Eq. (8.7.9), we obtain

$$\lambda_{ni}, \quad u_{ni} = \sum_{j=1}^{n} a_{ij}\psi_j, \quad i = 1, 2, \ldots, n. \tag{8.7.16}$$

λ_{n1} and u_{n1} calculated with the Ritz method are approximation of the accurate solution to the minimization problem of the Rayleigh quotient $R(u)$, which is given in Eq. (8.7.7); while λ_{ni} and u_{ni} $(i = 2, 3, \ldots, n)$ represent approximations of the second to the n-th eigenpairs of the minimization problem (8.7.8).

If the Ritz basis functions $\{\psi_i\}$ are taken from the domain D_A of the operator A, which is the set of comparison functions, it will result in faster convergence.

Given below is the result regarding the convergence of the Ritz method, when it is used to solve modal problems:

Theorem 8.19 If the operator A is positive definite and the embedding operator from the space H_A to the space H is compact, then approximations of eigenvalues and eigenfucntions, which are obtained by minimizing the Rayleigh quotient $R(u) = (Au, u)/(u, u)$ with the Ritz method, converge to solutions of the eigenvalue problem $Au = \lambda u$. The convergence is defined with respect to the metric on H_A and H.

In summary, when the Ritz method is applied to the following four types of problems, all approximate solutions obtained converge to the respective generalized solutions in spaces L_2 and H_A, which is the space with the strain energy norm. These four types of problems are listed below:

(1) Suppose the specified elastic coefficients, boundary shapes, and boundary constraints ensure the positive definiteness of the elasticity operator. Find the generalized solution of the static deformation in Elasticity.
(2) Suppose the elasticity operator to be positive definite and the energy embedding operator to be compact. Find generalized solutions to the modal problem in Elasticity.
(3) When the operator in a structural theory is positive definite, find the generalized solution of the static deformation in this structural theory,
(4) When the operator in a structural theory is positive definite and the energy embedding operator is compact, find generalized solutions to the modal problem in this structural theory.

References

1. Agmon S (1965) Lectures on elliptic boundary value problems. D. Van Nostrand, New York
2. Benadou M, Ciarlet PG (1975) Sur l' ellipticite du models linéaire de cogues de W. T. Koiter. Lecture notes in economics and mathematics systems 134, Computing methods in applied science and engineering, 2nd international symposium, Dec 15–19, 1975. Springer, New York, 1976, pp 89–136
3. Benadou M, Lalanne B (1985) Sur l'approximation des coques minces, par des méthods B-splines et éléments finis. In: Grellier JP, Campel GM (eds) Tendances Actuelles en Calcul des structures. Editions Pluralis, Paris, pp 939–958
4. Benadou M, Ciarlet PG, Miara B (1994) Existence theorems for two-dimensional linear shell theories. J Elast 34:111–138
5. Ciarlet PG, Miara B (1992) Justification of the two-dimensional equations of a linearly elastic shallow shell. Comm Pure Appl Math 45:327–360
6. Эйдус Д М (1951) О смешанной задаче теории упругости. ДАН СССР, 76(2) (in Russian)
7. Feng K (1979) Elliptic equations on composite manifold and composite elastic structures. Math Num Sinica 1(3):199–208
8. Feng K, Shi ZC (1984) The mathematical theory of elastic structure. Science Press, Beijing, China (in Chinese)
9. Fichera G (1965) Linear elliptic differential systems and eigenvalue problem. Lecture notes in mathematics. Springer, Berlin, Heidelberg, New York
10. Fichera G (1972) Existence theorems in elasticity. In: Flugge S (ed) Encyclopedia of physics, vol a/2, no 6, pp 347–389
11. Friedrichs K (1947) On the boundary-value problems of the theory of elasticity and Korn's inequality. Ann Math 48(2):441–471
12. Gilbarg D, Trudinger Neil S (1983) Elliptic partial differential equations of second order, 2nd edn. Springer, Berlin, Heidelberg
13. Gordegiani DG (1974) On the solveabillity of some boundary value problems for a variant of the theory of thin shells. Dokl. Akad, Nauk SSSR, p 215
14. Gurtin ME (1964) Variational principles for linear elastic-dynamics. Arch Ration Mech Anal 16(1):34
15. Hu HC (1990) Necessary and sufficient condition for correct use of generalized variational principle of elasticity in approximate solutions. Sci China Set A 33(2):196–205
16. Hu HC (1984) Variational principles of theory of elasticity with applications. Science Press, Beijing, China; Gordon, Breath Science Publisher, New York
17. Korn A (1908) Solution générale du problémé déguilibre dans la théorie de lélasticité dans le cas oń les efforts sont donnés á la surface. Ann. Université Toudouse
18. Kreyszig E (1978) Introductory functional analysis with applications. Wiley, New York, Senta Barbare, London, Sydney, Toronto
19. Kupradze VD (1965) Potential methods in the theory of elasticity. Israel Program for Scientific Translations, Jerusalem
20. Kupradze VD (1979) Three-dimensional problems of the mathematical theory of elasticity and thermoelasticity. North-Holland Publishing Company, Amsterdam
21. Leung AYT, Wang DJ, Wang Q (2004) On concentrated mass and stiffness in structural theories. Int J Struct Stab Dyn 2004(4):171–179
22. Meirovitch L (1980) Computational methods in structural dynamics. SIJTHOFF & NOORD-HOFF, Rockville, Maryland, USA
23. Міхлин С Г (1950) Прямые Методы в Математической Физике. Государственное Издательство Технико - Теоретической Литературы, Москва (in Russian)
24. Міхлин С Г (1952) Проблема Мнимума Квадратичного Функционала. Государственное Издательство Технико-Теоретической Литературы, Москва (in Russian)
25. Y Norio (2008) Oscillation theory of partial differential equations. World Sci Publ, Singapore

26. Payne LE, Weinberger HF (1961) On Korn's inequality. Arch Ration Mech Anal 8(2):89–98
27. Shoikhet BA (1974) On existence theorems in linear shell theory. PMM 38:567–571
28. Sun BH (1988) Application and analysis theory of combined elastic structure. Lanzhou University, Lanzhou (in Chinese)
29. Sun BH, Ye ZM (2009) Formulation of elastic multi-structures. Sci China Ser G Phys Mech Astron 52(6):935–953
30. Sun BH (2012) On existence theorem in theories of elastic structures. Adv Mech 42(5):538–546 (in Chinese)
31. Valid R (1995) The nonlinear theory of shells through variational principles: from elementary algebra to differential geometry. Wiley, New Jersey
32. Wang DJ, Hu HC (1982) A unified proof for the positive definiteness and compactness of two kinds of operators in the theory of elastic structure. Acta Mech Sin 14(2):111–121 (in Chinese)
33. Wang DJ, Hu HC (1982) A unified proof of the general properties of the linear vibrations in the theory of elastic structures. J Vib Shock 1(1):6–16 (in Chinese)
34. Wang DJ, Hu HC (1983) A unified proof for the positive-definiteness and compactness of two kinds of operators in the theories of elastic structures. In: Proceedings of the China-France symposium on finite element methods. Science Press China, New York, pp 6–16
35. Wang DJ, Hu HC (1985) Positive definiteness and compactness of two kinds of operators in theory of elastic structures. Scientia Sinica Ser A 28(7):727–739
36. Wang DJ, Wang WQ (1989) The reasonableness problem of theories of structures carrying concentrated masses, springs and supports in vibration problems. Acta Mech Solida Sin 2(2):247–251 (in Chinese)
37. Wang Q, Wang DJ (1993) Singularity under a concentrated force in elasticity. Appl Math Mech 14(8):707–711
38. Weinberger HF (1974) Variational methods for eigenvalue approximation, 2nd edn, 1987. Society for Industrial and Applied Mathematics
39. Wu JK (1981) A proof of ellipticity of the thin shell equations. Acta Mech Solida Sin 2(4):435–444 (in Chinese)

Index

D. Wang et al., *Qualitative Theory in Structural Mechanics*,
https://doi.org/10.1007/978-981-13-1376-9

Printed by Printforce, the Netherlands